Kyuzo Aoki (Ed.)

Essential Hypertension 2

With 164 Figures

Springer Japan KK

Dr. Kyuzo Aoki
2nd Department of Internal Medicine
Nagoya City University Medical School
Nagoya 467, Japan

Library of Congress Cataloging-in-Publication Data
Essential hypertension 2/Kyuzo Aoki, ed.
Proceedings of the Second International Symposium on Mechanism and Treatment in Essential Hypertension, held May 30-June 1, 1988, in Nagoya. Includes bibliographical references.
1. Essential hypertension—Pathophysiology—Congresses. 2. Calcium—Physiological effect—Congresses. I. Aoki, K. (Kyuzo), 1933- . II. International Symposium on Mechanism and Treatment in Essential Hypertension (2nd : 1988 : Nagoya-shi, Japan) III. Title: Essential hypertension two. RC685.H8E872 1989 616.1'32—dc20
ISBN 978-4-431-68092-5 ISBN 978-4-431-68090-1 (eBook)
DOI 10.1007/978-4-431-68090-1

Originally published by Springer-Verlag Tokyo in 1989
Softcover reprint of the hardcover 1st edition 1989

Typesetting: Asco Trade Typesetting Ltd., Hong Kong

Preface

The First International Symposium on Mechanism and Treatment in Essential Hypertension was held on October 23 and 24, 1985 in Nagoya. The Second International Symposium, which was held on May 30, 31, and June 1, 1988 in Nagoya, was a success thanks to the endeavors of all the participants. These symposiums were made possible by the generous support of the city of Nagoya, Aichi Prefecture, and various contributions.

Understanding of the mechanisms of gene (major gene, essential) hypertension is impossible unless we comprehend the physiological properties of arterial smooth muscle. Since Langendorff's discovery of the role of calcium ions in muscle contraction in 1895, we have made immense progress in our knowledge of the role of the calcium ion in the excitation-contraction coupling of the arterial smooth muscle. Investigation of the molecular mechanisms controlling the cellular basis of calcium ion action has been established with the discovery of the calcium binding protein (Ebashi 1963). The participation of the calcium induced calcium mechanism in the contraction of muscle has been directly demonstrated by using chemically skinned smooth muscle fibers (Endo et al. 1977). Methods for intracellular free calcium contents and calcium channel activity have been developed and contribute to the study of the role of the calcium ion in arterial smooth muscle contractility. Such recent progress promotes study of the basic physiological, biochemical, and molecular properties of the calcium channel, and, in the field of hypertension research, the sodium and potassium channels.

The common hemodynamic indicator of a rise in pressure in essential hypertension is elevation of the total peripheral vascular resistance. Abnormalities in transmembrane calcium ion movement at plasma membrane and sarcoplasmic reticulum levels have been reported as a possible cause of the increased contraction in the arterial smooth muscle. In view of the important role of the calcium ion in the regulation of arterial smooth muscle contractility, confirmation of the possibility that essential hypertension reflects abnormalities in the membrane calcium handling of the arterial smooth muscle, has been anticipated in the last two decades. Discovery of the spontaneously hypertensive rat (Aoki 1959 in introduction session) established the genetic basis of essential hypertension. New classification divided hypertension into the three categories of gene (essential) hypertension, environment (accessory gene with environmental factor) hypertension, and disease (nongene) hypertension, all of which could be applied to both experimental animal models and human hypertension (Aoki 1985 in introduction session). Calcium agonists and antagonists have become available for use in research of hypertension (1972). The history of the calcium membrane theory of gene hypertension is appeared in the last session.

This volume provides a pathophysiology of gene (essential) hypertension at the molecular, cellular, organic, and systemic levels. The 30 chapters are divided into five sections:

I Spontaneously hypertensive rats and classification of hypertension
II Calcium movement during contraction and relaxation in arterial smooth muscle
III Membrane calcium-handling abnormalities of Vascular Smooth Muscle in spontaneously hypertensive rats
IV Calcium, magnesium, and calcium antagonists in human hypertension
V Mechanism of blood pressure elevation in gene (essential) hypertension

The authors here provide new findings, an overview of their own work, related studies from other laboratories, and their scientific hypotheses. I would like to thank the contributors for their excellent papers and all the participants for their outstanding scientific achievements and cooperation.

It is hoped that the second symposium proceedings "Essential Hypertension 2" as well as the first symposium proceedings "Essential Hypertension" will stimulate the discovery of hypertension mechanisms in addition to preventive and therapeutic measures. I wish to acknowledge the staff of Springer-Verlag Tokyo for their highly professional and skillful efforts in making this attractive publication and thank the organizers, Drs. Fritz R. Bühler, Koichi Sato, and Masahiko Yamamoto, and the dedicated Ms. Junko Ito (Kusunoki) and Hiroko Hotta (Yokoyama) for administrative work.

Nagoya, July 1989 KYUZO AOKI

List of Contributors

Table of Contents

I Introduction:
Spontaneously Hypertensive Rats and Classification of Hypertension

Discovery and Development of the Spontaneously Hypertensive Rat

KYUZO AOKI[1]

Summary. In 1959, during the testing of the accuracy of a new blood pressure measurement apparatus for rats, a spontaneously hypertensive rat (Aoki SHR) was discovered among ten Wistar rats supplied from the Animal Center of Kyoto University. It was thought that the hypertension may be of genetic orign. In order to prove this hypothesis I began breeding the hypertensive male rat with a normotensive female rat to obtain genetically hypertensive rats. Many of the rats in the first generation had a blood pressure elevation with age and subsequently developed hypertension. The rats from the first generation with the highest blood pressure were selected and inbred to obtain genetically hypertensive rats. After the third generation, all rats obtained by the selective inbreeding method developed hypertension without any special manipulation or experimental maneuvers. The fact that the SHR strain was obtained by only three repetitive selective inbreeding suggests that there are a few major hypertension genes and the genes are the primary cause of the blood pressure elevation in SHR.

Key words: SHR—Spontaneously hypertensive rat—Establishment of SHR strain—Major hypertension gene—Gene hypothesis

Discovery of the Spontaneously Hypertensive Rat

In 1959, I joined the Department of Pathology, School of Medicine, Kyoto University as a graduate student. Professor Kozo Okamoto in the Department had organized his research on the production of animal models for the study of diabetes mellitus, endocrine diseases, obesity, arteriosclerosis, and hypertension through experimental maneuvers [1].

Professor Okamoto kindly gave me a research subject which was the production of renal hypertension following renal infarction and the mechanisms in the renal hypertension in rats. There was no reliable apparatus for measuring blood pressure in rats in the laboratories of the Department. It was attempted to produce a reliable apparatus for measuring blood pressure in rats because blood pressure should be measured in the rats as part of the hypertension research.

[1]Second Department of Internal Medicine, Nagoya City University Medical School, Mizuho-ku, Nagoya 467, Japan

Fig. 1. Tail water-plethysmographic apparatus for measurement of rat blood pressure. After Aoki [7, 9–11].

A new tail water-plethysmographic device was designed and made by modification from the methods of Byrom and Wilson [2], William et al. [3], Sobin [4], and Umezawa [5]. A plethysmographic chamber was made of a transparent acrylic tube with an exhaust pipe in the tap of the chamber to let out the air. The condition of rat tail was seen through the acrylic tube and the air in the tube could easily be drawn out from the exhaust pipe for setting the measurement of blood pressure. The chamber, cuff, and rat holder were arranged in a blood pressure measurement box having a temperature control system. For the measurement, the rat was warmed in the other box with a temperature control system at 38°C for 10 min. Then, the rat was putted in the holder. The tail of the rat was placed in the plastic chamber through the cuff in the blood pressure measurement box with temperature of 37°C, and the blood pressure was measured without anesthesia [6–11] (Fig. 1).

To examine the accuracy of a new blood pressure measurement apparatus, I used ten Wistar Kyoto rats which were kindly supplied by Tamio Katsuya from the Animal Center of Kyoto University. In the process of testing this apparatus in 1959, it was discovered that one male rat was consistantly hypertensive [6–11]. The rat was designated a spontaneously hypertensive rat (SHR). The systolic blood pressure of this SHR was in the range of 150 to 175 mmHg at the age of 7 weeks.

It was found among the Wistar rats supplied by the Animal Center that some rats were normotensive and a few rats were hypertensive. Such wide variations could be ascribed to either statistical limits of a homogeneous population or to variations in inheritance based on genetic differences. If the genetic difference hypothesis holds true, then by selective inbreeding it should be possible to separate the strains which differ in their genes to develop hypertension. The following experiment was planned and executed to prove the gene hypothesis in essential hypertension.

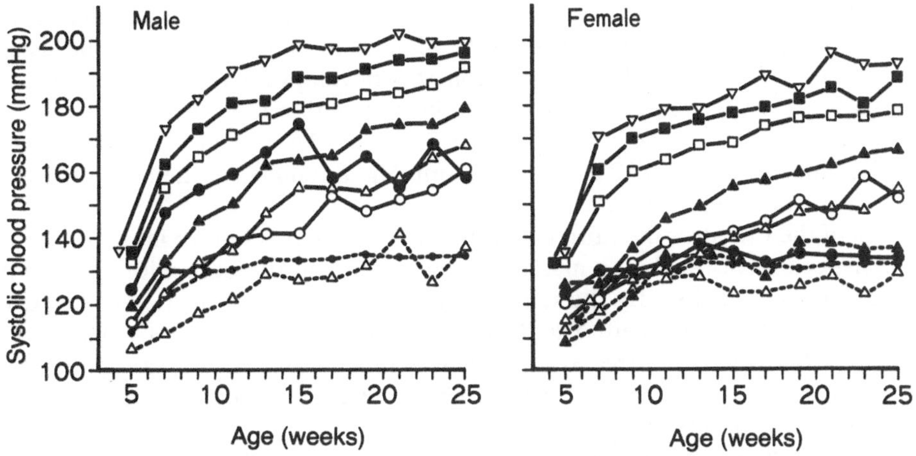

Fig. 2. Systolic blood pressure measurement of control rats and in male and female spontaneously hypertensive rats (SHR). Normotensive rats (···●···) and hypertensive rats (-○-) made up the control group. A male SHR (-●-) was bred with a female normotensive rat (-●-). Normotensive rats in the first (···△···) and second (···▲···) generation. Hypertensive rats in the first (-△-), second (-▲-), third (-□-), fourth (-■-), and fifth (-▽-) generations [6, 10, 11]

Process of the Development of the Strain of Spontaneously Hypertensive Rats

The male discovered to be SHR with blood pressure in the range of 150–175 mmHg was bred with a normotensive female rat with blood pressure in the range of 130–140 mmHg. The first generation obtained consisted of 36 animals from four litters. Hypertension was defined as systolic blood pressure over 150 mmHg. In the first generation, 15 of 16 males and 16 of 20 females were hypertensive. Thus, the incidence of hypertension was 94% in males and 80% in females. The blood pressure of the hypertensive rats was 115 ±4 and 169 ± 13 mmHg (M ± SD) in the males, and 115 ± 3 and 156 ± 14 mmHg in the females at the age of 5 and 25 weeks, respectively, in the first generation. The blood pressure of the one male normotensive rat was 108 and 138 mmHg, and that of the 4 female normotensive rat was 111 ± 3 and 131 ± 12 mmHg at the age of 5 and 25 weeks, respectively (Fig. 2).

Rats in the first generation with the highest blood pressure at the same age were selected for further inbreeding. All 20 males and 26 of 32 females in the second generation were hypertensive. The incidence of hypertension was 100% in males and 81% in females. The blood pressure of hypertensives was 119 ± 8 and 179 ± 14 mmHg in males, and 125 ± 12 and 167 ± 16 mmHg in females at the age of 5 and 25 weeks, respectively. The blood pressure of the female normotensive rats was 111 ± 3 and 131 ± 12 mmHg at the age of 5 and 25 weeks (Fig. 2).

The third generation, which consisted of 56 males and 64 females, was obtained by inbreeding the rats with the highest blood pressure from the second generation. All of the third generation rats developed hypertension, thus the incidence of hypertension was 100% in this generation in both sexes. The blood pressure of rats from the hypertensive strain was 133 ± 14 and 192 ± 18 mmHg in males, and it was 133 ± 12 and 179 ± 17 mmHg in females at the age of 5 and 25 weeks, respectively (Fig. 2).

All of the rats after the third generation delivered by selective inbreeding from the hypertensive rats had a markedly elevated blood pressure with age and developed hypertension. By the selective inbreeding method, a strain of SHR was developed and established. The development of the SHR strain by this method suggests that rats from the strain have the genetic tendency to develop hypertension and the tendency is transferred by a few major hypertension genes from parents to offspring (Fig. 2).

For a study on the development of the strain of spontaneously hypertensive rats, 66 young Wistar Kyoto rats to be used as controls for the SHR were supplied by the Animal Center of Kyoto University. It was confirmed in the control rats that some rats developed hypertension spontaneously, but some did not. Ten of 34 males and 5 of 32 females in the control rats developed hypertension. The incidence of hypertension was 29% and 16% in the males and females, respectively. The blood pressure of the normotensive rats was 114 ± 12 and 135 ± 8 mmHg in males, and 115 ± 11 and 133 ± 7 mmHg in females, at the age of 5 and 25 weeks, respectively. The results were added to the findings of the selective inbreeding to develop the strain of SHR, and were published in 1963 [6].

Stroke-Prone Spontaneously Hypertensive Rats

Stroke-prone SHR [12] were developed by the method [10] of selective inbreeding. The stroke-prone SHR showed a high incidence of cerebrovascular lesions due to high blood pressure. The blood pressure of Nagaoka stroke-prone SHR (Nagaoka SHRSP) (A_3 in Fig. 3) was significantly higher than that of stroke-resistant SHR (Nagaoka SHRSR) (C in Fig. 3), which was suggestive of a positive correlation between high blood pressure and the incidence of cerebrovascular lesions [12] (Fig. 3). The average blood pressure of the F_1 (Nagaoka SHRSP + Nagaoka SHRSR) breeding with A_3 and C was roughly the mean between both parental SHR. In addition, the blood pressure in the F_2, the offspring of F_1 mated with F_1, showed a continuous distribution within the range of blood pressure of A_3 and C. The same kind of genetic data were demonstrated in the hybrid groups between A_3 and WKY rats. The F_1 was intermediate in blood pressure level between the levels of the parents, and the F_2 ranged from a blood pressure as high as that of A_3 to that of WKY rats (left column in Fig. 3). (See Fig. 6 in Three-Way Classification of Hypertension, p 20) These results suggest that a few major hypertension genes transmit the hypertension in stroke-prone SHR.

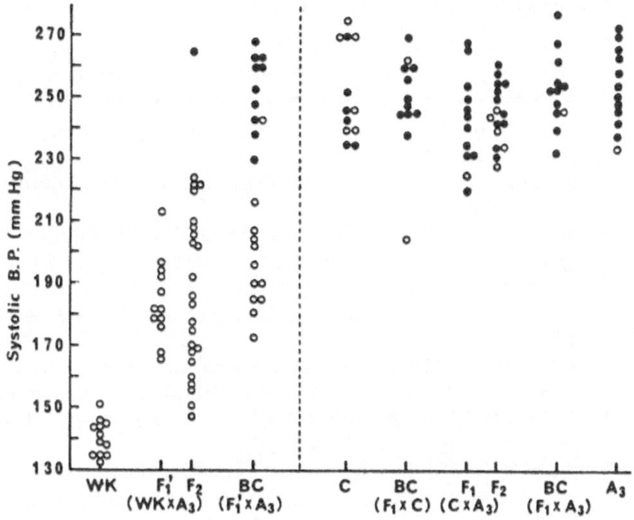

Fig. 3. Blood pressure measurements of rats with a cerebrovascular lesion were in range of 220–280 mmHg. Rats with a cerebrovascular lesion (●) and rats a without cerebrovascular lesion (○), *WK*, normotensive Wistar Kyoto rat. A_3 Nagaoka SHRSP, *C* Nagaoka SHRSR. From Nagaoka et al. [12]

Fig. 4. A spontaneously hypertensive rat (SHR) [10, 11]

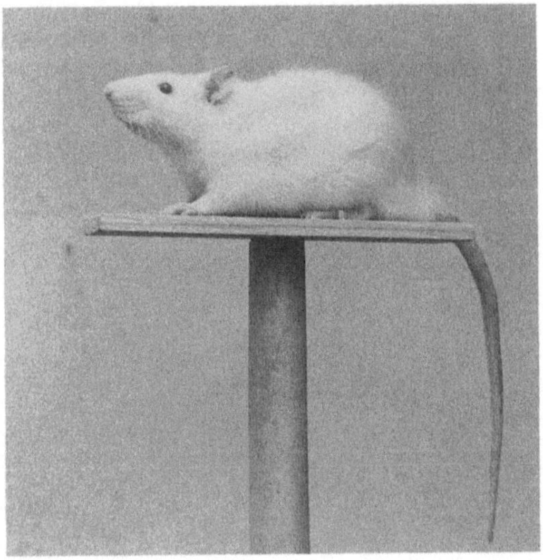

Conclusion

The spontaneously hypertensive rat (Aoki SHR) (Fig. 4) as well as stroke-prone SHR were developed and established by selective inbreeding of rats with spontaneous hypertension. It may be concluded that the hypertension is transmitted by a few major hypertension genes.

Reference

1. Okamoto K (1955) Experimental pathology of diabetes mellitus (Report II). III. Studies on rabbits from ancestors diabetic for several successive generations especially on spontaneous occurrence of diabetes in F_4 and F_5 rabbits. Tohoku J Exp Med 61 (Suppl III): III–62–112
2. Byrom FB, Wilson C (1938) A plethsmographic method for measuring systolic blood pressure in the intact rat. J Physiol 93: 301–304
3. William JR, Harrison TR, Grollman A (1939) A Simple method for determining the systolic blood pressure of the unanesthetized rat. J Clin Invest 18: 373–376
4. Sobin SS (1946) Accuracy of indirect determinations of blood pressure in the rat: Relation to temperature of plethysmograph and width of cuff. Am J Physiol 146: 179–186
5. Umezawa H (1952) An indirect method for measuring blood pressure of the unanesthetized rat (in Japanese). Tokyo Jikeikai Med J 66: 193–199
6. Okamoto K and Aoki K (1963) Development of a strain of spontaneously hypertensive rats. Jpn Circ J 27: 282–293
7. Aoki K (1983) Spontaneously hypertensive rat, SHR. In: Hypertensiology, essential and secondary hypertension; concepts, nature, diagnosis, and treatment. Shinkoh Igaku Shuppan, Tokyo, pp 144–169 (in Japanese)
8. Aoki K (1985) Essential hypertension and secondary hypertension in humans and rats. Asian Med J 28: 529–548
9 Aoki K (1985) Memories of discovery and development of spontaneously hypertensive rats. The 25th anniversary of the discovery of spontaneously hypertensive rats (in Japanese). Coronary 2: 97–100
10. Aoki K (1986) Discovery of the spontaneously hypertensive rat. In: Aoki K (ed) Essential hypertension, calcium mechanisms and treatment. Springer, Tokyo Berlin Heidelberg New York London Paris, pp 3–7
11. Aoki K (1988) The spontaneously hypertensive rat: Evidence of the genetic hypothesis in essential hypertension. In: Aoki K and Frohlich ED (eds) Calcium in essential hypertension. Academic, Tokyo San Diego New York London, pp 3–8
12. Nagaoka A, Iwatsuka H, Suzuki Z, Okamoto K (1976) Genetic predisposition to stroke in spontaneously hypertensive rats. Am J Physiol 230: 1354–1359

Three-Way Classification of Hypertension: Gene Hypertension, Environment Hypertension, and Disease Hypertension

KYUZO AOKI[1]

Summary. Bright described in 1827 [1] a patient in whom urine protein was associated with edema and hypertrophy of the heart, which greatly progressed the research of renal hypertension. In 1872, Gull and Sutton [2] demonstrated arterio-capillary fibrosis in patients with hypertension and separated the hypertensive disease from kidney disease. Frank designated in 1911 [3] this hypertensive disease as "*essentielle hypertonie*", which has been translated into English as essential hypertension and recognized as an arterial disease. In 1914, Volhard and Fahr [4] classified hypertension quantitatively into benign and malignant depending on the severity of hypertension. Pickering in 1955 [5] qualitatively differentiated hypertension into essential hypertension and secondary hypertension, which is a two class classification. Essential hypertension was characterized by high blood pressure and hypertensive cardiovascular hypertrophy of the heart and arteries. Secondary hypertension was defined as hypertension occuring as the prominent phenomenon of a disease. Renovascular hypertension, primary aldosteronism hypertension, pheochromocytoma hypertension, and others are considered as secondary hypertension.

Aoki proposed in 1985 [6], an etiological classification of hypertension, the following three-way classification: (1) gene (essential) hypertension, (2) environment (accessory gene) hypertension, and (3) disease (non-gene) hypertension. Gene hypertension, inherited through the major hypertension gene, may correspond to Frank's essential hypertension. Environment hypertension is characterized by the development of hypertension due to an interaction of environmental factors on an accessory hypertension genes. Salt-sensitive hypertension, obesity-sensitive hypertension, alcohol hypertension, and others of environment hypertension are considered. Disease hypertension, defined as high blood pressure resulting from a disease, corresponds to Pickering's secondary hypertension. This three-way classification according to hypertension gene, applicable to both experimental and clinical cases of hypertensive disease, may open the door to a wide range of further research on the mechanisms, therapy, and prevention of hypertensive diseases.

Key words: Three-way classification—Gene hypertension—Environment hypertension—Disease hypertension—Essential hypertension—Salt hypertension—Obesity hypertension—Renovascular hypertension—Pheochromocytoma hypertension—Aldosteronism hypertension—Major hypertension gene—Accessory hypertension gene

[1] Second Department of Internal Medicine, Nagoya City University Medical School, Mizuho-ku, Nagoya 467, Japan

History of Hypertension Classification

One-Way Classification

Bright [1] in 1827 recognized hypertension as a disease entity. He described the cases of patients in whom urine protein was associated with edema and hypertrophy of the heart. This evidence suggested that hypertrophy of the heart was induced by hypertension due to renal disturbance. Bright's description and findings promoted the discussion of the concept and mechanisms of hypertensive diseases. These discussions led to the hypothesis that a chemical substance was secreted from a certain organ into circulating blood causing an elevation in blood pressure. Tiegerstedt and Bergman [7] found in 1898 that an intravenous injection of saline extracts from the kidney induced a rise in blood pressure. The pressor substance was named renin. In 1934, Goldblatt and colleagues [8] found that constriction of renal artery produced a persistent rise in blood pressure. This study brought about the reexamination of renin as a possible cause of hypertension. Pickering and Prinzmetal [9] demonstrated in 1838, a prolonged rise in blood pressure by intravenous injection of kidney extracts. These studies promoted the progress of research on renal hypertension.

Gull and Sutton [2] observed in 1872 "arterio-capillary fibrosis" in patients with hypertension. The hypertensive disease was distinguished from kidney disease as well as from nephritis. The hypertensive disease was characterized by hypertension associated with hypertrophy of the heart and small arteries. In 1889, Huchard [10] used the term "presclerosis" for the hypertensive disease. von Basch found in 1893 that a majority of the patients with hypertensive disease were not associated with kidney disease; in 1893 von Basch [11] called the hypertension "latent arteriosclerosis." Allbutt [12] in 1895, called the raised blood pressure without proteinuria "senile plethora" and later revised the name to "hyperpiesis". In patients with hyperpiesis, blood pressure rose excessively around 40 years of age. In 1911, Frank [3] modified the term of "hyperpiesis" to "*essentielle hypertonie*". The term has been translated into English as essential hypertension, which is widely used as the diagnostic name of a type of hypertension. He [3] described that "*Der erhöhte Tonus der arteriolen Ring-muskulatur bei der essentiellen Hypertonie bleibt ätiologish zunächst ganz unklar.*"

Gull and Sutton [2] reported in 1872 "arterio-capillary fibrosis" as a possible cause of hypertension by inducing arterial vessel abnormalities which contribute to a generalized narrowing of the arteries in patients with essential hypertension. The narrowing of arteries elevates total peripheral vascular resistance, which raises arterial blood pressure. Thus, Gull and Sutton [2], von Basch [11], Allbutt [12], and Frank [3] have established the concept of essential hypertension as an arterial disease.

Two-Way Classification

Volhard and Fahr [4] in 1914 classified hypertension into two types, benign and malignant, depending on severity of high blood pressure. Pickering [5, 13], in 1955, proposed a new classification, a two-way classification, according to the degree and type of hypertension. He [5, 13] differentiated hypertension quantitatively and qualitatively. Regarding the quantitative classification, according to the level of blood pressure, hypertension was classified into benign and malignant [4]. According to qualitative differences, hypertension was classified into essential hypertension and secondary hypertension. Essential hypertension was characterized by high blood pressure with hypertensive cardiovascular hypertrophy. Pickering [5, 13] proposed that essential hypertension was dependent on inheritance and environment, and concluded that the cause of essential hypertension was polygenic and multifactorial. Pickering [5, 13] defined secondary hypertension as hypertension occuring as the prominent symptom of a disease. By this definition, secondary hypertension included renovascular hypertension, primary aldosteronism hypertension, pheochromocytoma hypertension, and other forms.

Three-Way Classification

Aoki [6, 14, 15], in 1985, proposed an etiological classification of hypertension, the following three-way classification of hypertension:

1. Gene (essential) hypertension
2. Environment (accessory gene) hypertension
3. Disease (non-gene) hypertension

Gene (major hypertension gene) hypertension, inherited through the major hypertension gene, may correspond to Frank's essential hypertension. The hypertension is caused by abnormalities in arterial smooth muscle. The blood pressure in patients with essential hypertension increases steeply at age from 30 to 50 years without any specific environmental factors. The hypertension induces hypertensive cardiovascular diseases in the patients.

Environment (accessory hypertension gene with environmental factor) hypertension is developed by the interaction of environmental factors on an accessory hypertension gene. There are many types of environment hypertension, such as salt, alcohol, obesity, low exercise, excessive exercise, mental stress, physical stress, anxiety, and doctor-induced hypertension. Salt hypertension [16, 17] and obesity hypertension [18–20] have been well defined.

Disease (non-gene, disease-induced) hypertension [6, 14, 15] is defined as high blood pressure resulting from some known or unknown disease. This type of hypertension is independent of hypertension genes. Disease hypertension is caused by different types of diseases, including renal, endocrine, nervous, and aortic valve diseases. Renovascular, renal parenchymal, renin-secreting tumor, primary aldosteronism, pheochromocytoma hypertension, are considered as disease hypertension.

Table 1. Three-way classification of hypertension. After Aoki [14, 15]

		Causes	Origin	Clinical diagnosis and abnormalities	Animal models
Type 1	Gene (essential) hypertension	Major gene inheritance	Arterial vessel	Arterial or essential hypertension	Smirk GH rat, Aoki SHR
				Abnormality of arterial vessel, arterial smooth muscle, membrane calcium transport	
Type 2	Environment hypertension	Environment with accessary gene inheritance	Environment	Salt (excess intake) hypertension	Dahl salt-hypertension rat
				Alcohol hypertension	
				Obesity hypertension	
				Low-exercise hypertension	
				Excess-exercise hypertension	
				Excess mental stress hypertension	
				Excess physical stress hypertension	
				Anxiety hypertension	
				Doctor-induced hypertension	
Type 3	Disease hypertension	Underlying disease	Underlying disease		
			Endocrine	Primary aldosteronism	Skelton hypertension
				Pheochromocytoma	
				Cushing's syndrome	
				Hyperthyroidism	
				Iatrogenic hypertension	
			Renal	Renal parenchymal disease	Goldblatt hypertension
				Renovascular hypertension	
				Primary reninism	
			Nervous	Sympathetic nerve hypertension	
				Parasympathetic nerve hypertension	
				Cerebral nerve hypertension	
				Neurogenic hypertension	
			Connective tissue	Arteriosclerotic hypertension	
			Drug origin	Liquorice hypertension	DOCA-salt hypertension
				Contraceptive hypertension	
				Steroid hypertension	
			Other origin	Hematologic hypertension	
				Immunologic hypertension	

GH, genetically hypertensive; SHR, spontaneously hypertensive rats; DOCA, deoxycorticosterone acetate

Inheritance of Hypertension in Humans

Weitz's Family Origin Hypothesis

In 1923, Weitz [21] proposed a hypothesis of family origin of essential hypertension. He reported that the family data were compatible with an autosomal dominant of inheritance in essential hypertension.

Pickering's Polygenic Theory

The mode of hypertension inheritance in humans was investigated in normotensive and hypertensive populations. Pickering [5, 13] reported that the frequency distribution curve of arterial blood pressure in population samples did not demonstrate a bimodal curve and thus concluded that the separation of the disease of essential hypertension from the population was an artifact. He [5, 13] believed that there was no particular cause for elevated arterial blood pressure in essential hypertension. In 1954, Hamilton et al. [22] pointed out that the differences between subjects with essential hypertension and normotension were quantitative, not qualitative. They proposed that the term essential hypertension should be designated for a group of subjects with high blood pressure [22–26]. Pickering [5, 13], in 1955, proposed a polygenic theory based on an alternative quantitative hypothesis, and that the cause of essential hypertension was polygenic and the high blood pressure was the result of a number of factors. The three major factors in order were: age, inheritance, and environment. These factors elevate blood pressure in essential hypertension. Pickering [5, 13] concluded that blood pressure is inherited by multifactorial or polygenic inheritance as graded over the whole range from normotension to hypertension (Fig. 1).

Platt's Single Gene Hypothesis

In 1959, Platt [27–30] demonstrated that the frequency distribution curves of arterial pressure of siblings showed a dip at 150 mmHg systolic and 90 mmHg diastolic pressure (Fig. 2). The curves showed a division line between normotension and hypertension. Platt [27–30] postulated the existence of two populations: a population which consists of hypertensive subjects, in whom blood pressure rose steeply with increasing age, and a population with no rise according to age (Fig. 3).

From the blood pressure measurements of employees of the Metropolitan Life Insurance Company of New York, Thomson [31] noted in 1950 that the age of onset of diastolic hypertension was usually between 45 and 54 years and less common after that age, with hypertension defined as a diastolic blood pressure of 90 mmHg or over. Cruz-Coke [32], in 1959, demonstrated that the rise in diastolic blood pressure per year was nearly 6 mmHg in those who developed hypertension between the ages of 30 and 49, and 3.5 mmHg after the age of 50, but the rise in subjects with normotension was less than 1 mmHg between the age of 30 and 69. In 1959, Morrison and Morris [33] showed that the rise of pressure with age in the relatives of hypertensives was greater than in the rela-

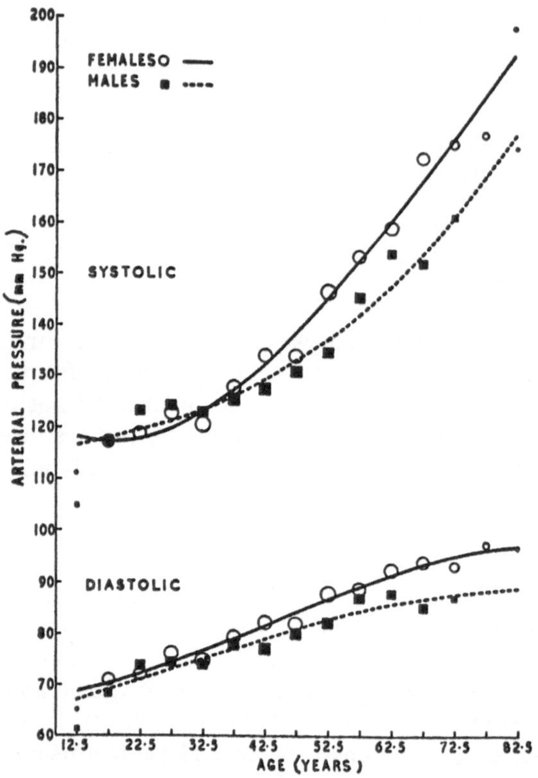

Fig. 1. Systolic and diastolic pressures for females (○) and males (■) for 5-year age groups of a population sample. Blood pressure increases with age. From Hamilton et al. [22]

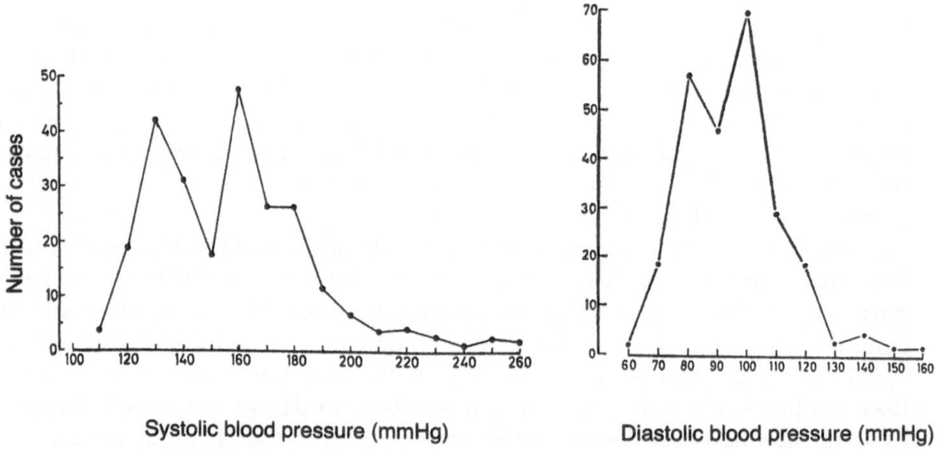

Fig. 2. The frequency distribution of systolic and diastolic blood pressures of siblings aged 45–60 years and of propositi aged 45–60 years with essential hypertension. The distribution curve is bimodal. From Platt [28]

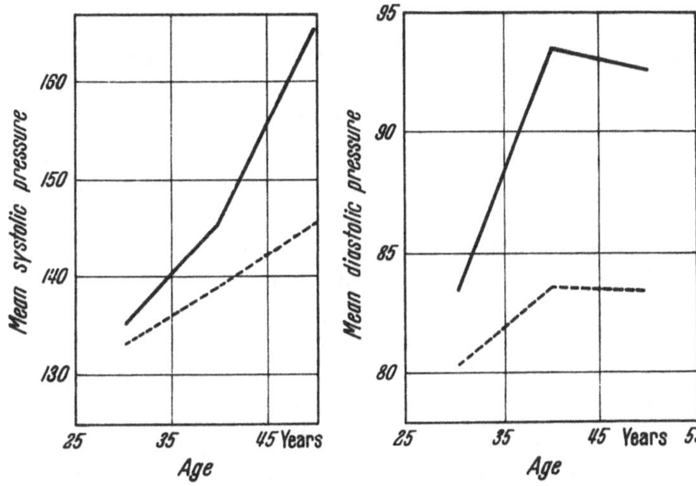

Fig. 3. The rise in blood pressure with age in the relatives of hypertensive (—) and normotensives (- - -). The increase is greater in relatives of hypertensives than in relatives of normotensives. From Platt [30]

tives of normotensives (Fig. 3). They suggested that the different rates of rise in blood pressure were dependent upon genetic, not environmental factors.

The evidence suggesting two distinct populations, normotensive and hypertensive, was very clear, indicating that essential hypertension is the manifestation of single gene inheritance as a Mendelian dominant [27–30].

Aoki's Major Gene Hypothesis

In 1985, Aoki [6, 14, 15] proposed that essential hypertension is inherited in the major hypertension gene which is transmitted from parents to children. The characteristic of the hypertension gene is to manifest hypertension around the age of 30.

The major hypertension gene induces an alteration in the cell membrane and sarcoplasmic membrane of arterial smooth muscle, which may consist of abnormalities of membrane potential, calcium channel, and calcium binding sites, etc. The abnormalities increase the calcium concentration in the cytosol of arterial smooth muscle through an increase in the influx and release of calcium or a reduction in the efflux of calcium. The increased concentration of calcium induces an excessive overlapping and a prolongation of interaction of actin and myosin filaments, producing overcontraction and incomplete relaxation of arterial smooth muscle. The abnormalities in the contractile process of arterial smooth muscle bring about shortening of the long length of the muscle and widening of the short length of the muscle, which reduces the size of the arterial lumen. Reduction of the diameter of the arterial lumen results in an increase in vascular resistance, which elevates blood pressure and brings about the development and persistence of hypertension in patients with gene (essential) hypertension [6, 34, 35].

The Three-Way Classification: Gene Hypertension, Essential Hypertension, and Disease Hypertension

In 1827, Bright [1] reported that hypertension is induced by renal disease. On the contrary, Frank [3], in 1911, defined essential hypertension as an arterial smooth muscle disease, and Platt [27–30] concluded that essential hypertension is an inherited disease. Weitz [21] proposed in 1923, the family origin of essential hypertension inherited by an autosomal dominant hypertension gene. On the contrary, Murphy [36] speculated in 1973 that the family component in the blood pressures was corrected, but the component might be environmental in origin, and concluded that essential hypertension was due to environmental and non-genetic factors.

In the biogenetic natural science area, selective inbred strains of rats [37–44], mice [45, 46], and rabbits [47, 48] have provided evidence that genetic factors determine the level of blood pressure. Selective breeding for blood pressure in animals with high × high and low × low concentrates genes for high and low blood pressure in the respective strains. This selective breeding process tends to "fix" the genes in the homozygous state [49]. The spontaneously hypertensive rats (Aoki SHR) have two to six loci controlling blood pressure. Since the number of loci is small, the identification of the major genes may be possible in SHR [49].

Since most genes (DNA) are translated via RNA into proteins, genetic differences must be expressed physiologically and biochemically through "mutant" proteins in the genetically hypertensive rat, SHR, as well as in patients with hypertension induced by the major hypertension gene.

Major Gene Hypertension

Rat Gene Hypertension

Smirk genetically hypertensive rats. Smirk and Hall [37] in 1958 and Phelan and Smirk [38] in 1960 reported a colony of genetic hypertensive (GH) rats by breeding rats with above-average blood pressure. The average blood pressure of the rats, which consisted of some hypertensive and some normotensive rats, was 132 mmHg. In 1958, the rats of the Otago hypertensive colony [40] had an average systolic blood pressure of approximately 140 mmHg, and the incidence of hypertension was 30%. In 1983, the GH rats of the Department of Medicine, University of Otago were in their 59th generation of selective inbreeding [40]. By the age of 10 days, the blood pressure in the GH rats was 5 mmHg higher than that in the noromotensive rats, and at the age of 6 weeks it was 25 mmHg higher. The blood pressure increased with age in both normotensive and GH rats, but the rate of increases was greater in GH rats than in normotensive rats [39]. The blood pressure of GH rats at 7–15 months was in the range of 152–220 mmHg with an average of 170 mmHg; that of the normotensive rats was in the range of 115–137 mmHg with an average of 130 mmHg. A number of modified diets, such as low-salt and high-salt diets, did not affect the blood pressure of GH rats [40].

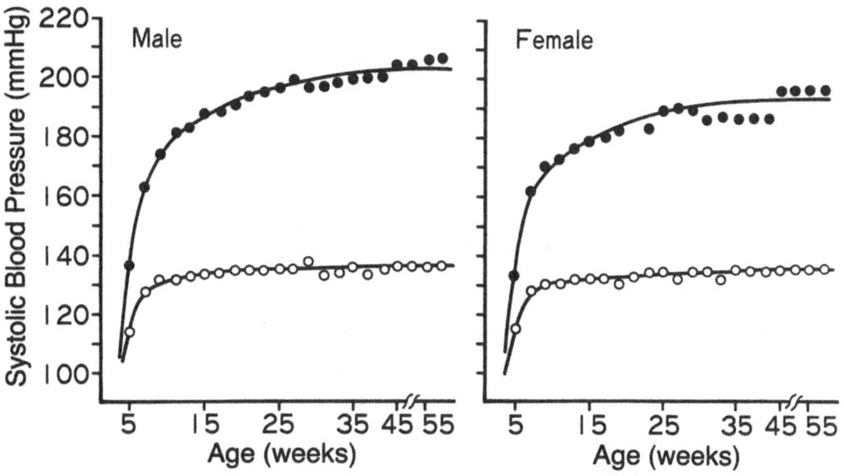

Fig. 4. The rise in systolic blood pressure with age in F_4 of SHR (●) (30 males and 19 females at 30 weeks of age) and normotensive rats (○) (24 males and 27 females). The rise in SHR is greater than in normotensive rats. After Okamoto and Aoki [41]

Hypertension in the GH rats was inherited from parents to offspring. The inheritance mode of high blood pressure could be revealed by crossbreeding, mating the GH rats with another pure line of rats. The gene analysis for blood pressure in the offspring showed that the inheritance of the level of blood pressure was additive. The number of hypertensive genes was at least five. Simpson and Phelan [40] concluded that blood pressure in GH rats was controlled by a large number of genes each with separate, distinct small effects. However, it is not clear in GH rats if hyertension is induced by a small or major hypertension gene.

Aoki spontaneously hypertensive rats. Aoki [41, 42] developed a strain of spontaneously hypertensive rats (SHR) by mating a male SHR with a normotensive female rat. All rats obtained from offspring of SHR developed hypertension after the F_3 generation (see Aoki, Discovery and Development of the Spontaneously Hypertensive Rat, pp 3–8). The average systolic blood pressure of F_3 SHR was 133, 180, and 192 mmHg at 5, 15, and 25 weeks of age in males, and 133, 168, and 179 mmHg in females, respectively [41, 42]. SHR showed a striking increase in blood pressure at the age of 5–15 weeks. The rate of the rise in blood pressure was greater in SHR than in normotensive Wistar rats (Fig. 4).

The low-sodium (0.079% sodium), standard-sodium (0.276%), and high-sodium (2.76%) diets did not affect blood pressure in the SHR. The blood pressure was 200, 192, and 193 mmHg in low-, standard-, and high-sodium diets in SHR, respectively, at the 25th week of the diet experiment. There was no difference in blood pressure among the varying sodium diets. These results indicated that hypertension in SHR was independent of sodium intake [43] (Fig. 5).

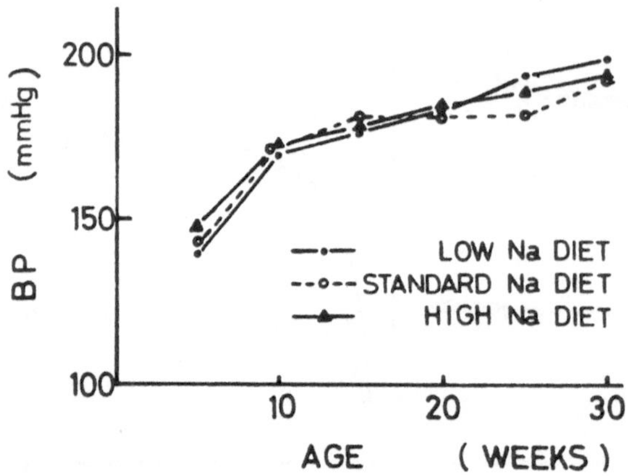

Fig. 5. Effects of salt intake on blood pressure in SHR. Blood pressures of the low-salt (0.079% sodium) (-●-), standard-salt (0.276%) (-○-), and high-salt (2.76%) (-▲-) diet-fed SHR are given. From Aoki et al. [43]

Random brother-sister inbreeding resulted in a uniform population of SHR, and there was no significant difference in average blood pressure between F_2, F_3, F_4 SHR in NIH [50]. The result of F_1 hybrid crosses between normotensive and hypertensive rats showed that there was no significant difference between the blood pressures of the progeny of the two reciprocal crosses. The average blood pressure of the groups of F_1 hybrids was intermediate between the values of normotensive and hypertensive rats [50]. Failure to obtain a uniform population of either hypertensive or normotensive rats in the F_1 hybrid group is not in accordance with either a simple dominant or recessive form of Mendelian inheritance. However, the possibility still remains that inheritance of hypertension in SHR is an incomplete dominant form. Louis et al. [50] have suggested that the progeny of SHR and normotensive rats favors a polygenic inheritance.

The genetic determination of blood pressure was analyzed by breeding three normotensive strains of rats with SHR [51]. The average blood pressure in F_1 hybrids of SHR and Wistar-Imamichi (WI) rats, was intermediate between that of the two parents. The F_2 hybrids showed an intermediate blood pressure level between that of the parent strains. The blood pressure mean values of the F_2 and backcross generations were lower than the expected values of an assumption of completely additive inheritance. There was no evidence either for bimodality of blood pressure distribution or backcross F_2 generations for the 1:2:1 ratio in normotension, the intermediate level of parents blood pressure, and hypertension. The values of genetic determination were calculated from the variance of the F_1 and F_2 generations. The values showed that this trait had a heritable characteristic and that the genetic determination of blood pressure in SHR was approximately 70% [51].

By cross-analysis of SHR and normotensive Wistar-Kyoto (WKY) rats, the F_1 hybrids showed an intermediate blood pressure level between SHR and WKY rats [51]. The degree of genetic determination derived from the variances of the F_1 and F_2 was 90%. These data indicated that hypertension was highly heritable and that the blood pressure was determined essentially by genetic factors. The calculated dominant ratio was greater than 90%. From these results, Tanase et al. [51] suggested the possibility of incomplete dominant inheritance of major hypertension genes. The calculated value of the dominant ratio from the genetic parameters showed that dominance did exist but was incomplete. All rats in SHR strain after the third generation developed hypertension, suggesting that the number of the major hypertension genes is three.

Tanase [52] demonstrated, in 1979, unequivocal Mendelian segregation of blood pressure in backcross and F_2 populations, and a single major locus influencing blood pressure in SHR. The blood pressure of the two homozygous genotypes at the ht locus differed by 32 mmHg in females and 38 mmHg in males. Other genes may modify the expression of the ht gene. The SHR allele at the ht locus was designated ht [52].

The vascular muscle from SHR responded to the cations of Co^{2+}, La^{3+}, Sr^{2+}, and Mn^{2+} with marked contraction, but that from normotensive rats did not [53]. The response of arterial muscle to Co^{2+} was controlled by a single autosomal locus with inheritance by partial dominance [49]. Genes controlling the arterial muscle response to Co^{2+} were the two homozygous types at the locus and differed in blood pressure by 15 mmHg. Cobalt possibly interacts with the regulatory functions of Ca^{2+} in the arterial muscle; the use of Co^{2+} unmasked some alterations in Ca^{2+} handling properties in the cell membrane of arterial muscle. The muscle protein involved in the SHR could be identified [49]. The locus controlling the arterial muscle response to Co^{2+} was named Hyp-2, for hypertension locus number 2.

The blood pressure of Nagaoka stroke-prone SHR (Nagaoka SHRSP, A_3 in Fig. 6) was significantly higher than that of stroke-resistant SHR (Nagaoka SHRSR, C in Fig. 6), which was suggestive of a positive correlation between high blood pressure and the incidence of cerebrovascular lesions [44].The average blood pressure of the F_1 breeding with A_3 and C was roughly intermediate between both parental SHR. In addition, the blood pressure in the F_2 obtained by F_1 mated with F_1 showed a continuous distribution within the range of blood pressure in A_3 and C. Similar genetic data were demonstrated in the hybrid groups between A_3 and WKY rats. The F_1 was intermediate from a blood pressure as high as that of A_3 to that of WKY rats (Fig. 6). These results seem to be compatible with the findings by Tanase et al. [51], who indicated that three major genes transmit the hypertension in SHR.

Human Gene Hypertension

In humans, gene (essential) hypertension is transmitted by the major hypertension gene. The characteristics of the gene hypertension are: (1) the chronic elevation of diastolic blood pressure, (2) blood pressure markedly increases during the ages of 30–50 years, and (3) an association with hypertensive cardiovascular hypertrophy. In most cases the systolic pressure is above the range

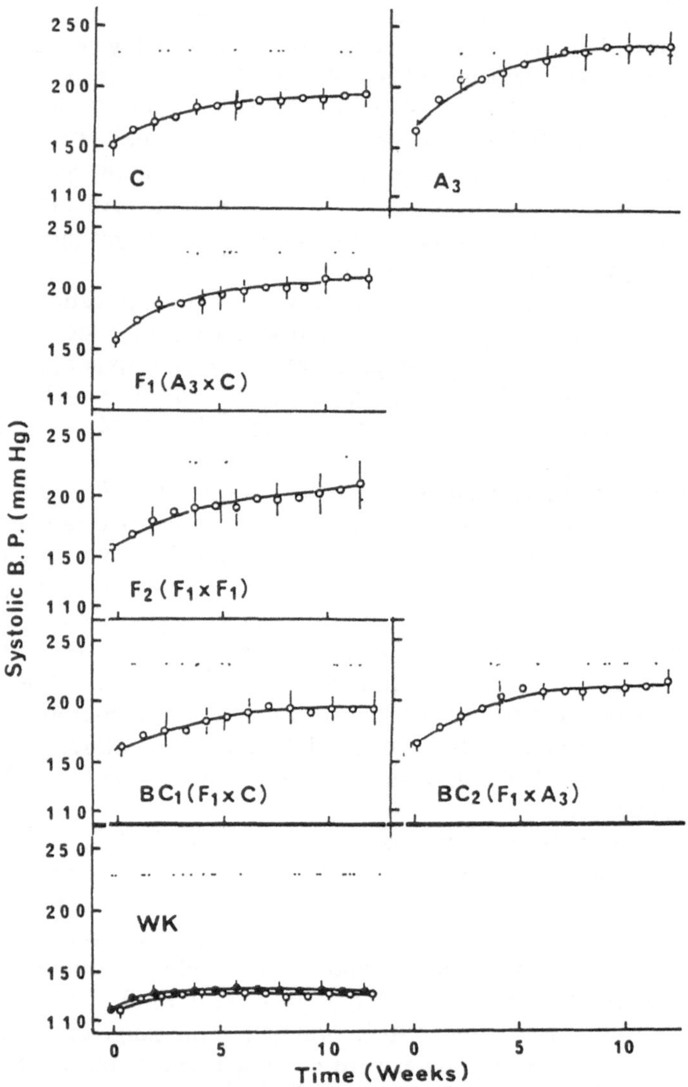

Fig. 6. Systolic blood pressure in Nagaoka stroke-prone SHR (A_3), stroke-resistant SHR (*C*), F_1, F_2, BC_1, BC_2, and Wistar-Kyoto noromotensive rats (*WK*). The average blood pressure of the F_1 from A_3 bred with C was roughly the intermediate between both SHR parents. From Nagaoka et al. [44]

generally encountered in the patients with the hypertension. Variations in blood pressure are greater in the patients with hypertension than in normotensive subjects. The majority of patients with hypertension have no symptoms referable to the blood pressure elevation in the early stage of hypertension. A typical symptom of elevated pressure in severe hypertension is headache, which is commonly localized to the occipital region.

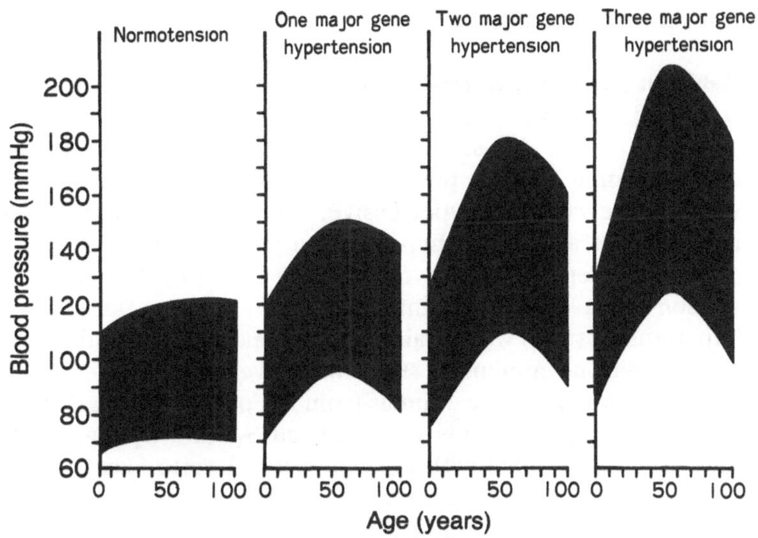

Fig. 7. Theoretical blood pressure is represented in non-hypertension gene normotension (normotension) and gene (essential) hypertension with one to three major hypertension genes. The rate of blood pressure increase with age in subjects with gene hypertension is greater than that in subjects with normotension. The blood pressure of the three-gene hypertension is higher than that of the two-gene hypertension. The blood pressure of two-gene hypertension is higher than that of the one-gene hypertension. From Aoki [14, 15]

The levels of blood pressure in gene hypertension may be determined by the effects of major genes which are a completely additive inheritance. By the major gene hypothesis, gene hypertension can be divided into three groups: one-gene hypertension, two-gene hypertension, and three-gene hypertension, which may correspond to mild, moderate, and severe hypertension, respectively [14, 15] (Fig. 7).

Environment (Accessory Gene) Hypertension

Environment hypertension is developed by the interaction of an environmental factor on an accessory hypertension gene. Thus, this hypertension is inherited through the accessory gene. Environment hypertension includes salt-sensitive hypertension, alcohol hypertension, mental stress hypertension, and physical stress hypertension [6, 14, 15]. Salt hypertension develops in subjects with an accessory hypertension gene which are susceptible to the hypertensive effect of an excessive intake of salt. Diminution of salt intake in patients with salt hypertension decreases blood pressure. Obesity hypertension is induced by overweight in subjects with an accessory hypertension gene which is susceptible to the pressor effect of obesity.

Rat Environment Hypertension

Dahl salt-hypertensive rats. Dahl et al. [54], in 1962, observed that some rats remained normotensive despite the fact that they were fed chronically large amounts of salt (11% sea salt equivalent to 7% NaCl in the diet). They pointed out that some animals never developed hypertension after the high-salt diet whereas a few became hypertensive. Dahl et al. [54] reasoned that if the sensitivity to salt as a cause of hypertension was genetically controlled, it should be possible to separate the two strains, a strain of rats which developed hypertension from excess salt intake, and another strain which did not. A salt-hypertensive strain was obtained by the selective inbreeding of salt-hypertension rats; the salt-normotensive strain was developed by selective inbreeding of salt-normotension rats. The animals from the parents with salt hypertension and the subsequent 2 generations were labeled S_1, and S_2, and S_3, respectively. The animals from parents with normotension despite excess salt ingestion and the subsequent two generations were labeled R_1, R_2, and R_3. The Response of blood pressure to salt intake differed between salt-sensitive S_3 and salt-resistant R_3 rats. None of the 39 R_3 rats showed hypertension with high-salt feeding. Neither the presence nor absence of excess dietary salt affected the blood pressure of the R_3 rats. Whereas 49 of the 60 S_3 rats were hypertensive with the high-salt diet. The mean systolic blood pressure of salt-fed S_3 males was 180 mmHg. When the S_3 rats were maintained on the control diet, hyertension did not develop. These S strain rats might have the gene to develop hypertension under high-salt intake which unmasked the effect of the hypertension gene and caused hypertension to develop in Dahl S rats (Fig. 8).

By demonstrating two strains of rats, S and R, which differ in their susceptibility to excess salt ingestion, Dahl et al. [54] concluded that the sensitivity of the pressor response to excess salt intake was genetically transmitted in S rate but not in R rats.

Human Environment Hypertension

We have observed that an excessive intake of salt, calories, or alcohol elevates blood pressure in some subjects, but not in others. A similar phenomenon may hold true in pyelonephritis. Some patients with pyelonephritis develop hypertension, but many patients with pyelonephritis do not. Pyelonephritis may have manifested a genetic predisposition to hypertension. Our hypothesis would predict that environment hypertension could develop hypertension by the interaction of an environmental factor on a genetic disposition. This genetic disposition is transmitted by the accessory hypertension gene.

Kawasaki's salt hypertension. The effects of dietary sodium intake on blood pressure have been investigated in various types of human hypertension. The results obtained demonstrate that factors other than dietary or intrinsic sodium play a key role in the development of hypertension, since salt loading does not

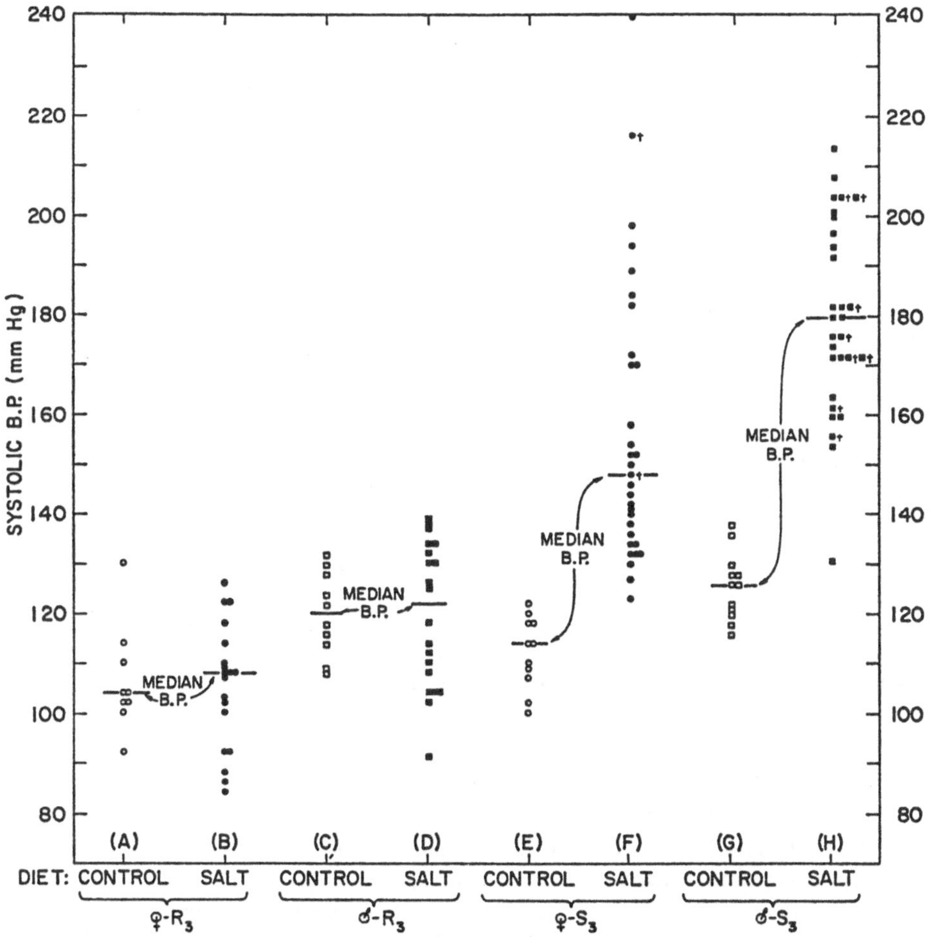

Fig. 8. Evidence of genetic variation in susceptibility to a high-salt diet. Effect of dietary salt alone, on systolic blood pressure of rats inbred for resistance (R_3) and sensitivity (R_3) to triiodothyronine-salt regimen. Blood pressures of R_3 animals were normotensive after 6 months, whereas those of S_3 animals were hypertensive after only 3 months on the same diets. + Animal died before 3rd month; B.P. shown was recorded after only 1–2 months on the regimen. From Dahl et al. [54]

increase blood pressure in some subjects (55–57). Kawasaki et al. [17], in 1978 have studies the effects of dietary sodium intake on blood pressure in idiopathic hypertension. The patients received the basic diet containing 100 mEq sodium per day for one week. Then they received a low-sodium diet of 9 mEq sodium per day for one week. Finally, these patients were put on a high-sodium diet of 249 mEq sodium per day for one week. Patients whose average mean blood pressure value on day 6 of the high-sodium diet exceeded more than 10%, and those with less than 10% were defined as "salt-sensitive" and "salt-

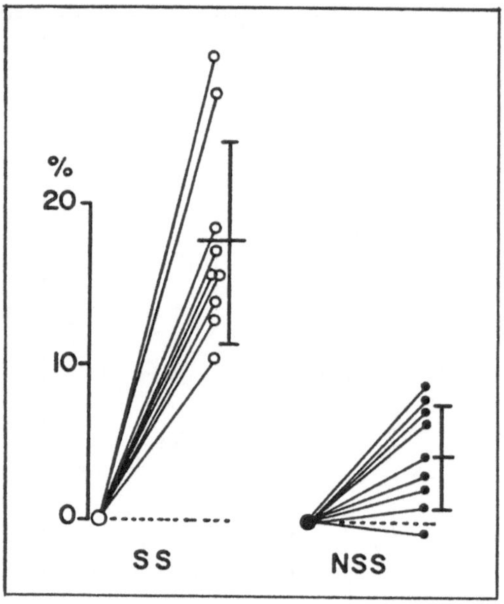

Fig. 9. Percent increase of mean arterial pressure in salt-sensitive (*SS*) and salt-nonsensitive (*NSS*) patients with hypertension when the diet was changed from low sodium to high sodium. Values = M ± SD. From Kawasaki et al. [17]

nonsensitive", respectively. The mean blood pressure in the average-sodium, low-sodium, and high-sodium diet was 120 ± 5 (M ± SEM), 105 ± 4, and 124 ± 4 mmHg in the salt-sensitive patients, and 114 ± 4, 110 ± 3 and 113 ± 4 mmHg in the salt-nonsensitive patients, respectively. The average increase in mean blood pressure was significantly greater in the salt-sensitive group (19 ± 2 mmHg) than in the salt-nonsensitive group (3 ± 1 mmHg) (Fig. 9). The salt-nonsensitive patients excreted more sodium and gained less weight during the high-sodium diet than the salt-sensitive patients, which indicated a defect in sodium excretion from the kidney in the salt-sensitive patients. Patients with idiopathic hypertension may be divided into "salt-sensitive" and "salt-nonsensitive" hypertension from the changes in blood pressure from low-sodium to high-sodium intake. The factors of the salt-sensitive or salt-nonsensitive are probably of genetic origin as in Dahl's S or R rats. The environmental factor of high-sodium intake may interact with the sodium-sensitive accessory gene, which elevates blood pressure in "salt-sensitive" hypertension. Thus, salt hypertension is induced by the environmental factor of salt intake.

Obesity hypertension. Obesity may be defined as an excess of body fat. The association of obesity and hypertension is a constant epidemiological finding [18–20, 58, 59]. We have observed that some subjects with a body weight gain become hypertensive, but others remain normotensive. A relatively small reduction in weight of approximately 5 kg in Japanese obesity hypertensive patients reduces the blood pressure by 30/20 mmHg. Hypotensive effects of weight loss have been consistently observed in obesity hypertension [18–20]. Young adults are vulnerable to the hypertensive effects of weight gain or the hypotensive effects of weight loss. From the family history, the obese population can be separated into two classes: obesity with an increase in blood pressure, and obesity with no increase in blood pressure. There are "obesity-sensitive" hypertensive and "obesity-nonsensitive" normotensive subjects. The sensitivity to obesity as a cause of hypertension may be genetically controlled. This hypertension gene is called the obesity-sensitive accessory hypertension gene.

Alcohol hypertension. Chronic intake of excessive amounts of alcohol leads to a transient and persistent elevation in blood pressure [59–65]. For example, when a group of 132 patients who consumed ethanol were admitted to the hospital, 52% were hypertensive. Those who remained abstinent over the next 6 months tended to remain normotensive [63]. Hypertension due to alcohol intake is a reversible hypertension. The question remains whether chronic alcohol consumption causes blood pressure elevation or whether those most prone to drink have a predisposition to hypertension. Reduction in blood pressure in alcoholics who abstain, indicates that a combination of alcohol with predisposition is the cause of hypertension [64]. Studies of Twins can clarify the contribution of genetic factors to the association of alcohol consumption and blood pressure. The data for mono- and dizygous twins shows that a significantly greater proportion of high alcohol consumers have a history of hypertension than the low alcohol consumers. There is a significant association between alcohol consumption and blood pressure, with differences mainly in to the larger dizygous group [65]. The resutls suggest that alcohol intake may interact with an alcohol-sensitive accessory hypertension gene to elevate blood pressure and develop hypertension.

Disease Hypertension

In 1914, Volhard and Fahr [4] proposed that all hypertension was due to renal disease. This view was widely accepted until the 1940's. It is now generally accepted that renal damage is the result of hypertension rather than the cause, and that a defect in renal function is involved in the late stage of essential hypertension. It is true that some diseases cause hypertension. Therefore, hypertension is often associated with chronic renal disease, primary aldosteronism, pheochromocytoma, hyperthyroidism, and other diseases. We [6, 14, 15] have named this hypertension "disease hypertension", "disease-induced hypertension" or "disease-dependent hypertension".

Animal Disease Hypertension

Since the first animal model of hypertension was induced by renal arterial constriction in dogs by Goldblatt et al. [8] in 1934, it has been known that several kinds of diseases are associated with hypertension.

Goldblatt's renovascular hypertension. Goldblatt et al. [8] suggested that ischemia in the kidneys is the initial condition for the development of this hypertension. Either almost complete occulusion of the unilateral renal artery or moderate constriction of both main renal arteries causes the persistent rise in blood pressure. These observed results indicate that renal ischemia from the obstruction of renal arteries is a condition leading to persistent elevation of blood pressure. Goldblatt hypertension is a renal vascular disease- or renal ischemic disease-induced hypertension. This mode of hypertension falls in the category of disease hypertension.

Neurogenic hypertension. Since the discovery of the aortic depressor nerves by Cyon and Ludwig [66] in 1866, the role of baroreceptor in hypertension has been investigated [67]. However, there is no definitive evidence for the development of hypertension due to sinoaortic denervation [68]. External stimuli produce an elevation of blood pressure in animals with sinoaortic denervation [69], but the pressure in the animal may be normal or low during rest and sleep. Lowe [70] has found that patients with bilateral common carotid occlusion do not have particulary high blood pressure. Many studies have confirmed that sinoaortic denervation does not produce persistent hypertension [68]. Stimuli that evoke the defense reflex, produce cardiovascular effects similar to those of hypothalamic stimulation. The evidence demonstrates that the chief role of the reflex is to buffer the short term and quick changes in blood pressure. Experimental neurogenic hypertension shows an elevation of blood pressure during stimuli and low blood pressure during rest. Neurogenic hypertension is a transient hypertension, which is induced by nervous diseases.

The importance of the brain in hypertension is often neglected [71] despite experimental and clinical evidence of involvement [72]. Lesions of the bilateral nucleus of the tractus solitarii (NTS) induced by an electric method can produce acute and transient hypertension [72]. The lesions abolish arterial baroreceptor reflexes. NTS lesions in rats result in a profound, immediate, and partially differentiated excitation of sympathetic neurons, leading to an elevation in peripheral vascular resistance and a rise in blood pressure. The chronic state after NTS lesioning is characterized by marked lability of blood pressure, tachycardia, and exaggerated responses of blood pressure to evoked behaviors. The state can result from lack of inhibition of the sympathetic vasomotor outflow. Defects of the neurotransmitter systems, particularly the noradrenergic innervation of the NTS, could over time, possibly lead to marked lability of blood pressure due to stimuli. Thus, the neurogenic hypertension by electric-induced lesioning of NTS is characterized by a marked lability of blood pressure due to stimuli and a normal or low blood pressure during rest and sleep. These observations point out that there is no evidence for an organic brain disorder as the cause of the persistent elevation of blood pressure [73].

Skelton's adrenal regeneration hypertension. Skelton [74], in 1955, studied the effect of bilateral adrenalectomy on the production of hypertension by methylandrostenediol in the rat. He found that one control animal developed hypertension. The right adrenal gland of this rat had been completely removed, but a large regeneration adrenal cortex was present on the left side attached to the adrenalectomy scar. He considered that the hypertension was the result of excessive hormonal secretion by the regenerated adrenal cortical tissue. Skelton [74] demonstrated that the effects of unilateral nephrectomy, unilateral adrenalectomy with contralateral adrenal enucleation and 1 % NaCl solution as drinking water, in young female rats bring about the development of hypertension. The capsule and some adherent cells of the zona glomerulosa were left in situ. Systolic blood pressure of the adrenal-enucleated rats reached 217 ± 9 mmHg at 7 weeks after the operation. After the hypertension had been established, removal of the regenerated adrenal or substitution of water for saline drinking fluid caused a reduction in blood pressure to normotension. A similar syndrome of adrenal-regeneration hypertension was produced by hormone overdosage [75]. The results suggested that the adrenal-regeneration hypertension may be the result of the interaction of salt with oversecretion of some hormones from the regenerating adrenal cortex. It was concluded that the increased function of adrenal cortex is of significance in the pathogenesis of salt-adrenal-regeneration hypertension. [74–76].

Human Disease Hypertension

Renovascular hypertension. Renal arterial obstruction causes renovascular hypertension in humans. The arterial obstruction leads to a decrease in 50% of the cross-sectional area of the lumen, which brings about a systolic pressure gradient of 30 mmHg. The pressure gradient stimulates juxtaglomerular cells, which liberates renin in plasma. Renin from the obstructed kidney induces renovascular hypertension. The renin acts on angiotensinogen to liberate angiotensin I. The two terminal amino acids histidyl (radical of histidine) and leucine, of angiotensin I are split by the conventing enzyme to yield angiotensin II which is the most powerful pressor substance. The increased angiotensin II leads to increased peripheral vascular resistance, which induces hypertension in the initial stages of renovascular hypertension.

Renovascular hypertension is developed by a decrease in the cross-sectional area of the lumen of the renal artery. Therefore, renovascular hypertension is one of the most well-known types of disease hypertension.

Primary aldosteronism hypertension. Since Conn (1955) [77] described aldosteronism, primary aldosteronism has been discussed as an endocrine mechansim of essential hypertension. The true primary aldosteronism is Conn's syndrome, which occurs in less than 0.5% of unselected hypertensive patients. Primary aldosteronism, characterized by the overproduction of aldosterone, induces hypokalemia, metabolic alkalosis, sodium retension, and hypertension. Thus, the patients may have the symptoms of hypokalemia, such as muscle weakness, polyuria, nocturia, polydipsia, paresthesia, tetany, muscle paralysis, and headache. Plasma sodium is toward the upper limit of the normal range or above. Total exchangeable sodium is increased, and total body sodium is elevated. Plasma volume, extracellular fluid volume, and total body water and expanded.

Cardiac output is high; vascular resistance is within the normal range. The treatment of Conn's syndrome is the removal of the adenoma, or inhibition of the sodium and potassium transport in the distal renal tubules by the use of the spironolactone or amiloride. Removal of the adenoma is the choice of treatment, which decreases blood pressure to normal.

The hypertensinogenic action of aldosterone may be related to its effects of sodium retention via its action on "mineralcorticoid" receptor. However, the mechanisms of hypertension remain unexplained. The hypertension is a clinical manifestation of the disease of primary aldosteronism. Thus, this hypertension is in the category of disease hypertension.

Pheochromocytoma hypertension. Pheochromocytoma is a tumor which secretes excessive catecholamines into circulation. These catecholamines are norepinephrine, epinephrine, and dopamine. The incidence of pheochromocytoma is less than 0.1% of patients with newly diagnosed hypertension. The clinical features of the disease in accordance with the level of catecholamines. Fifty percent of these patients have paroximal hypertension. A typical attack of hypercatecholaminemia is characterized by the marked elevation of blood pressure with severe headache, profuse sweating, palpitation, pallor, nausea, abdominal discomfort, chest discomfort, paresthesia, and anxiety. The attack appears spontaneously, but occasionally it is brought by emotion, smoking, exercise, change of posture, abdominal palpitation, defecation, sexual activity, medical examination, administration of β-blocker, or anesthesia. An attack is induced by drugs, such as synephrine in orange juice, amobarbital, or nitroglycerine. The crisis is short, less than 15 min, and after the crisis patients experience profound fatigue. Postural hypotension is frequently observed. The hypertension arises from the catecholamine-induced vasoconstriction which causes an elevation of vascular resistance. Hypertension and symptoms disappear after removal of the pheochromocytoma. Pheochromocytoma hypertension is a type of disease hypertension.

Mechanisms of Elevation of Blood Pressure in Hypertensive Diseases

A schema of the mechanisms of elevation of blood pressure in gene, environment, and disease hypertension is shown in Fig. 10. In gene hypertension, the major hypertension gene induces alteration of arterial smooth muscle, which produces contraction of arterial smooth muscle. The contraction brings about vasoconstriction of arteries, which leads to an increase in vascular resistance. The increased vascular resistance results in the elevation of blood pressure, which is hypertension. Therefore, the major hypertension gene is the trigger in the elevation of blood pressure in gene (essential) hypertension (Fig. 10).

Environment hypertension is developed by the interaction of an accessory hypertension gene with an environmental factor. Salt hypertension, obesity hypertension, and alcohol hypertension are within the category of environment hypertension. An interaction of the salt-sensitive gene with an excessive intake

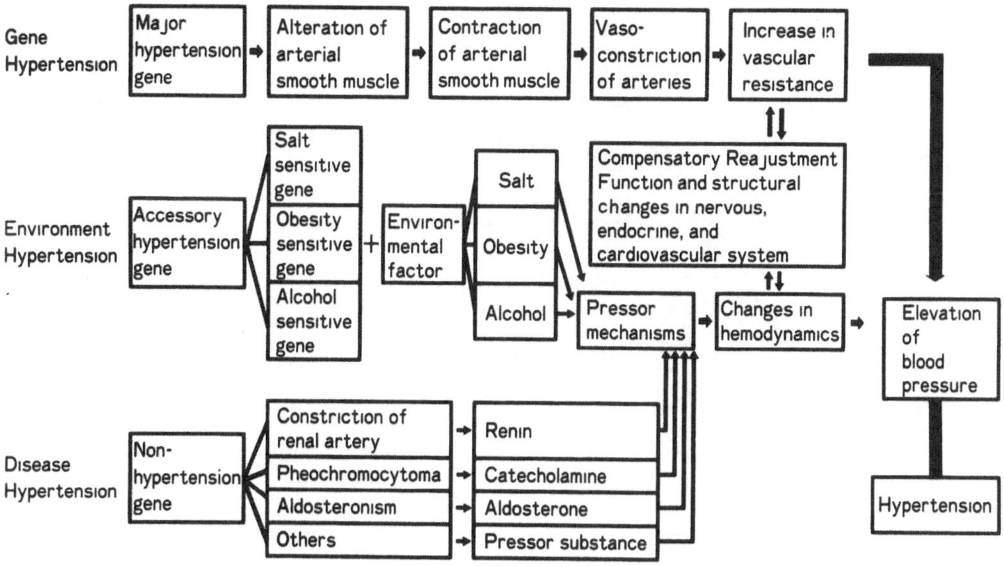

Fig. 10. A schematic demonstration of the mechanisms of elevation of blood pressure in gene, environment, and disease hypertension

of salt, the obesity-sensitive gene with obesity, and the alcohol-sensitive gene with excessive alcohol intake effect pressor mechanisms, which alters the hemodynamics. The hemodynamic change brings about elevated blood pressure and causes hypertension (Fig. 10).

Disease hypertension is caused by different types of diseases, including renal arterial constriction, pheochromocytoma, aldosteronism, and others. The diseases effect the pressor mechanisms mostly through the pressor substance, which changes the hemodynamics. The hemodynamic changes bring about the elevation of blood pressure and thus hypertension (Fig. 10).

Compensatory mechanisms for the increase in vascular resistance and changes in hemodynamics induce functional and structural changes in the nervous system, and endocrine system as well as the heart and arteries.

Conclusion

The etiological three-way classification of hypertension consists of three types of hypertension: Type 1 is gene hypertension (major hypertension gene, essential), Type 2 is environment hypertension (accessary gene with environment factor), and Type 3 is disease hypertension (non-gene). This classification is applicable for both experimental and clinical cases of hypertension. The new classification may aid the practical research on the mechanisms, therapy, and prevention of hypertension (Fig. 11).

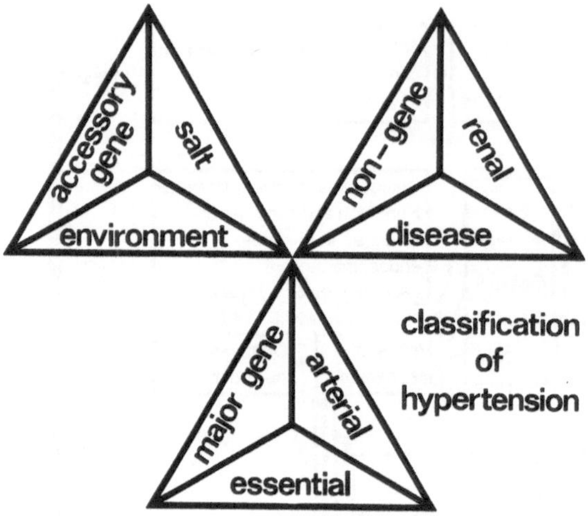

Fig. 11. The three-way classification of hypertension

References

1. Bright R (1827) Reports of medical cases, selected with a view of illustrating the symptoms and cure of diseases by a reference to morbid anatomy. Longman, London
2. Gull WW, Sutton HG (1872) On the pathology of the morbid state commonly called chronic Bright's disease with contracted kidney: Arterio-capillary fibrosis. Med Chir Trans 55: 273–326
3. Frank E (1911) Bestehen Beziehungen zwischen chromaffinem System und der chronischen Hypertonie des Menschen? Ein kritischer Beitrag zu der Lehre von der physio-pathologischen Bedeutung des Adrenalines. Dtsch Arch klin Med 103: 397–412
4. Volhard F, Fahr T (1914) Die Brightsche Nierenkrankheit. Klinik, Pathologie und Atlas. Springer, Berlin
5. Pickering GW (1955) The classification of hypertension. In: High blood pressure. Grune and Stratton, New York, J and A Churchill, London, pp 122–130
6. Aoki K (1985) Essential hypertension and secondary hypertension in humans and rats. Asian Med J 28: 529–548
7. Tiegerstedt R, Bergman PG (1898) Niere and Kreislauf. Scand Arch Physiol 8: 223–240
8. Goldblatt H, Lynch J, Hanzal RF, Summerville WW (1934) Studies on experimental hypertension. J Exp Med 59: 347–379
9. Pickering GW, Prinzmetal M (1938) Some observations on renin, a pressure substance contained in normal kidney, together with a method for its biological assay. Clin Sci 3: 211–227
10. Huchard H (1889) Maladies du coeur et des vaisseaux. Doin, Paris
11. von Bach S (1893) Ueber latente Arteriosclerose und deren Beziehung zu Fettleibigkeit, Herzerkrankungen und anderen Begleiterscheinungen. Urban & Schwartzenberg, Vienna
12. Allbutt TC (1895) Diseases of the arteries, including angina pectoris. Macmillan, London
13. Pickering GW (1961) The aetiology of essential hypertension, the genetic factor. In: The nature of essential hypertension. J and A Churchill, London, pp 22–57
14. Aoki K (1986) Etiological classification of hypertension: Essential hypertension, environment hypertension, and disease hypertension. In: Aoki K (ed) Essential hyper-

tension, calcium mechanisms and treatment, Springer, Tokyo Berlin Herdelberg New York London Paris, pp 11–24
15. Aoki K (1988) The three-way classification of hypertension: Essential hypertension, environment hypertension, and disease hypertension. In: Aoki K, Frohlich ED (eds) Calcium in essential hypertension. Academic Press, Tokyo San Diego New York London, pp 9–36
16. Weinberger MH, Miller JZ, Luft FC, Grim CE, Fineberg NS (1986) Definitions and characteristics of sodium sensitivity and blood pressure resistance. Hypertension 8 (Suppl II): 127–134
17. Kawasaki T, Delea CS, Bartter FC, Smith H (1978) The effect of high sodium and low sodium intake on arterial pressures and other related variables in humans subjects with ideopathic hypertension. Am J Med 64: 193–198
18. Reisin E (1983) Obesity and hypertension: Effect of weight reduction. In: Robertson JIS (ed) Clinical aspects of essential hypertension. Handbook of hypertension, vol 1. Elsevier, Amsterdam, pp 30–43
19. Reisin E, Abel R, Modern M, Silverberg DS, Eliahou ME, Modan B (1978) Effect of weight loss without salt reduction on the reduction of blood pressure in overweight hypertensive patients. N Engl J Med 298: 1–6
20. Reisin E, Frohlich ED, Messerli FH, Dreslinski GR, Dunn FG, Jones MM, Batson HM Jr. (1983) Cardiovascular changes after weight reduction in obesity hypertension. Ann Intern Med 98: 315–319
21. Weitz W (1923) Zur Atiologie der genuinen order vasculären Hypertonie. Z Klin Med 96: 151–181
22. Hamilton M, Pickering GW, Roberts JAF, Sowry GSC (1954) The aetiology of essential hypertension. 1. The arterial pressure in the general population. Clin Sci 13: 11–35
23. Hamilton M, Pickering GW, Roberts JAF, Sowry GSC (1954) The aetiology of essential hypertension. 2. Scores for arterial blood pressures adjusted for differences in age and sex. Clin Sci 13: 37–49
24. Hamilton M, Pickering GW, Roberts JAF, Sowry GSC (1954) The aetiology of essential hypertension. 4. The role of inheritance. Clin Sci 13: 273–304
25. Oldham PD, Pickering GW, Roberts JAF, Sowry GSC (1960) The nature of essential hypertension. Lancet I: 1085–1093
26. Pickering WG (1960) Inheritance of high blood pressure. In: Bock KD, Cottier PT (eds) Essential hypertension. Springer, Berlin, pp 30–38
27. Platt R (1947) Heredity in hypertension. Quart J Med NS 16: 111
28. Platt R (1959) The nature of essential hypertension. Lancet II: 55–57
29. Platt R (1959) The nature of essential hypertension. Lancet I: 1189–1190
30. Platt R (1960) The nature of essential hypertension. In: Bock KD, Cottier PT (eds) Essential hypertension. Springer, Berlin, pp 39–44
31. Thomson KJ (1950) Proceedings of the 38th annual meeting of the medical section of the american life convention
32. Cruz-Coke R (1959) The nature of essential hypertension. Lancet II: 853
33. Morrison SL, Morris JN (1959) Epidemiological observations on high blood-pressure without evident cause. Lancet II: 864–870
34. Aoki K (1986) Calcium membrane theory of essential hypertension. In: Aoki K (ed), Essential hypertension, calcium mechanisms and treatment. Springer, Tokyo Berlin Heidelberg New York London Paris, pp 223–242
35. Aoki K (1988) The calcium membrane theory of essential hypertension. In: Aoki K, Frohlich ED (eds) Calcium in essential hypertension. Academic, Tokyo San Diego New York London, pp 623–653
36. Murphy EA (1973) Genetics in hypertension, a perspective. Circ Res 32; 33 (Suppl I): I–129–137
37. Smirk FH, Hall WH (1958) Inherited hypertension in rats. Nature 182: 727–728
38. Phelan EL, Smirk FH (1960) Cardiac hypertrophy in genetically hypertensive rats. J Pathol Bacteriol 80: 445–448

39. Jones DR, Dowd DA (1970) Development of elevated blood pressure in young genetically hypertensive rats. Life Sciences 9 (I): 247–250
40. Simpson FO, Phelan EL (1984) Hypertension in the genetically hypertensive strain. In: De Jong W (ed) Experimental and genetic models of hypertension. Handbook of Hypertension, vol 4. Elsevier, Amsterdam, pp 200–223
41. Okamoto K, Aoki K (1963) Development of a strain of spontaneously hypertensive rats. Jpn Circ J 27: 282–293
42. Aoki K (1986) Discovery of the spontaneously hypertensive rats. In: Aoki K (ed), Essential hypertension, calcium mechanisms and treatment. Springer, Tokyo Berlin Heidelberg New York London Paris, pp 3–7
43. Aoki K, Yamori Y, Ooshima A, Okamoto K (1972) Effects of high or low sodium intake in spontaneously hypertensive rats. Jpn Circ J 36: 539–545
44. Nagaoka A, Iwatsuka H, Suzuki Z, Okamoto K (1976) Genetic predisposition to stroke in spontaneously hypertensive rats. Am J Physiol 230: 1354–1359
45. Schlager G (1974) Selection for blood pressure levels in mice. Genetics 76: 537–549
46. Schlager G, Weibust RS (1967) Genetic control of blood pressure in mice. Genetics 55: 497–506
47. Alexander N, Hinshaw LB, Drury DR (1954) Development of a strain of spontaneously hypertensive rabbits. Proc Soc Exp Biol Med 86: 855–858
48. Fox RR, Schlager G , Laird CG (1969) Blood pressure in thirteen strain of rabbits. J Hered 60: 312–314
49. Rapp JP (1983) Genetics of experimental and human hypertension. In: Genest J, Kuchel O, Hamet P, and Cantin M (eds), Hypertension, pathophysiology and treatment, 2nd edn. McGraw-Hill, New York, pp 582–598
50. Louis WJ, Tabei R, Sjoerdsma A, Spectors (1969) Inheritance of high blood-pressure in the spontaneously hypertensive rats. Lancet I: 1035–1036
51. Tanase H, Suzuki Y, Ooshima A, Yamori Y, Okamoto K (1970) Genetic analysis of blood pressure in spontaneously hypertensive rats. Jpn Circ J 34: 1197–1212
52. Tanase J (1979) Genetic control of blood pressure in spontaneously hypertenive rats (SHR). Exp Anim 28: 519–530 (in Japanese)
53. Shibata S, Kurahashi K, Kuchii M (1973) A possible etiology of contractility impairment of vascular smooth from spontaneously hypertensive rats. J Pharmacol Exp Ther 185: 406–417
54. Dahl LK, Heine M, Tassinari L (1962) Effects of chronic excess salt ingestion. Evidence that genetic factors play an important role in susceptibility of experimental hypertension. J Exp Med 115: 1173–1190
55. Hatch FT, Wertheim AR, Eurman GH, Watkin DM, Froeb HF, Epstein HA (1954) Effects of diet in essential hypertension. Am J Med 17: 499–513
56. Tobian L (1974) Hypertension and the kidney. Arch Intern Med 133: 959–967
57. Miall WE, Oldham PD (1963) The hereditary factor in arterial blood pressure. Br Med J I-1 (No 5323) 75–80
58. Stamler R, Stamler J, Riedlinger WF, Algera G, Roberts FH (1978) Weight and blood pressure findings in hypertension screening of 1 million Americans. J Am Med Assoc 240: 1607–1610
59. Memorandum from a WHO/ISH Meeting (1986) 1986 guidelines for the treatment of mild hypertension. J Hypertension 4: 383–386
60. Puddey IB, Beilin LJ, Vandongen R, Rogers P (1985). Evidence for a direct effect of alcohol on blood pressure in normotensive men—a randomised controlled trial. Hypertension 7: 703–713
61. Puddey IB, Beilin LJ, Vandongen R (1987) Regular alcohol use raises blood pressure in treated hypertensive subjects: A randomised controlled trial. Lancet I: 647–651
62. Saunders JB, Beevers DJ, Paton A (1981) Alcohol-induced hypertension. Lancet II: 653–656
63. Arkwright PD, Beilin LJ, Vandongen RV, Rowse IA, Labor C (1982) The pressor effect of moderate alcohol consumption in man: A search for mechanisms. Circulation 66: 515–519

64. Beilin JF, Arkwright PD (1983) Alcohol and hypertension. In: Robertson JIS (ed), Clinical aspects of essential hypertension. Handbook of Hypertension, vol 1. Elsevier, Amsterdam, pp 44–63
65. Myrhed M (1974) Alcohol consumption in relation to factors associated with ischemic heart disease: A co-twin control study. Acta Med Scand (Suppl) 567: 1–93
66. Cyon E, Ludwig C (1886) Die Reflexe eines der sensiblen Nerven des Herzens auf die motorischen der Blutgefasse. Verh Kgl Ges Wiss, Leipziar, 18: 307–328
67. Pickering TG, Sleight P (1977) Baroreceptors and hypertension. In: De Jong W, Provoost AP, Shapiro AP (eds) Hypertension and brain mechanisms. Elsevier, Amsterdam, pp 43–60
68. Heymans C, Neil E (1958) Reflexogenic areas of the cardiovascular system, Churchill, London
69. Masson GMC, Aoki K, Page IH (1966) Effects of sinoaortic denervation on renal and adrenal hypertension. Am J Physiol 211: 99–104
70. Lowe RD (1961) Ischaemia of the brain as a cause of chronic hypertension in man. Clin Sci 21: 403–407
71. De Jong W, Zandberg P, Palkovists M, Bohus B (1977) Acute and chronic hypertension after lesions and transections of the rat brain stem. In: De Jong W, Provoost AP, Shapiro AP (eds) Hypertension and brain mechanisms. Elsevier, Amsterdam, pp 189–197
72. Doba N, Reis DJ (1973) Acute fulminating neurogenic hypertension produced by brainstem lesions in the cat. Circ Res 32: 584–593
73. Magnus O, Koster M, Vander Drift JHA (1977) Cerebral mechanisms and neurogenic hypertension in man, with special reference to baroreceptor control. In: De Jong, Provoost AP, Shapiro AP (eds) Hypertension and brain mechanisms. Elsevier, Amsterdam, pp 199–218
74. Skelton FR (1955) Development of hypertension and cardiovascular-renal lesions during adrenal regeneration in the rat. Proc Soc Exp Biol Med 90: 342–346
75. Floulkes R, Gardiner SM, Bennett T (1987) Adrenal regeneration in the rat. J Hypertens 5: 637–644
76. Selye H, Hell CE, Rowley EM (1943) Malignant hypertension produced by treatment with desoxycorticosterone acetate and sodium chloride. Can Med Assoc J 49: 88–92
77. Conn JW (1955) Primary aldosteronism. J Lab Clin Med 45: 661–664

II Calcium Movement During Contraction and Relaxation in Arterial Smooth Muscle

Membrane Potential and Calcium Channels in Arterial Smooth Muscle Cells

MARK T. NELSON[1], NICHOLAS B. STANDEN[2], JOSEPH E. BRAYDEN[1], and JENNINGS F. WORLEY III[3]

Summary. Noradrenaline and membrane depolarization decrease the diameters of arteries. We found that noradrenaline and membrane depolarization can contract rabbit mesenteric arteries by a common mechanism, namely by opening voltage-dependent calcium channels. The effects of noradrenaline and membrane depolarization on calcium channels were examined directly at the single channel level using the patch clamp technique. Membrane depolarization increased calcium entry by elevating the open state probability (*P*o) approximately 2.7 fold for an 8 mV depolarization over the physiological range of membrane potentials. Noradrenaline also increased *P*o in the absence of a membrane depolarization. The effect of noradrenaline may be mediated by a second messenger, because it exerts its influence without directly contacting the channels. These results predict that steady state contractions to noradrenaline would depend on calcium entry through voltage-dependent calcium channels. Indeed, noradrenaline contractions increased with membrane depolarization or with the calcium channel agonist, Bay R 5417, and were inhibited by organic and inorganic calcium channel blockers. In all cases, maintained contractions increased about 2.7 fold for an 8 mV depolarization, suggesting that the voltage-dependence of calcium channels was responsible for this voltage-dependence. Therefore, in the presence of noradrenaline, calcium channels can open over the physiological range of membrane potentials. These results suggest that any agent that changes membrane potential can affect force and blood vessel diameter.

Key words: Arterial smooth muscle—Calcium channels—Calcium channel blockers—Membrane potential—Noradrenaline

Introduction

Calcium entry into smooth muscle cells in the arterial wall regulates force development and, thus, the diameter of arteries. Calcium enters smooth cells

[1] Department of Pharmacology, University of Vermont College of Medicine, Burlington, VT 05405, USA
[2] Department of Physiology, University of Leicester, Leicester LE1 7RH, UK
[3] Department of Pharmacology and Toxicology, West Virginia University, Morgantown, WV 26506, USA

through voltage-dependent calcium channels. Noradrenaline and membrane depolarization contract arteries by increasing calcium influx. Calcium entry has been proposed to occur through two pathways: (1) voltage-dependent calcium channels and (2) noradrenaline-activated calcium-permeable channels that are voltage-insensitive [1]. We demonstrate here that membrane depolarization and noradrenaline can contract mesenteric arteries by a common mechanism; namely, by opening voltage-dependent calcium channels [2]. Furthermore, we show that in the presence of noradrenaline, calcium entry through voltage-dependent calcium channels can regulate force over the physiological range of membrane potentials.

Materials and Methods

Single smooth muscle cells were isolated from rabbit mesenteric arteries as previously described [3]. Isolation involves perfusion of a section of artery (approx. 5 mm) with a low calcium (0.2 m*M*) Hank's solution containing collagenase (Sigma, Type IV, 1.5 mg/ml) and elastase (Sigma, Type II, 0.4 mg/ml) for about 1 h at 37°C, after which single cells are obtained by mechanical agitation.

Single calcium channels were measured in cell-attached membrane patches [2]. Calcium channel recordings have been made with barium (80 m*M* $BaCl_2$, 10 m*M* Hepes, pH 7.4) as the permeant ion (this improves the signal to noise ratio of the small calcium channel currents). Single channel data recorded on video tape were analyzed by PDP 11/73 and Compaq 286 computers.

Contractions of ring segments of mesenteric artery were measured on a myograph while simultaneously measuring the membrane potential of a smooth muscle cell in the artery with a microelectrode [2] (37°C, Hank's solution with 1.8 m*M* $CaCl_2$, 4.17 m*M* HCO_3, gassed with 95% O_2, 5% CO_2, pH 7.2).

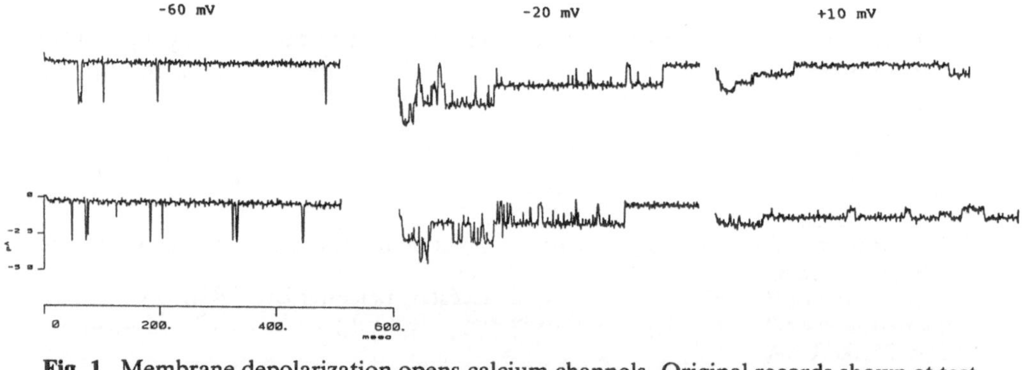

Fig. 1. Membrane depolarization opens calcium channels. Original records shown at test potentials of *−60 mV*, *−20 mV*, *+10 mV*. On-cell recording. Filtered at 500 Hz, digitized at 2 kHz

Results

Membrane Depolarization Opens Calcium Channels in Arterial Smooth Muscle

Figure 1 shows that membrane depolarization decreases the single channel current as expected for a reduction in driving force and causes a dramatic increase in the fraction of time a calcium channel is open. The unitary current-voltage relationship was nonlinear with 80 m*M* barium as the charge carrier, with the slope conductance decreasing from 37 pS at −60 mV to about 18 pS to 0 mV [2]. Over the physiological range of membrane potentials (−70 to −40 mV), *P*o increased e-fold for 7–8 mV depolarization [2]. These results predict that, in the physiological membrane potential range, a membrane depolarization of 5 mV should increase calcium influx through voltage-dependent calcium channels about twofold.

Noradrenaline Opens Voltage-Dependent Calcium Channel

β-adrenergic stimulation increases cardiac contractility by opening voltage-dependent calcium channels [4]. This effect appears to be mediated by cyclic AMP-dependent phosphorylation of the calcium channel protein. We explored the possibility that alpha-adrenergic stimulation of arterial smooth muscle could open voltage-dependent calcium channels. Indeed, noradrenaline when applied to an intact single smooth muscle cell increased *P*o of calcium channels in an on-cell patch (Fig. 2) [2]. At the test potential shown here (0 mV), noradrenaline increased *P*o 1.9-fold. At more negative potentials, noradrenaline increased *P*o as much as fourfold and had no effect on the single channel conductance [2]. Since noradrenaline was applied to the cell and not directly to the calcium channels, these results suggest the involvement of an intracellular second messenger.

These results indicate that noradrenaline and membrane depolarization can increase calcium entry by opening voltage-dependent calcium channels. Maintained force in arteries depends on calcium entry. Therefore, these results would also predict that maintained contractions to noradrenaline should increase as calcium channels are opened by membrane depolarization. This possibility was tested in the following experiments.

Opening Calcium Channels by Membrane Depolarization Increases Noradrenaline Contractions

Membrane depolarization increased noradrenaline contractions (Fig. 3). Elevation of external potassium from 5 m*M* to 15 m*M* and from 5 m*M* to 20 m*M* depolarized the membrane potential by about 14.8 mV and 20.1 mV, respectively. In the absence of noradrenaline, elevation of external potassium to 20 m*M* increased force only slightly. These results are consistent with noradrenaline contracting mesenteric arteries by activating voltage-dependent calcium channels.

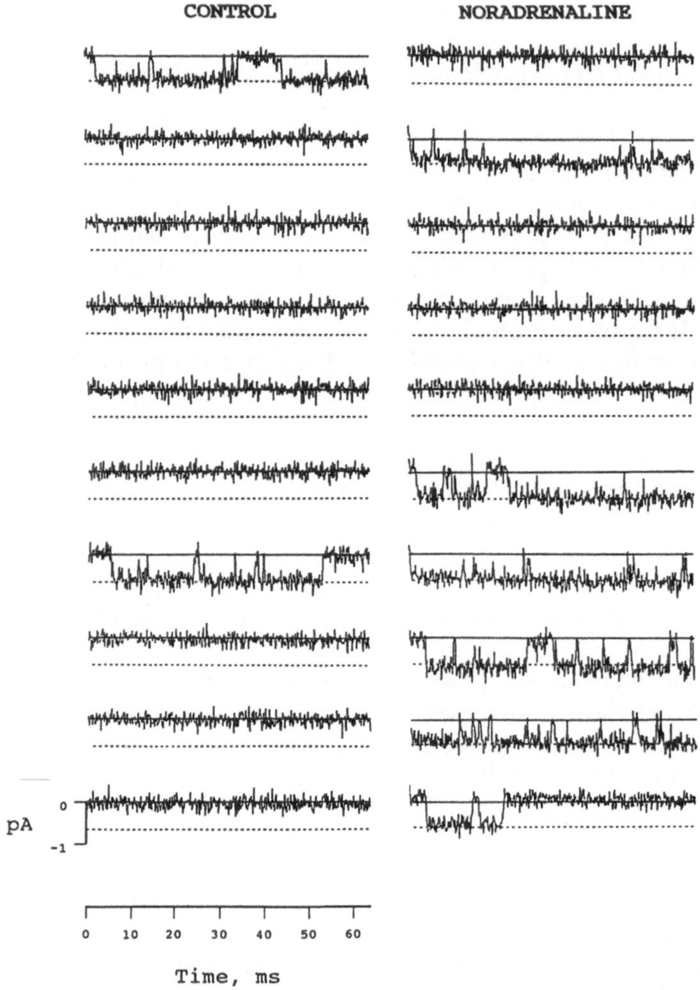

Fig. 2. Noradrenaline opens voltage-dependent calcium channels. Original records filtered at 2 kHz, digitized at 8 kHz. Test potential was 0 mV. On-cell recording. Noradrenaline (10 μM) was added directly to the bath

Blocking Voltage-Dependent Calcium Channels Reduces Noradrenaline Contractions

These results predict that inhibitors of voltage-dependent calcium channels should reduce noradrenaline contractions, providing that the noradrenaline-induced change in the calcium channel protein does not affect the binding of the calcium channel inhibitors to the channel. The organic calcium channel inhibitors, diltiazem and verapamil, reduced contractions to noradrenaline, with 5 μM verapamil almost completely relaxing the artery (Fig. 4). Nisoldipine, a dihyd-

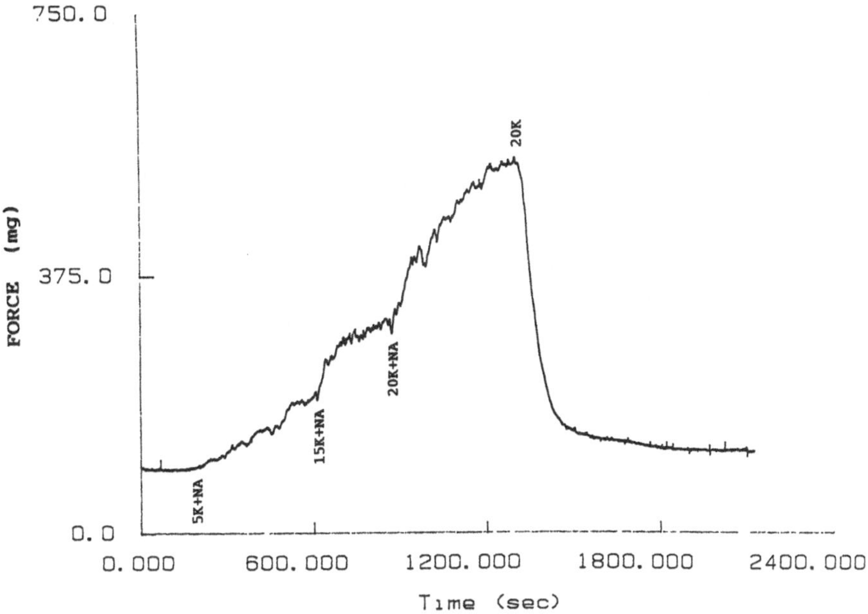

Fig. 3. Membrane depolarization increases noradrenaline contractions. Mesenteric artery was superfused at a flow rate of 2 ml/min. At time 0, the artery was bathed in a standard physiological solution with 5 m*M* external potassium (*5K*). Noradrenaline (*NA*) (2 μM) was then added to the superfusate and force developed. At approximately 600 s, external potassium was increased to 15 m*M* while keeping noradrenaline constant. At about 900 s, external potassium was increased from 15 to 20 m*M*. At about 1300 s, noradrenaline was removed from the superfusate while keeping external potassium constant at 20 m*M*

ropyridine calcium channel inhibitor, also inhibited noradrenaline contractions [2]. Dihydropyridine inhibition depends on the state of the calcium channel with membrane depolarization promoting inhibition [5]. When the effects of nisoldipine on noradrenaline and potassium contractions were compared under conditions of similar channel states, i.e., at the same membrane potential; nisoldipine inhibited both types of contractions with a similar apparent affinity [2].

Inorganic divalent cations such as cadmium and cobalt are well-known blockers of voltage-dependent calcium channels, with cadmium usually being the most potent [6]. We found that cadmium and cobalt block single calcium channels from mesenteric artery with dissociation constants of 36 μM and 325 μM, respectively [7]. Cadmium and cobalt cause interruptions in the unitary current that are consistent with these ions binding in the channel pore and blocking movement of the permeant ion, in this case, barium. The single channel results predicted that both cadmium and cobalt should reduce noradrenaline contractions, with cadmium having a higher affinity than cobalt. Indeed, as shown in Fig. 5, cadmium and cobalt reduce noradrenaline contractions, with apparent inhibition constants of about 50 μM and 500 μM, respectively.

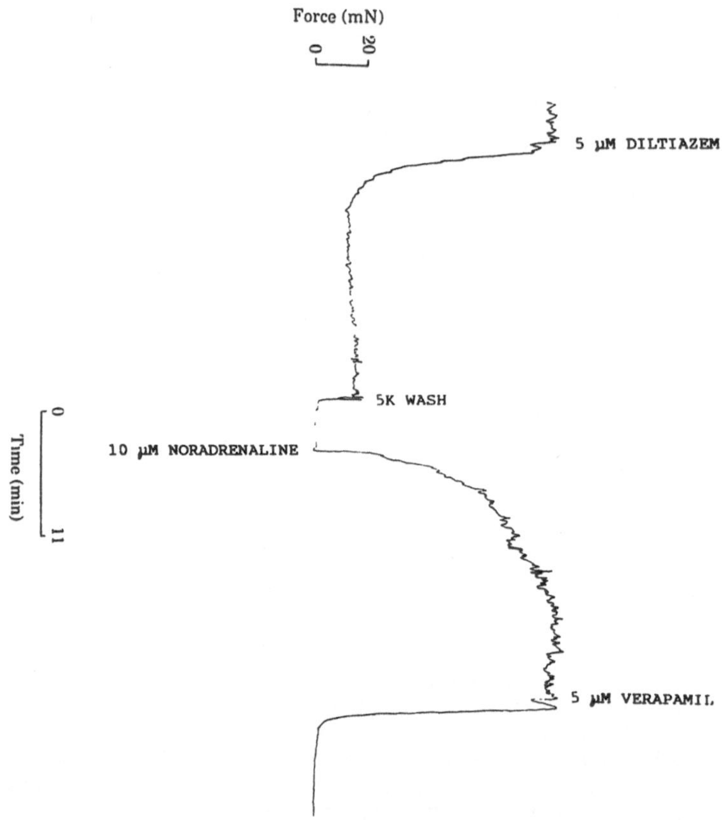

Fig. 4. Organic calcium channel blocking drugs, diltiazem and verapamil, inhibit noradrenaline contractions. diltiazem (5 μM); verapamil (5 μM); noradrenaline (10 μM)

Discussion

Maintained Contractions of Rabbit Mesenteric Arteries Depend on Calcium Entry Through Voltage-Dependent Calcium Channels

Our results are consistent with noradrenaline contracting mesenteric arteries by opening voltage-dependent calcium channels. This conclusion is based on the observations that agents that open voltage-dependent calcium channels (membrane depolarization and Bay K 8644) increase noradrenaline contractions [2], and blockers of calcium channels (nisoldipine, verapamil, cadmium, cobalt, and diltiazem) reduce noradrenaline contractions. Finally, we demonstrate directly at the single channel level that noradrenaline can open voltage-dependent calcium channels.

Voltage-dependent calcium channels in on-cell patches are opened by noradrenaline when it is added to the bath solution, i.e., noradrenaline does not directly contact the channels. This suggests that noradrenaline is acting through

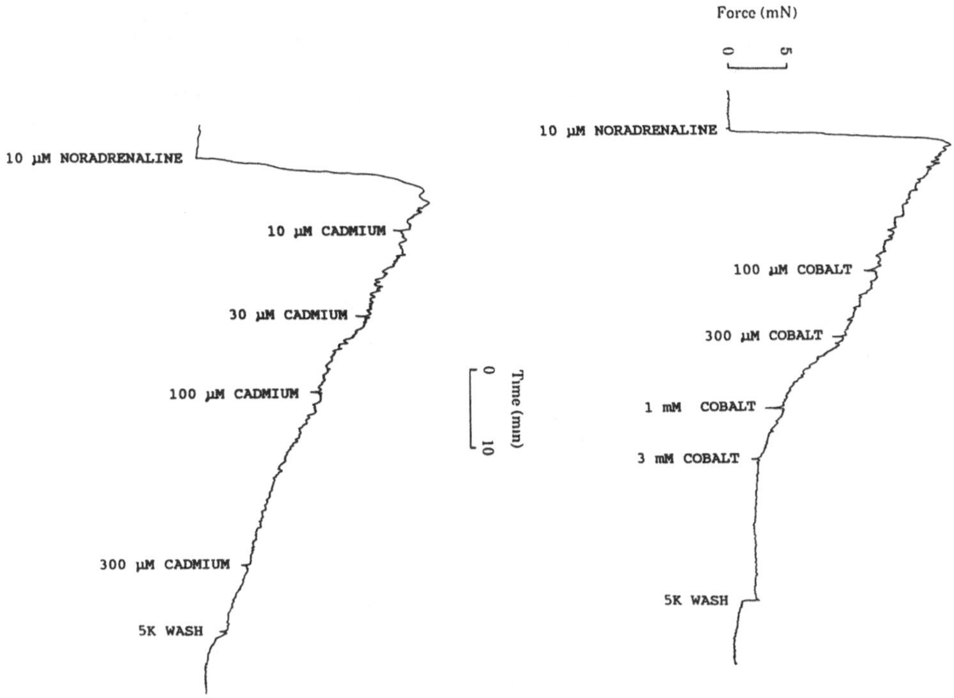

Fig. 5. Inorganic calcium channel blockers, cadmium and cobalt, inhibit noradrenaline contractions. Noradrenaline (10 μM)

a second messenger system. The effects of noradrenaline on mesenteric arteries are blocked by traditional alpha-1 blockers, e.g., prazosin. Alpha receptor stimulation has been shown to increase inositol trisphosphate (IP_3) and diacylglycerol (DAG) levels [8]. IP_3 has been shown to cause a transient contraction in this artery by releasing calcium from an internal store, most likely the sarcoplasmic reticulum [9]. DAG can stimulate protein kinase C, which might lead to a phosphorylation of the calcium channel and, thus, an increase in *P*o [10]. Calcium channel phosphorylation appears to be responsible for the increase in calcium currents in heart observed during β-adrenergic stimulation [4].

Link Between Membrane Hyperpolarization and Vasodilation

Pharmacological and endogenous agents that hyperpolarize are potent vasodilators. Our results suggest that in the presence of some vasoconstrictors, calcium entry through voltage-dependent calcium channels regulates tone. Thus, membrane hyperpolarization would vasodilate by closing calcium channels. We found that a membrane hyperpolarization of about 2 mV reduces force by 30%. Such a decrease in force would theoretically be expected to lead to a substantial increase in blood flow under physiological conditions.

Acknowledgments. This work was supported by the National Science Foundation, American Heart Association, Miles Laboratories, Burroughs Wellcome, Wellcome Trust, Medical Research Council, NIH, and the Pharmaceutical Manufacturers Association Foundation. JFW was supported by a fellowship from the Muscular Dystrophy Association; NBS is supported by a Wellcome Trust Research Leave award; and MTN is an Established Investigator of the American Heart Association.

References

1. Bulbring E, Tomita T (1987) Catecholamine action on smooth muscle. Pharmacol Rev 39: 50–86
2. Nelson MT, Standen NB, Brayden JE, Worley JF (1988) Noradrenaline contracts arteries by activating voltage-dependent calcium channels. Nature 336: 382–385
3. Worley JF, Deitmer J, Nelson MT (1986) Single nisoldipine-sensitive calcium channels in smooth muscle cells isolated from rabbit mesenteric artery. Proc Natl Acad Sci USA 83: 5746–5750
4. Kameyama M, Hescheler J, Hofmann F, Trautwein W (1986) Modulation of Ca current during the phosphorylation cycle in the guinea pig heart. Pflugers Arch 407: 123–128
5. Nelson MT, Worley JF (1989) Dihydropyridine inhibition of single calcium channels and contraction in rabbit mesenteric artery depends on voltage. J Physiol (Lond) 412: 65–91
6. Hagiwara S, Byerly L (1981) Calcium channel. Annu Rev Neurosci 4: 69–125
7. Nelson MT, Standen NB, Worley JF, Huang Y (1989) Effects of external and internal cadmium on single calcium channels in isolated smooth muscle cells from rabbit mesenteric artery. (abstract) J Physiol (Lond) 410: 11
8. Hashimoto T, Hirata M, Itoh T, Kanmura Y, Kuriyama H (1986) Inositol 1,4,5-trisphosphate activates pharmacomechanical coupling in smooth muscle of the rabbit mesenteric artery. J Physiol (Lond) 370: 605–618
9. Somlyo AV, Bond M, Somlyo AP, Scarpa A (1985) Inositol trisphosphate-induced calcium release and contraction in vascular smooth muscle. Proc Natl Acad Sci USA 82: 5231–5235
10. Flockerzi V, Oeken H-J, Hofmann F, Pelzer D, Cavalie A, Trautwein W (1986) Purified dihydropyridine-binding sites from skeletal muscle t-tubules is a functional calcium channel. Nature 323: 66–68

Excitatory Agonists and Ca-Permeable Channels in Arterial Smooth Muscle Cells

INGRID DECLERCK, GUY DROOGMANS, and RIK CASTEELS[1]

Summary. The contribution of Ca entry to the activation of vascular smooth muscle cells of rabbit ear artery by the agonists norepinephrine and ATP is discussed. Stimulation with norepinephrine is not accompanied by a depolarization of the cell membrane. The opening of voltage-gated Ca channels by depolarization can therefore be excluded. The existence of receptor-operated Ca channels, which are activated by a direct interaction of the agonist with its receptor, is also unlikely. The stimulation with norepinephrine is not accompanied by a change in transmembrane current as measured by means of the whole-cell patch clamp technique in single smooth muscle cells. On the other hand, a rather unexpected inhibition of the voltage-gated Ca channels by α-agonists was observed by means of this technique.

These findings, as well as the limited increase of $[Ca^{2+}]_i$ observed during the tonic component of contraction, are consistent with a limited role of Ca entry during norepinephrine stimulation. ATP ($10^{-7} - 10^{-5}$ *M*) activates a transient inward current at a holding potential of -60 mV in chemically dispersed, single smooth muscle cells. The reversal potential of this current and its modification by ionic substitutions indicate that this channel is mainly permeable to monovalent and also to divalent cations.

Key words: ATP — Ca-influx — Norepinephrine — Receptor-operated Ca channels — Voltage-gated Ca channels

Introduction

A change in the intracellular level of free Ca ions ($[Ca^{2+}]_i$) is an essential step in the excitation-contraction coupling of vascular smooth muscle, linking the physiological stimulus to the activation of the contractile proteins. Excitatory agonists can induce such changes by activating Ca channels in the cell membrane. Receptor activation may modify the permeability of the cell membrane for monovalent cations or anions, which might induce a depolarization of the cell membrane and the concomitant activation of voltage-gated Ca channels. On the other hand, agonists can also generate an influx of extracellular calcium by mod-

[1]Laboratorium voor Fysiologie, Campus Gasthuisberg KU Leuven, B-3000 Leuven, Belgium

ulating the voltage-gated Ca channels directly (e.g., by shifting their activation to more negative potentials) or by activating a separate population of Ca channels, the so-called receptor-operated Ca channels.

Recent investigations have shown that norepinephrine (NE) and adenosine triphosphate (ATP) play an important role in the sympathetic neurotransmission of vascular smooth muscle [1, 2]. In this vascular paper we will discuss the effects of both agonists on the entry of calcium ions in vascular smooth muscle cells of rabbit ear artery.

Effects of α-Adrenergic Agonists

The increase of cytoplasmic free calcium during stimulation with α-agonists is supposed to occur in two phases: an initial transient release of sequestered calcium from an internal pool, which is responsible for the phasic component of the contraction, would be followed by a sustained entry of Ca ions from the extracellular space, giving rise to the tonic component of the contraction. A discussion of the mechanism of Ca release induced by α-agonists is beyond the scope of this review. We will limit the discussion to the entry of calcium associated with agonist stimulation.

Indirect evidence indicates that the tonic component of the NE-induced contraction depends on an influx of extracellular calcium ions. Removal of extracellular Ca or blocking passive Ca entry with inorganic Ca-entry blockers (La, Mn) relaxes vascular smooth muscle tissues which have been precontracted with NE. On the other hard, tissues exposed to either of these solutions respond to α-receptor activation with a transient contraction. This contraction, which is believed to be due to Ca release, can be induced only once, and it cannot be repeated by subsequent agonist applications.

Membrane Potential and α-Receptor Activation

Exogenously applied NE in the concentration range 10^{-9}–10^{-6} M induces a dose-dependent contractile response which is not accompanied by a significant change of the membrane potential in the rabbit ear artery [3]. This observation suggests that the contractile response induced by α-receptor activation does not depend at least in these vascular smooth muscle cells on an influx of extracellular Ca ions activated by a depolarization of the cell membrane. Similar effects of agonist stimulation on the resting potential have also been reported for other vascular smooth muscle tissues [4, 5], but agonist-induced depolarizations and hyperpolarizations have also been observed. The degree of depolarization is variable and less than the threshold to induce a contractile response with increased $[K^+]_o$ in some tissues, whereas in others a close correlation between contraction and depolarization has been observed [6–12]. Depending on the degree of depolarization induced by the agonist, an influx of Ca ions through voltage-activated Ca channels is likely to occur, and this might partly explain why the organic Ca-entry blockers exert such diverse effects on the contractions evoked by agonists in different vascular smooth muscle tissues.

These variable effects of α-receptor activation on the resting potential might be related to the modulation of ionic permeabilities of the cell membrane by the concomitant changes in $[Ca^{2+}]_i$: that is, the opening of Ca activated K channels would hyperpolarize the membrane and thereby compensate a direct depolarizing action by α-receptor activation due e.g., to the opening of Na or Cl channels. This hypothesis has been tested by studying the effects of α-receptor activation on the membrane potential of vascular smooth muscle cells of rabbit ear artery in which the K channels were blocked by adding 10–15 m*M* tetraethylammonium (TEA) to the perfusion medium. This latter substance causes a depolarization to values between −40 and −30 mV, which is accompanied by a tonic contraction. Higher concentrations of TEA fail to induce larger depolarizations. If the depolarization induced by TEA reaches a level of around −35 mV spontaneous action potentials occur (Fig. 1). If the membrane potential remains more negative than −35 mV, action potentials can still be induced by applying depolarizing current pulses or by increasing $[K^+]_o$. Application of NE to tissues exposed to TEA results in a dose-dependent depolarization of the cell membrane (Fig. 2). These observations support the hypothesis that the opening of Ca activated K channels might counteract any agonist-induced depolarization by "clamping" the membrane potential at a value close to the resting potential.

The effects of α-receptor activation on the passive membrane permeability to monovalent ions have been further examined with ^{42}K or ^{86}Rb, ^{24}Na, and ^{36}Cl flux experiments. Agonist stimulation significantly enhances the exchange rate of each of these ions in the rabbit ear artery and in other vascular smooth muscle tissues [3, 13–16]. The membrane potential can, depending on the relative amplitudes of these increases in ionic permeabilities, remain unchanged, as has been observed in the rabbit ear artery. However, the membrane may also de-

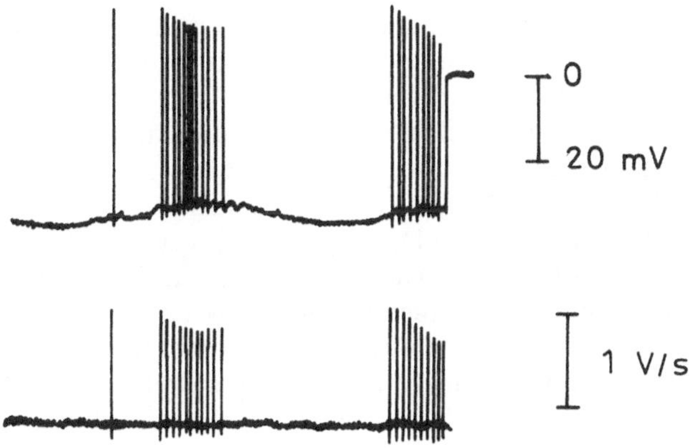

Fig. 1. Bursts of spontaneous action potentials occurring during exposure of rabbit ear artery to a solution containing 10 m*M* tetraethyl ammonium. The *top trace* shows the action potentials and the *bottom trace*, their time derivatives

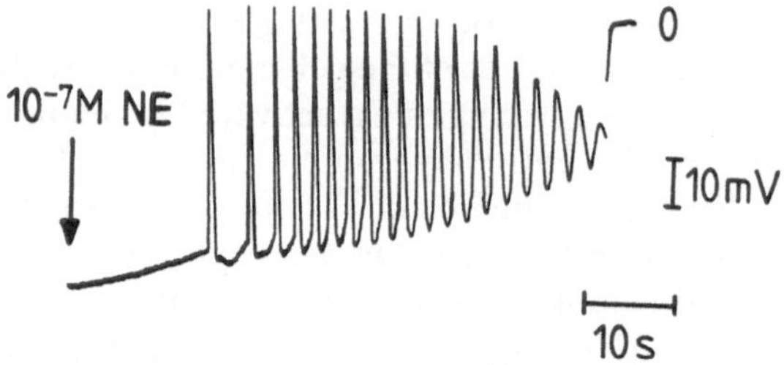

Fig. 2. Effect of 10^{-7} *M* norepinephrine (*NE*) on the membrane potential and on the spike discharge of rabbit ear artery exposed to a solution containing 10 m*M* tetraethyl ammonium

polarize or hyperpolarize to a variable extent, as observed in other smooth muscle tissues. In addition, changes in the Na and Cl permeability might be responsible for the depolarizing effect of NE in solutions containing TEA. Moreover, it has been reported that the increase in cytoplasmic Ca induced by NE activates K and Cl channels in single smooth muscle cells [17, 18].

It can be concluded from these observations that the effect of NE on the membrane potential, if any at all, is indirect and due to changes of the membrane permeability of monovalent ions caused by the release of Ca from internal stores. If the membrane is depolarized above the threshold for activation of voltage-gated Ca channels, it will be associated with an influx of calcium ions through these channels, which can be inhibited by Ca-entry blockers. On the other hand, this depolarization and the concomitant activation of Ca channels may also be inhibited by K-channel activators.

Does α-Receptor Activation Open Receptor-Operated Channels?

The preceding observations suggest that the maintained component of the contractile response induced by α-receptor activation in the rabbit ear artery does not depend on the opening of voltage-gated Ca channels. The term receptor-operated or ligand-gated channels has been introduced to designate the putative Ca channels which are closely linked to membrane receptors and activated by an agonist-receptor interaction [19, 20].

This classification of Ca channels into voltage-gated and receptor-operated channels is largely based on their presumed different sensitivities to organic Ca channel blockers. These substances strongly inhibit the voltage-gated Ca channels and the contraction caused by excess K. In contrast, the contractions induced by NE are only slightly affected, which could be explained by assuming a large contributioin of receptor-operated channels to the increase of $[Ca^{2+}]_i$ and to force development. However, the contribution of intracellular Ca release to the contraction could also lead to an apparent insensitivity to the Ca channel blockers. On the other hand, the finding that the tonic component of the agonist-induced contraction depends to a larger extent on extracellular calcium than the early phasic component is in agreement with a contribution of receptor-operated channels to this tonic component of contraction.

The effects of α-receptor activation on Ca entry has also been studied by means of unidirectional ^{45}Ca influx and net ^{45}Ca uptake measurements. In rat aorta [21] and in rabbit ear artery [3] the ^{45}Ca uptake measured after short loading periods is increased, whereas the total exchangeable ^{45}Ca fraction measured after a prolonged loading period is not affected. In rabbit aorta it has been reported that NE either does not have an effect on the ^{45}Ca influx [22] or that it increases the net uptake of ^{45}Ca [23]. The interpretation of ^{45}Ca uptake data during α-receptor activation is, however, not straightforward. Agonist stimulation causes a redistribution of cellular Ca, and also modifies the kinetics of the Ca exchange between the different cellular compartments. This could give rise to an apparent increased rate of ^{45}Ca uptake and/or to an increased exchangeable ^{45}Ca pool. In addition, the contribution of ^{45}Ca entry through Ca channels activated by a possible depolarization of the cell membrane during agonist stimulation has to be ruled out.

This uncertainty about the role of Ca influx during agonist stimulation can only be resolved by an approach that allows an unequivocal demonstration of the existence of receptor-operated channels. By means of the whole-cell patch-clamp technique, it is possible to measure and compare transmembrane ionic currents at rest and during α-receptor activation in single vascular smooth muscle cells of the rabbit ear artery [24]. These isolated cells contract upon activation of α-receptors, suggesting that the isolation procedure does not affect the receptors and the putative receptor-operated channels. Dialysis of the cell with pipette solution allows suppression of contaminating K currents by using CsCl in the pipette solution, and buffering of the changes in $[Ca^{2+}]_i$ by adding 11 mM EGTA to the pipette solution, thereby circumventing currents that would be activated by Ca. Ca ions in the bathing solutions are replaced by Ba ions at a concentration of 10 mM in order to maximize the Ca channel currents [24]. Stimulation with 10 μM NE does not cause under these conditions a detectable change of the holding current measured at a potential close to the normal resting potential of the cell. This observation is not consistent with the activation of receptor-operated Ca channels. However, it cannot be excluded that some soluble, endogenous second messenger or essential protein structure is lost during the dialysis of the cell interior with the pipette solution, and that NE therefore fails to activate receptor-operated channels. A decreased level of GTP cannot explain the absence of an agonist effect, because α-agonists do not evoke a current when 0.1 mM GTP or either of its nonhydrolyzable analogues Gpp(NH)p or GTP-γ-S are added to the patch pipette solution.

The changes of the free intracellular calcium concentration, as measured by means of aequorin and fluorescent dyes (quin-2, fura-2), show a remarkable pattern during agonist stimulation [25, 26], suggesting that extracellular calcium plays at least quantitatively a limited role in the activation of the contractile proteins of smooth muscle cells during exposure to agonists: after a fast initial increase that depends on release of Ca from internal pools, $[Ca^{2+}]_i$ declines to levels close to the resting value, while the contraction is sustained. Such a small increase of $[Ca^{2+}]_i$ can be achieved by a limited Ca influx. Moreover, its effectiveness for increasing $[Ca^{2+}]_i$ is enhanced because the buffering action of the intracellular Ca stores is largely abolished during the stimulation with the agonist. This influx may not be observed in patch-clamp experiments because it may be below the detecion level for electrical measurements.

Effect of α-Receptor Activation on Voltage-Gated Ca Channels

The above findings do not substantiate the hypothesis that NE opens receptor-operated channels. We have therefore investigated the alternative hypothesis, i.e., the modulation of voltage-activated Ca channels by α-receptor activation, by comparing the currents through these channels in the presence and in the absence of NE [24]. The records in Fig. 3A show that the agonist appreciably inhibits the current through voltage-activated Ca channels. This rather unexpected observation cannot be due to a time-dependent run-down of the Ca channels because the currents recover after washing out the agonist. This recovery was complete in some cells, but in most cells it was only partial, probably because washing out the agonist requires a relatively long time (at least 10 min) which overlaps with the concomitant run-down of the Ca channels. However, a second application of NE also induces similar inhibitory effects on the voltage-activated Ca channels. All these data are summarized in Fig. 3B. It shows the current-voltage relationship before and during application of the agonist and after its washing out. The peak inward current is reduced at all potentials and more specifically by 44.1% $\pm$ 3.1% in the voltage range between -10 and $+20$ mV. However, the curve is not significantly shifted along the voltage axis. Because the patch pipette contains 11 mM EGTA, which prevents pronounced changes of the free intracellular Ca concentration, it is unlikely that this reduction is due to an increase of $[Ca^{2+}]_i$ by the release of Ca from internal stores.

The α_1-agonist phenylephrine mimics the effects of NE, while the α_2-agonist clonidine does not affect the voltage-activated Ca channel currents. A concomitant activation of β-receptors is prevented in these experiments by the presence of 1μM propranolol. These findings indicate that the current inhibition is caused by activation of α_1-receptors.

The observed inhibition is not unique to stimulation with α_1-agonists, because it also occurs when the cells are stimulated with histamine. More recently, a similar inhibitory effect has been observed in rat portal vein [27], and it has also been described in cultured rat aortic smooth muscle cells during stimulation with [Arg^8]vasopressin [28]. These latter observations indicate that the inhibition is probably not a direct effect of the agonist, but rather an effect mediated by a second messenger, which is common to all these agonists. We have therefore tried to find out whether the down modulation of Ca channels is mediated by changes of the intracellular levels of cyclic nucleotides. Patch pipettes filled with either 10 μM cAMP or 10 μM cGMP were used, as dialysis of the cells with this pipette solution increases the intracellular level of either nucleotide and thereby buffers the putative receptor-mediated changes. However, the inhibitory effects of α-receptor activation on the voltage-activated Ca channels persist under these conditions, suggesting that the effect of α-receptor activation is not mediated by changes in cyclic nucleotide levels.

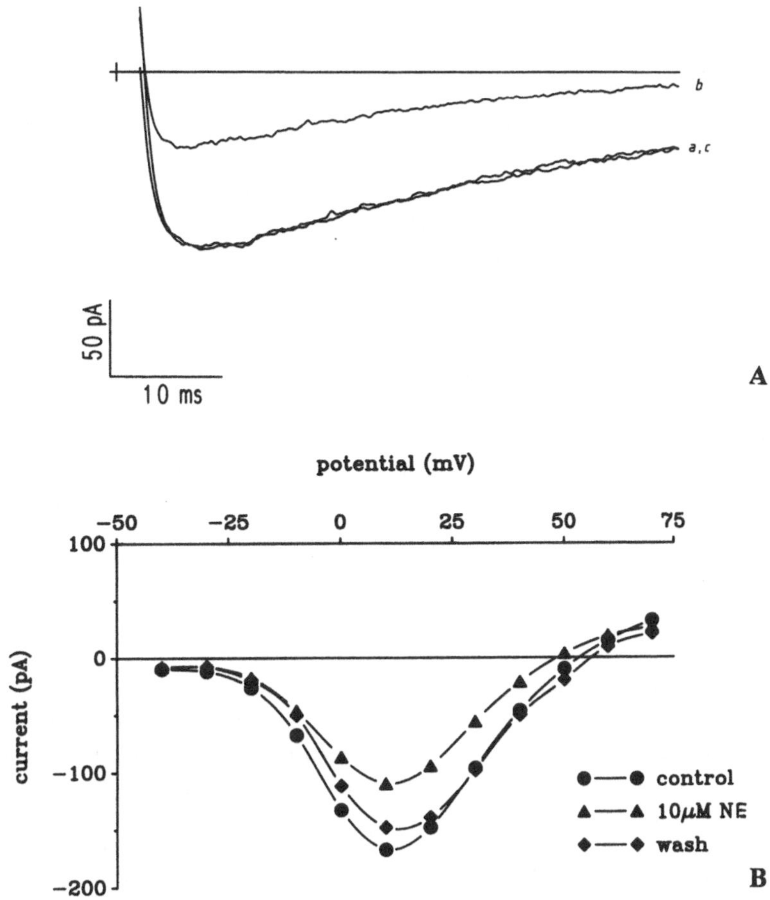

Fig. 3. **A** The effect of 10 μM norepinephrine (*NE*) on the Ba current in single smooth muscle cells from rabbit ear artery induced by a depolarizing voltage step from a holding potential of −50 mV to +10 mV. Traces *a* and *c* correspond to control currents measured before and after application of the agonist, whereas trace *b* shows the current measured in the presence of the agonist. **B** Current-voltage relationship of the peak inward Ba current before (circles), during (triangles) and after (*diamonds*) application of 10 μM norepinephrine. From [24]

It has recently been suggested that the α_1-adrenergic receptor is coupled to its signal transduction mechanisms by guanine-nucleotide binding proteins (G-proteins). In order to establish the role of G-proteins in the down modulation of Ca channels by α-adrenoceptor activation, the cells were internally perfused with pipette solutions containing either 0.1 mM GTP or one of its non hydrolyzable analogues Gpp(NH)p or GTP-γ-S. The amplitude of the peak inward currents recorded with these solutions is of the same order of magnitude as those observed with electrodes filled with the standard pipette solution, suggesting

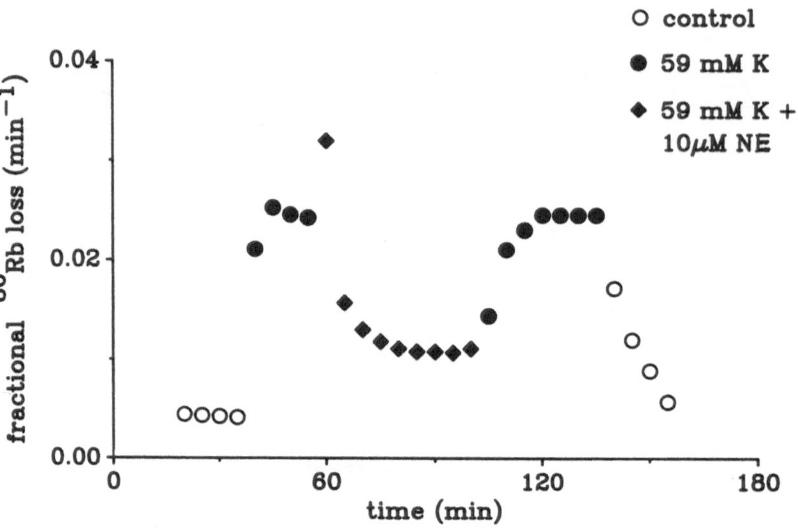

Fig. 4. Stimulation of the ^{86}Rb efflux rate from rabbit ear artery by K depolarization (59 m*M* K), and the effect of 10 μ*M* norepinephrine (*NE*) on this stimulation

that these substances do not exert a direct effect on these currents. The inhibition of the Ca channel currents induced by α-receptor activation also occurs with these pipette solutions. After washing out the agonist the currents return to the control value when the cells are dialyzed with GTP. However, the inhibition becomes irreversible when the pipette solution contains a non hydrolyzable GTP analogue [24]. These observations are consistent with a mediation of this inhibition by G-proteins, but it cannot be excluded that the run-down of the Ca channels is accelerated by these non hydrolyzable GTP analogues.

This inhibitory effect of NE on the voltage-activated Ca channels also occurs in intact preparations. In order to substantiate this hypothesis the effect of NE on the ^{86}Rb-efflux stimulated with K depolarization has been studied [14]. Depolarization of the cell membrane by 59 m*M* $[K^+]_0$ results in an increase of the ^{86}Rb-efflux rate, which can be blocked by Ca entry blockers. This enhanced efflux therefore reflects the opening of Ca activated K channels as a consequence of the increase of $[Ca^{2+}]_i$ due to the activation of voltage-gated Ca channels by the K depolarization. NE, applied to these K-depolarized tissues, causes an additional, transient increase of the ^{86}Rb-efflux which is followed by a maintained reduction of the efflux rate below its level in 59 m*M* $[K^+]_0$ in the absence of the agonist (Fig. 4). The initial stimulation of the efflux is consistent with the release of calcium, which causes a further increase in $[Ca^{2+}]_i$. The subsequent decline of the efflux rate suggests that $[Ca^{2+}]_i$ is reduced below its value in the absence of NE. The finding can be explained by assuming that NE inhibits the voltage-activated Ca channels, as observed in single cells.

The physiological significance of this inhibition is not clear, but it may curtail excessive increases of intracellular Ca levels that could be deleterious for cell function.

Effects of ATP

Perivascular nerve stimulation induces excitatory junction potentials (e.j.p.) in the rabbit ear artery [2]. During repetitive stimulation at a sufficiently high frequency these e.j.ps can summate and reach threshold for eliciting an action potential. They are believed to be caused by the release of a cotransmitter, most likely ATP, from the adrenergic nerve terminal. Exogenously applied ATP mimics the time course of the e.j.p. and of the phasic component of the contractile response to nerve stimulation [2].

In order to study the nature of the ionic current responsible for the e.j.p., we applied the whole-cell patch-clamp technique to chemically dispersed single smooth muscle cells of rabbit ear artery. Because depolarization of the cell membrane activates large outward K currents that may interfere with the determination of the ATP-induced current, we employed pipette solutions in which K ions were replaced by Cs ions. Bath application of ATP ($10^{-7} - 10^{-5}$ *M*) activates at a holding potential of −60 mV a dose-dependent, transient, inward current (Fig. 5). However, a second stimulation with ATP causes a much smaller, if any, current.

The voltage dependence and reversal potential of this ATP-activated current were measured by applying at its peak value a voltage ramp starting from the holding potential of −60 mV and changing at a rate of 0.2 mV/ms to +40 mV. From the resulting current record in the presence of ATP we subtracted the current record observed during a similar voltage change in the absence of ATP in order to correct for capacitive and leakage current, as well as for other ionic currents activated by depolarization (Fig. 6).

The reversal potential of the ATP-activated current lies around +5mV and does not correspond to the equilibrium potential of any cation or anion present in the bathing or pipette solution. This observation suggests that the channels activated by ATP are not selective for any particular ion.

Substitution of Na ions by a large cation, such as *N*-methyl-D-glucamin, shifts the reversal potential of the ATP-activated channels to negative potentials (−32 mV), whereas replacement of half the extracellular Cl by gluconate does not

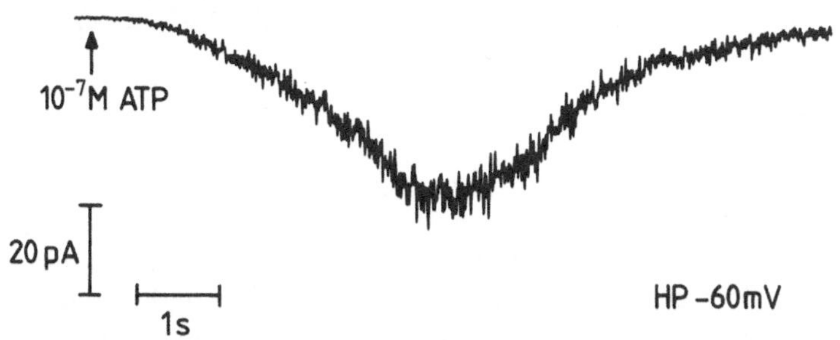

Fig. 5. Effect of 10^{-7} *M* ATP on the transmembrane ionic current recorded at a holding potential of −60 mV in single cells from rabbit ear artery

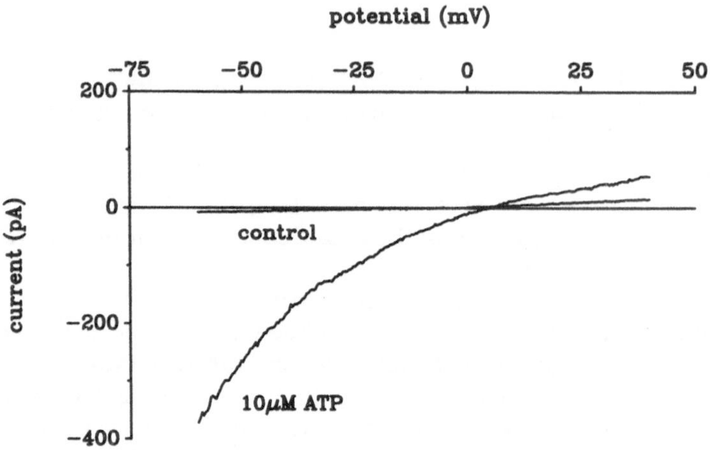

Fig. 6. Current-voltage relationships of single cells of rabbit ear artery in a normal physiological solution and in the presence of 10 μM ATP obtained by applying a voltage ramp from −60 to +40 mV. The difference between both curves gives the voltage dependence of the ATP-activated current. The pipette solution contained 140 mM Cs in order to suppress outward K currents

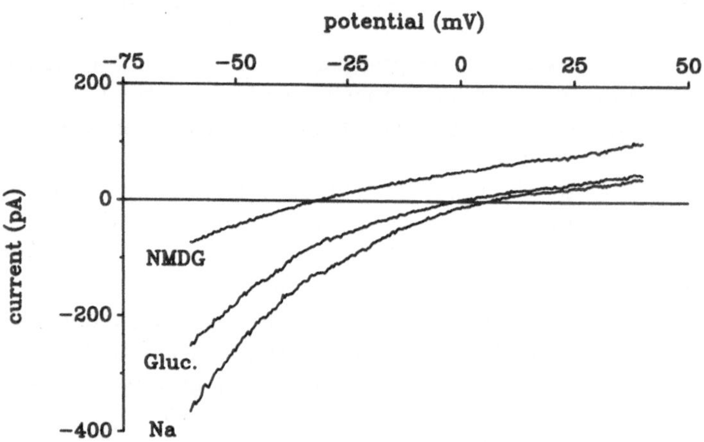

Fig. 7. Voltage dependence of the current activated by 10 μM ATP in cells exposed to a normal physiological solution (*Na*), or to a solution in which half of the Cl was replaced with gluconate (*Gluc.*) or to a solution in which Na ions were replaced by an impermeable cation (*NMDG*)

significantly modify the reversal potential (Fig. 7). These observations suggest that the ATP-activated channel is a cationic channel that excludes anions.

In order to find out whether the ATP-activated channel is also permeable for divalent cations, extracellular Na ions were replaced by 92 mM Ca. Under these conditions ATP still activates an inward current at negative potentials, indicating that the ATP-activated channels are also permeable to divalent cations. The

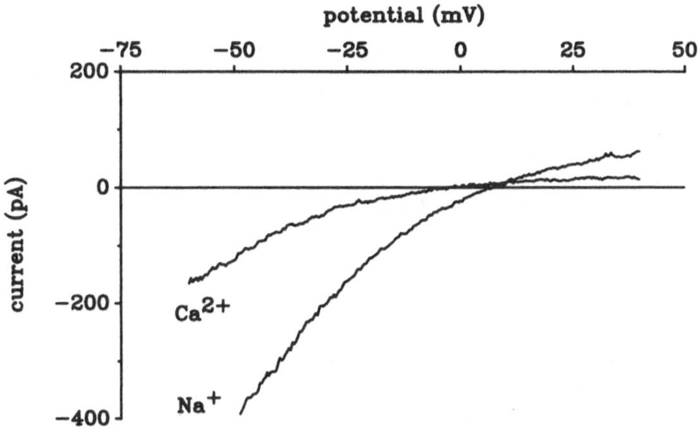

Fig. 8. Effect of the substitution of extracellular Na ions by Ca ions (*Ca^{2+}*) on the current-voltage relationship of the current induced by 10 μM ATP (*Na^{+}*)

reversal potential is, however, shifted to a value around −6 mV (Fig. 8).

These ATP-activated channels can explain the non adrenergic e.j.ps evoked by nerve stimulation and can contribute to the entry of activator Ca either by depolarizing the cell membrane, which results in the opening of voltage-activated Ca channels, or by directly allowing the entry of Ca ions. The contribution of the latter mechanism will be limited because of the much higher concentration of Na ions in the extracellular medium. These data are similar to those reported recently by other investigators [29, 30]

References

1. Burnstock G, Kennedy C (1985) A dual function for adenosine 5′-triphosphate in the regulation of vascular tone; excitatory cotransmitters with noradrenaline from perivascular nerves and locally released inhibitory intravascular agent. Circ Res 58: 319–330
2. Suzuki H (1985) Electrical responses of smooth muscle cells of the rabbit ear artery to adenosine triphosphate. J Physiol 359: 401–431
3. Droogmans G, Raeymaekers L, Casteels R (1977) Electro-and pharmacomechanical coupling in smooth muscle cells of the rabbit ear artery. J Gen Physiol 70: 129–148
4. Su C, Bevan J, Ursillo RC (1964) Electrical quiescence of pulmonary artery smooth muscle during sympathomimetic stimulation. Circ Res 15: 20–27
5. Holman ME, Surprenant AM (1979) Some properties of the excitatory junction potentials recorded from saphenous arteries of rabbits. J Physiol 287: 337–351
6. Casteels R, Kitamura K, Kuriyama H, Suzuki H (1977) The membrane properties of the smooth muscle cells of the rabbit main pulmonary artery. J Physiol 271: 41–61
7. Haeusler G (1978) Relationship between noradrenaline induced depolarization and contraction in vascular smooth muscle. Blood Vessels 15: 46–54
8. Kuriyama H, Makita Y (1983) Modulation of noradrenergic transmission in the guinea-pig mesenteric artery: an electrophysiological study. J Physiol 335: 609–627

9. Bolton T, Lang RJ, Takewaki T (1984) Mechanism of action of noradrenaline and carbachol on smooth muscle of guinea-pig anterior mesenteric artery. J Physiol (London) 351: 549–572
10. Holman ME, Surprenant AM (1980) An electrophysiological analysis of the effects of noradrenaline and α-receptor antagonists on neuromuscular transmission in mammalian muscular arteries. Br J Pharmacol 71: 651–661
11. Mulvany MJ, Nilsson H, Flatman JA (1982) Role of membrane potential in the response of rat small mesenteric arteries to exogenous noradrenaline stimulation. J Physiol 332: 363–373
12. Makita Y, Kanmura Y, Itoh T, Suzuki H, Kuriyama H (1983) Effects of nifedipine derivatives on smooth muscle cells and neuromuscular transmission in the rabbit mesenteric artery. Naunyn-Schmiedebergs Arch Pharmacol 324: 302–312
13. Bolton TB, Clapp LH (1984) The diverse effects of noradrenaline and other stimulants on ^{86}Rb and ^{42}K efflux in rabbit and guinea-pig arterial muscle. J Physiol 355: 43–64
14. Casteels R, Droogmans G (1985) Dependence on calcium of potassium- and agonist-induced changes in potassium permeability of the rabbit ear artery. J Physiol 364: 151-167
15. Smith J, Jones AW (1985) Calcium-dependent fluxes of potassium-42 and chloride-36 during norepinephrine activation of rat aorta. Circ Res 56: 507–516
16. Aaronson PI, Jones AW (1985 Calcium regulation of potassium fluxes in rabbit aorta during activation by noradrenaline or high potassium medium. J Physiol 367: 27–43
17. Benham C, Bolton T (1986) Spontaneous transient outward currents in single visceral and vascular smooth muscle cells of the rabbit. J Physiol 381: 385–406
18. Byrne NG, Large WA (1987) Action of noradrenaline on single smooth muscle cells freshly dispersed from the rat anococcygeus muscle. J. Physiol (London) 389: 513–525
19. Bolton TB (1979) Mechanism of action of transmitters and other substances on smooth muscle. Physiol Rev 59: 606–718
20. van Breemen C, Aaronson P, Loutzenhiser R (1979) Sodium-calcium interactions in mammalian smooth muscle. Pharmacol Rev 30: 167–208
21. Godfraind T (1976) Calcium exchange in vascular smooth muscle, action of noradrenaline and lanthanum. J Physiol 260: 21–35
22. Deth R, van Breemen C (1974) Relative contributions of Ca influx and cellular Ca release during drug induced activation of rabbit aorta. J Membrane Biol 30: 363–380
23. Meisheri KD, Hwang O, van Breemen C (1981) Evidence for two separate Ca pathways in smooth muscle plasmalemma. J Membrane Biol 59: 19–25
24. Droogmans G, Declerck I, Casteels R (1987) Effect of adrenergic agonists on Ca^{2+} channel currents in single vascular smooth muscle cells. Pflugers Arch 409: 7–12
25. Morgan J, Morgan K (1984) Stimulus-specific patterns of intracellular calcium levels in smooth muscle of ferret portal vein. J Physiol 351: 155–167
26. Himpens B, Casteels R (1987) Measurement by Quin2 of changes of the intracellular calcium concentration in strips of the rabbit ear artery and of the guinea-pig ileum. Pflugers Arch 410: 32–37
27. Pacaud P, Loirand G, Mironneau C, Mironneau J (1987) Opposing effects of noradrenaline on the two classes of voltage-dependent calcium channels of single vascular smooth muscle cells in short-term primary culture. Pflugers Arch 410: 557–559
28. Galizzi JP, Qar J, Fosset M, Van Renterghem C, Lazdunski M (1987) Regulation of calcium channels in aortic smooth muscle cells by protein kinase C activators (diacylglycerol and phorbol esters) and by peptides (casopressin and bombesin) that stimulate phosphoinositide breakdown. J Biol Chem 262: 6947–6950
29. Benham CD, Bolton TB, Byrne NG, Large WA (1987) Action of externally applied adenosine triphosphate on single smooth muscle cells dispersed from rabbit ear artery. J Physiol 387: 473–488
30. Benham CD, Tsien RW (1987) A novel receptor-operated Ca^{2+} permeable channel activated by ATP in smooth muscle. Nature 328: 275–278

Arterial Smooth Muscle Contraction Is Dependent on Sustained Increases in Myoplasmic Ca^{2+}

CHRISTOPHER M. REMBOLD and RICHARD A. MURPHY[1]

Summary. Agonist stimulation of vascular smooth muscle produces a transient rise in myoplasmic $[Ca^{2+}]$ as detected by both aequorin and fluorescent dyes. There are reports that $[Ca^{2+}]_i$ may subsequently fall to near resting values with substained stimulation and contraction. We tested the hypothesis that cellular injury (with artifactually high resting $[Ca^{2+}]_i$) may obscure sustained elevations in $[Ca^{2+}]_i$. Aequorin-loaded swine carotid media preparations were separated into three groups based on the stress induced by maximal doses of histamine (10–100 μM). Agonist-induced steady-state (30-min) $[Ca^{2+}]_i$ increases were attenuated and resting $[Ca^{2+}]_i$ values were higher in tissues that produced less active stress. Tissue damage, with elevated resting $[Ca^{2+}]_i$, appears to mask agonist-dependent increases in aequorin-estimated $[Ca^{2+}]_i$ responsible for contraction in vascular smooth muscle.

Key words: Aequorin—Cross-bridge phosphorylation—Excitation-contraction coupling —Latch bridges—Swine carotid media

Introduction

$[Ca^{2+}]$-stimulated phosphorylation of the 20K myosin regulatory light chain has been proposed to be the primary regulator of cross-bridge cycling and stress development in smooth muscle [1]. Supporting this hypothesis is the finding that estimates of intracellular $[Ca^{2+}]$ ($[Ca^{2+}]_i$) and myosin phosphorylation can transiently rise to high values upon agonist stimulation [2–4]. The regulation of force maintenance is more controversial. Stress maintenance is associated with small but significant increases in both $[Ca^{2+}]_i$ and myosin phosphorylation [4–7]. We hypothesized that the high levels of force are maintained by the accumulation of attached, dephosphorylated cross-bridges (termed latch bridges) which are formed by the dephosphorylation of attached cross-bridges [8–10]. Small elevations in $[Ca^{2+}]_i$ and myosin phosphorylation can thus produce a high force. This hypothesis explains many of the mechanical characteristics of smooth muscle

[1]Division of Cardiology, Departments of Medicine and Physiology, University of Virginia Health Sciences Center, Charlottesville, VA 22908, USA

such as the direct dependence of the unloaded shortening velocity on phosphorylation and the high economy of force maintenance [10, 11].

An alternative hypothesis for stress maintenance has been proposed [12–14]. Several studies suggested that intracellular [Ca^{2+}] decreased to near resting levels with sustained agonist stimulation [3, 12, 15–19], while other studies found sustained increases in [Ca^{2+}]$_i$ [4, 6, 7, 20–23]. Phorbol diesters, activators of protein kinase C, can induce smooth muscle contraction [13, 24]. Rasmussen et al. [13] proposed that protein kinase C phosphorylates some unknown cytoskeletal or myofibrillar protein to regulate the tonic phase of contraction (latch), independent of sustained increases in [Ca^{2+}]$_i$.

The issue we address is whether Ca^{2+}-dependent phosphorylation is obligatory for stress maintenance, or whether Ca^{2+}-independent mechanisms are involved in contraction of smooth muscle. We hypothesized that cellular injury can increase [Ca^{2+}]$_i$ (presumably in superficial cells). Such increase in resting [Ca^{2+}]$_i$ estimates may artifactually obscure small, sustained elevations in [Ca^{2+}]$_i$. Our index of cellular damage was stress (force/tissue cross-sectional area) generated by stimulation with maximal doses of histamine.

Materials and Methods

Tissue Preparations and Aequorin Loading

Swine common carotid arteries were obtained from a slaughterhouse and transported at 0°C in physiological salt solution (PSS). Dissection of medial strips, mounting and determination of the optimum length for stress development were performed as illustrated by Driska et al. [2]. The intimal surface was mechanically rubbed to remove the endothelium. PSS contains NaCl 140 m*M* KCl 5 m*M* 3-[*N*-morpholino]propanesulfonic acid (MOPS) 2 m*M* $CaCl_2$ 1.6 m*M* $MgCl_2$ 1.2 m*M* Na_2HPO_4 1.2 m*M* D-glucose 5.6 m*M* (pH adjusted to 7.4 at 37°C). Agonist stimulation was performed by injecting an appropriate volume of 10 m*M* stock histamine into the tissue bath. Agonist stock solutions were prepared daily.

Aequorin (batch 2, obtained from Dr. John Blinks, Mayo Medical School, Rochester, MN) was loaded as described by Rembold and Murphy [4], which was based on the method of Morgan and Morgan [3]. This procedure involves incubation of free-floating tissues in a series of Ca^{2+}-free solutions at 2°C. Solution 1 chelates extracellular Ca^{2+} with high EGTA; solution 2 contains aequorin; solution 3 contains higher [Mg^{2+}] which may help to "reseal" the membrane; and solution 4 is Ca^{2+}-free PSS with high $MgCl_2$. The tissues were mounted isometrically, stretched to a length which produced a stress of approximately 0.5×10^5 N/m^2, and warmed to 22°C. Extracellular $CaCl_2$ was slowly increased to 1.6 m*M*, and the tissues were equilibrated overnight at 37°C. The next morning, the tissues were stretched to within 5% of their optimal length for stress development. The loading procedure does not affect maximal stress development or the time course of myosin phosphorylation [6].

Light Detection

Light measurements were made in a light-tight enclosure modeled on the apparatus of Blinks [25] as described previously [6]. The apparatus was modified by connecting the photomultiplier output to a C-10 photon counter (Thorn EMI). The digital output was counted by a Metrabyte DASH16 AD board in an AT&T 6300 computer. The dark count of the photomultiplier tube was collected for at least 5 min. The photon count and analogue force signal were stored by the computer every second throughout the experimental period. Force was measured with a Cambridge Technology 300H servo in the isometric mode and displayed on a rectilinear recorder in synchronization with the light signal. Stress was calculated as force per cross-sectional area, which was estimated from measured length, weight, and a density of 1.050 g/cm^3.

Aequorin light signals are presented in the form log L/L_{max}, where L is the photon count (in counts per second) and L_{max} is an estimate of the peak light intensity that would be recorded if all of the aequorin in the tissue were instantaneously exposed to 5 m*M* $CaCl^2$ [26]. L_{max} can be considered a measure of the total undischarged aequorin present in the tissue. Hence, the ratio L/L_{max} will be invariant with respect to the efficiency of aequorin loading in different tissues. L_{max} was calculated according to the method of Allen and Blinks [26]. Changes in log L/L_{max} and active stress were compared with resting values using Student's paried *t*-test, and significance was defined at $P < 0.05$.

Aequorin light emission was calibrated in a series of Ca^{2+}/ethyleneglycol-bis-(β-aminoethyl ether) *N*, *N*, *N'*, *N'*-tetraacetic acid (EGTA) buffers with 0.5 m*M* Mg^{2+} at 37°C as described in Rembold and Murphy [4].

Results

Representative tracings of tissues stimulated with maximal doses of histamine are shown in Fig. 1. Tissues that produced high stress values (solid line) typically had the lowest resting $[Ca^{2+}]_i$, higher transient peaks, and larger changes in $[Ca^{2+}]_i$ observed 30 min after stimulation. In contrast, tissues that produced low stress (dotted line) had higher resting and smaller changes in $[Ca^{2+}]_i$.

In all, 32 aequorin-loaded swine carotid media preparations were stimulated with 10 or 100 μM histamine. Increasing histamine from 10 to 100 μM did not increase the steady-state stress. These preparations were divided into three groups based on the steady-state stress (30 min after histamine addition). Stress values exceeding 2×10^5 N/m^2 were exceptional (12%), and these tissues exhibited the lowest resting $[Ca^{2+}]_i$ and the largest change in $[Ca^{2+}]_i$ (Table 1). Stress values below 1×10^5 N/m^2 were observed in 34% of the preparations. This prevalence of damaged tissues was similar to that observed in tissues that were not loaded with aequorin (29%, $n = 187$) suggesting that dissection and storage were responsible for tissue injury, rather than the aequorin loading procedure. These tissues exhibited significantly higher resting values and smaller steady-state increases in $[Ca^{2+}]_i$ when compared with the high stress group.

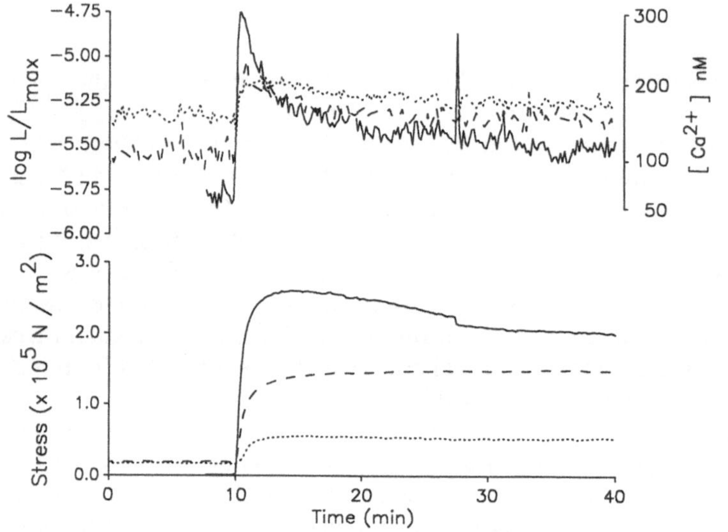

Fig. 1. Representative tracings of log L/L_{max} (myoplasmic $[Ca^{2+}]$) and active stress after stimulation with 10 or 100 μM histamine in three preparations that produced high (*solid line*), intermediate (*dashed line*), and low (*dotted line*) levels of stress. Histamine was added to the bathing solution at 10 min. The *perturbations* in the solid lines at 27 min were due to a solution flush. The lines are plotted as 10 sec averages of log L/L_{max} and stress values. The $[Ca^{2+}]_i$ scale was calculated from calibration in 0.5 mM $MgCl_2$ at 37°C

Table 1. Estimated resting and histamine-induced sustained $[Ca^{2+}]$ in tissues generating different levels of stress

Group ($\times 10^5$ N/m^2)	Stress ($\times 10^5$ N/m^2)	Resting Log L/L_{max} (nM Ca^{2+})	Sustained Log L/L_{max} (nM Ca^{2+})	Change Log L/L_{max} (nM Ca^{2+})	n
<1	0.65 ± 0.06	−5.38 ± 0.04 (145 ± 8)	−5.22 ± 0.02 (179 ± 6)	0.15 (34)	11
1–2	1.40 ± 0.07*	−5.53 ± 0.04 (109 ± 9)*	−5.31 ± 0.03 (156 ± 8)	0.22 (47)*	17
>2	2.42 ± 0.23*	−5.74 ± 0.04 (70 ± 9)*	−5.38 ± 0.06 (143 ± 14)*	0 36 (73)*	4
>1	1.60 ± 0.11*	−5.57 ± 0.04 (104 ± 7)*	−5.32 ± 0.03 (157 ± 7)*	0.25 (53)*	21

* $P < 0.05$ in comparison with those tissues that produced $< 1.0 \times 10^5$ N/m^2.

Table 2. Estimated resting $[Ca^{2+}]$ in tissues producing different levels of stress before and after 10 min in a Ca^{2+}-free physiological salt solution with 1 mM EGTA

Group ($\times 10^5$ N/m^2)	Stress ($\times 10^5$ N/m^2)	Resting Log L/L_{max} (nM Ca^{2+})	Ca^{2+}-free Log L/L_{max} (nM Ca^{2+})	Change Log L/L_{max} (nM Ca^{2+})	n
<1	0.71 ± 0.13	−5.56 ± 0.07 (105 ± 14)	−5 77 ± 0.08 (65 ± 15)	−0.20 (40)	4
1–2	1.29 ± 0.08*	−5.61 ± 0.03 (94 ± 6)	−5.76 ± 0.03 (65 ± 6)	−0.15 (29)	9
>2	2.44 ± 0.18*	−5.74 ± 0.07 (70 ± 14)	−5.86 ± 0.07 (48 ± 13)	−0.12 (22)*	5

* $P < 0.05$ in comparison with those tissues that produced $< 1.0 \times 10^5$ N/m^2

A second group of tissues were exposed to PSS containing 1 m*M* EGTA with no added $CaCl_2$ for 10 min to remove extracellular Ca^{2+} and deplete permeable cells. These tissues were also divided into three groups based on the stress produced in response to histamine or 109 m*M* KCl (substituted for NaCl in PSS). Those tissues with less injury (i.e., higher stresses) had smaller decreases in resting log L/L_{max} on removal of extracellular $CaCl_2$ (Table 2).

Discussion

We have proposed that Ca^{2+}-dependent myosin light chain phosphorylation is the primary regulator of both stress development and maintenance (latch) in smooth muscle [4, 9, 10, 23]. One characteristic of this covalent regulation is that small increases in $[Ca^{2+}]_i$ and cross-bridge phosphorylation can induce large amounts of active stress due to the accumulation of attached, dephosphorylated cross-bridges, termed latch bridges [2, 5, 8, 9, 23]. Estimates of both basal $[Ca^{2+}]_i$ and basal phosphorylation values are subject to significant errors. We hypothesized that such errors may be responsible for observations of $[Ca^{2+}]_i$-[3, 15, 17, 18] or phosphorylation-[27, 28] independent contractions.

A likely source of error in estimates of basal $[Ca^{2+}]_i$ is the presence of superficial cell injury. Light emission by aequorin is proportional to $[Ca^{2+}]^{2.5}$. Therefore, a small population of injured cells with higher resting $[Ca^{2+}]_i$ could signifcnatly increase resting $[Ca^{2+}]_i$ estimates. This could mask or reduce the true average increases in steady-state $[Ca^{2+}]_i$ which produce a contraction.

The maximum active stress induced by a supramaximal stimulus is a good index of tissue viability. Dead cells would not respond to stimulation and injured cells with elevated $[Ca^{2+}]_i$ would be supercontracted in the unstimulated preparation and not add to the force induced by the agonist. We previously established a minimum active stress of 1×10^5 N/m^2 as the criteria for acceptance of an experiment [4]. By comparing tissues generating low, normal, or high stress as a proxy for injury, we could assess the effect on basal or agonist-induced increases in $[Ca^{2+}]_i$.

Reductions in tissue viability, as assessed by stress generation, were associated with significant increases in estimated basal myoplasmic $[Ca^{2+}]$. This hypothesis is supported by the response to removal of extracellular $CaCl_2$ (Table 2). Damaged preparations (i.e., low stress) had larger decreases in intracellular $[Ca^{2+}]$ when extracellular $CaCl_2$ was removed.

The small selected population of the "best" tissues had a significantly lower basal $[Ca^{2+}]_i$ than the average population of acceptable tissues generating more than 1×10^5 N/m^2 (Table 1). This implies that reported average values [4, 6] may be artifactually high. The magnitude of histamine-induced initial $[Ca^{2+}]_i$ transients was attenuated in tissues generating low stress (Fig. 1). Steady-state $[Ca^{2+}]_i$ estimates during histamine-induced contractions were lower in the more viable tissues, but the net increases over basal values were higher (Table 1). The response of injured cells to histamine is not predictable and may be quite variable. Some cells might show an increase in light emission, and others might not respond. However, dead cells or in general any cells in which Ca^{2+} is elevated for long periods of time would be depleted of active aequorin and unable to generate a light signal on stimulation [6]. Thus, the presence of dead cells would attenuate the stimulus-induced increases in estimated $[Ca^{2+}]_i$. We conclude that in vivo basal $[Ca^{2+}]_i$ values are probably lower than the average estimates in isolated tissues using aequorin. Initial transients and the magnitude of the increase in steady-state $[Ca^{2+}]_i$ may be underestimated. Nevertheless, agonist-induced steady-state $[Ca^{2+}]_i$ may be overestimated in preparations with cell injury. These effects are quite striking in the histamine-stimulated carotid media, in which steady-state $[Ca^{2+}]_i$ and phosphorylation values are higher than can be elicited with most agonists in the majority of tissues. The slow, submaximal contractions reported as being $[Ca^{2+}]$-independent may be associated with $[Ca^{2+}]_i$ increases that were undetectable in preparations that had an injury-induced, elevated basal $[Ca^{2+}]_i$ signal.

Our results are quantitatively applicable only to the swine carotid media preparation. The carotid media may be more subject to tissue damage because the adventitia is dissected away and the endothelium is rubbed off. Nevertheless, the swine carotid media is much thicker than most vascular smooth muscle preparations, which would minimize edge effects. The magnitude of artifacts will be proportional to the fraction of injured cells. We believe this is comparatively low in the swine carotid media because the average active stresses (based on smooth muscle cell cross-sectional area) are higher than generally reported for other tissues.

Fluorescence indicators do not have the marked increase in signal as a function of $[Ca^{2+}]$ that characterizes aequorin. This would serve to decrease the effect of a small proportion of damaged cells on the resting fluorescence signal. However, fluorescent dyes are not inactivated by high $[Ca^{2+}]$ as is aequorin. Thus, dead cells will still contribute to the resting $[Ca^{2+}]$ fluorescence signal. Significant artifacts may be present and should be evaluated. Our results indicate that all estimates of myoplasmic $[Ca^{2+}]$ should be accompanied by stringent criteria for tissue viability. Reports of Ca^{2+}-independent contractions can reasonably be explained by the inability to detect the small increases in $[Ca^{2+}]_i$ (and phosphorylation) that are responsible for the observed contraction.

Acknowledgements. We thank Barbara Weaver for technical assistance and Dana Hannum-Baugh for manuscript preparation. Arteries were donated by Smithfield Co., Smithfield, VA. Dr. C.M. Rembold is a Lucille P. Markey Scholar, and this work was supported by a grant from the Lucille P. Markey Charitable Trust in addition to support from PHS grants 1RO1-HL38918 and 5PO1-HL19242.

References

1. Kamm KE, Stull JT (1985) The function of myosin and myosin light chain kinase phosphorylation in smooth muscle. Ann Rev Pharmacol Toxicol 25: 593–620
2. Driska SP, Aksoy MO, Murphy RA (1981) Myosin light chain phosphorylation associated with contraction in arterial smooth muscle. Am J Physiol 240: C222–C233

3. Morgan JP, Morgan KG (1982) Vascular smooth muscle: the first recorded Ca^{2+} transients. Pflügers Arch 395: 75–77
4. Rembold CM, Murphy RA (1988) Myoplasmic $[Ca^{2+}]$ determines myosin phosphorylation in agonist stimulated swine arterial smooth muscle. Circ Res 63: 593–603
5. Ratz PH, Hai C-M, Murphy RA (1989) Dependence of stress on cross-bridge phosphorylation in vascular smooth muscle. Am J Physiol 256: C96–C100
6. Rembold CM, Murphy RA (1986) Myoplasmic calcium, myosin phosphorylation, and regulation of the crossbridge cycle in swine arterial smooth muscle. Circ Res 58: 803–815
7. Rembold CM, Murphy RA (1988) $[Ca^{2+}]$-dependent myosin phosphorylation in phorbol diester stimulated smooth muscle contraction. Am J Physiol 255: C719–C723
8. Dillon PF, Aksoy MO, Driska SP, Murphy RA (1981) Myosin phosphorylation and the cross-bridge cycle in arterial smooth muscle. Science 211: 495–497
9. Hai C-M, Murphy RA (1988) Crossbridge phosphorylation and regulation of the latch state in smooth muscle. Am J Physiol 254: C99–C106
10. Hai C-M, Murphy RA (1988) Regulation of shortening velocity by crossbridge phosphorylation in smooth muscle. Am J Physiol 255: C86–C94
11. Hai C-M, Murphy RA (1989) Ca^{2+}, Cross-bridge phosphorylation, and contraction. Ann Rev Physiol 51: 285–298
12. Alkon DL, Rasmussen H (1988) A spatial-temporal model of cell activation. Science 239: 998–1005
13. Rasmussen H, Takuwa Y, Park S (1987) Protein kinase C in the regulation of smooth muscle contraction. FASEB J 1: 177–185
14. Small JV, Fürst DO, De Mey J (1986) Localization of filamin in smooth muscle. J Cell Biol 102: 210–220
15. Bradley AB, Morgan KG (1987) Alterations in cytoplasmic calcium sensitivity during porcine coronary artery contractions as detected with aequorin. J Physiol (London) 385: 437–448
16. Caponi AM, Lew PD, Vallotton MB (1985) Cytosolic free calcium levels in monolayers of cultured rat aortic smooth muscle cells. J Biol Chem 260: 7836–7842
17. DeFeo TT, Morgan KG (1986) Comparison of two different indicators: quin 2 and aequorin in isolated single cells and intact strips of ferret portal vein. Pflügers Arch 406: 427–429
18. Morgan JP, Morgan KG (1984) Stimulus-specific patterns of intracellular calcium levels in smooth muscle of ferret portal vein. J Physiol (London) 351: 155–167
19. Sumimoto K, Kuriyama H (1986) Mobilization of free Ca^{2+} measured during contraction-relaxation cycles in smooth muscle cells of the procine coronary artery using quin 2. Pflügers Arch 406: 173–180
20. Bitar KN, Bradford P, Putney JW, Makhlouf GM (1986) Cytosolic calcium during contraction of isolated mammalian gastric muscle cells. Science 232: 1143–1145
21. Brock TA, Alexander RW, Ekstein LS, Atkinson WJ, Gimbrone MA (1985) Angiotensin increases cytosolic free calcium in cultured vascular smooth muscle cells. Hypertension 7: I105–I109
22. Matsumoto T, Kanaide H, Nishimura J, Shogakiuchi Y, Kobayashi S, Nakamura M (1986) Histamine activates H1-receptors to induce cytosolic free calcium transients in cultured vascular smooth muscle cells from rat aorta. Biochem Biophys Res Comm 135: 172–177
23. Rembold CM, Murphy RA (1988) Myoplasmic $[Ca^{2+}]$ determines myosin phosphorylation and isometric stress in agonist stimulated swine arterial smooth muscle. Eur J Pharmacol 12 (Suppl 5): 538–542
24. Danthuluri NR, Deth RC (1984) Phorbol-ester-induced contractions of arterial smooth muscle and inhibition of α-adrenergic response. Biochem Biophys Res Comm 125: 1103–1109
25. Blinks JR (1982) The use of photoproteins as calcium indicators in cellular physiology. Techniques Cell Physiol 126: 1–38

26. Allen DG, Blinks JR (1979) The interpretation of light signals from aequorin-injected skeletal and cardiac muscle cells: a new method of calibration. In: Ashley CC, Campbell AK (eds) Detection and measurements of free Ca^{2+} in cells. Elsevier/North-Holland, Amsterdam, pp 159–174
27. Moreland S, Moreland RS (1987) Effects of dihydropyridines on stress, myosin phosphorylation, and V_o in smooth muscle. Am J Physiol 252: H1049–H1058
28. Park S, Rasmussen H (1986) Carbachol-induced protein phosphorylation changes in bovine tracheal smooth muscle. J Biol Chem 261: 15734–15739

Role of Second Messengers on Mobilization of Calcium and Related Events in Vascular Smooth Muscles

Kenji Kitamura[1], Takeo Itoh[1], Masato Hirata[2], Matthew C.E. Gwee[3], and Hirosi Kuriyama[1]

Summary. In vascular smooth muscles the contraction-relaxation cycles are regulated by changes in the intracellular free Ca concentration; second messengers (signal transductors) play an essential role in the regulation of the cell activity through modulation of free Ca in the cytosol. The Ca homeostasis in vascular smooth muscles is controlled by various factors: increases in the intracellular Ca concentration are mediated by Ca influx through the activation of voltage-dependent and receptor-operated channels and the release of Ca from the sarcoplasmic reticulum; decreases in the free Ca concentration are caused by Ca efflux through the activation of the Ca active transport system and through active Ca accumulation into the sarcoplasmic reticulum. Furthermore, the Na-Ca exchange diffusion mechanism is thought to contribute partly to the reduction in the free Ca. In this chapter, the role of second messengers such as cyclic AMP, cyclic GMP, diacylglycerol (including the action of phorbol esters) and inositol 1,4,5-trisphosphate on Ca regulation in vascular smooth muscles is reviewed.

Key words: Ca mobilization—Cyclic nucleotides—Diacylglycerol—Inositol 1,4,5-trisphosphate—Vascular smooth muscles

Introduction

In vascular smooth muscles the contraction-relaxation cycle is regulated by the intracellular concentration of free Ca, i.e., at concentrations over 0.1 μM, Ca promotes the phosphorylation of myosin light chain (20K D proteins, MLC_{20}) and initiates cross-bridge cycling between actin and myosin which mediates the contraction. At 10 μM Ca, the maximum amplitude of contraction is achieved as estimated from mechanical responses in skinned smooth muscle tissues. Although the role of Ca on the contraction-relaxation process has been estab-

Department of Pharmacology, Faculty of Medicine[1], and Department of Biochemistry, Faculty of Dentistry[2], Kyushu University, Fukuoka 812, Japan
[3]Department of Pharmacology, Faculty of Medicine, National University of Singapore, Singapore 0511

lished [75, 37] the underlying mechanisms responsible for mediating the contraction are not yet fully understood. For example, stimulation of the membrane by an agonist increases the free Ca in the cytosol and phosphorylates MLC_{20} through the formation of Ca-calmodulin-myosin light chain kinase complex. As a consequence, the cycling rate of cross-bridges between actin and myosin is increased. When the Ca concentration is subsequently reduced, the phosphorylation of MLC_{20} and the cycling rate of cross-bridges between actin and myosin are reduced, whereas the amplitude of contraction induced by the agonist is maintained. This phenomenon has been referred to as the "latch bridge" or "latch state" [17, 64]. Thus, to maintain prolonged muscle contraction in the presence of a low concentration of Ca, a mechanism to matintain high Ca sensitivity (2nd Ca sensitivity) of the contractile proteins has been postulated [64]. However, details of such a mechanism have not yet been clarified.

The amount of free Ca required to regulate a contraction is obtained through the influx of Ca across the cell membrane and through the release of Ca from the intracellular store site, mainly the sarcoplasmic reticulum (SR). Influx of Ca is due to activation of the voltage-dependent and the receptor-operated channels [9]. However, the nature of the latter needs further clarification. Release of Ca from the SR is thought to be Ca-induced as well as through activation by inositol 1,4,5-trisphosphate ($InsP_3$). The latter is synthesized from the breakdown of phosphatidyl inositol 4,5-bisphosphate (PI-P_2) through the activation of various membrane receptors by the appropriate agonist [8, 50, 1]. The mutual relationship between Ca-induced and $InsP_3$-induced Ca release mechanisms is now under further investigation. Reduction in the free Ca in the cytosol is caused by extrusion of Ca and reaccumulation into the SR. The former is thought to be due to activation of the active Ca transport which requires ATP and, in part, of the Na-Ca exchange diffusion mechanism [19, 3]. However, the physiological role of the Na-Ca exchange diffusion mechanism in vascular smooth muscle cells is still controversial. In contrast to this, the active Ca transport system in the SR is generally accepted to play an important role in modulating the free Ca in the cytosol.

Mobilization of Ca in the smooth muscle cell is regulated by second messengers (signal transductors) such as the cyclic nucleotides (cyclic AMP and cyclic GMP), $InsP_3$, and diacylglycerol (DG). In relation to Ca mobilization, both cyclic AMP and cyclic GMP tend to reduce the free Ca in the cytosol through phosphorylation of cyclic AMP-dependent protein kinase (A-kinase) and cyclic GMP-dependent protein kinase (G-kinase) [75, 37]. $InsP_3$ increases the amount of free Ca via activation of the Ca channel distributed in the SR. Inositol 1,3,4, 5-tetrakisphosphate ($InsP_4$) [35], a metabolite of $InsP_3$ synthesized by the activation of $InsP_3$-3 kinase, is reported to modify the receptor-operated ion channel in neuroblastoma-glioma hybrid cells [28]. On the other hand, it was reported that the actions of $InsP_4$ require the presence of $InsP_3$ [35]. DG is known to activate protein kinase C which phosphorylates many proteins in cell organelles.

In the chapter, we would like to review some recent findings concerning the actions of the second messengers in relation to the mobilization and function of Ca in vascular smooth muscles. The review will deal with studies measuring the mechanical response of vascular smooth muscle cells and corresponding changes in the intracellular Ca concentration estimated using quin-2 or fura-2. The results of studies using biochemical procedures and electrophysiological techniques are also discussed. Here, the number of references cited is necessarily limited and, therefore, references to our work are not included in the reference list.

Actions of Cyclic Nucleotides

Cyclic AMP on Vascular Smooth Muscles

Activation of the β_1-adrenoceptor subtype distributed in the coronary artery and β_2-adrenoceptor subtype distributed in many other vascular beds causes muscle relaxation. The nature of this relaxation is thought to be due to an increase in the amount of cyclic AMP. However, the potency of cyclic AMP in inducing relaxation of smooth muscle cells seems to be much weaker than its positive inotorpic action induced by activation of β_1-adrenoceptors in cardiac muscles. In the vascular system β_2-adrenoceptor blockade lowers the blood pressure due to inhibition of the β-adrenoceptor-related mechanism rather than by direct actions on smooth muscles. Thus, the β-adrenoceptor activation in vascular smooth muscles may play a minor role in the regulation of blood pressure. Nevertheless, from a physiological viewpoint the mechanism of β-adrenoceptor activation has received much attention in relation to the synthesis of cyclic AMP.

Synthesis of cyclic AMP, after formation of the agonist-β-adrenoceptor complex, requires activation of GTP-binding proteins (G_s or N_s, which is composed of the α-, β- and γ-subunits; activation of the α-subunit is thought to be an essential process for activation of the catalytic subunit—adenylate cyclase). The activity of the GTP-binding protein (G-protein) and adenylate cyclase may be modified by various factors such as cholera toxin, forskolin, and others. In addition, the synthesis of cyclic AMP is inhibited by activation by the addition of pertussis toxin (islet cell activating factor; IAP), which can, therefore, enhance the synthesis of cyclic AMP [52].

The role of cyclic AMP in vascular smooth muscle tissues as observed from studied with cyclic AMP and a permeable cyclic AMP analogue, dibutyryl cyclic AMP, can be summarized as follows:

1. Cyclic AMP activates the A-kinase which phosphorylates MLCK with consequent inhibition of the phosphorylation reactions induced by Ca-calmodulin-MLCK complexes; as a consequence, contractions evoked in intact and saponin-treated skinned muscle tissues are inhibited.
2. Cyclic AMP is thought to activate the active Na-K transport mechanism [62], i.e., activation of the β-adrenoceptor causes stimulation of the cyclic AMP-dependent phosphorylation of Na-K ATPase (activation of Na-K pump); the subsequent changes in the ionic distribution activate the Na-Ca exchange diffusion mechanism.
3. Cyclic AMP accelerates Ca accumulation into the SR by activation of Ca-ATPase as determined from studies on the caffeine-induced contraction of skinned muscle tissue in Ca-free solution and also from the results of the

studies with SR-rich vesicles. As a consequence free Ca in the cytosol is partially reduced [60].

4. Cyclic AMP accelerates the active extrusion of Ca by activation of Ca-ATPase as shown in studies on Ca accumulation into the sarcolemmarich vesicle.
5. In cardiac muscle cyclic AMP activates the voltage-dependent Ca channel as reported in studies using the voltage clamp procedure. This can explain the positive inotropic action of cyclic AMP mediated through activation of the β-adrenoceptor [4]. However, in smooth muscles, cyclic AMP applied as dibutyryl cyclic AMP or injected intracellularly did not modify the voltage-dependent Ca current. These observations confirm the previous findings obtained from studies on the guinea-pig urinary bladder by Klöckner and Isengerg [40]. Droogmans et al. [18] also reported that norepinephrine (NE) and isoprenaline inhibit the voltage-dependent Ca channels in rabbit ear artery independent of the action of cyclic AMP.
6. Cyclic AMP modifies the Ca-dependent K-channel following its injection into smooth muscles. However, it is not yet certain whether the hyperpolarization induced by activation of the β-adrenoceptor has a causal relation with activation of the Ca-dependent K-channel through increased cyclic AMP synthesis.

   From the above observations, cyclic AMP seems to reduce the free Ca in the cytosol and to cause relaxation of the vascular tissues. This reduction in the intracellular free Ca concentration requires the activation of the active Ca extrusion pump to a greater extent than that of the Na-Ca exchange diffusion. These actions of cyclic AMP may explain the differences observed in smooth muscle and cardiac muscle tissues; cyclic AMP relaxes the former and contracts the latter.
7. The β-adrenoceptor which regulates the action of vascular smooth muscle tone also contributes to the activity of perivascular nerve terminals, i.e., activation of the β_2-adrenoceptor distributed in nerve terminals enhances the release of the chemical transmitter. Isoprenaline also increases the release of the chemical transmitter at the nerve terminals, and this action is thought to be due to the action of cyclic AMP synthesized at the nerve terminals.

Cyclic GMP

Cyclic GMP plays a more important role than cyclic AMP in the regulation of the contraction-relaxation cycles in vascular smooth muscles. Synthesis of cyclic GMP is modulated by various substances such as endogenous substances released from nerve terminals, endothelium-derived relaxing factor, and many drugs. Thus putative substances released from the atria promote cyclic GMP synthesis; such substances include atrionatriuretic factor (ANF), atriopeptin, and α-human atrial natriuretic polypeptide (α-hANP) [58]. Furthermore, endothelium-derived relaxing factor (EDRF), released from endothelial cells by various stimulants such as acetylcholine (ACh), histamine, ATP, bradykinin, and angiotensin II, also increases the amount of cyclic GMP [32, 48]. It is a well-known fact that nitro-compounds such as trinitroglyceride (nitroglycerin), sodium nitroprusside, isosorbide dintrate, and nicorandil also increase the amount of cyclic GMP [43, 31, 48, 20].

Endothelium-derived relaxing factor. In 1980, Furchgott and Zawadzki [23] found that ACh releases a substance from the endothelium that produces relaxation of tissues precontracted by NE and high K. Since then numerous publications have appeared concerning the nature of this EDRF and the underlying mechanism of the EDRF-induced relaxation [22, 73]. EDRF has been reported to increase the amount of cyclic GMP by activation of guanylate cyclase distributed in the cytosol in a manner similar to that observed after the application of nitro-compounds [42, 57, 33]. The synthesis of cyclic GMP is inhibited by methylene blue or hemoglobin as observed in studies using the nitro-compounds [42, 33]. Recently, an elegant investigation was made to elucidate the nature of EDRF [53]; nitric oxide (NO) was found to play a key role in the endothelium-dependent relaxation induced by EDRF as postulated by Furchgott [22].

ANF and α-hANP. Newly found peptide hormones released from the cardiac atrium possess a potent vasodilating action which has also been attributed to the synthesis of cyclic GMP [14, 38, 44, 48]. These peptide hormones show some variation in structure in the different species studied. The synthesis of cyclic GMP by ANF or α-ANP vtilizes a different metabolic pathway from that produced by EDRF and requires the presence of G-protein and guanylate cyclase at the sarcolemma. This metabolic pathway is not inhibited by methylene blue or hemoglobin [48].

Actions of cyclic GMP that are mediated through the activation of G-kinase to produce vascular relaxation are thought to be due to the following mechanisms:

a) Cyclic GMP phosphorylates MLC_{20} via activation of the G-kinase as does cyclic AMP through the activation of A-kinase [48, 49]. Thus, the phosphorylation of MLC through activation of the Ca-calmodulin-MLCK complex is inhibited with consequent relaxation of the precontracted tissues. However, this action has been reported to be weaker than that of cyclic AMP or may even be absent [37].
b) Cyclic GMP activates the active Ca extrusion process; this action requires the presence of calmodulin [55, 56]. However, cyclic GMP does not activate Ca accumulation into the Sr through the Ca pump, and it also does not have any effect on the voltage-dependent Ca channels in vascular smooth muscle cells. In cardiac muscles, however, cyclic GMP synthesized by the action. In cardiac muscles, however, cyclic GMP synthesized by the action of ACh or exogenously applied as the dibutyryl cyclic GMP inhibits the voltage-dependent Ca current [21].

The main action of cyclic GMP synthesized in vascular smooth muscle cells in response to α-hANP, EDRF, or the nitro compounds is, therefore, thought to be the activation of the active Ca extrusion process leading to a reduction in the free Ca in the cytosol; fluorescent dyes such as fura-2 or quin-2 are used for the estimation of Ca in such studies.

EDRF related events. Endothelial cells release many substances which regulate smooth muscle cell activity. Together with the release of EDRF, an endothelium-derived hyperpolarizing factor (EDHF) is also released; i.e., EDHF hyperpolarizes the membrane by a different mechanism from that involving the EDRF released [41]. In particular, stimulation of the muscarinic M_2-receptor in the endothelium releases EDRF and that of the M_1-receptor releases EDHF as determined by using pirenzepine. Furthermore, the release of EDRF is inhibited by methylene blue, but not the release of EDHF. The action of EDHF is transient, and the hyperpolarization occurs within only a few minutes, but this short duration of action of EDHF is not due to desensitization of the smooth muscle cells. The hyperpolarization induced by EDHF accompanies the increase in the ionic conductance and, therefore, an increase in the K-conductance is postulated. However, apamine, a bee venom and an inhibitor of one subtype of the Ca-dependent K channel, has no effect on this hyperpolarization [41]. Although EDRF produces prolonged relaxation of precontracted tissues, responses to both EDRF and EDHF do not occur in the endothelium-denuded tissues. Thus, substances released from endothelial cells relax precontracted tissues by an increase in the amount of cyclic GMP and hyperpolarize the membrane to reduce its excitability through inhibition of the voltage-dependent Ca channels.

Interactions of agonists on the synthesis of cyclic nucleotides. As described previously, activation of the β-adrenoceptor results in the synthesis of cyclic AMP, and this action is inhibited by stimulation of the α_2-adrenoceptor, i.e., binding of the agonist to the α_2-adrenoceptor activates the G_i-protein which inhibits the activation of adenylate cyclase involved in the synthesis of cyclic AMP. ACh is also reported to increase the amount of cyclic GMP in smooth muscle cells via the release of EDRF from endothelial cells. Therefore, more precise investigations are required to clarify whether ACh increases cyclic GMP in endothelium-denuded vascular smooth muscle tissues or not.

Synthesis of cyclic nucleotides requires the presence of ATP and GTP. ATP plays an essential role in the activation of smooth muscle cells, i.e., it is required to promote the activation of contractile proteins, to accumulate Ca into the SR, to extrude Ca from smooth muscle cells, and also to activate the voltage-dependent Ca channel. In the sarcolemma the voltage-dependent Ca current, as measured by the whole cell clamp method, produces a rapidly developing rundown phenomenon in the absence of ATP; intracellular perfusion of ATP prevents the rundown. The nonhydrolyzable adenosine derivatives (AMP-PNP) cannot substitute for the action of ATP in preserving the activity of this channel. These derivatives of adenosine also activate the Ca-dependent K-channel as measured by the whole cell voltage clamp procedure.

GTP also has multiple roles in vascular smooth muscle cells. Activation of G-protein requires the presence of GTP. High concentrations of GTP (above 100 μM) also act on the contractile proteins, and low concentrations (above 1 μM) of GTPγS more potently evoke the contractions in skinned muscle tissues prepared by saponin treatment. However, GTPγS and GTP do not act on the Ca-independent contraction evoked by Mg-ATP following application of rigor solution (lack of ATP and Ca in the solution). Thus, GTP increases the Ca sensitivity of the contractile proteins. Furthermore, GTP and GTPγS release Ca from the SR as observed during the application of both agents in the resting state following accumulation of Ca into the SR and the subsequent caffeine-induced contraction in skinned muscle strips. in the rabbit mesenteric artery, Saida and Van Breemen [61] reported that the release of Ca by $InsP_3$ requires the presence of GTP . However, in the same tissues, GTP acts additively with $InsP_3$-induced Ca release from the SR, but not with that of caffeine. However, it is not yet clear how much GTP is required in smooth muscle cells for physiological functions.

Actions of Inositol 1,4,5-Trisphosphate

The application of ACh to the coronary arteries of various species produces varying responses e.g., in the guinea-pig coronary artery ACh consistently hyperpolarizes the membrane if the endothelium is intact, but not after the endothelium is denuded. Much the same is observed after application of ACh to porcine coronary artery [7] and to the rat aorta [66]. However, in the rabbit intact coronary artery, ACh transiently hyperpolarizes the membrane and subsequently generates a depolarization whereas, after removal of the endothelium, the response is only depolarization. The hyperpolarization is due to the release of EDHF as noted previously, and in some, but not all, tissues ACh depolarizes the membrane as a direct action. However, ACh consistently produces contractions in the above three tissues. In the guinea-pig mesenteric artery, activation of the α_1-adrenoceptor with a low concentration of NE (below 1 μM) does not modify the membrane potential, yet tissue contraction occurs. Therefore, the depolarization of the membrane is a prerequisite for the activation of the voltage-dependent Ca channel with increases in the ca influx, but it is not an essential process to produce contraction. This agonist-induced contraction evoked without depolarization of the membrane has been termed pharmaco-mechanical coupling mechanism [65]. Such responses have been confirmed in various vascular tissues (rabbit pulmonary, ear, and saphenous arteries, guinea-pig tail and basilar arteries, and others).

The pharmaco-mechanical coupling occurs not only with exogenously applied NE or ACh but also with the endogenous release of NE by perivascular nerve stimulation, e.g., in the guinea-pig mesenteric artery, perivascular nerve stimulation provokes excitatory junction potential (e.j.p.) and, upon repetitive stimulation, enhances the amplitude of e.j.p. and triggers the spike when the depolarization exceeds the threshold. The e.j.p. is not affected by prazosin, an α_1-adrenoceptor blocker, but yohimbine (an α_2-adrenoceptor blocker) and phentolamine (an α_1- and α_2-adrenoceptor blocker) enhance the amplitude of e.j.p; however, the contraction evoked by perivascular nerve stimulation is blocked by prazosin. The nature of the transmitter involved in the generation of the e.j.p. is unknown, i.e., whether this results from the activation of a specific NE receptor γ-adrenoceptor) by released NE [29] or from activation of the purinergic receptor (P_1) by released ATP, a co-transmitter of NE [10].

The pioneering work concerning inositol phospholipid metabolism in relation to Ca mobilization was done by Hokin and Hokin [30] using pancreatic slices. These workers have shown that ACh accelerated incorporation of ^{32}P into phosphatidyl inositol and phosphatidic acid. Michell [46] subsequently proposed the Ca gate theory which suggests that activation of phosphatidylinositide metabolism triggers the opening of the Ca gate at the surface membrane and increases the influx of Ca. Streb et al. [67] then reported that, in the rat pancreatic acinar cells, $InsP_3$ releases Ca stored in the cell (endoplasmic reticulum). It was also found that not only ACh but 5-hydroxytryptamine synthesizes $InsP_3$, which preceded the monophosphate [8]. During the course of research in various laboratories, the inositol phospholipid metabolism was established and the cascade of events deduced: activation of various receptors (α_1-adrenergic, H_1-histaminergic, M_2-muscarinic, and others) hydrolyses PI-P_2 (and partly PI-P_1) resulting in the synthesis of $InsP_3$ and DG. DG is an unstable substance and forms phosphatidic acid. More details concerning these biochemical studies have already been reviewed by many investigators [8, 50, 76, 1].

The effects of ACh on the porcine coronary artery or of NE on the rabbit mesenteric artery, in relation to the actions of $InsP_3$, were briefly reviewed. It should be noted here that ACh has no effect on the electrical and mechanical responses of the rabbit mesenteric artery, and NE also has no effect on the porcine coronary artery. The action of $InsP_3$ on the Ca release mechanism in relation to the agonist action can be summarized as follows:

1. ACh and NE produce contraction of individual tissues in Ca-free solution containing 0.1 μM EGTA. Procaine (above 1 mM) blocks the contraction evoked by either ACh or NE in the presence or absence of Ca. The minimum concentration of procaine required to inhibit the contraction evoked by ACh was lower than that required for caffeine, partly due to the inhibition of $InsP_3$ synthesis by procaine. Following the application of a low concentration (10–100 nM) of A23187, a Ca ionophore, ACh and NE produced contraction in both tissues with a very slow time course.
2. In both tissues, the Ca transient observed by the application of ACh or NE is blocked by pretreatment with either atropine or prazosin, respectively. Ca release from the SR was estimated by fluorescence intensities using fura-2 or quin-2, in the presence or absence of extracellular Ca. Acceleration of Ca efflux by ACh was also detected in porcine coronary artery using labelled Ca.
3. Application of either ACh or NE reduces PI-P_2 and increases $InsP_3$ and phosphatidic acid (PA) in a concentration-dependent manner.
4. When the activity of phosphatidylinositol 4, 5-bisphosphate phosphodiesterase (PDE, phospholipase C) is assessed from the production of $InsP_3$, the cytosol fraction shows about a 10 times higher activity than that observed from the particulate fraction in smooth muscle cells.
5. In vascular tissues, $InsP_3$-3 phosphatase plays a role in the synthesis of inositol 1,3,4,5-tetrakisphosphate ($InsP_4$) which is dephosphorylated to InsP1, 3,4-trisphosphate ($InsP_3'$) by $InsP_4$-5-phosphatase in the cytosol [35]. $InsP_3$ is dephosphorylated by the respective phosphatases to $InsP_2$ and InsP inositol, respectively [1]. When the activity of $InsP_3$ phosphatase was estimated

from the reduction in the amount of $InsP_3$, much the same potency was observed in both particulate and cytosolic fractions. Increased concentrations of Ca enhance the activity of the phosphatase.

6. In the porcine aorta, $InsP_3$ releases Ca from the SR-rich fraction to a greater extent than that observed in the sarcolemma-rich fraction.
7. In permeabilized, dispersed cells prepared by saponin treatment (skinned muscle cells), $InsP_3$ releases Ca from the SR. The K_3 value for Ca release from the SR by various concentrations of $InsP_3$ was estimated to be 0.7 μM. The release of Ca from the SR was also estimated using the caged $InsP_3$ procedure [74].
8. In skinned smooth muscle tissues of the rabbit mesenteric artery, a contraction is obtained upon application of $InsP_3$ after the accumulation of Ca into the SR. Consequently, the amount of Ca stored in the SR will be reduced, and subsequently applied caffeine will produce a smaller amplitude of contraction than that evoked by caffeine alone. When $InsP_3$ was applied in the presence of 0.3 μM Ca, repetitively applied caffeine at 20-min intervals produced the same amplitude of contraction without any attenuation. In skinned muscle tissues, the release of Ca from the SR by $InsP_3$ to evoke a contraction requires a much higher concentration than that expected from the K_m value measured in extracted SR fragments. These differences may be caused by buffering of cytosolic Ca by EGTA, i.e., to evoke the contraction from the resting level, Ca was buffered with relatively high concentrations of EGTA (2–4 mM) so that the released Ca may be chelated quickly rather than bind with the calmodulin.
9. The $InsP_3$-induced Ca release can also be estimated from the activation of the Ca-dependent K current. In dispersed smooth muscle cells of the rabbit portal vein, the Ca-dependent K channel is activated when the intracellular free Ca is increased. The released Ca from the SR activates the Ca-dependent K channel to produce the oscillatory outward current. Appearance of the oscillatory Ca-dependent K current is also increased in proportion to the concentrations of extracellular Ca. When $InsP_3$ was perfused into the cell, the threshold required to produce the oscillatory current was lowered (shifted to a more hyperpolarized direction), and this increased the number of occurrences and the amplitude of oscillatory currents. When pretreatment with low concentrations of A23187 was applied, the oscillatory current ceased. The direct action of $InsP_3$ on the Ca channels was not observed, but the amplitude of the Ca current was inhibited by a marked increase in the Ca concentration following the intracellular perfusion of high concentrations of $InsP_3$; the inhibition is due to acceleration of the inactivation process of the Ca current indirectly.
10. $InsP_4$ has no direct actions on the Ca current but accelerates the development of the oscillatory Ca-dependent K current. Since $InsP_4$ is not able to release Ca from the SR in $InsP_3'$ showed a weak ability to release Ca from the SR [35]. $InsP_4$ probably acts after its dephosphorylation into $InsP_3'$ by $InsP_4$ 5-phosphatase located in the cytosol.
11. $InsP_3$ itself releases Ca from the SR and does not require the presence of GTP. Both agents additively release Ca from the SR. GTPγS showed a more potent action on Ca release than GTP.

12. The site of $InsP_3$ action in the SR has been estimated using the photoaffinity labelling technique. Incubation of $InsP_3$ with *para*-azidobenzoyl γ-alanine (ABγA) in saponin-treated, dispersed smooth muscle cells of the dog trachea results in the $InsP_3$-induced Ca release being irreversibly blocked after photo-radiation; the three $InsP_3$ binding sites were determined. When cells were pretreated with high concentrations of cold $InsP_3$, such $InsP_3$-ABγA binding to the receptor protein of SR was inhibited. $InsP_3$-ABγA complex leaked out Ca stored in the SR and prevented any accumulation of Ca into the $InsP_3$-releasable site. Elucidation of the amino acid sequences for the three peaks is urgently required for understanding of the $InsP_3$-induced Ca release mechanism.

From our present knowlegee, it is difficult to estimate the precise amount of $InsP_3$ required to mobilize Ca from the SR in various regions of vascular beds in different species. Distribution densities of various agonist receptors in relation to voltage-dependent Ca channels may also be important to solve the problem.

Actions of Protein Kinase C Activated by a Phorbol Ester (Instead of Diacylglycerol) on Vascular Smooth Muscles

Protein kinase C is widely distributed in many tissues including vascular smooth muscle tissues. The biochemical events involved in the formation of protein kinase C and its likely phsiological functions have been reviewed by Nishizuka [50], Williamson et al. [76], and Abdel-Latif [1]. Protein kinase C is activated by DG in the presence of phosphatidyl serine (PS) and low concentrations of Ca. Since DG is an unstable substance, phorbol esters such as phorbol 12,13-didecanoate (PDD), 12-0-tetradecanoylphorbol-13-acetate (TPA), and phorbol 12,13-dibutyrate (PDBu), or an analogue of DG, 1,2-diolein, are used as a substitute for DG. Phorbol esters are more potent activators of protein kinase C than the physiological product DG. These substances distribute into many cells and permanently activate protein kinase C [69, 12, 39]. In vascular smooth muscles, the role of DG and protein kinase C have been evaluated from studies on the actions of TPA or PDD. The results obtained are summarized as follows:

1. TPA (0.1 μM) has a dual action on the K-induced contraction with initial enhancement followed by inhibition during prolonged exposure. The enhancement occurs to a greater extent with a low (39 mM) than with a high (128 mM) K. TPA produces much the same effect on the contraction evoked by Ca in skinned muscle tissues; it enhances the contraction evoked by low Ca and inhibits the contraction evoked by high Ca. 1,2-Diolein also produces much the same response as that obtained with TPA, although with a lower potency.
2. TPA with PS has no effect on the Ca-independent contraction evoked in a relaxing solution (Ca-free solution) containing 10 mM EGTA and 4 mM MgATP following the application of adenosine-5-*o*-3-thiophosphate (ATPγS) and 0.3 μM Ca. The TPA-induced enhancement of the contraction evoked by high K has no causal relation with the amount of free Ca as estimated using quin-2 in intact muscle tissues.

3. From the above observations on the action of TPA, it is concluded that TPA has a dual action on the K- and Ca-induced contraction in intact and skinned muscle tissues, respectively, with the actions being dependent on the amount of free Ca in the cytosol; the TPA-induced enhancement of the contraction has no causal relation with functions of the sarcolemma and SR but requires the presence of Ca. Protein kinase C inhibits the actin-activated ATPase and also phosphorylates the MLC at a different site from that at which MLCK is phosphorylated by the Ca-calmodulin complex, thus relaxing smooth muscle tissues [34]. Presumably, the inhibitory effects of TPA observed in the presence of high concentrations of K (above 60 m*M*) in intact tissues or high CA (above 0.5 μM) in skinned muscle tissues may be related to this phosphorylating action of phorbol esters.
4. In the presence or absence of extracellular Ca, low concentrations of TPA (0.1–1 n*M*) slightly enhance the contraction evoked by ACh, but high concentrations of TPA (above 100 n*M* inhibit the ACh-induced contraction. TPA dose-dependently inhibits the reduction in the amount of PI-P_2 and increases the amount of PA in muscles under treatment with ACh.
5. When the contraction and Ca transient evoked by ACh are compared, TPA markedly reduces the Ca transient measured using fura-2 with no significant change in the amplitude of contraction. TPA and PS enhance the contraction evoked by $InsP_3$ in the presence of 0.3 μM Ca in skinned muscle tissues.
6. The above (4 and 5) indicate that TPA inhibits the ACh-induced increase in the intracellular free Ca by inhibiting the hydrolysis of PI-P_2, but it enhances the Ca sensitivity of the contractile proteins; i.e., the ACh-induced contraction is controlled by a negative feed-back regulation of PI-P_2 hydrolysis together with a positive feed-back regulation of the Ca sensitivity of the contractile proteins. The former may partly explain the generation of the agonist-induced desensitization of the receptor.
7. TPA enhances the contraction evoked by high K (below 60 m*M*) in intact tissues and by Ca (below 0.5 μM) in skinned muscle tissues [59, 13]. To investigate further the effects of TPA on intact and skinned smooth muscle tissues, force development, unloaded shortening velocity measured by the slack test (V_{max}; an indicator of cycling rate of cross-bridges between actin and myosin) [15, 16, 54, 25], and phosphorylation of MLC_{20} are compared. TPA enhances the force development and V_{max} more than the corresponding increase in phosphorylation of MLC_{20} on the K-induced contraction. The enhancing actions of TPA on the above parameters occur to a greater extent on the tonic contractions than those observed on the phasic contractions evoked by 128 m*M* K. These actions of TPA with PS are inhibited by preteatment with 1-(5-isoquinolinesulphonyl)-2-methyl-piperazine (H-7), a potent and selective inhibitor of protein kinase C [27].
8. In skinned muscle tissues, TPA with PS time-dependently enhance the amplitude of contraction induced by 0.3–1.0 μM Ca and V_{max}. At concentrations of Ca above 0.5 μM, the phosphorylation of MLC_{20} is enhanced to a lesser extent than the force development and V_{max}. Therefore, TPA may enhance the contraction by phosphorylation-dependent and -independent processes.

9. Using the whole cell voltage clamp procedure, the inward Ca current, measured in dispersed smooth muscle cells of the guinea-pig portal vein, is inhibited by extra- and intra-cellular perfusion of TPA or protein kinase C with or without occurrence of transient enhancement.

There is evidence that factors other that, or in addition to, myosin phosphorylation are involved in regulating the shortening velocity of smooth muscle tissues. Siegman et al. [63] suggested that in addition to the phosphorylation of MLC_{20}, the cycling rate of phosphorylated cross-bridges between actin and myosin may also be directly regulated by Ca in the intact guinea-pig taenia coli. Caldesmon [36] may also regulate actin through the process referred to as the caldesmon-actin flip flop model, or through some unknown processes. Furthermore, the role of leiotonin, a substance resembling troponin C located in smooth muscles, must not be ruled out [45].

In vascular smooth muscle tissues (mesenteric and coronary arteries) ACh produces a sustained contraction while the Ca transient, as measured by aequorine, quin-2, fura-2, rapidly declines and then remains just above the basal Ca level [47]. Aksoy et al. [2] noted that the maintenance of contraction is associated with a reduction in the isotonic shortening velocity and in the phosphorylation ofMLC_{20}. Therefore, the concept of the latch mechanism was introduced. Our basic assumption concerning the actions of protein kinase C, on the basis of the actions of TPA (with PS), is that activation of force development occurs with an increase in the V_{max}, with a slight increase in the phosphorylation of MLC_{20}, and without any change in the amount of free Ca in the cytosol. Thus, the actions of TPA may not have a direct relation to the latch mechanism, but protein kinase C activation may partly be related to the force maintenance during the decrease in the concentration of Ca.

Calcium Influx by Activation of the Voltage-Dependent and Receptor-Operated Calcium Channels

Features of the Ca influx which regulate the contraction-relaxation cycle are now becoming clear. Here we would like to review briefly the properties of the Ca channels mainly in vascular smooth muscle cells. Based on the patch clamp method, the voltage-dependent Ca channels of vascular smooth muscle cells are classified into two subtypes, T (transient or fast) and L (Long-lasting or slow) depending on the properties of the unit current as classified in cardiac muscles [51, 71]. The slop conductances of L- and T-types are 25 pS and 8 pS, respectively, measured in 110 m*M* Ba (in a pipette) and 130 m*M* Cs (in a bath) for the rabbit ear artery [6], 15 pS and 8 pS (80 m*M* Ba in a pipette and 120 m*M* NaCl in a bath) for the rabbit mesenteric artery [77], 18–24 pS and 7–9 pS (90 m*M* Ba in a pipette and 130, m*M* K asparate in a bath) for the dog saphenous vein [78], and 25 pS and 12 pS (50 m*M* Ba in a pipette and 145 m*M* KCl in a bath) for the guinea-pig taenia coli [79]. The T-type of channel possesses a low threshold (−70 mV), low sensitivity to Ca antagonist especially 1,4-dihydropyridine derivative), and undergoes very fast inactivation (20–50 ms under sustained command pulse). The L-type of channel possesses a high threshold (−10 mV), is sensitive to Ca antagonists, and slowly accelerates the inactivation process (time constant longer than 700 ms). However, it is unlikely that all smooth muscle cell membranes possess two subtypes of Ca channels; for example, the longitudinal muscle cells of rabbit ileum, stomach cells of frog or amphyma, and guinea-pig aorta possess only the L-subtype of voltage-dependent Ca channel [11].

The chemical configuration of the 1,4-dihydropyridine-sensitive voltage-dependent Ca channels (L-subtype) distributed in the tubular structure of skeletal muscles (T-system together with triad) have now been analyzed and found to consist of α_1, α_2, β-, γ-, and δ-subtypes; the α_1-subtype plays an essential role in channel [68, 70]. The characteristics of the 1,4-dihydropyridine-sensitive voltage-dependent Ca channels distributed in smooth muscle membrane have not yet been established; moreover, the physiological characteristics seem to differ from those of cardiac muscles. In cardiac muscle cells, Ca antagonists such as verapamil and gallopamil act on an intracellular site [26]. However, in smooth muscle cells, intracellular perfusion of these antagonists does not produce any effect; both agents act only if they are applied extracellularly. Furthermore, cyclic AMP together with protein kinase A accelerates the voltage-dependent Ca channels in cardiac muscles but not those in smooth muscles. Thus, the properties of the L-subtypes of Ca channels in cardiac and smooth muscle cells may differ, at least in part.

In sarcolemmal vesicles prepared from vascular smooth muscles, the density of the voltage-dependent Ca channel (1,4-dihyropyridine-sensitive Ca channel) has been estimated to be 0.2 pmol/mg protein (in a T-system of skeletal muscle, 65 pmol/mg protein; in cardiac muscle, 1–12 pmol/mg protein) [24]. In dispersed smooth muscle cells prepared from porcine coronary artery, the B_{max} value of [^{3}H] nitrendipine is 93 fmol/10^6 cells, which is equivalent to 56×10^3 binding sites per cell; this value is approximately the same as that obtained using the electrophysiological method in the guinea-pig aorta and other excitable cells. With the assumption that the surface area of a smooth muscle cell is 4000–6000 μm^2, the density of nitrendipine-binding sites was calculated to be about 10 sites/μm^2. The number of voltage-dependent Ca channels in smooth muscle cells seems to be very low in comparison with that of skeletal and cardiac muscles. Nevertheless, the action of Ca antagonists occurs to a greater extent in smooth muscle cells than in the other muscle systems. Ca antagonists may act on the hydrophobic site located in the midregion of the α_1-subtype Ca channel. However, more detailed experiments would be required to clarify the nature of the voltage-dependent Ca-channels in vascular smooth muscle cells.

Many pharmacologists are keen to establish the presence and nature of the receptor-operated Ca channels in smooth muscles to explain Ca mobilization as observed in studies involving the mechanical and also electrical responses [9, 72]. If such Ca channels are present, interpretations of mechanical responses obtained from smooth muscles become easier. ACh depolarizes the membrane of smooth muscle cells of the guinea-pig ileum placed in Na-free and Cl-free solution but not in Ca-free solution. IN dispersed smooth muscle cells of the guinea-pig ileum, ACh generates the inward current by changes in permeability to Na and K, but it has no effect on the voltage-dependent Ca channel. Presumably the presence of Ca may inhibit the opening probability of the channels without changing the amplitude of the ACh-induced single channel current. In the rabbit ear artery, Benham and Tsien [5] reported that ATP activates the purinergic receptor, thus causing the activation of the Na- and Ca-permeable channel. At 100 m*M* Ca or Ba, the unit conductance of this channel is calculated to be below 5 pS, and the selectivity for Ca and Na are calculated to be 3:1. This was the first evidenceto elucidate the receptor-operated, Ca-permeable channel in smooth muscles. Further experiments are required to clarify the nature of such a receptor-operated Ca channel; it is not known whether this channel has the property of a selective Ca channel or a non-selective cation channel.

Conclusion

A limited volume of body fluid regulates the homeostasis of cells, tissues, and organs in living beings: thus, responses of vascular beds to various regulating factors may differ one region to the other. Such factors include innervation (noradrenergic, cholinergic, dopaminergic, histaminergic, purinergic, non-cholinergic-nonadrenergic, and so on), receptors (for multiple humoral transmitters, autacoids, putative peptides, hormones), and various ion channels distributed in vascular smooth muscle cells. Furthermore, the regulation of smooth muscle cell activity by the endothelial cells should also be considered. It is surprising that so much information received from the extracellular environment through the first messengers of vascular smooth muscle cells is transferred to only five second messengers (two cyclic nucleotides, $InsP_3$, DG, and Ca). To understand the regional differences in functions of the vascular bed, more detailed investigations should be carried out on the molecular mechanisms of action of the second messengers in relation to the action of Ca during the contraction-relaxation process of smooth muscles in various regions of the vascular bed.

Acknowledgements. This work is supported by a grant from the Ministry of Education of Japan, and one of the authors (MCEG) wishes to acknowledge gratefully sponsorship from the Japanese Society for the Promotion of Science.

References

1. Abdel-Latif AA (1986) Calcium-mobilizing receptors, polyphospho-inositides, and the generation of second messengers. Pharmacol Rev 38: 227–272
2. Aksoy MO, Murphy RA, Kamm KE (1982) Role of Ca^{2+} and myosin light chain phosphorylation in regulation of smooth muscle. Am J Physiol 242: C109–C116
3. Ashida T, Blaustein MP (1987) Regulation of cell calcium and contractility in mammalian arterial smooth muscle: the role of sodium-calcium exchange. J Physiol 392: 617–635
4. Bean BP, Nowcky MC, Tsien RW (1984) β-Adrenergic modulation of calcium channels in frog ventricular heart cells. Nature 307: 371–375
5. Benham CD, Tsien RW (1987) A novel receptor-operated Ca^{2+}-permeable channel activated by ATP in smooth muscle. Nature 328: 275–278

6. Benham CD, Hess P, Tsien RN (1987) Two types of calcium channels in single smooth muscle cells from rabbit ear artery studied with whole-cell and single-channel recordings. Circ Res 61 (Suppl I): 10–16
7. Beny JL, Brunett PC, Huggel H (1986) Effects of mechanical stimulation, substance P and vasoactive intestinal polypeptide on the electrical and mechanical attivities of circular smooth muscles from pig. Coronary arteries contracted with acetylcholine; role of endothelium. Pharmacology 33: 61–68
8. Berridge MJ (1984) Inositol trisphosphate and diacylglycerol as second messengers. Biochem J 220: 345–360
9. Bolton TB (1979) Mechanisms of action of transmitters and other substances on smooth muscle. Physiol Rev 59: 606–718
10. Burnstock G (1981) Purinergic receptors—receptor and recognition- series B, vol 12. Chapman and Hall, London
11. Caffrey JM, Josephson IR, Brown AM (1986) Calcium channels of amphibian stomach and mammalian aorta smooth cells. Biophys J 49: 1237–1242
12. Castagna M, Takai Y, Kaibuchi K, Sano K, Kikkawa U, Nishizuka Y (1982) Direct activation of calcium-activated, phospholipid-dependent protein kinase by tumor-promoting phorbol esters. J Biol Chem 257: 7847–7851
13. Chatterjee M, Tejada M (1986) Phorbol ester-induced contraction in chemically skinned vascular smooth muscle. Am J Physiol 251: C356–C361
14. Currie MG, Geller DM, Cole BM, Boylon JG, Sheng WY, Holmberg SW, Needleman P (1983) Bioactive cardiac substances: potent vasorelaxant activity in mammalian atria. Science 221: 71–73
15. Dillon PF, Murphy RA (1982) Tonic force maintenance with reduced shorting velocity in arterial smooth muscle. Am J Physiol 242: C102–C108
16. Dillon PF, Murphy RA (1982) High force development and crossbridge attachment in smooth muscle from swine carotid arteries. Circ Res 50: 799–804
17. Driska SP, Aksoy MO, Murphy RA (1981) Myosin light chain phosphorylation associated with contraction in arterial smooth muscle. Am J Physiol 240: C222–C233
18. Droogmans G, Declerck I, Casteels R (1987) Effect of adrenergic agonists on Ca^{2+}-channel currents in single vascular smooth muscle cells. Pflügers Arch Eur J Physiol 409: 7–12
19. Eggermont JA, Vrolix M, Raeymaekers L, Wuytack F, Casteels R (1988) Ca^{2+}-transport ATPases of vascular smooth muscle. Circ Res 62: 266–278
20. Endoh M, Taira N (1983) Relationship between relaxation and cyclic GMP formation caused by nicorandil in canine mesenteric artery. Naunyn-Schmiedeberg's Arch Pharmacol 322: 319–321
21. Fishmeister R, Hartzell HÇ (1987) Cyclic guanosine 3′,5′-monophosphate regulates the calcium current in single cells from frog ventricle. J Physiol 387: 453–472
22. Furchgott RF (1984) The role of endothelium in the response of vascular smooth muscle to drugs. Ann Rev Pharmacol Toxicol 24: 175–197
23. Furchgott RF, Zawadzki JR (1980) The obligatory role of endothelial cells in the relaxation of arterial smooth muscle by acetylcholine. Nature 288: 373–376
24. Godfraind T, Miller R, Wibo M (1986) Calcium antagonism and calcium entry blockade. Pharmacol Rev 38: 321–416
25. Hellstrand P, Arner A (1985) Myosin light chain phosphorylation and the cross bridge cycle at low substrate concentration in chemically skinned guinea-pig *Taenia coli*. Pflügers Arch Eur J Physiol 405: 323–328
26. Hescheler J, Pelzer D, Trube G, Trautwein W (1982) Does the organic calcium channel blocker D600 act from inside or outside on the cardiac cell membrane? Pflügers Arch Eur J Physiol 393: 287–291
27. Hidaka H, Inagaki M, Kawamoto S, Sasaki Y (1984) Isoquinolinesulfonamides, novel and potent inhibitors of cyclic nucleotide dependent protein kinase and protein kinase c. Biochemistry 23: 5036–5041
28. Higashida H, Brown DA (1986) Two polyphosphyatidyl inositide metabolites control two k^{+} currents in a neuronal cell. Nature 232: 333–335

29. Hirst GDS, Neild TO (1980) Evidence for two populations of excitatory receptors for noradrenaline on arteriolar smooth muscle. Nature 283: 767–768
30. Hokin MR, Hokin LE (1953) Enzyme secretion and the incorporation of P32 into phospholipids of pancreas slices. J Biol Chem 203: 967–977
31. Ignarro LJ, Kadwitz PJ (1985) The pharmacological and physiological role of cyclic GMP in vascular smooth muscle relaxation. Ann Rev Pharmacol Toxicol 25: 171–191
32. Ignarro LJ, Burke TM, Wood KS, Wolin MS, Kadowitz PJ (1984) Association between cyclic GMP accumulation and acetylcholine-elicited relaxation of bovine intrapulmonary artery. J Pharmacol Exp Ther 228: 682–690
33. Ignarro LJ, Byrns RE, Buga GM, Wood KS (1987) Endothelium-derived relaxing factor from pulmonary artery and vein possesses pharmacologic and chemical properties identical to those of nitric oxide radical. Circ Res 61: 866–879
34. Inagaki M, Yokokura H, Itoh T, Kanmura Y, Kuriyama H, Hidaka H (1987) Purified rabbit protein kinase C relaxes skinned vascular smooth mucle and phosphorylates myosin light chain. Arch Biochem Biophys 254: 136–141
35. Irvine RF, Letcher AJ, Heslop JP, Berridge MJ (1986) The inositol tris/tetrakis phosphate pathway-demonstration of Ins (1,4,5) P_3 3-kinase activity in animal tissues. Nature 320: 631–634
36. Kakiuchi S, Sobue K (1981) Ca^{2+}- and calmodulin-dependent flip-flop mechanism in microtubule assembly-disassembly. FEBS Lett 14: 141–143
37. Kamm KE, Stull JT (1985) The function of myosin and myosin light chain kinase phosphorylation in smooth muscle. Ann Rev Pharmacol Toxicol 25: 593–620
38. Kangawa K, Matsuo H (1984) Purification and complete amino acid sequence of alpha-human atrial natriuretic polypeptide (alpha-hANP). Biochem Biophys Res Commun 118: 131–139
39. Kikkawa U, Takai Y, Tanaka Y, Miyake R, Bishizuka Y (1983) Protein kinase C as a possible receptor protein of tumor-promoting phorbol esters. J Biol Chem 258: 11442–11445
40. Klöchner U, Isenberg G (1985) Calcium currents of cesium loaded isolated smooth muscle cells (urinary bladder of the guinea pig). Pflügers Arch EurJ Physiol 405: 340–348
41. Komori K, Suzuki H (1987) Electrical responses of smooth muscle cells during cholinergic vasodilation in the rabbit saphenous artery. Circ Res 61: 586–593
42. Kukovetz WR, Holzmann S (1983) Mechanism of nitrate-induced vasodilation and tolerance. Zeitschrift für Kardiologie 72 (Suppl 3): 14–19
43. Kukovetz WR, Holzmann S, Wurm S, Pöch G (1979) Evidence for cyclic GMP-mediated relaxant effects of nitro-compounds in cornary smooth muscle. Naunyn-Schmiedeberg's Arch Pharmacol 310: 129–138
44. Matsuoka H, Ishii M, Sugimoto T, Hirata Y, Sugimoto T, Kangawa K, Matsuo H (1985) Inhibition of aldosterone production by alpha-human natriuretic polypeptide is associated with an increase in cGMP production. Biochem Biophys Res Commun 127: 1052–156
45. Mikawa T, Toyooka T, Nonomura Y, Ebashi S (1977) Essential factor of gizzard 'troponin' fraction. A new type of regulatory protein. J Biochem 81: 273–275
46. Michell RH (1975) Inositol phospholipids and cell surface receptor function. Biochim Biophys Acta 415: 81–147
47. Morgan JP, Morgan KG (1984) Stimulus-specific patterns of intracellular calcium levels in smooth muscle of ferret portal vein. J Physiol 351: 156–167
48. Murad F (1986) Cyclic guanosine monophosphate as a mediator of vasodilation. J Clin Invest 78: 1–5
49. Murad F, Rapoport RM, Fiscus R (1985) Role of cyclic-GMP in relaxation of vascular smooth muscle. J Cardiovasc Pharmacol 7 (Suppl 3): S111–S118
50. Nishizuka Y (1986) Studies and perspectives of protein kinase C. Science 233: 305–312
51. Nowycky MC, Fox AP, Tsien RW (1985) three types of neuronal calcium channel with different calcium agonist sensitivity. Nature 316: 440–443

52. Okajima F, Katada T, Ui M (1985) Coupling of the guanine nucleotide regulatory protein to chemotactic peptide receptors in neutrophil membranes and its uncoupling by islet-activating protein, pertussis toxin. J Biol Chem 260: 6761–6768
53. Palmer RMJ, Ferrige AG, Moncada S (1987) Nitric oxide release accounts for the biological activity of endothelium-derived relaxing factor. Nature 327: 524–526
54. Paul RJ, Doerman G, Zeugner C, Rüegg JC (1983) The dependence of unloaded shortening velocity on Ca^{2+}, calmodulin, and duration of contraction in "chemically skinned" smooth muscle. Circ Res 53: 342–351
55. Popescu LM, Ignat P (1983) Calmodulin-dependent Ca^{2+}-pump ATPase of human smooth muscle sarcolemma. Cell Calcium 4: 219–235
56. Popescu LM, Panoiu C, Hinescu M, Nutu O (1985) The mechanism of cGMP-induced relaxation in vascular smooth muscle. Eur J Pharmacol 107: 393–394
57. Rapoport RM, Murad F (1983) Agonist-induced endothelium-dependent relaxation in rat thoracic aorta may be mediated through cGMP. Circ Res 52: 372–377
58. Rapoport RM, Ginsburg R, Waldman SA, Murad F (1986) Effects of atriopeptin on relaxation and cyclic GMP levels in human coronary artery in vitro. Eur J Pharmacol 124: 193–196
59. Rasmussen H, Barrett PQ (1984) Calcium messenger system; an integrated view. Physiol Rev 64: 938–984
60. Ross EM, Gilman AG (1980) Biochemical properties of hormone-sensitive adenylate cyclase. Annu Rev Biochem 49: 533–564
61. Saida K, Van Breemen C (1987) GTP requirement for inositol-1,4,5-trisphosphate-induced Ca^{2+} release from sarcoplasmic reticulum in smooth muscle. Biochem Biophys Res Commun 144: 1313–1316
62. Scheid CR, Honeyman TW, Fay FS (1979) Mechanisms of β-adrenergic relaxation of smooth muscle. Nature 227: 32–36
63. Siegman MJ, Butler TM, Mooers SU (1985) Energetics and regulation of crossbridge states in mammalian smooth muscle. Experientia 41: 1020–1025
64. Singer HA, Kamm KE, Murphy RA (1986) Estimates of activation in arterial smooth muscle. Am J Physiol 251: C465–C473
65. Somlyo AV, Somlyo AP (1968) Electromechanical and pharmacomechanical coupling in vascular smooth muscle. J Pharmacol Exp Ther 156: 129–145
66. Southerton JS, Taylor SG, Weston AH (1987) Comparison of the effects of BRL 34915 and acetylcholine-liberated EDRF on rat isolated aorta. J Physiol 382: 50
67. Streb H, Irvine RF, Berridge MJ, Schulz I (1983) Release of Ca^{2+} from a non mitochondrial intracellular store in pancreatic acinar cells by inositol-1,4,5-trisphosphate. Nature 306: 67–69
68. Takahashi M, Seagar MJ, Jones JF, Reber BFH, Catteral WA (1987) Subunit structure of dihydropyridine-sensitive calcium channels from skeletal muscle. Proc Natl Acad Sci USA 84: 5478–5482
69. Takai Y, Kishimoto A, Kikkawa N, Mori T, Nishizuka Y (1979) Unsaturated diacylglycerol as a possible messenger for the activation of calcium-activated, phospholipid-dependent protein kinase system. Biochem Biophys Res Commun 91: 1218–1224
70. Tanabe T, Takeshima H, Mikami A, Flockerzi V, Takahashi H, Kangawa K, Kojima M, Matsuo H, Hirose T, Numa S (1987) Primary structure of the receptor for calcium channel blockers from skeletal muscle. Nature 328: 313–318
71. Tsien RW (1983) Calcium channels in excitable cell membranes. Annu Rev Physiol 45: 341–358
72. van Breemen C, Mangel A, Fahim M, Meisheri K (1982) Selectivity of calcium antagonistic action in vascular smooth muscle. Am J Cardiol 49: 507–510
73. Vanhoutte PM, Rubanyi GM, Miller VM, Houston DS (1986) Modulation of vascular smooth muscle contraction by the endothelium. Ann Rev Physiol 48: 307–320
74. Walker JW, Somlyo AV, Goldman YE, Somlyo AP, Trentham DR (1987) Kinetics of smooth and skeletal muscle activation by laser pulse photolysis of caged inositol 1,4,5-trisphosphate. Nature 327: 249–252

75. Walsh MP (1985) Calcium regulation of smooth muscle contraction. In: Marimé D (ed) Calcium and cell physiology. Springer, Berlin Heidelberg New York Tokyo
76. Williamson JR, Cooper RH, Joseph SK, Thomas AP (1985) Inositol trisphosphate and diacylglycerol as intracellular second messengers in liver. Am J Physiol 248: C203–C216
77. Worley JF III, Deitmer JW, Nelson NT (1986) Single nisoldipine-sensitive calcium channels in smooth muscle cells isolated from rabbit mesenteric artery. Proc Natl Acad Sci USA 83: 5746–5750
78. Yatani S, Seidel CL, Allen J, Brown AM (1987) Whole-cell and single-channel calcium currents of isolated smooth muscle cells from saphenous vein. Circ Res 60: 523–533
79. Yoshino M, Someya T, Nishio A, Yabu H (1988) Whole-cell and unitary Ca channel currents in mammalian intestinal smooth muscle cells: evidence for the existence of two types of Ca channels. Pflügers Arch Eur J Physiol 411: 229–231

Myogenic Contraction and Relaxation of Arterial Smooth Muscle

Koichi Nakayama and Yoshio Tanaka[1]

Summary. The present study was undertaken to determine, using fura-2, the relation between alterations of the intracellular Ca^{2+} signals and smooth muscle tone produced by stretch, agonistic stimuli, and high K^+ in helical strips of canine cerebral artery. Simultaneous measurements of the Ca^{2+} signals and tension development were made with a fluorimeter. The autofluorescence, motion artifact, and Ca^{2+} buffering action, considered as main disadvantages when fluorescent dye is used, seemed to be practically negligible in the measurement of Ca^{2+} signals and the stretch-induced contraction. The onset of the delayed contraction in response to stretch was always preceded by increasing strength of Ca^{2+} signals, while the Ca^{2+} signals were relatively constant during the maintenance of tone. Ca^{2+} antagonists such as nimodipine and diltiazem inhibited both mechanical response and Ca^{2+} signals in a concentration-dependent manner. A putative inhibitor of calmodulin, W-7, and one of myosin light chain kinase inhibitor, ML-9, showed an uncoupling action on the mechanical activity and Ca^{2+} signals. W-7 and ML-9 inhibited the mechanical response to stretch to a greater extent than it affected Ca^{2+} signals. The results suggest that fura-2 is useful for the simultaneous monitoring of myogenic tone and Ca^{2+} signals at the whole tissue level in cerebral arteries.

Key words: Ca^{2+} antagonist—Ca^{2+} signals—Cerebral artery—Endothelium—Myogenic tone

Introduction

The idea that mechanical deformation such as stretch, stress, or strain of the vascular wall enhances vascular tone originated from the observation by Baylis [1] and is the basis of the myogenic mechanism of regulation of blood flow. The stretch-induced contraction has also been considered to be related to the etiology of vasospastic contractions, for instance, hypertension and vasospasm. Our previous reports [2–4] have demonstrated the Ca^{2+}-dependent contraction of cerebral arteries of dogs and rabbits in response to dynamic stretch (quick stretch). The contraction in response to stretch is greatly modified by pro-

[1]Department of Pharmacology, School of Pharmaceutical Sciences, University of Shizuoka, Yada, Shizuoka 422, Japan

moters and inhibitors of Ca^{2+} signaling such as Ca^{2+} agonist (Bay k 8644), Ca^{2+} antagonists like nimodipine and diltiazem [4, 5], and putative inhibitors of calmodulin (W-7, [N-(6-aminohexyl)-5-chloro-1-naphthalenesulfonamide hydrochloride]), myosin light chain kinase (ML-9, [1-(5-chrolonaphthalene-1-sulfonyl)-1H-hexahydro-1,4-diazepine hydrochloride)], and protein kinase C (H-7, [1-(5-isoquinolinesulfonyl)-2-methylpiperazine dihydrochloride]) [6]. We have previously reported a reliable method of simultaneous recording of mechanical activity and cytosolic Ca^{2+} signals, using fura-2 in the cerebral artery at the tissue level [6]. The myogenic tone of vascular tissue seems to be a good alternative for contractions due to pharmacological and electric stimuli in the study of Ca^{2+} signals transduction at the multicellular or tissue level. The present study was undertaken to investigate the effects of several Ca^{2+} modifiers on the relation between alterations in myogenic tone and cytosolic Ca^{2+} signals in canine cerebral artery. In the study, we also examined the role of endothelium in the initiation of the stretch-induced contraction of cerebral artery.

Methods

A helical strip of canine cerebral artery (basilar and middle cerebral arteries) was loaded with fura-2, and mounted in a quartz glass chamber and perfused with Tyrode's solution. Figure 1 shows a schematic diagram of the experimental setup. Ca^{2+} signals and isometric tension development were simultaneously recorded with a fluorimeter (CAF-100, JASCO, Tokyo, Japan) (Fig. 1). The length of the preparation was adjusted to 130% of the initial muscle length (100%), and the preparation was equilibrated for about 2 h. The artery strip was stretched with an electromagnetic puller at a rate of 10 cm/s, the amount of stretch being 140% of the initial muscle length (100%), so that the total length during the period of stretching for 30s reached 170% of the initial length. Stretches were repeated at 20-min intervals to produce reproducible stretch-induced responses (Fig. 1). UV light for excitation (340 and 380 nm) was focused on the artery; and the corresponding emission signals (500 nm) as well as the ratio signals (340/380), a measure of cytosolic Ca^{2+} signals, were monitored. The artery segment was intermittently exposed to UV light, and the observation period was limited to about 2 h when the actual experiments were started. To calculate intracellular Ca^{2+} concentration $[Ca^{2+}]_i$, the following equation was used: $[Ca^{2+}]_i = K_D (R - R_{min}) \beta/(R_{max}\text{-}R)$ where the dissociation rate constant (K_D) equals 224 nM [7]. R is the measured ratio at each point, and R_{max} is the ratio at the maximum fluorescence, which was obtained by adding Ca^{2+} ionophore ionomycin (10 μM) and 80 mM KCl to the organ bath containing Tyrode's solution. To obtain R_{min}, 4 mM EGTA was added to quench the dye fluorescence. β is the ratio of the 500 mM fluorescence at 380 nm excitation for $Ca^{2+} = 0$ to that in Ca^{2+}-containing solution (2 mM) in the presence of ionomycin. All measurements were recorded on a pen-writing oscillograph. Further experimental details were described in earlier publications [2–6].

Bay k 8644 and nimodipine were supplied by Bayer Yakuhin, Osaka, Japan.

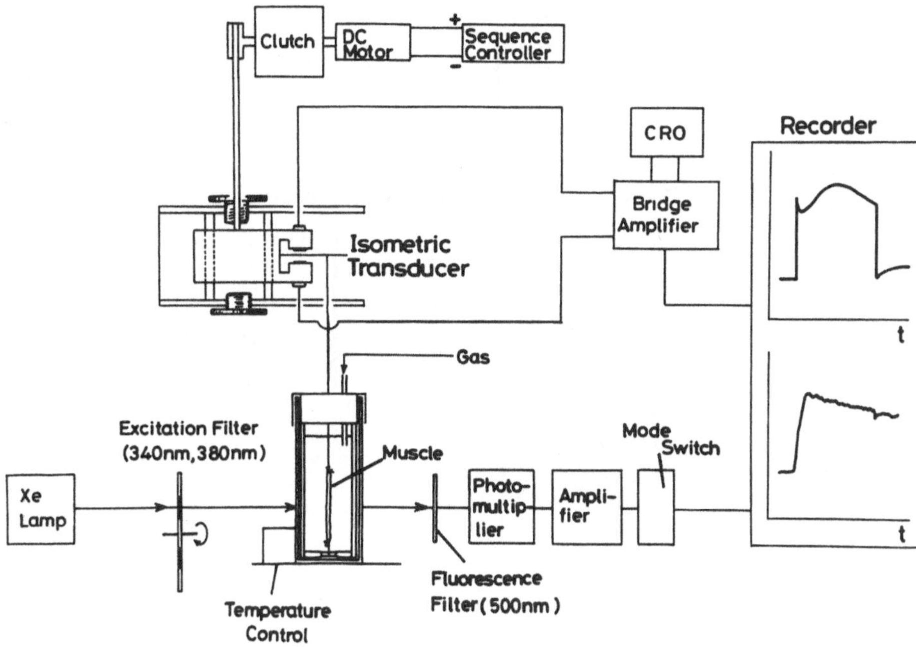

Fig. 1. Experimental setup of simultaneous recording of mechanical activity and Ca^{2+} signals with schematic diagram of electromagnetic puller for stretching the muscle

Other drugs used in the present study were commercially available. All drugs were expressed on a molar basis in the organ bath.

Data were expressed as the mean ± SE. Statistical analysis was made, using paired or unpaired Student's *t*-test, and P values less than 0.05 were considered significant.

Results

Simultaneous Recording of Stretch-Induced Contraction and Ca^{2+} Signals

Figure 1 shows a typical example of mechanical activity and the corresponding Ca^{2+} signals in response to quick stretch. The initial tension rise coincident with stretch and subsequent fall in tension at the completion of stretch was followed by a delayed contraction, which reached a maximum within about 10 s (Fig. 2). The contractile response was maintained or gradually decreased during stretch for 30 s. The 500-nm emission signal of fluorescence excited at 340 nm was increased, while that at 380 nm was decreased. As a consequence, the Ca^{2+} signal ratio (340/380), a measure of intracellular Ca^{2+}, began to increase 1–2 s before the tension rise of the delayed contraction and reached a maximum within 5 s. $[Ca^{2+}]_i$ at the control resting tension was 113 ± 52 n*M* ($n = 4$); while it

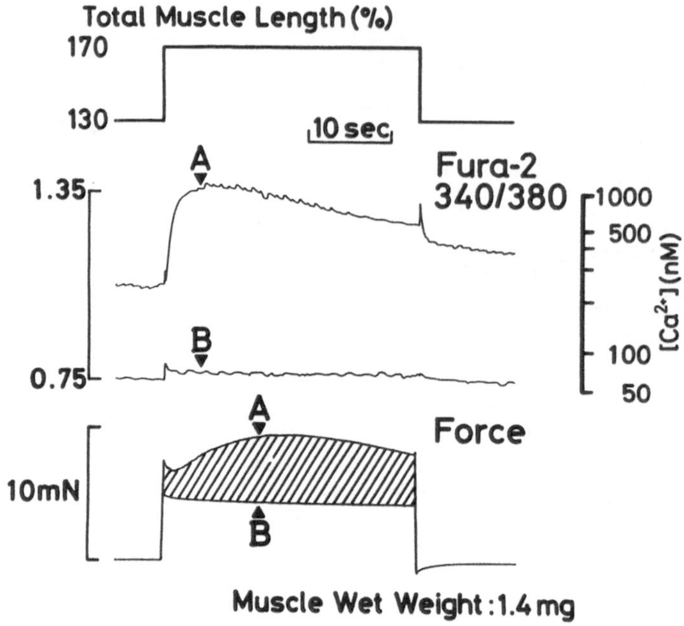

Fig. 2. Typical mechanical response to quick stretch and Ca^{2+} signals of a canine cerebral artery loaded with Fura-2. Active tension was produced by quick stretch at a rate of 10 cm/s to 140% of the initial (100%) muscle length and a stimulus period of 30 s. The corresponding ratio Ca^{2+} signals (340/380 nm) from artery bathed in normal (*A*) or Ca^{2+}-free Tyrode's solution containing 0.2 mM EGTA (*B*) were monitored. *Hatched area* of the mechanogram shows active tension. The muscle strip was initially stretched to a standardized length, i.e., 130% of the initial length, so that the total length during the period of stretching reached 170% of the initial length

reached 764 ± 252 n*M* ($n = 4$) during stretch-induced contraction. When the artery strip was immersed in a Ca^{2+}-free medium containing 0.2 m*M* EGTA, the tension developed by quick stretch decreased with an increase in number of stretches applied to the muscle, and the active mechanical response to stretch disappeared (Fig. 2). The corresponding Ca^{2+} signal was strongly attenuated in Ca^{2+}-free medium. Thus, the hatched area of the mechanogram on the recording chart showed an active mechanical response to quick stretch with a dimension of force (mN) × time (s).

In a separate series of experiments, we examined the role of endothelium in the initiation of the stretch-induced contraction. Fig. 3 shows a typical example of the time course of stretch-induced contraction in the cerebral artery with and without endothelium. The cerebral artery with endothelium produced the contraction in response to quick stretch reproducibly during an observation period of 2 h (Fig. 3a). The responsiveness to stretch of the endothelium-rubbed artery descreased slightly with time. However, there is no significant difference in the mechanical responses to stretch between endothelium-rubbed and normal arteries (Fig. 3b).

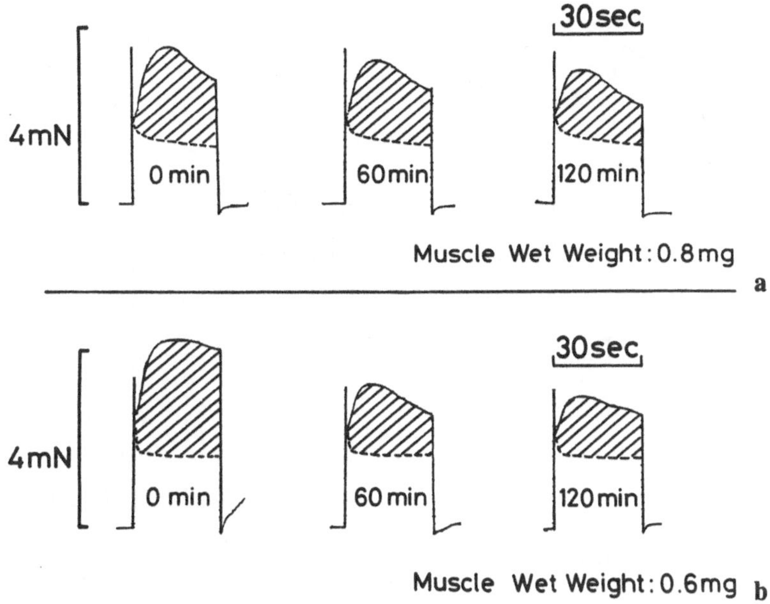

Fig. 3a, b. Time course of tension development upon quick stretch of a pair of cerebral strips with **a** or without **b** endothelium. Endothelium was mechanically rubbed. The artery was stretched every 20 min during an observation period of 120 min. *Hatched area* shows the active contractile component which was superimposed on the passive one

Effects of Drugs

Promoters of Ca^{2+} Influx

Our previous studies demonstrated that the contractile response to stretch was strongly potentiated by promoters of Ca^{2+} influx [2–6]. Figure 4 shows the effects of Bay k 8644 (10^{-7} *M*), a dihydropyridine Ca^{2+} agonist, on the mechanical activity and the Ca^{2+} signals in response to qucik stretch. Bay k 8644 potentiated both the stretch-induced contraction and the Ca^{2+} signals, while Ca^{2+}-withdrawal or Ca antagonist effectively inhibited the potentiation. Table 1 summarizes the relation between alterations in mechanical activity and Ca^{2+} signals augmented by quick stretch, 80 m*M* K^+, and agonistic stimuli such as 5-hydroxytryptamine (3×10^{-6} *M*) and prostaglandin $F_{2\alpha}$ (10^{-5} *M*). The stretch-induced contraction was divided into two components: The initial phase of contraction produced by stretch was correlated with the increasing strength of Ca^{2+} signals, while tone maintained in the presence of relatively low Ca^{2+} signals. Eighty m*M* K^+ and prostaglandin $F_{2\alpha}$ produced phasic and tonic contractions, while 5-hydroxytryptamine produced only phasic contraction. The Ca^{2+} signals remained at the increased level during phasic and tonic contractions produced by 80 m*M* K^+ and prostaglandin $F_{2\alpha}$. The Ca^{2+} signals were transiently increased with concomitant phasic contraction when 5-hydroxytryptamine was given.

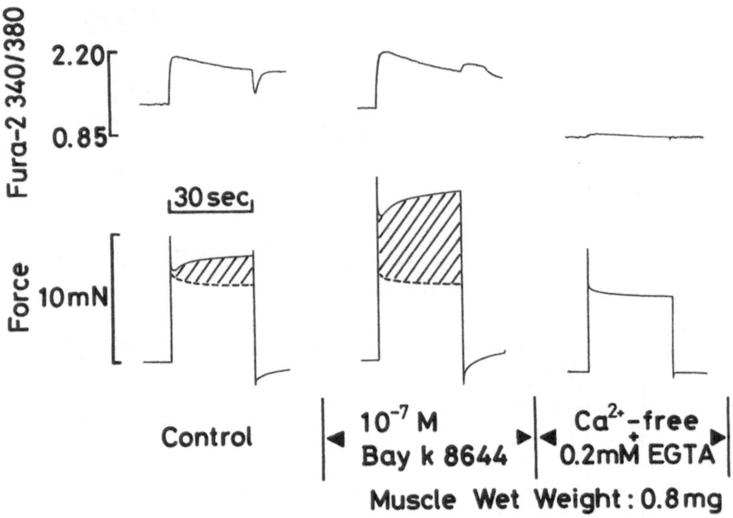

Fig. 4. Potentiating effects of Bay k 8644 on the contractile response to quick stretch and the corresponding Ca^{2+} signals. Ca^{2+} withdrawal inhibited both mechanical activity and fluorescence

Table 1. The relation between alterations in smooth muscle tone and Ca^{2+} signals of canine cerebral artery

Stimulus	No. of preparations	Tension developed (mN)	Ca^{2+} signal/tension
Quick stretch	10		
Initial phase		1.8 ± 0.4	[a]0.21 ± 0.05
Maintenance phase		2.7 ± 0.4	0.07 ± 0.02
80 mM K^+-induced contracture	4		
Phasic		6.1 ± 1.4	0.05 ± 0.01
Tonic		8.3 ± 0.6	0.03 ± 0.004
5-HT (3×10^{-6} M)	5		
Phasic		4.6 ± 1.7	0.05 ± 0.01
Prostaglandin $F_{2\alpha}$ (10^{-5} M)	5		
Phasic		3.9 ± 1.3	0.06 ± 0.02
Tonic		9.1 ± 3.0	0.02 ± 0.01

Each value represents mean ± SE.
[a] Statistically significant difference ($P < 0.05$) from the value of the maintenance phase of quick stretch.

Effects of Inhibitors

When the length of arterial strip was adjusted to 130% of its slack length, which was defined as the standard length [2], the cerebral artery showed an intrinsic basal tone. To begin with, we tested the effects of drugs on the basal tone and Ca^{2+} signals. Figure 5 shows typical examples. The Ca^{2+} antagonist diltiazem

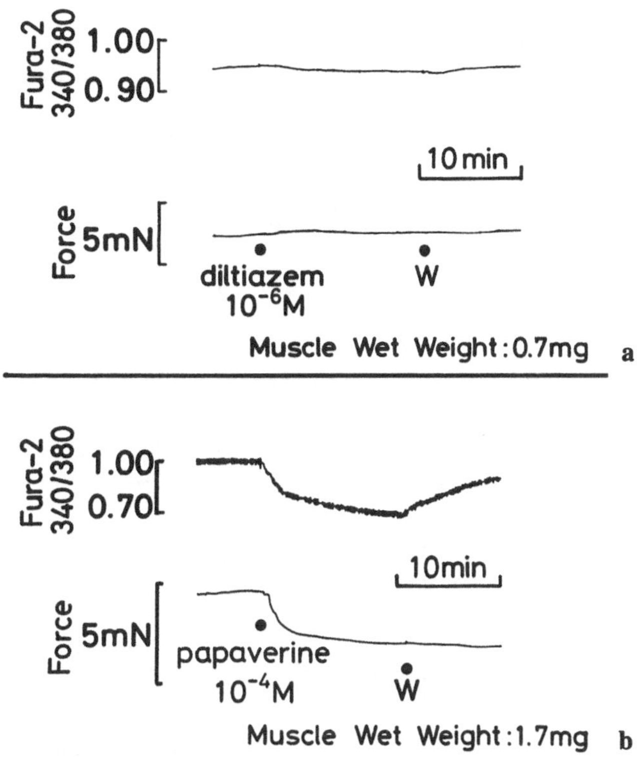

Fig. 5a, b. Different effects of diltiazem **a** and papaverine **b** on basal tension and Ca^{2+} signals in canine cerebral artery strips. *W*, washout

(Fig. 5a), nimodipine at 10^{-6} *M*, a putative inhibitor of calmodulin (W-7) at 10^{-6} *M*, and one of myosin light chain kinase (ML-9) at 10^{-6} *M* (data not given) showed no apparent effect on basal tone and Ca^{2+} signals. Papaverine (10^{-4} *M*), however, totally abolished both basal tone and Ca^{2+} signals (Fig. 5b). Incubation of the artery in Ca^{2+}-free medium for 30 min only depressed Ca^{2+} signals without any apparent effect on basal tone.

Differential inhibition of stretch-induced contraction and Ca^{2+} signals was produced by Ca^{2+} antagonists (Ca^{2+} channel blockers) and intracellular Ca^{2+} antagonists. Figure 6 shows the relation between Ca^{2+} signals and stretch-induced contraction. Ca^{2+} antagonists such as nimodipine (Fig. 6a) and diltiazem inhibited both Ca^{2+} signals and the contraction in response to quick stretch in a concentration-dependent manner (Fig. 6b). A putative inhibitor of calmodulin, W-7, and one of myosin light chain kinase (ML-9) showed an uncoupling effect on mechanical activity and Ca^{2+} signals. These drugs inhibited the mechanical response to stretch to a greater extent than they affected Ca^{2+} signals (Fig. 6b).

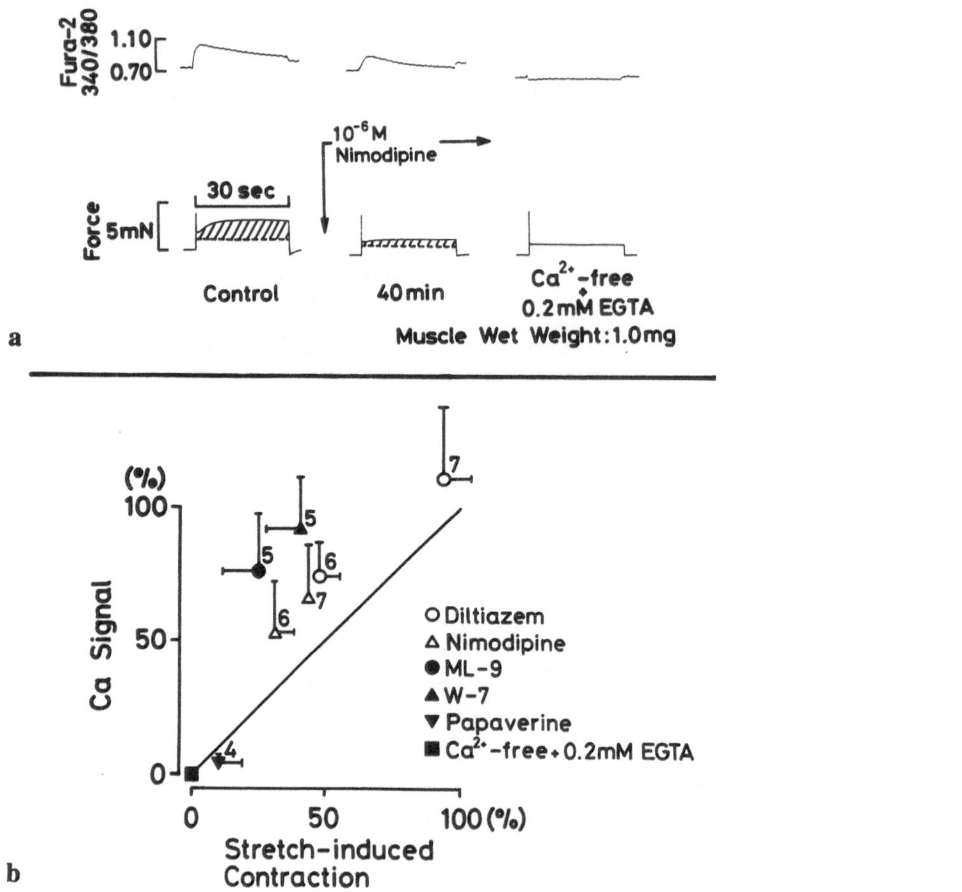

Fig. 6a, b. The relation between alterations in contractile response to quick stretch and Ca^{2+} signals. **a** Effects of nimodipine on the stretch-induced contraction and the corresponding Ca^{2+} signals. **b** Summarized data representing the different inhibitory effects of various drugs on the stretch-induced contraction and the Ca^{2+} signals. Each figure in the inset shows logalismic dose (*M*). Each point with bars represents the mean ± SE of 4–6 preparations

Discussion

Since the stretch activation of vascular tissue is essentially a contractile response to a rapid change in muscle length, it is important to know how the mechanical movement affects the recording of Ca^{2+} signals. We have already reported that motion artifact, autofluorescence, and Ca^{2+} buffering action (which have been considered as the main disadvantages for the measurement of Ca^{2+} signals using fluorescence dye such as fura-2 and quin 2) were practically negligible when the contractile response of canine cerebral artery to quick stretch and the Ca^{2+} signals were simultaneously recorded [6, 7]. Furthermore, we exposed the artery strip intermittently to UV light and limited the observation period to about 2 h so that the deteriorating effects, such as dye-bleaching and accumulation of byproducts of fura-2 [8] during UV exosure on the recording of the Ca^{2+} signals, might also be in part avoided.

The present study confirmed our previous reports that the stretch activation of cerebral artery was myogenic [2–5]. Recently, it has been reported that the endothelium accelerates cell to cell signal transduction [9] or liberates constrictor substances such as cyclooxygenase products of arachidonic acid, since the response to stretch was prevented by pretreatment with indomethacin [10]. We stimulated the muscle strip mechanically with quick stretch, a dynamic stretch [11]; but others stretched the muscle slowly, i.e., with a slow stretch or static stretch [10]. Stretch-activated ion channels have been identified in vascular endothelial cells [12]. These channels may act as mechanoreceptors responding to forces such as flow and shear stress. However, the sensory mechanism of endothelium may be rate-dependent, which may be one of the reasons why our results are different from those of others. Our previous study demonstrated that the potentiating effects of hemolysate/oxyhemoglobin obtained from canine autologous blood on the stretch-induced contraction was in part dependent on endothelium [13]. Thus, the contraction of cerebral artery upon stretch produced by the present procedure is very likely to be attributed to myogenic mechanism, i.e., direct stimulation of arterial smooth muscle by stretch. However, it seems also possible that endothelium modulates vascular contractility, which depends on various factors such as type of stimulus, animal species, and vascular regions.

As to the possible cause of the stretch-induced contraction, the involvement of the transmembrane influx of extracellular Ca^{2+} through voltage-dependent and/or receptor-operated channels [2–6] or a specialized pathway such as stretch-sensitive channel [14] as well as the release of Ca^{2+} from some intracellular store sites possibly near the sarcolemma, has been suggested [3]. In the present study, we confirmed the crucial rôle of Ca^{2+} by monitoring both mechanical activity and cytosolic Ca^{2+} signals. The contractile component due to stretch which depended on extracellular Ca^{2+} was enormously potentiated by Bay k 8644, a promoter of influx of Ca^{2+}, and was susceptible to Ca^{2+} withdrawal and Ca antagonist. Interestingly, the phase of initial tension rise upon stretch was preceded by a greater increase in the Ca^{2+} signals over their level seen during maintenance or the later slow decline of the tension, which seems to indicate that the Ca^{2+} sensitivity of contractile elements is altered during stretch activation. Although, it has been reported that static stretch did not affect the pCa -tension relation in smooth muscle skinned fibers such as taenia caecum and pulmonary artery of guinea pig [15], in fact, stretch is known to sensitize the contractile elements to Ca^{2+} in skeletal and cardiac muscles [16]. Thus, the maintenance phase of stretch-induced contraction may be related to a contractile state which needs relatively low $[Ca^{2+}]_i$ and energy expenditure such as latch [17].

In the present study, we found different inhibition of mechanical activity and Ca^{2+} signals by several inhibitors of Ca^{2+} kinetics. Although Ca^{2+} antagonists inherently have a slight effect on basal/intrinsic tone, these drugs are very effective once the tone is potentiated by promoters of Ca^{2+} influx [4]. We showed clearly the different vasodilating properties between diltiazem, a Ca^{2+} antagonist, and papaverine, a nonspecific vasodilator. Diltiazem slightly affected both basal tone and Ca^{2+} signals while inhibiting the potentiated tone by lowering Ca^{2+} signals. On the other hand, papaverine totally abolished both mechanical activity and Ca^{2+} signals at all times. This action seems to be one of the main reasons why Ca^{2+} antagonists are beneficial in the treatment of hypertension and vasospasm without causing circulatory disorders such as orthostatic hypotension and coronary steal syndrome. We observed an uncoupling effect of putative intracellular Ca^{2+} antagonists such as W-7 and ML-9 on the mechanical activity and Ca^{2+} signals. Although details of the pharmacological properties of these drugs are still controversial, it seems possible that these drugs inhibit the stretch-induced contraction by affecting intracellular contractile process without a dominant lowering of cytosolic Ca^{2+} signals.

The myogenic mechanism/theory of control of blood flow under physiological and pathological conditions has had a long history. However, studies on myogenic contraction and relaxation in vascular tissue are still important with respect to (1) the mechanisms of transmembrane or cell to cell signal transduction; (2) the interaction between blood and/or endothelium with medial smooth muscle or other vascular components; and (3) the specific role of Ca^{2+}, i.e., extra- and intracellular Ca^{2+} components or stretch-sensitive Ca^{2+} channels. the present method for the simultaneous recording of stretch-induced contraction and Ca^{2+} signals of cerebral artery at the whole tissue level seems to be useful not only for understanding the myogenic mechanism of vascular contractility, but also for testing new types of vasodilator and vasoconstrictor drugs.

Acknowledgements. The present study was supported in part by a Grant-in-Aid from the Ministry of Education, Science, and Culture of Japan (No. 63571051) and by Sankyo Life Science Foundation, Tokyo, Japan. We express our thanks to Dr. Aoki, the president of the symposium.

References

1. Bayliss WM (1902) On the local reactions of the arterial wall to change of internal pressure. J Physiol (Lond) 28: 220–231
2. Nakayama K (1982) Calcium-dependent contractile activation of cerebral artery produced by quick stretch. Am J Physiol 242: H760–H768
3. Nakayama K, Suzuki S, Sugi H (1986) Physiological and ultrastructural studies on the mechanism of stretch-induced contractile activation in rabbit cerebral artery smooth muscle. Jpn J Physiol 36: 745–760
4. Nakayama K (1986) Myogenic activity of vascular tissues and the effects of calcium agonists and antagonists. In: Aoki K (ed) Essential hypertension: calcium mechanisms and treatment. Springer, Tokyo, pp 95–105
5. Nakayama K, Ishii K, Kato H (1983) Effect of Ca-antagonists on the contraction of cerebral and peripheral arteries produced by electrical and mechanical stimuli. Gen Pharmacol 14: 111–113
6. Nakayama K, Tanaka Y (1988) Calcium transients and stretch-induced myogenic tone in vascular tissue. In: Halpern W (ed) Proceedings of Second International

Symposium on Resistance Arteries at Stowe, Vermont. Perinatology Press, Ithaca, pp 212–218
7. Grynkiewicz G, Poenie M, Tsien RY (1985) A new generation of Ca^{2+} indicators with greatly improved fluorescence properties. J Biol Chem 260: 3440–3450
8. Becker PL, Fay FS (1987) Photobleaching of fura-2 and its effect on determination of calcium concentration. Am J Physiol 253: C613–C618
9. Harder D (1987) Pressure-induced myogenic activation of cat cerebral arteries is dependent on intact endothelium. Circ Res 60: 102–107
10. Katusic ZS, Shepherd JT, Vanhoutte PM (1987) Endothelium-dependent contraction to stretch in canine basilar arteries. Am J Physiol 252: H671–H673
11. Johansson B, Mellander S (1975) Static and dynamic changes in the vascular myogenic response to passive changes in length as revealed by electrical and mechanical recordings from the rat portal vein. Circ Res 36: 76–83
12. Lansman JB, Hallam TJ, Rink TJ (1987) Single stretch-activated ion channels in vascular endothelial cells as mechanotransducers? Nature 325: 811–813
13. Nakayama K, Hashimoto K (1984) Blood components and cerebral vasospasm. Bibl Cardiol 38: 148–160
14. Endo M, Kitazawa S, Yagi S, Iino M, Kakuta Y (1977) Some properties of chemically skinned smooth muscle fibers. In: Casteels R, Godfraind T, Rueegg JC (eds) Elsevier/North-holand, Amsterdam New York Oxford, pp 199–209
15. Bevan JA (1983) Diltiazem selectively inhibits cerebrovascular extrinsic but not intrinsic myogenic tone. A review. Circ Res 52 (Suppl I): I104–109
16. Pagani ED, Alousi AA (1987) New approaches for the treatment of the failing myocardium: myofibrillarproteins as potential targets for pharmacotherapy. In: Beamish RE, Panagia V, Dhalla NS (eds) Pharmacological aspects of heart disease. Martinus Nijhoff, Boston, pp 341–352
17. Chatterjee M, Murphy RA (1986) Calcium-dependent stress maintenance without myosin phosphorylation in skinned smooth muscle. Science 221: 464–466

Relationship Between Cytosolic Calcium Level and Contractile Tension in Vascular Smooth Muscle

HIDEAKI KARAKI and HIROSHI OZAKI[1]

Summary. In rat aortic strips loaded with a fluorescent Ca^{2+} indicaotr, fura-2, muscle contraction was recorded simultaneously with cytosolic Ca^{2+} levels ($[Ca^{2+}]_{cyt}$). High-K^+, norepinephrine, prostaglandin $F_{2\alpha}$, and ionomycin induced sustained contraction following sustained increase in $[Ca^{2+}]_{cyt}$, whereas Bay k 8644 and tetraethylammonium caused rhythmic contractions following the rhythmic changes in $[Ca^{2+}]_{cyt}$. Tension changes were always preceded by the changes in $[Ca^{2+}]_{cyt}$. KCl, norepinephrine, prostaglandin $F_{2\alpha}$, and phorbol esters induced concentration-dependent increase in both muscle tension and $[Ca^{2+}]_{cyt}$ although receptor-agonists and phorbol esters induced a greater contraction than high K^+ at a given $[Ca^{2+}]_{cyt}$. Phorbol esters, at the concentrations which do not change muscle tension or $[Ca^{2+}]_{cyt}$, potentiated the high K^+-induced contraction with little effect on high K^+-stimulated $[Ca^{2+}]_{cyt}$. Verapamil inhibited the changes due to high K^+. Although verapamil inhibited the $[Ca^{2+}]_{cyt}$ stimulated by receptor-agonists and phorbol esters to the resting level, a portion of the sustained contraction was not inhibited. In Ca^{2+}-free solution, norepinephrine, prostaglandin $F_{2\alpha}$, caffeine, or ionomycin induced only a transient increase in $[Ca^{2+}]_{cyt}$. Caffeine and ionomycin induced a transient contraction although norepinephrine and prostaglandin $F_{2\alpha}$ induced a transient contraction followed by relatively sustained contraction in Ca^{2+}-free solution. Phorbol ester at a higher concentration induced sustained contraction in a Ca^{2+}-free solution without changing $[Ca^{2+}]_{cyt}$. Sodium nitroprusside inhibited the stimulated muscle tension more strongly than it reduced $[Ca^{2+}]_{cyt}$. In Ca^{2+}-free solution, sodium nitroprusside inhibited the changes due to receptor-agonists and phorbol esters. Forskolin showed similar effects to sodium nitroprusside. These results indicate that vasoconstrictors and vasodilators modulate smooth muscle contraction by changing $[Ca^{2+}]_{cyt}$ and also by changing the sensitivity of contractile elements of Ca^{2+}. Furthermore, at least a part of the receptor-mediated sustained contraction may be due to activation of protein kinase C.

Key words: Vascular smooth muscle—Calcium regulation—Fura-2-Ca^{2+} fluorescence—Vasoconstrictors—Vasodilators

[1]Department of Veterinary Pharmacology, Faculty of Agriculture, The University of Tokyo, Bunkyo-ku, Tokyo 113, Japan

Introduction

Contraction of vascular smooth muscle can be initiated either by binding of stimulatory transmitters or other substances to specific receptors or by directly changing membrane electrical properties [1, 2]. Receptor activation may stimulate turnover of membrane phosphoinositides [3, 4]. These changes open Ca^{2+} channels located in the membrane to permit influx of external Ca^{2+} and/or release of cellular bound Ca^{2+} [1, 5, 6]. Resulting increase in the concentration of cytosolic Ca^{2+} ($[Ca^{2+}]_{cyt}$), together with calmodulin, activates myosin light chain kinase, phosphorylates myosin light chain, and induces muscle contraction [7]. Ca^{2+} in the cytoplasm is either sequestered to cellular binding sites, presumably sarcoplasmic reticulum [8], or extruded from the cell by an active pumping mechanism on the cell membrane [9, 10]. Drugs affecting vascular smooth muscle contraction may modify either of these processes [11]. Critical evaluation of the effects of stimulants and relaxants in smooth muscle requires quantitative measurement of $[Ca^{2+}]_{cyt}$ and comparison with the contractile response. Recently, fluorescent dyes such as quin2 and fura-2 have become available to measure the change in $[Ca^{2+}]_{cyt}$ in various biological preparations. This technique has been applied to smooth muscle strips with simultaneous measurement of contractile tension [12–18]. In this article, we briefly summarize recent results obtained in smooth muscle with this method.

Methods

The thoracic aorta was isolated from a male Wistar rat weighing 250–300 g and cut into a spiral strip, 2 mm wide and 7 mm long. The endothelium was removed by gently rubbing the intimal surface with a finger moistened with normal physiological salt solution in order to avoid the possible complication of the results by the release of the endothelium-derived relaxing factor [19]. The normal physiological salt solution contained: NaCl 136.9 m*M*, KCl 5.4 m*M*, $CaCl_2$ 1.5 m*M*, $MgCl_2$ 1.0 m*M*, *N*-2-hydroxyethylpiperazine-*N'*-2-ethanesulfonic acid (HEPES) 20 m*M*, ethylenediamine-tetraacetic acid (EDTA) 0.01 m*M*, and glucose 5.5 m*M*. The solution was continuously bubbled with 100% O_2 at 37°C and pH 7.4. High K^+ solution was made by substituting NaCl with KCl to make final concentration of K^+ 72.7 m*M*. Ca^{2+}-free solution was made by removing $CaCl_2$ and adding 0.5 m*M* ethyleneglycol bis (β-aminoethyl ether)-*N*,*N*,*N'*,*N'*-tetraacetic acid (EGTA). In some experiments, external free Ca^{2+} concentration was reduced by adding 4 m*M* EGTA to the normal solution. The decrease in pH resulting from the EGTA-Ca^{2+} interaction was adjusted with tris(hydroxymethyl) aminomethane.

The membrane permeable acetoxymethyl ester of fura-2 (fura-2/AM) is sparingly soluble in water. Fura-2/AM dissolved in dimethylsulfoxide disperses in physiological salt solution as small particles up to the concentration of only 1 μM or less [18]. Single cells such as platelets are loaded with fura-2 using this solution. However, since the particle size is bigger than matrix of extracellular space and since effective concentration of fura-2/AM is low, smooth muscle tissue incubated in this solution does not take up fura-2/AM. In order to decrease the

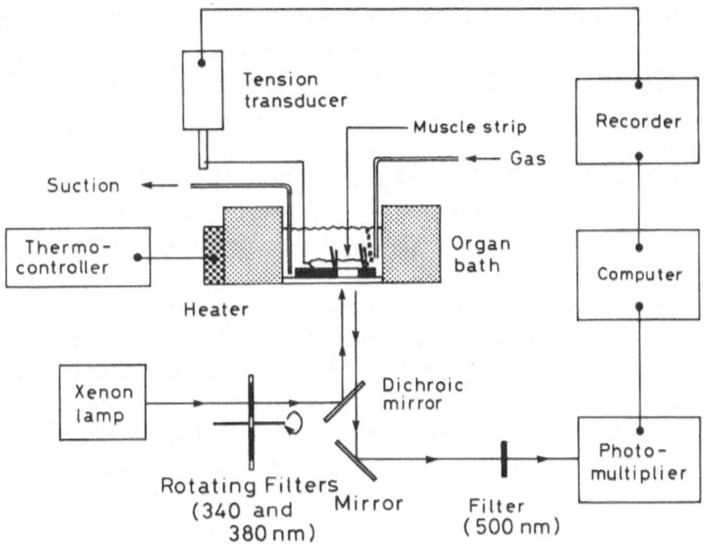

Fig. 1. Block diagram of the fluorimeter designed to measure fluorescence in smooth muscle strip

particle size and to increase the solubility of fura-2, noncytotoxic detergent, cremophor EL (0.02 %), was added. Muscle strips were treated with 5 μM fura-2/AM for 3–4 h at room temperature. Fura-2/AM is changed to fura-2 in the cell by esterases. After the fura-2-loading, the muscle was rinsed with normal solution for 15–30 min in order to remove uncleaved fura-2/AM from the tissue before starting the experiment.

Experiments were performed with a fluorimeter (CAF-100, Japan Spectroscopic, Tokyo, Japan) (Fig. 1). The muscle strip was held horizontally in a temperature controlled organ bath. One end of the muscle strip was connected to a strain gauge transducer to monitor the mechanical activity. The excitation light was obtained from a xenon high pressure lamp (75 W) equipped with a rotating filter wheel. The wheel contained 340 ± 5.5 nm and 380 ± 5.5 nm interference filters. Thus, the muscle strip was excited by these two different wavelengths alternatively at 48 Hz. The 500 ± 10 nm lights emitted from muscle strip excited at 340 nm (F_{340}) and that excited at 380 nm (F_{380}) were collected at the photomultiplier. The ratio of these two fluorescences ($R_{340/380}$) was calculated by a computer in the fluorimeter.

Autofluorescence

In the aorta without fura-2-loading, resting autofluorescence was detected at both F_{340} and F_{380} [12, 14, 20]. Addition of high K^+, norepinephrine, or ionomycin increased both F_{340} and F_{380} with an increase in tension (Fig. 2). The amounts of the increase in F_{340} and F_{380} were 5%–6% of the respective resting autofluorescence. The increase in the autofluorescence was detected 1–2 s after the beginning of the high K^+-induced contraction, 3–5 s after the beginning of norepinephrine-induced contraction, or 5–8 s after the beginning of ionomycin-induced contraction. Addition of EGTA decreased the fluorescence to the respective resting levels. This autofluorescence may be due to the increase in the amount of reduced pyridine nucleotides (NADH, NADPH); it changes in correlation with mechanical activity [12, 14] or $[Ca^{2+}]_{cyt}$ [20]. Since F_{340} and F_{380} increased proportionally, $R_{340/380}$ stayed constant during the muscle contraction in rat aorta although $R_{340/380}$ changed in other smooth muscle such as guinea-pig taenia caeci.

Fura-2 loading increased the resting fluorescence to approximately 250% of the autofluorescence measured before the fura-2-loading. Addition of high K^+ or norepinephrine rapidly increased F_{340}, decreased F_{340}, and increased $R_{340/380}$ (Fig. 3). Addition of ionomycin, induced slow increase in both $[Ca^{2+}]_{cyt}$ and muscle tension. Addition of EGTA rapidly decreased both muscle tension and $[Ca^{2+}]_{cyt}$ due to high K^+, norepinephrine, and ionomycin.

The increase in F_{340} was 4–6 times greater than that in the fura-2-unloaded muscle. Further, both F_{340} and F_{380} changed before the changes in muscle tension. Thus, the fura-2-Ca^{2+} signals are different from the NAD(P)H fluorescence, which was observed in the fura-2-unloaded muscle, regarding the rate and amount of the increase in the F_{340} and F_{380}, and the direction of the change in the F_{380}. Thus, the fluorescent signal of fura-2-loaded rat aorta may be indicative of $[Ca^{2+}]_{cyt}$ only when the direction of change of in F_{380} is opposite to that of F_{340} which returned to original level or gradually increased above the resting level. These results seem to be due to the insufficient fura-2-loading.

Absolute Ca^{2+} Concentration

The absolute cytosolic Ca^{2+} concentration was calculated from $R_{340/380}$ using the equation described by Grynkiewictz et al. [21]:

$$[Ca^{2+}]_{cyt} = \frac{R - R_{min}}{R_{max} - R} \times \mathrm{Kd} \times \mathrm{s}$$

Kd and R represent the dissociation constant of fura-2 for Ca^{2+} and experimentally determined $R_{340/380}$, respectively. This equation can be used only in the absence of autofluorescence. However, rat aortic strips have basal, constant autofluorescence, and Ca^{2+}-dependent, variable autofluorescence at F_{340} and F_{380} (Fig. 2). In order to apply this equation to the tissue with autofluorescence, we obtained R_{max} and R_{min} by the addition of ionomycin and EGTA, respectively, assuming that ionomycin maximizes membrane permeability to Ca^{2+} [22]. As shown in Fig. 4, application of ionomycin to the tissue previously contracted by high K^+ further increased $R_{340/380}$. Increase in the external concentration of Ca^{2+} from 1.5 to 10 mM did not induce any change in $[Ca^{2+}]_{cyt}$. Therefore, $R_{340/380}$ in the presence of ionomycin was considered as R_{max}. Absolute $[Ca^{2+}]_{cyt}$ calculated by assuming the Kd value as 224 nM at 37°C [21] is also shown. Compared to the large increase in $[Ca^{2+}]_{cyt}$ due to ionomycin, increase in muscle tension was relatively small. This is not because high K^+ induces maximum ten-

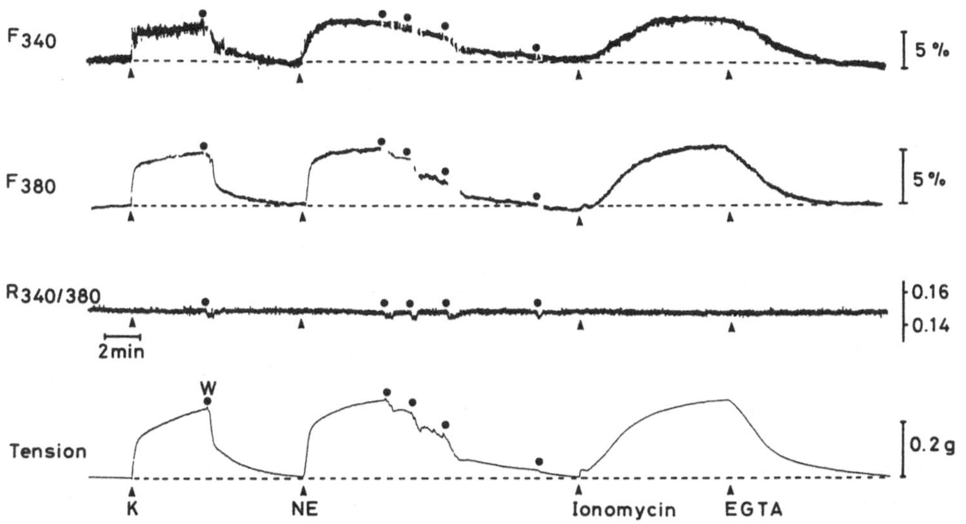

Fig. 2. Changes in fluorescence measured simultaneously with muscle tension in a strip of rat aorta without fura-2 loading stimulated by 72.7 mM KCl (K), 0.3 μM norepinephrine (NE), or 10 μM ionomycin or relaxed by 4 mM EGTA. Changes in F_{340} and F_{380} are shown by relative value taking resting fluorescene level as 100%. $R_{340/380}$ is also shown. *Solid circles*, washing the muscle with normal solution

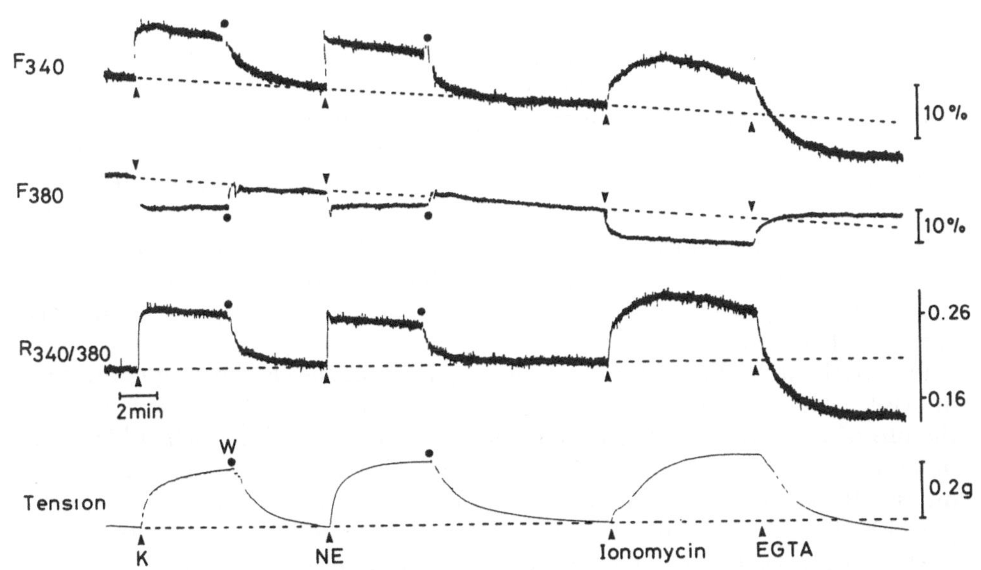

Fig. 3. Changes in fluorescence measured simultaneously with muscle tension in a fura-2 loaded rat aorta stimulated by 72.7 mM KCl (K), 0.3 μM norepinephrine (NE), or 10μM ionomycin. Changes in F_{340} and F_{380} are shown by relative value taking resting fluorescence level of the fura-2 loaded aorta as 100%. $R_{340/380}$, which indicates relative $[Ca^{2+}]_{cyt}$, is also shown. *Solid circles*, washing the muscle with normal solution

sion development, but may because ionomycin somehow uncouples muscle tension from the increase in $[Ca^{2+}]_{cyt}$. Application of EGTA to the high K^+-contracted tissue decreased $R_{340/380}$ below the resting level and this level was considered as R_{min}. The constant value, s, is the ratio of F_{380} of the tissue measured in Ca^{2+}-free solution to that measured in Ca^{2+} containing solution in the presence of ionomycin.

An example of a calibration curve is shown in Fig. 5. This figure indicates that the $R_{340/380}$ may be used as an indicator of $[Ca^{2+}]_{cyt}$ when $R_{340/380}$ is in a particular range from 0.18 to 0.26 in this case. $R_{340/380}$ values are variable depending on the fluorimeter because of the difference in their light paths for 340 and 380 n*M* lights. We did not take account of the Ca^{2+}-dependent, variable autofluorescence because it is less than 2.5% of the total fluorescence. From these assumptions, the resting aorta seems to contain 228 ± 25 n*M* ($n = 4$) Ca^{2+} which increased to 1748 ± 153 n*M* ($n = 4$) and to 1528 ± 180 n*M* ($n = 4$) during high K^+- and norepinephrine-induced contractions, respectively. However, fura-2 binds to soluble proteins in cytoplasm and both Kd of fura-2 for Ca^{2+}, and the fluorescence characteristics of fura-2 in cytoplasm may be changed from those measured in vitro [18, 23]. Thus, it is not possible to calculate the absolute $[Ca^{2+}]_{cyt}$ without knowing the Kd value in cytoplasm. For this reason, we decided not to use the absolute value of $[Ca^{2+}]_{cyt}$ but to use the ratio $R_{340/380}$ as an indicator of relative $[Ca^{2+}]_{cyt}$.

Vasoconstrictors

Stimulation of the muscle with high K^+ increased $[Ca^{2+}]_{cyt}$, which remained elevated as long as the contractile force was maintained. Stimulation with norepinephrine and ionomycin also increased $[Ca^{2+}]_{cyt}$, which decreased only slightly during sustained contraction (Fig. 3). 12-Deoxyphorbol 13-isobutyrate (DPB, 1 μM) and other phorbol esters such as 12-deoxyphorbol 13-isobutyrate 20-acetate (DPBA) and phorbol 12 ,13-dibutyrate (PDBu) also increased $[Ca^{2+}]_{cyt}$ and muscle tension in rat aorta. Rat aorta showed rhythmic changes in muscle tension and $[Ca^{2+}]_{cyt}$ immediately after the addition of tetraethylammonium, a K^+ channel blocker, or 1–2 min after the addition Bay k 8644, a Ca^{2+}-channel activator [24] (Fig. 6).

Previously, it has been reported that receptor agonists caused a rapid increase in cytosolic Ca^{2+} concentration which then declined to a very low level with the use of aequorin in ferret portal vein [25, 26] and bovine trachea [27]. The difference may be explained by the different characteristics of fura-2 and aequorin. It has been shown that aequorin luminescence over-represents the regional highest $[Ca^{2+}]_{cyt}$ because the intensity of this luminescence corresponds to the third power of Ca^{2+} concentration [28]. Consequently, presence of an area in which Ca^{2+} level is higher than average $[Ca^{2+}]_{cyt}$ increases the total aequorin luminescence to the level higher than that due to average $[Ca^{2+}]_{cyt}$. Thus, receptor-mediated Ca^{2+} release produces a regional, high Ca^{2+} area which may result in the initial large and transient increase in the aequorin signal.

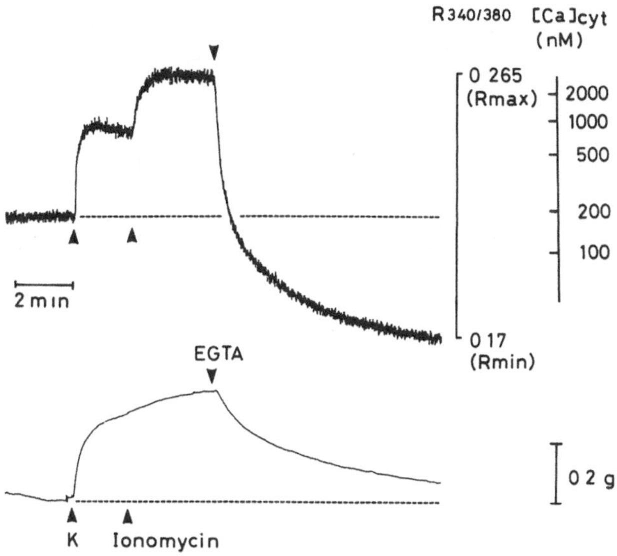

Fig. 4. Method to obtain R_{max} and R_{min}. After the muscle was contracted by 72.7 m*M* KCl (*K*) (in the presence of 1.5 m*M* Ca^{2+}), 1 μ*M* ionomycin was added to determine the maximum level of $R_{340/380}$ (R_{max}). Subsequently, 4 m*M* EGTA was added to determine the minimum level of the $R_{340/380}$ (R_{mim}). $R_{340/380}$ (*R*) was converted to cytosolic Ca^{2+} concentrations by the method of Grynkiewictz et al. [21] modified by Scanlon et al. [22]

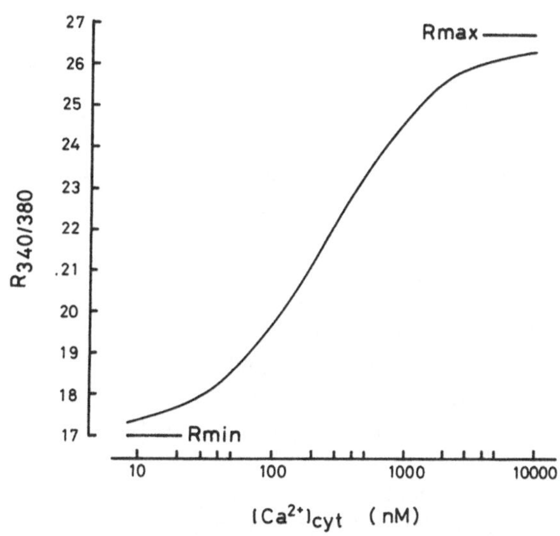

Fig. 5. Correlation between fluorescence ratio ($R_{340/380}$) and $[Ca^{2+}]_{cyt}$ in the presence of autofluorescence. R_{max} and R_{min} were measured as shown in Fig. 4. Ca^{2+} concentration was calculated using Kd value of 224 n*M* [21]

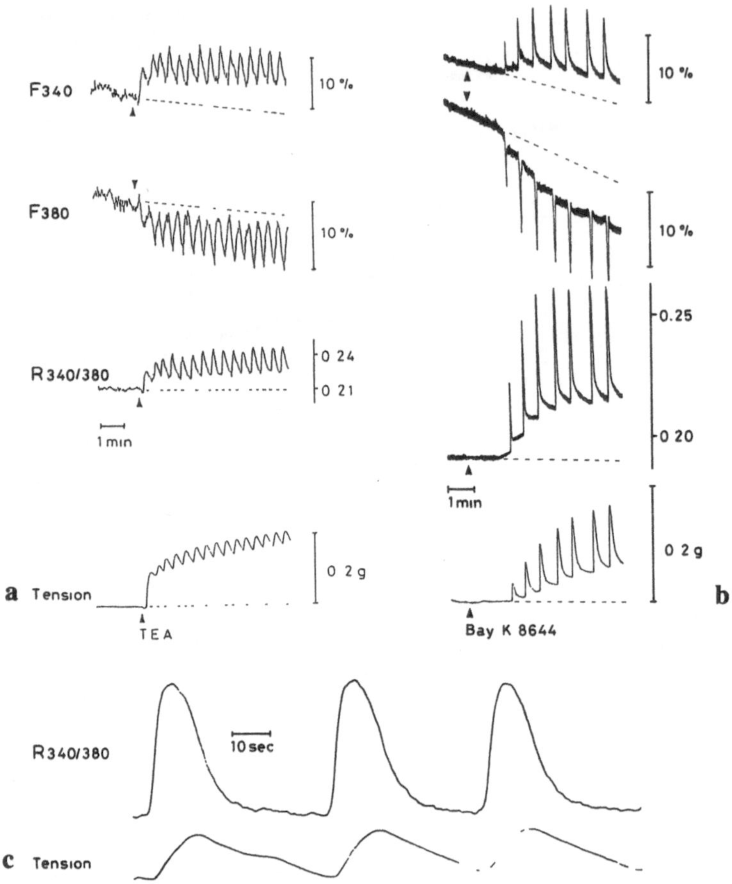

Fig. 6a–c Fluctuations in fluorescence and tension caused by **a** 10 m*M* tetraethylammonium (*TEA*) and **b**, **c** 1 μM Bay k 8644. Changes in F_{340}, F_{380}, and $[Ca^{2+}]_{cyt}$ (shown by $R_{340/380}$) are shown. Changes in F_{340} and F_{380} are shown by relative value taking resting fluoprescence level as 100%. In **c** fluctuations in $R_{340/380}$ and muscle tensin caused by Bay k 8644 are shown with expanded time scale

Subsequent diffusion of this Ca^{2+} to cytoplasm establishes uniformity of $[Ca^{2+}]_{cyt}$. Simultaneously, opening of Ca^{2+} channels contributes to an increase of the average $[Ca^{2+}]_{cyt}$, which is detected as a steady level of aequorin signal. In contrast to this, fura-2 forms a simple 1:1 complex with Ca^{2+} [21]. Thus, fura-2 signals may represent the average $[Ca^{2+}]_{cyt}$ or major portion of $[Ca^{2+}]_{cyt}$, including the area where the contractile elements exist. Receptor-agonists also induce only a transient increase in $[Ca^{2+}]_{cyt}$ measured with quin 2 in dispersed toad stomach cells [29] and dispersed porcine coronary arterial cells [30] and with fura-2 in the cultured aortic smooth muscle cells of rat [31]. The difference between smooth muscle tissue and single cells may be due to the lack of cell-to-cell contact [32] and/or changes in the characteristics of the cells during dispersion or culture [33].

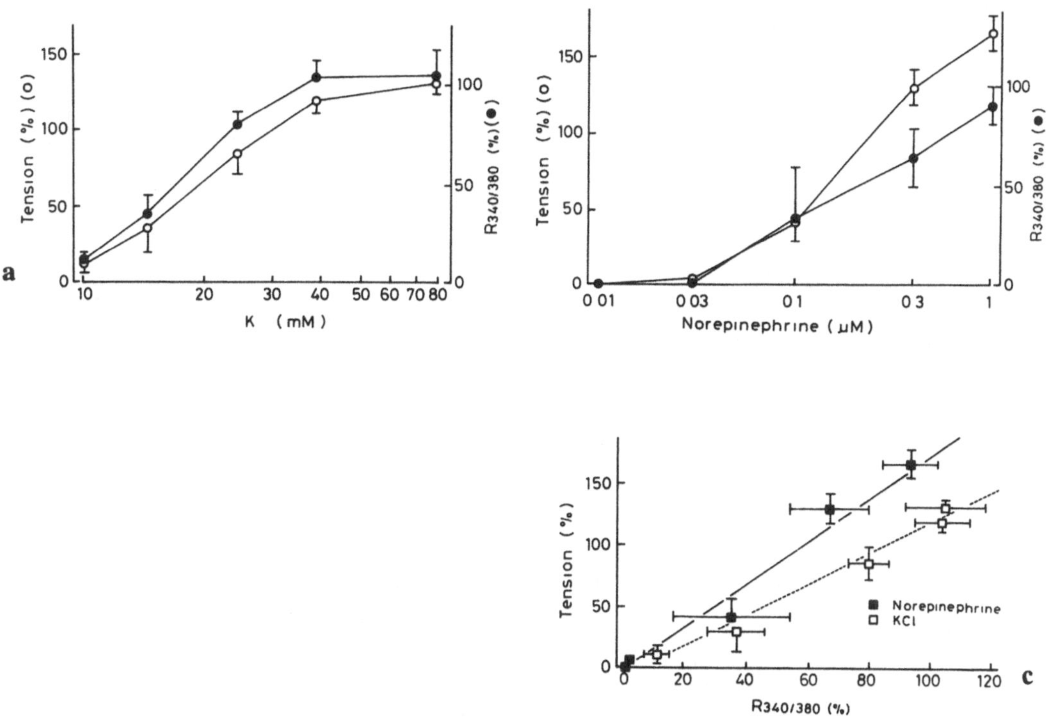

Fig. 7a–c. Effect of cumulative application of **a** KCl and **b** norepinephrine on tension and $[Ca^{2+}]_{cyt}$ (shown by $R_{340/380}$). The 72.7 mM KCl-induced increases in $[Ca^{2+}]_{cyt}$ and tension measured before starting the experiment represent 100%. **c** Relationship between $[Ca^{2+}]_{cyt}$ (shown by $R_{340/380}$) and tension development in the presence of different concentrations of KCl or norepinephrine. Mean ± SE of 4–6 experiments are shown

Cumulative addition of KCl or norepinephrine induced concentration-dependent increases in both muscle tension and $[Ca^{2+}]_{cyt}$ (Fig. 7). If these data are plotted on the $[Ca^{2+}]_{cyt}$ (ordinate)-tension (abscissa) graph (Fig. 7c), it is clearly shown that a positive correlation exists between $[Ca^{2+}]_{cyt}$ and tension development in the presence of high K^+ or norepinephrine. However, the slope of the $[Ca^{2+}]_{cyt}$-tension curve due to high K^+ is smaller than that due to norepinephrine, indicating that greater contraction is induced at a given $[Ca^{2+}]_{cyt}$ in the presence of norepinephrine than in the presence of high K^+ [14, 16, 34]. Prostaglandin $F_{2\alpha}$ and phorbol esters showed similar effects as norepinephrine [34]. Furthermore, phorbol esters at concentrations which did not change resting muscle tension or $[Ca^{2+}]_{cyt}$ potentiated the high K^+-induced contraction with little effect on the high K^+-stimulated $[Ca^{2+}]_{cyt}$.

Sustained increments in both muscle tension and $[Ca^{2+}]_{cyt}$ due to high K^+ or ionomycin were rapidly inhibited by removing external Ca^{2+}. This result suggests that the sustained contractions induced by high K^+ and ionomycin are attributable to Ca^{2+} influx. Ca^{2+}-removal also inhibited the $[Ca^{2+}]_{cyt}$ stimulated by norepinephrine, prostaglandin $F_{2\alpha}$, and phorbol esters. However, a portion of

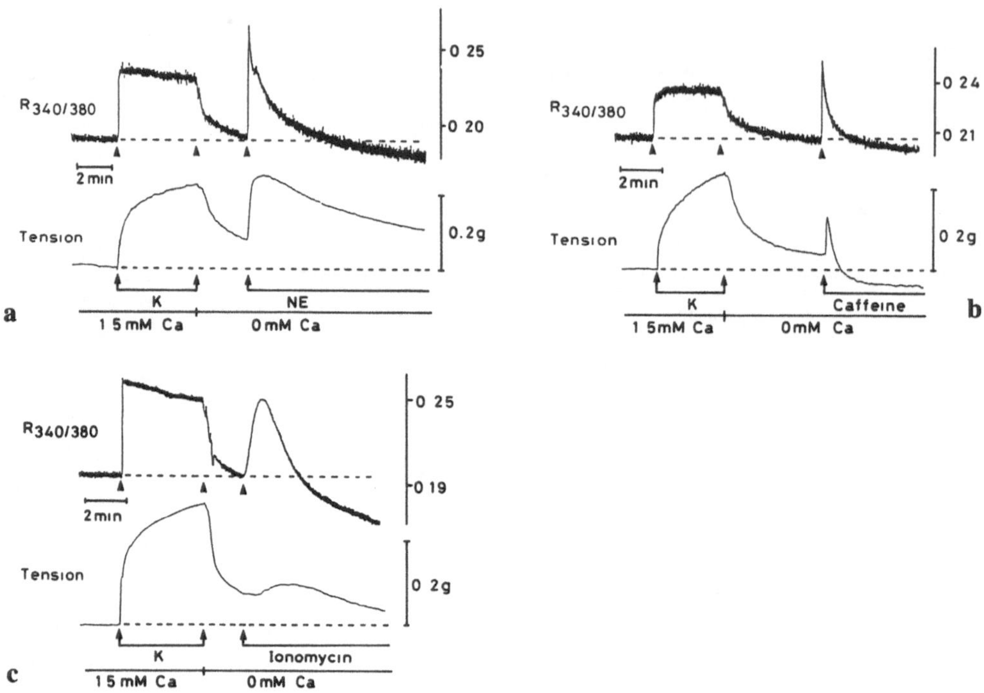

Fig. 8a–c. Effect of **a** norepinephrine **b** caffeine, and **c** ionomycin on $[Ca^{2+}]_{cyt}$ (shown by $R_{340/380}$) and tension in the absence of external Ca^{2+}. After treatment of the muscle for 5 min in 72.7 m*M* KCl (*K*) solution in the presence of 1.5 m*M* Ca^{2+}, muscle was washed with Ca^{2+} free solution (with EGTA 0.5 m*M*). After $[Ca^{2+}]_{cyt}$ decreased to the resting level, 10 μM norepinephrine (*NE*), 20 m*M* caffeine, or 10 μM ionomycin was added

the contraction induced by these agonists were resistant to Ca^{2+}-removal. These results suggest that receptor-mediated contractions are due to Ca^{2+} influx and also due to the mechanism which is less dependent on the increase in $[Ca^{2+}]_{cyt}$.

In a Ca^{2+}-free solution containing 0.5 m*M* EGTA, both $[Ca^{2+}]_{cyt}$ and muscle tension decreased (Fig. 8). When the $[Ca^{2+}]_{cyt}$ approached the resting level, 10 μM norepinephrine was added. Norepinephrine induced a transient increase in $[Ca^{2+}]_{cyt}$ and muscle tension followed by sustained contraction. This result supports the suggestion that the first transient increase in the $[Ca^{2+}]_{cyt}$ induced by norepinephrine is attributable to Ca^{2+} released from an intracellular store [35]. In Ca^{2+}-free solution, caffeine (20 m*M*) and ionomycin (10 μM) also caused a transient increase in $[Ca^{2+}]_{cyt}$, indicating that caffeine and ionomycin release cellular Ca^{2+}. The effect of caffeine may be attributable to Ca^{2+}-induced Ca^{2+} release from the sarcoplasmic reticulum [15, 36, 37]. In contrast to this, ionomycin may increase the permeability to Ca^{2+} of the membranes in the cell and release Ca^{2+} from both the sarcoplasmic reticulum and other storage sites [15].

In Ca^{2+}-free solution, although norepinephrine induced a similar increase in $[Ca^{2+}]_{cyt}$ to that of caffeine or ionomycin, the norepinephrine-induced contraction was markedly greater and more sustained than that induced by caffeine or ionomycin. These results again indicate that receptor-agonists increase the Ca^{2+}-sensitivity of the contractile elements. Caffeine-induced transient increase in $[Ca^{2+}]_{cyt}$ and muscle tension was followed by a rapid decrease in msucle tension. This decrease may be due to the decrease in the sensitivity of contractile elements to Ca^{2+} [15, 37]. Ionomycin induced a large transient increase in $[Ca^{2+}]_{cyt}$ followed by a small transient contraction. Ionomycin may also inhibit the contractile elements although the mechanism of this inhibition remains to be clarified. A possible explanation for the increase in Ca^{2+}-sensitivity due to receptor-agonists is that diacylglycerol, which is produced by the receptor mediated breakdown of membrane phosphoinositides, activates protein kinase C which phosphorylates contractile elements to change their Ca^{2+} sensitivity [38]. This mechanism may also be responsible for the sustained contraction induced by receptor-agonists and phorbol esters in Ca^{2+}-free solution. In cardiac muscle, stimulation of α-adrenoceptors slightly increased $[Ca^{2+}]_{cyt}$ measured with aequorin and greatly increased muscle tension, possibly because the positive inotropic effect of α-adrenoceptor stimulation is in large part of the result of an increase in myofibrillar sensitivity to Ca^{2+} [39].

Vasodilators

Figure 9 shows the effect of sodium nitroprusside on high K^+-induced increase in $[Ca^{2+}]_{cyt}$ and muscle tension. Low concentrations of sodium nitroprusside decreased the high K^+-stimulated $[Ca^{2+}]_{cyt}$ and muscle tension proportionally, with little effect on the slope of the $[Ca^{2+}]_{cyt}$-tension curve. Figure 10 shows that low concentration of sodium nitroprusside reduced the norepinephrine-induced increase in $[Ca^{2+}]_{cyt}$ and muscle tension with little effect on the the slope of the $[Ca^{2+}]_{cyt}$-tension curve. These results indicate that the decrease in $[Ca^{2+}]_{cyt}$ is responsible for the inhibitory effect of sodium nitroprusside. Since sodium nitroprusside lowered the norepinephrine-stimulated $[Ca^{2+}]_{cyt}$ more strongly than that stimulated by high K^+, it seems likely that sodium nitroprusside inhibits receptor-linked Ca^{2+} channels more strongly than voltage-dependent Ca^{2+} channels, supporting the previous suggestion [6].

Higher concentrations of sodium nitroprusside strongly inhibited the contraction induced by norepinephrine and less strongly inhibited the contraction induced by high K^+ (Figs. 9 and 10). The increments in $[Ca^{2+}]_{cyt}$ were also reduced, however to a lesser extent than the contractions, resulting in the decrease in the slope of the $[Ca^{2+}]_{cyt}$-tension curve [17]. Sodium nitroprusside completely inhibited the norepinephrine-induced transient contraction in Ca^{2+}-free solution, whereas it only partially inhibited the transient increase in $[Ca^{2+}]_{cyt}$ (Fig. 11). These results indicate the possibility that sodium nitroprusside inhibit these contractions by the decrease in $[Ca^{2+}]_{cyt}$ and also by the increase in the threshold of $[Ca^{2+}]_{cyt}$ for tension development [17].

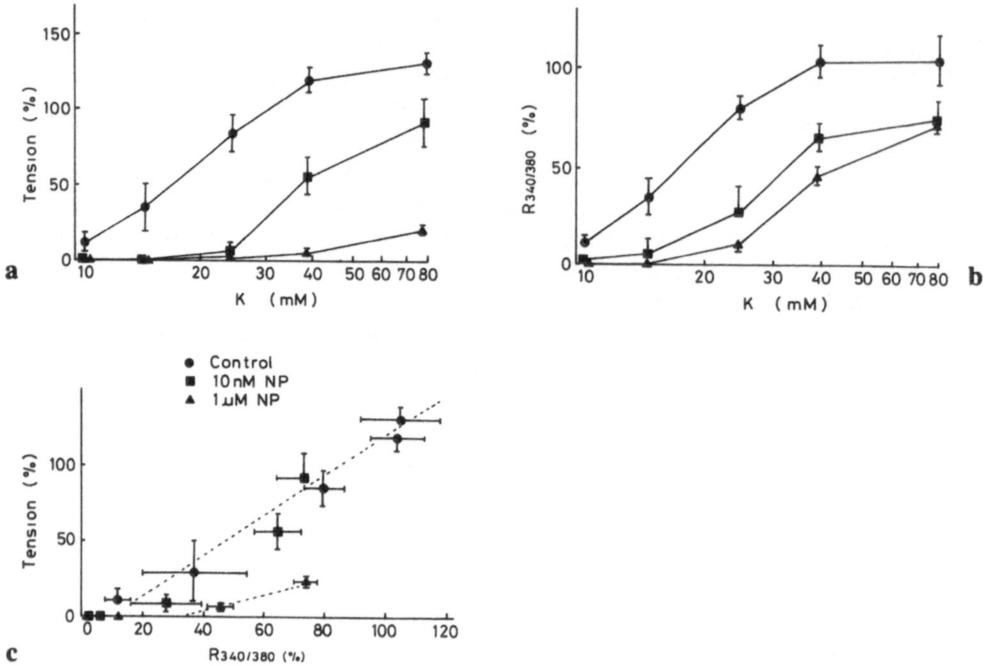

Fig. 9a–c. Effect of sodium nitroprusside on KCl (*K*)-induced increases in **a** tension and **b** $[Ca^{2+}]_{cyt}$ (shown by $R_{340/380}$). The 72.7 m*M* KCl-induced increases in $[Ca^{2+}]_{cyt}$ and tension measured before starting the experiment represent 100%. Sodium nitroprusside (*NP*) (10 n*M* or 1 μ*M*) was added 5 min before the addition of the stimulants. **c** Effects of sodium nitroprusside on the relationship between $[Ca^{2+}]_{cyt}$ and tension development in the presence of different concentrations of KCl. Mean ± SE of 4–6 experiments are shown

The inhibitory effects of sodium nitroprusside may be mediated by the increase in cyclic GMP level [40, 41]. Cyclic GMP has been shown to inhibit the contractile response in permeabilized smooth muscle through the activation of cyclic GMP-dependent protein kinase [42, 43]. Therefore, the dual effects of sodium nitroprusside may be attributable to its effects on different stages of the excitation-contraction coupling in smooth muscle cells.

Forskolin either did not change or slightly decreased the resting $[Ca^{2+}]_{cyt}$ and muscle tone [18]. This result is not consistent with the unexpected finding with aequorin that forskolin increases resting $[Ca^{2+}]_{cyt}$ [25–27]. These results also support the possibility that aequorin signals represent the regional, high Ca^{2+} area, whereas fura-2 signals represent the average $[Ca^{2+}]_{cyt}$.

In the presence of 0.1 μ*M* forskolin, the norepinephrine- and high K^+-stimulated $[Ca^{2+}]_{cyt}$ and muscle tension were inhibited without changing the slope of the $[Ca^{2+}]_{cyt}$-tension curve. A higher concentration (1 μ*M*) inhibited muscle contraction more strongly than $[Ca^{2+}]_{cyt}$. Norepinephrine-induced changes were more strongly inhibited than those induced by high K^+ [18]. These

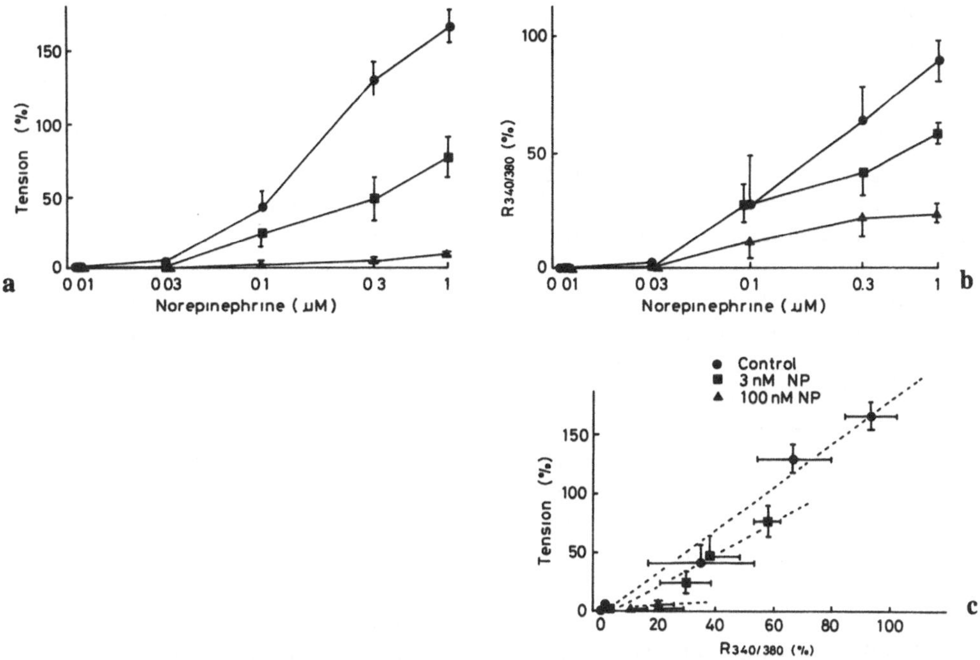

Fig. 10a–c. Effect of sodium nitroprusside on norepinephrine-induced increases in **a** tension and **b** $[Ca^{2+}]_{cyt}$ (shown by $R_{340/380}$). The 72.7 m*M* KCl-induced increases in $[Ca^{2+}]_{cyt}$ and tension measured before starting the experiment represent 100%. Sodium nitroprusside (*NP*) (3 and 100 n*M*) was added 5 min before the addition of the stimulants. **c** Effects of sodium nitroprus-side on the relationship between $[Ca^{2+}]_{cyt}$ and tension development in the presence of different concentrations of norepinephrine. Mean ± SE of 4–6 experiments are shown

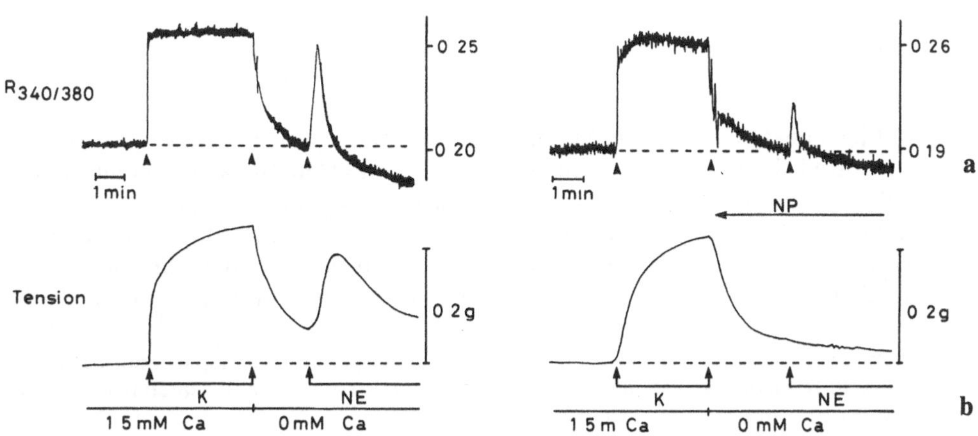

Fig. 11a,b. Effect of sodium nitroprusside (*NP*) on norepinephrine (*NE*)-induced phasic contraction and $[Ca^{2+}]_{cyt}$ (shown by $R_{340/380}$) in the absence of external Ca^{2+}. After treatment of the muscle for 4 min in 72.7 m*M* KCl (*K*) solution in the presence of 1.5 m*M* Ca^{2+} solution, external solution was replaced with Ca^{2+}-free solution (with EGTA 0.5 m*M*) **a** without or **b** with 1 µ*M* sodium nitroprusside. After $[Ca^{2+}]_{cyt}$ decreased to the resting level, 10 µ*M* norepinephrine was added results indicate that the inhibitory effects of forskolin were quite similar to those of sodium nitroprusside. Forskolin has been shown to increase cyclic AMP [44], which inhibits various processes in the smooth muscle cells [11, 45]. In cultured smooth muscle cells, it has been suggested that cyclic AMP activates cyclic GMP-dependent protein kinase and decreases $[Ca^{2+}]_{cyt}$ [33].

Addition of 1 μM verapamil did not change the resting $[Ca^{2+}]_{cyt}$ or the resting tension. In the presence of verapamil, however, the increases in both $[Ca^{2+}]_{cyt}$ and muscle tension induced by the cumulative addition of KCl were lowered. Since verapamil proportionally decreased $[Ca^{2+}]_{cyt}$ and muscle tension, the slope of the $[Ca^{2+}]_{cyt}$-tension relationship did not change, indicating that the inhibitory effect is due to a decrease in $[Ca^{2+}]_{cyt}$. Verapamil also inhibited the increments in $[Ca^{2+}]_{cyt}$ and muscle tension induced by norepinephrine. Higher concentration (10 μM) of verapamil inhibited the norepinephrine-stimulated $[Ca^{2+}]_{cyt}$ almost completely. Verapamil has been shown to inhibit high K^+-stimulated Ca^{2+} influx by inhibiting voltage-dependent Ca^{2+} channels and to inhibit norepinephrine-induced Ca^{2+} influx by inhibiting receptor-operated Ca^{2+} channels [1, 6, 11, 46]. The decrease in $[Ca^{2+}]_{cyt}$ induced by verapamil may be attributable to this effect. It was again found, however, that a portion of the norepinephrine-induced sustained contraction was not inhibited by verapamil. Verapamil showed similar inhibitory effects on $[Ca^{2+}]_{cyt}$ and contraction stimulated by prostaglandin $F_2\alpha$ and phorbol esters [47]. These results indicate that a portion of the contraction induced by receptor-agonists is maintained at the resting $[Ca^{2+}]_{cyt}$, again suggesting that norepinephrine increases the Ca^{2+} or contractile elements.

The transient increases in $[Ca^{2+}]_{cyt}$ and muscle tension due to norepinephrine were not modified by 1 μM verapamil. This effect indicates that verapamil does not inhibit the norepinephrine-induced release of stored Ca^{2+} or contractile elements.

Conclusion

Mechanical activity and changes in cytosolic free Ca^{2+} concentrations were measured simultaneously using a fura-2 dual-wavelength excitation method. Results suggest that vasoconstrictors and vasodilators modulate the smooth muscle contraction by changing the $[Ca^{2+}]_{cyt}$ and also by changing the sensitivity of the contractile system to Ca^{2+}. Furthermore, at least a part of the receptor-mediated sustained contraction may be due to activation of protein kinase C.

Acknowledgements. We are grateful to Dr. Koichi Sato of our department for his help. This study was supported in part by a Grant-in-aid for Scientific Research from the Ministry of Education, Science, and Culture, Japan.

References

1. Bolton TB (1979) Mechanisms of action of transmitters and other substances on smooth muscle. Physiol Rev 59: 606–718

2. Vanhoutte PM, Verbeuren TJ, Webb C (1981) Local modulation of adrenergic neuroeffector interaction in the blood vessel wall. Physiol Rev 61: 151–247
3. Rasmussen H, Barrett PO (1984) Calcium messenger system: an integrated view. Physiol Rev 64: 938–984
4. Abdel-Latif AA (1986) Calcium-mobilizing receptors, polyphospho-inositides, and the generation of second messengers. Pharmacol Rev 38: 227–272
5. Weiss GB (1977) Calcium and contractility in vascular smooth muscle. Adv Gen Cell Pharmacol 2: 71–154
6. Karaki H, Weiss GB (1984) Calcium channels in smooth muscle. Gastroenterology 87: 960–970
7. Hartshorne DJ (1982) The contractile apparatus of smooth muscle and its regulation by calcium. In: Grass MF III, Barnes CD (eds) Vascular smooth muscle: metabolic, ionic and contractile mechanisms. Academic Press, New York, pp 135–161
8. Somlyo AV, Bond M, Somlyo AP, Scarpa A (1985) Inositol trisphosphate-induced calcium release and contraction in vascular smooth muscle. Proc Natl Acad Sci USA 82: 5231–5235
9. Van Breemen C (1977) Calcium requirement for activation of intact aortic smooth muscle. J Physiol 272: 317–329
10. Daniel EE (1985) The use of subcellular membrane fractions in analysis of control of smooth muscle function. Experientia 41: 905–913
11. Karaki H (1987) Use of tension measurements to delineate the mode of action of vasodilators. J Pharmacol Methods 18: 1–21
12. Ozaki H, Sato K, Satoh T, Karaki H (1987) Simultaneous recordings of caleium signals and mechanical activity using fluorescent dye fura 2 in isolated strips of vascular smooth muscle. Jpn J Pharmacol 45: 429–433
13. Himpens B, Somlyo AP (1988) Free-calcium and force transients during depolarization and pharmacomechanical coupling in guinea-pig smooth muscle. J Physiol 395: 507–530
14. Sato K, Ozaki H, Karaki H (1988) Changes in cytosolic calcium level in vascular smooth muscle strip measured simultaneously with contraction using fluorescent calcium indicator fura 2. J Pharmacol Exp Ther 246: 294–300
15. Sato K, Ozaki H, Karaki H (1988) Multiple effects of caffeine on contraction and cytosolic Ca^{2+} levels in vascular smooth muscle of rat aorta. Naunyn Schmiedebergs Arch Pharmacol 338: 443–448
16. Karaki H, Sato K, Ozaki H (1988) Different effects of norepinephrine and KCl on the cytosolic Ca^{2+}-tension relationship in vascular smooth muscle of rat aorta. Eur J Pharmacol 151: 325–328
17. Karaki H, Sato K, Ozaki H, Murakami K (1988) Effects of sodium nitroprusside on cytosolic calcium level in vascular smooth muscle. Eur J Pharmacol 156: 259–266
18. Abe A, Karaki H (1989) Effect of forskolin on cytosolic Ca^{2+} level and contraction in vascular smooth muscle. J Pharmacol Exp Ther (in press)
19. Furchgott RF, Zawadzki JV (1980) The obligatory role of endothelial cells in the relaxation of arterial smooth muscle by acetylcholine. Nature (Lond) 288: 373–376
20. Ozaki H, Satoh T, Karaki H, Ishida Y (1988) Regulation of metabolisms and contraction by cytoplasmic calcium in the intestinal smooth muscle. J Biol Chem 263: 14074–14079
21. Grynkiewictz G, Poenie M, Tsien RY (1985) A new generation of Ca^{2+} indicators with greatly improved fluorescence properties. J Biol Chem 260: 3440–3450
22. Scanlon M, Williams DA, Fay SF (1987) A Ca^{2+}-insensitive form of fura-2 associated with polymorphonuclear leukocytes. J Biol Chem 262: 6308–6312
23. Konishi M, Olson A, Hollingworth S, Bayler SM (1988) Myoplasmic binding fura-2 investigated by steady-state fluorescence and absorbance measurement. Biophys J 54: 1089–1104
24. Schramm M, Thomas G, Towarrt R, Franckowiak G (1983) Novel dihydropyridines with positive inotropic action through activation of Ca^{2+} channels. Nature (Lond) 303: 535–537

25. Morgan JP, Morgan KG (1984) Stimulus-specific patterns of intracellular calcium levels in smooth muscle of ferret portal vein. J Physiol (Lond) 351: 155–167
26. Morgan JP, Morgan KG (1984) Alteration of cytoplasmic ionized calcium levels in smooth muscle by vasodilators in the ferret. J Physiol (Lond) 357: 539–551
27. Takuwa Y, Takuwa N, Rasmussen H (1988) The effects of isoproterenol on intracellular calcium concentration. J Biol Chem 263: 762–768
28. Blinks JR (1982) The use of photoproteins as calcium indicators in cellular physiology. Techniques Cell Physiol 126: 1–38
29. Williams DA, Fay SF (1986) Calcium transients and resting levels in isolated smooth muscle cells as monitored with quin 2. Am J Physiol 250: C779–C791
30. Sumimoto K, Kuriyama H (1986) Mobilization of free Ca^{2+} measured during contraction-relaxation cycles in smooth muscle cells of the porcine coronary artery using quin 2. Pflügers Arch 406: 173–180
31. Hassid A (1986) Atriopeptin II decreases cytosolic free Ca in cultured vascular smooth muscle cells. Am J Physiol 251: C681–C686
32. Ambler SK, Poenie M, Tsien RY, Taylor P (1988) Agonist-stimulated oscillations and cycling of intracellular free calcium in individual cultured muscle cells. J Biol Chem 263: 1952–1959
33. Lincoln TM, Wear CB, Cornwell TL, Taylor A (1989) Reduction of Ca^{2+} levels in smooth muscle cells by forskolin and isoproterenol is mediated by activation of cyclic GMP-dependent protein kinase. FASEB J 3: A881
34. Karaki H, Ozaki H, Ohyama T, Sato K (1989) Protein kinase C mediates the contraction induced by receptor agonists in vascular smooth muscle. Jpn J Pharmacol 49 (Suppl): P96
35. Karaki H, Weiss GB (1988) Calcium release in smooth muscle. Life Sci 42: 111–122
36. Endo M, Kitazawa T, Yagi S, Iino M, Kakuta Y (1977) Some properties of chemically skinned smooth muscle fibers. In: Casteels R, Godfraind T, Rüegg JC (eds) Excitation-contraction coupling in smooth muscle. Elsevier/North-Holland, Amsterdam, pp 199–209
37. Karaki H, Ahn HY, Urakawa N (1987) Caffeine-induced contraction in vascular smooth muscle. Arch Int Pharmacodyn Ther 285: 60–71
38. Rasmussen H, Takuwa Y, Park S (1987) Protein kinase C in the regulation of smooth muscle contraction. FASEB J 1: 177–185
39. Endoh M, Blinks JR (1988) Actions of sympathomimetic amines on the Ca^{2+} transients and contractions of rabbit myocardium: Reciprocal changes in myofibrillar responsiveness to Ca^{2+} mediated through α- and β-adrenoceptors. Circ Res 62: 247–265
40. Lincoln TM, Fisher-Simpson V (1983) A comparison of the effects of forskolin and nitroprusside on cyclic nucleotides and relaxation in the rat aorta. Eur J Pharmacol 101: 17–27
41. Ignarro LJ, Kadowitz PJ (1985) The pharmacological and physiological role of cyclic GMP in vascular smooth muscle relaxation. Ann Rev Pharmacol Toxicol 25: 171–191
42. Pfitzer G, Hofmann F, Di Salvo J, Rüegg JC (1984) cGMP and cAMP inhibit tension development in skinned coronary arteries. Pflügers Arch 401: 277–280
43. Pfitzer G, Merkel L, Rüegg JC, Hofmann F (1986) Cyclic GMP-dependent protein kinase relaxes skinned fibers from guinea-pig taenia coli but not from chicken gizzard. Pflügers Arch 407: 87–91
44. Seamon KB, Daly JW (1983) Forskolin, cyclic AMP and cellular physiology. Trend Pharmacol Sci 4: 120–123
45. Bülbring E, Tomita T (1987) Catecholamine action in smooth muscle. Pharmacol Rev 39: 49–96
46. Flaim SF (1982) Comparative pharmacology of calcium blockers based on studies of vascular smooth muscle. In: Flaim SF, Zelis R (eds) Calcium blockers. Urban and Schwarzenberg, New York, pp 155–178
47. Sato K, Ozaki H, Karaki H (1989) Inhibitory effect of verapamil on cytoplasmic Ca level and muscle tension in vascular smooth muscle. Jpn J Pharmacol 49 (Suppl): P97

Calcium Mobilization Mechanisms in Smooth Muscle

MAKOTO ENDO, MASAMITSU IINO, and TSUTOMU KOBAYASHI[1]

Summary. The role of intracellular calcium store in calcium mobilization in smooth muscle cells was examined using guinea pig taenia caeci, portal vein and pulmonary artery. By using saponin-skinned fiber bundles, two calcium release mechanisms, calcium-induced calcium release (CICR) and inositoltrisphosphate-induced calcium release (IICR), were found in the calcium store. Caffeine caused calcium release by enhancing CICR. In the absence of sensitizing drugs such as caffeine, CICR could only be evoked with levels of free calcium above 1 μM. Similar to CICR, IICR was accelerated by calcium ion and by adenine and related compounds, and inhibited by procaine. However, unlike CICR, the enhancing effect of calcium ion on IICR was apparent in the submicromolar range. The two calcium release mechanisms were distributed heterogeneously in the whole calcium store. A part of the store, called Sα, was with both CICR and IICR, but the remainder (Sβ) was only with IICR and not with CICR. This heterogeneity was confirmed by using ryanodine, which specifically acts on the CICR mechanism to fix the CICR channel in an open state. Thus, after ryanodine treatment, Sα no longer accumulated calcium but Sβ did. When intact smooth muscle cells were treated with ryanodine in the presence of caffeine, caffeine contractures as well as the initial phase of agonist-induced contractions were abolished, indicating the essential role of Sα in these contractions.

Key words: Calcium-induced calcium release (CICR)—Calcium release—Inositol-trisphosphate-induced calcium release (IICR)—Intracellular calcium store—Ryanodine

Introduction

The physiological contraction of smooth muscle cells is mediated by an elevated cytoplasmic free calcium ion concentration [1]. The source of calcium ion to activate the smooth muscle contractile system is known to be either the extracellular medium or an intracellular calcium store. The relative importance of each calcium source mentioned above during various types of contractions has not yet been precisely determined. It is considered that the component of contraction that is abolished when calcium ions in the extracellular medium are

[1]Department of Pharmacology, Faculty of Medicine, University of Tokyo, Bunkyo-ku, Tokyo 113, Japan

removed is the part evoked by calcium ion entering from the extracellular medium. However, this is not necessarily true, because if by some mechanism extracellular calcium ion is required for the process to cause a release of calcium from the intracellular calcium store, such a process should also be abolished by the removal of extracellular calcium ion. Furthermore, the amount of calcium in the store might be reduced by the removal of extracellular calcium, if the removal occurs for a prolonged period. Therefore, while it is quite clear that the part of contraction remaining after removal of extracellular calcium must be due to the release of calcium from the intracellular calcium store, it is not easy to determine, in the part abolished, whether the major calcium source is the extracellular medium or the intracellular store. We have obtained some answers to part of this problem by specifically underming the calcium-accumulating capacity of a part of the intracellular calcium store by the use of a plant alkaloid, ryanodine, which is known to fix a kind of calcium channel of the calcium store in an open state [2]. These experiments also provide independent evidence for the hypothesis that the part remaining after the removal of external calcium is indeed due to a release of calcium from the intracellular store.

Materials and Methods

Guinea pigs, mostly male, weighing 240 g–300 g, were stunned and bled. Thin bundles of smooth muscle fibers (150 μm–250 μm in diameter for taenia caeci and 70 μm in thickness and 300 μm in width for portal vein or pulmonary artery), were carefully obtained from the respective tissues in normal external solution under a stereomicroscope. The fiber bundles were held horizontally, one end being connected to a tension transducer (AE801, Akers) to record isometric tension. Contraction and relaxation were induced by changing the solution surrounding the bundles, which was accomplished by transferring the preparation from one well to another, each containing the appropriate solution.

For the skinned fiber experiments, the cells in the bundles were chemically skinned by immersing them in a relaxing solution containing saponin (50 μg/ml) for 30 min. Calcium release from the store was determined by procedures established in our laboratory [3, 4]. First calcium was loaded to the store to a fixed level by utilizing the calcium pump activity in the membrane of the store. Then, calcium and MgATP in the solution was withdrawn to stop the calcium pump activity to prevent reuptake of calcium during release experiments, and a calcium releasing stimulus to be tested was given for various periods of time in the presence of a highly concentrated calcium buffer (total 10 m*M* EGTA). The calcium buffer was necessary as the processes of calcium release are calcium ion concentration-dependent (see Results and Discussion section); if the buffer were absent, the result of calcium release would cause a rise in calcium ion concentration in the medium and facilitate further calcium release, which would obscure the magnitude of the direct effect of the calcium releasing stimulus to be tested. After termination of the calcium releasing stimulus, the amount of calcium remaining in the store was determined by applying a strong calcium releasing stimulus to discharge all of the calcium in the store, the amount released being determined by the use of fura-2 fluorescence.

To measure the fluorescence, the skinned fiber preparation was fixed in a small glass capillary cuvette (inner diameter about 400 μm) which was placed on the stage of a fluorescence microscope. The fluorescent light was collected onto a photomultiplier and the signal was fed into a microcomputer through a processor. Both 340 and 380 nm wavelengths of light were used to excite the dye and fluorescence at 530 nm was measured.

Normal external solution contained (m*M*): NaCl, 150; KCl, 4; Ca methanesulphonate $(Ms)_2$, 2; $Mg(Ms)_2$, 2; HEPES (N-2-hydroxyethyl-piperazine-N′-2-ethanesulphonic acid), 5, glucose, 5.6. pH was adjusted to 7.40 at 25°C with NaOH. In high-K solutions, NaCl was replaced by an equivalent amount of KCl by a specified amount. In calcium-free solutions, $Ca(Ms)_2$ was omitted and 5 m*M* EGTA (ethyleneglycol-bis-[β-aminoethylether]-N-N′-tetraacetic acid) was added. Various agents were dissolved in the appropriate solution. For compositions of solutions used for skinned fiber experiments, refer to the previously published papers [3, 4].

Results and Discussion

Responses of Smooth Muscles in the Absence of External Calcium Ion

It is well known that while high-K-induced contraction of smooth muscle is abolished by the removal of calcium ion from the extracellular medium, agonist (neurotransmitter)-induced contraction remains after the removal [5]. This was confirmed as is shown in Fig. 1, in which responses of smooth muscles from the pulmonary artery to three kinds of stimuli are given. In a calcium-free medium, high-K solution could no longer evoke any contraction but the initial part of agonist (noradrenaline)-induced contraction was almost unaltered, although the later part was strongly suppressed. Caffeine-induced contraction persisted in the calcium-free medium. Essentially the same results were obtained with smooth muscles from taenia caeci and portal vein. The only difference between the different smooth muscles was the exact time relationship between the initial and later phases of agonist-induced contractions. The later phase in noradrenaline-induced contractions of portal vein showed a clear and distinct rapidly rising phase making the separation of the two phases very clear, while the peak of the later phase in taenia caeci activated by acetylcholine or carbachol, was completely fused with the initial phase, the whole response constituting a single broad peak in the presence of external calcium (Fig. 4). Exactly the same results were obtained when instead of calcium-free solution, the organic calcium antagonists such as diltiazem or nifedipine were used in the presence of normal concentration of calcium ion in the external medium. These results quite clearly indicated that in caffeine-induced contractions and the initial phase of agonist-induced contraction of smooth muscles, the calcium ion comes from an intracellular calcium store, while the high-K-induced contraction and the later phase of agonist-

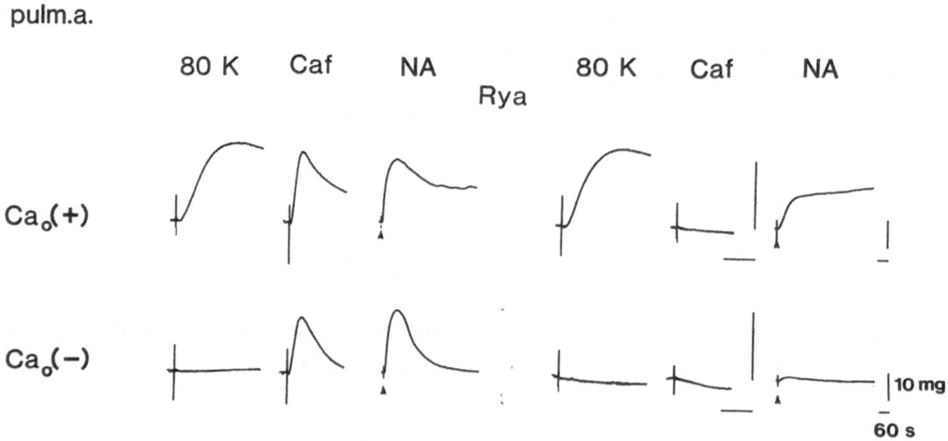

Fig. 1. Responses of smooth muscles of pulmonary artery (*pulm.a.*) of the guinea pig to 80 m*M* K (*80 K*), 25 m*M* caffeine (*Caf*) or 10 μM noradrenaline (*NA*) in the presence of 2 m*M* calcium in the external solution (*upper panel*) or in its absence (*lower panel*), before (*left*) and after (*right*) a treatment with 30 μM ryanodine (*Rya*) together with 25 m*M* caffeine. Note that *calibration bars* for 80 K and Caf are different from those for NA

induced contraction require calcium influx. What was not determined here is whether in the latter types of contractions calcium entering into the cell directly acts on the contractile proteins or indirectly acts by somehow causing calcium release from an intracellular calcium store.

Caffeine-induced contractions are considered to be due to calcium release from an intracellular calcium store, and the results shown in Fig. 1 are consistent with this idea. The magnitude of caffeine-induced contraction in the calcium-free solution, however, was slightly smaller than that in normal external medium, although the difference was not so conspicuous nor so consistent as in the case of the later phase of agonist-induced contractions. The difference may well be due to the difference in the amount of calcium in the store that must have been reduced in the calcium-free solution at least to some extent. The virtually unaltered responses of the initial phase of agonist-induced contraction in the calcium-free medium could then be interpreted as an indication of saturation of the effectiveness of the stimulus in this case. Although the dose of caffeine used in Fig. 1 was probably high enough to cause a saturating magnitude of calcium release, the contractile responses were clearly not saturated as indicated by the smaller tensions developed which was most likely due to the additional relaxing action of caffeine.

Two Different Calcium Release Mechanisms in Smooth Muscle Calcium Store

The properties of calcium release mechanisms from the intracellular calcium store were studied with skinned fiber preparations of smooth muscles. Two dif-

Table 1. Comparison of the properties of calcium release mechanisms of smooth muscle calcium store .

	Effective Ca concentration	Potentiators	Inhibitors	Ryanodine effects	Distribution
CICR	> 1 μM	Caffeine Adenine compounds	Mg ion Procaine	Open fixation	5%–80%
IICR	−0.1 μM	Adenine compounds	Procaine	No effects	Whole store

CICR, calcium-induced calcium release; IICR, inositoltrisphosphate-induced calcium release

ferent calcium release mechanisms were detected. One was a calcium release mechanism activated by applying calcium ion itself (calcium-induced calcium release: CICR), and the other was that activated by inositoltrisphosphate (IP_3-induced calcium release: IICR). The characteristics of these two calcium release mechanisms were examined and the results were summarized in Table 1. Some characteristics of the two release mechanisms are similar but others are quite different.

CICR was activated by calcium ion but only in concentrations above micromolar range unless sensitized by agents such as caffeine. IICR could be evoked in the practical absence of calcium ion, but its rate was strongly enhanced if calcium ion in a submicromolar range was present in the IP_3-containing calcium-releasing medium [3]. Since this concentration of calcium was too low to activate the CICR mechanism, this effect of calcium could not be due to the additional calcium release through the CICR mechanism, but to the direct enhancing effect of the calcium ion on the IICR mechanism. Adenine compounds such as ATP, nonhydrolyzable analogues of ATP, ADP, AMP, adenosine, and adenine enhanced both CICR and IICR. Procaine inhibited both CICR and IICR.

Caffeine stimulated the CICR mechanism by increasing the calcium sensitivity of the mechanism as well as by increasing the maximum rate of calcium release at the optimum calcium ion concentration. Magnesium ion showed the opposite effect; it decreased both the calcium sensitivity of CICR and the maximum rate of calcium release at the optimum calcium ion concentration. On the other hand, IICR was not affected by caffeine at all, and magnesium ion did not appear to exert any appreciable action on IICR either. Calcium release by IP_3 is actually inhibited by magnesium ion, but this is mainly due to the activation by magnesium ion of IP_3-destructing enzyme. In fact, in the presence of a sufficient concentration of an inhibitor of the IP_3-destructing enzyme, 2,3-bisphosphoglycerate, magnesium ion did not appreciably inhibit IICR.

The properties of the CICR mechanism in smooth muscle calcium store described above are very similar to those in skeletal muscles [6], in which CICR was first demonstrated and characterized.

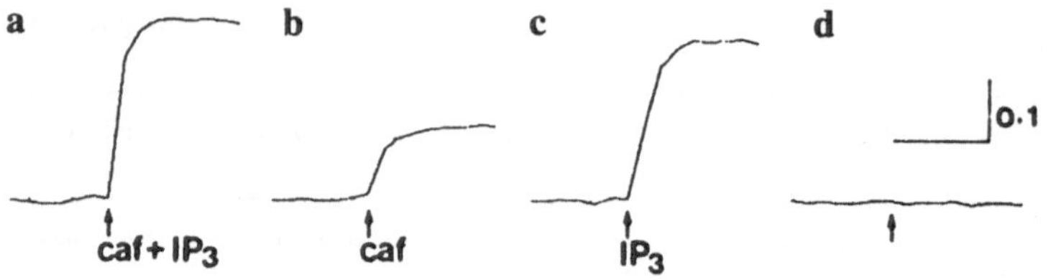

Fig. 2. Heterogeneity of smooth muscle calcium store in regard to the calcium release mechanisms. A skinned smooth muscle fiber bundle of taenia. Records of fura-2 fluorescence caused by calcium release from loaded store induced by 50 m*M* caffeine plus 25 m*M* AMP (*Caf*) or by 10 μM inositoltriphosphate (IP_3), or their combination. *Panel* **d**, application of the combination of the agents but without previous calcium loading. From [3]

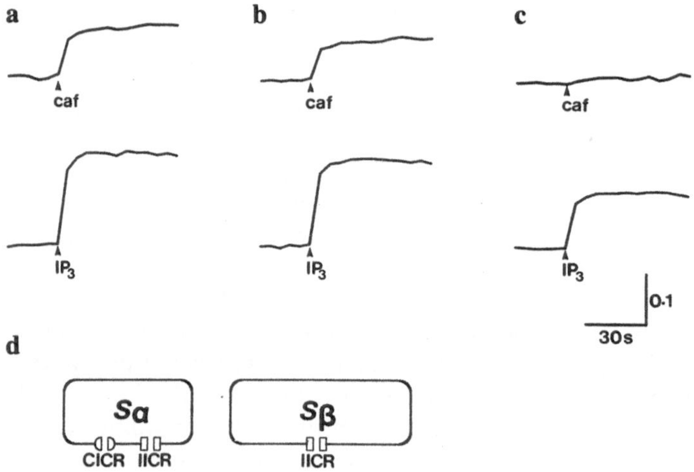

Fig. 3a–d. Effect of ryanodine (30 μM) on calcium-accumulating capacity of intracellular calcium store of a skinned fiber bundle of taenia. The capacity was determined by releasing calcium from the store as in Fig. 2. **a** Control. **b** After treatment with ryanodine together with 10 μM inositoltrisphosphate (IP_3) at pCa 7 for 2 min. **c** After additional treatment with ryanodine together with 50 m*M* caffeine (*caf*) for 2 min. **d** Scheme of compartments of the calcium store in smooth muscle cells. $S\alpha$ includes both the calcium-induced calcium release (*CICR*) mechanism and the inositoltrisphosphate-induced calcium release (*IICR*) mechanism. $S\beta$ is with only IICR. From [7]

The CICR and the IICR mechanisms were heterogeneously distributed in the smooth muscle calcium store [3]. Figure 2 shows the calcium release responses to maximal CICR and IICR stimulation of the calcium store in a skinned fiber bundle of taenia caecum. It can be seen that the maximal stimulation of IICR by applying 10 μM of IP_3 caused a much larger calcium release than the maximal stimulation of CICR by applying 50 m*M* caffeine. Simultaneous application of IP_3 and caffeine caused essentially the same amount of calcium release as IP_3 alone. After the maximal CICR stimulation, the activation of IICR still evoked a calcium release from the store without any reloading, although the amount of calcium released was smaller compared with the case without initial CICR stimulation by about the amount of calcium released during the preceding CICR (data not shown). However, after the maximal IICR stimulation, the activation of CICR no longer evoked a calcium release unless calcium was reloaded to the store (data not shown). These and other results are consistent with the hypothesis that the IICR mechanism distributes on the membrane of the whole smooth muscle calcium store, but the CICR mechanism distributes on only a part of it. In other words, a portion of the store, called Sα, is with both the CICR and the IICR mechanisms, while the remainder (Sβ) is with only IICR (Fig. 3d). The ratio of the calcium accumulating capacity of Sα to the capacity of the whole store differs in different smooth muscles. It is about 40%, 5%, and 60% of the whole store in taenia caeci, portal vein and pulmonary artery of the guinea pig, respectively [7].

This haterogeneity of calcium store was confirmed by using ryanodine [7], which is known to specifically act on the CICR mechanism of skeletal muscle SR only when the CICR channel is open, and fixes the channel in an open state [2]. It was shown that the alkaloid also acts on the CICR channel of the smooth muscle calcium store in a similar manner. Thus, as shown in Fig. 3, after ryanodine treatment during maximal CICR stimulation with caffeine, Sα no longer held calcium but Sβ did (Figs. 3b, c). When ryanodine was applied during the maximal IICR stimulation, no such effect was produced (Figs. 3a, b), indicating the specificity of the alkaloid action to the CICR channel.

Functional Removal of the Calcium Store in Smooth Muscle Cells with Ryanodine

The effect of ryanodine on intact smooth muscle preparations was then examined [7]. Ryanodine did not exert any effects even when the smooth muscle preparations were repeatedly activated by a high-K medium or by applying an appropriate agonist. However, when caffeine was repeatedly applied in the presence of ryanodine, the first contraction was exactly the same as that without the alkaloid, but the second and subsequent caffeine applications no longer induced any appreciable contractions (Fig. 1). This ryanodine effect persisted even after the removal of the alkaloid at least for a few hours at 20°C, but was reversed by incubation without the alkaloid at 37°C for about 40 min. These results can be interpreted as follows. In the case of skinned fibers, ryanodine specifically acted on the CICR mechanism only when the mechanism was activated by caffeine, and fixed the CICR channel in an open state, making the portion of the calcium store with CICR channels, Sα, leaky enough so that no appreciable amount of calcium was further accumulated in Sα. In other words the function of Sα was abolished by the alkaloid. Since ryanodine exerted no effects during high-K- or agonist-induced contractions, the CICR channels were not activated during these high-K- or agonist-induced contractions, certainly not to the level activated during the caffeine treatment, although the magnitude of high-K- or agonist-induced contractions was the same or even greater than that of caffeine-induced contractions. The absence of the effects of ryanodine when applied during high-K- and agonist-induced contractions also indicates that ryanodine does not have direct action on the contractile protein system.

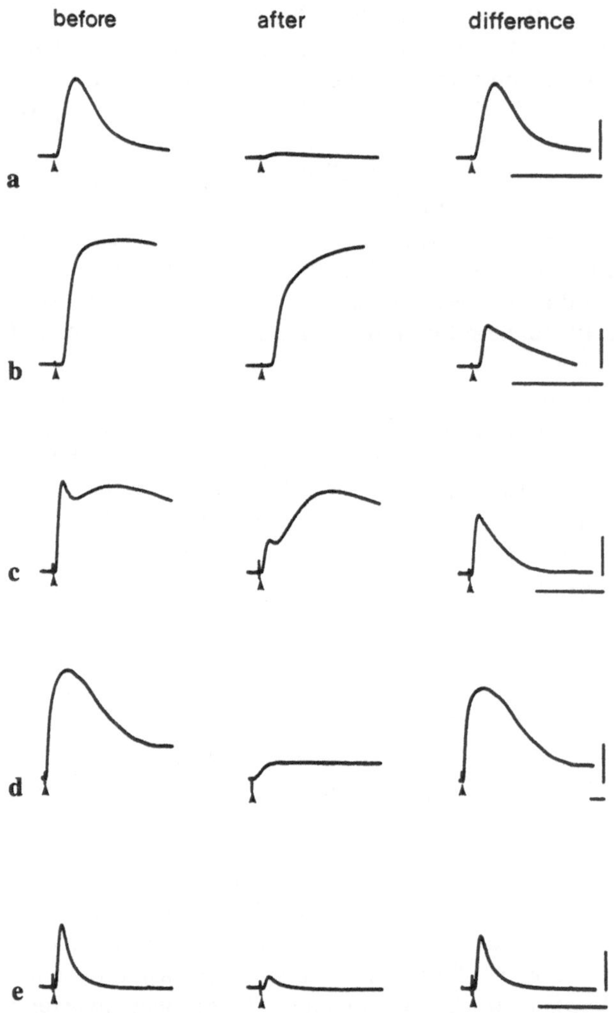

Fig. 4a–e. Tension responses of intact smooth muscle fiber bundles before (*left*) and after (*middle*) the treatment with 30 μM ryanodine together with 25 mM caffeine. Traces on the right are the differences of tension traces (*left–middle*). Contractions were evoked by 25 mM caffeine (**a** taenia), 10 μM carbachol (**b** taenia), or 10 μM noradrenaline (**c** and **e**, portal vein; **d**, pulmonary artery). **a–d** 2 mM external calcium; **e** calcium-free, 5 mM EGTA. *Horizontal bars* 60 s, *Vertical bars* 1 mN except **d** (0.1 mN). From [7]

Even after the effective ryanodine treatment to abolish further caffeine contractures, the magnitude of the high-K contractions was generally almost the same as before the treatment (Fig. 1), indicating the calcium release from Sα does not appreciably contribute to the high-K contractions. However, from time to time, high-K contractures were reduced after the effective ryanodine treatment. Therefore, the participation of calcium release from Sα might also be important under a certain condition yet to be specified.

As shown in Fig. 4, the initial phase of the agonist-induced contractions was strongly depressed after the effective ryanodine treatment, although the later phase was almost unaffected. This clearly indicates the essential role of $S\alpha$ in the initial phase and its lack of a role in the later phase of agonist-induced contractions. As already described, the calcium-accumulating capacity of $S\alpha$ in the portial vein is only about 5% of that of the whole store, but the functional removal of S_α abolished a major part of the initial phase of noradrenaline-induced contraction of this muscle (Figs. 4c, e). Similar observations were made in other kinds of smooth muscle as well. Therefore, it appears that $S\alpha$ contributes more than $S\beta$ in the release of calcium in response to agonists. If the second messenger to cause calcium release in response to agonist stimulation is IP_3 as generally assumed, why is the contribution of $S\beta$ in calcium release much smaller than that of $S\alpha$, while both parts of the store have the IICR mechanism? It might be more difficult for IP_3 to reach $S\beta$ than $S\alpha$, or the calcium ion concentration that has an enhancing effect on IICR might be higher in the vicinity of $S\alpha$ than $S\beta$. These and other possibilities need further exploration.

References

1. Fay FS, Shleven HH, Granger WC, Taylor SR (1979) Aequorin luminescence during activation of single smooth muscle cells. Nature 280: 506–508
2. Fleischer S, Ogunbunmi EM, Dixon MC, Fleer EAM (1985) Localization of Ca^{2+} release channels with ryanodine in junctional terminal cisternae of sarcoplasmic reticulum of fast skeletal muscle. Proc Natl Acad Sci USA 82: 7256–7259
3. Iino M (1987) Calcium dependent inositol trisphosphate-induced calcium release in the guinea-pig taenia caeci. Biochem Biophys Res Commun 142: 47–52
4. Endo M, Iino M (1988) Measurement of Ca^{2+} release in skinned fibers from skeletal muscle. Methods Enzymol 157: 12–26
5. van Breemen C, Hwang OK, Meisheri KD (1981) The mechanism of inhibitory action of diltiazem on vascular smooth muscle contractility. J Pharmacol Exp Ther 218: 459–463
6. Endo M (1985) Calcium release from sarcoplasmic reticulum. Curr Top Membr Transp 25: 181–230
7. Iino M, Kobayashi T, Endo M (1988) Use of ryanodine for functional removal of the calcium store in smooth muscle cells of the guinea-pig. Biochem Biophys Res Commun 152: 417–422

Sarcoplasmic Reticulum of Mesenteric Arteries Buffers Stimulated Ca^{2+} Entry

KOOICHI SAIDA[1,2], ZELIN SHENG[1], and CORNELIS VAN BREEMEN[1]

Summary. The effect of sarcoplasmic reticulum (SR) Ca^{2+} depletion on the efficacy of Ca^{2+} entry to cause contraction was investigated by sequentially exposing small mesenteric arteries to Ca^{2+}-free experimental solutions containing EGTA with or without agonist. The agonist-mediated Ca^{2+} depletion of the SR markedly reduced the rate and magnitude of the contraction due to Ca^{2+} influx through voltage-gated channels. This buffering of Ca^{2+} influx by the norepinephrine-sensitive SR fits in with a two compartment model for the SR in mesenteric arteries in which norepinephrine, through the intermediate action of inositol-1,4,5-trisphosphate, releases Ca^{2+} from the superficial SR, which can be replenished from the deeper SR. This functional model was deduced from earlier observations in small mesenteric arteries that norepinephrine in Ca^{2+}-free experimental solutions activated a biphasic contraction with an amplitude of half the magnitude of the rapid monophasic contraction induced by caffeine. It is concluded that norephinephrine potentiates the contractile effect of Ca^{2+} entry in mesenteric arteries by shunting out the superficial buffer barrier.

Key words: Ca^{2+} entry—Sarcoplasmic reticulum—Ca^{2+} buffering—Mesenteric artery

Introduction

The intracellular Ca^{2+} concentration in vascular smooth muscle is one of the most important factors for the regulation of vascular tone and blood pressure. Control of the intracellular Ca^{2+} concentration is accomplished mainly by Ca^{2+} channels and MgATP-dependent Ca^{2+} pumps in both the plasmalemma and sarcoplasmic reticulum (SR). In essential hypertension total peripheral resistance is abnormally elevated [1]. This is thought to be due to mishandling of the intracellular Ca^{2+} concentration in vascular smooth muscle. In spontaneously hypertensive rats, the most widely studied model of human hypertension, Sugiyama et al. [2] demonstrated a significant increase in the intracellular Ca^{2+} concentration of vascular smooth muscle as compared with normotensive rats. This increase appears to result from an increase in the Ca^{2+} entry through the Ca^{2+} channels as shown in our previous study [3]. In this paper, therefore, we will consider the role of SR in buffering this Ca^{2+} entry in mesenteric arteries.

[1]Department of Pharmacology, University of Miami School of Medicine, Miami, Florida 33101, USA
[2]Present address: Department of Pharmacology, Upjohn Research Laboratories, 23 Wadai, Tsukuba 300–42, Japan

Ca^{2+} Buffering by SR

The SR in smooth muscle is a membranous tubular system that has components closely underlying the surface membrane as well as deeper portions that are contiguous with the rough endoplasmic reticulum and the nuclear envelope [4]. These structures are present in sufficient abundance to serve as intracellular sources and sinks for Ca^{2+} [5]. The SR takes up Ca^{2+} to induce at least part of relaxation, and it releases Ca^{2+} during the initial phase of contraction in smooth muscle. In addition to these functions it has been proposed that the SR also acts as a superficial buffer barrier to Ca^{2+} entry from the extracellular space into the myoplasm [6]. This hypothesis has subsequently been confirmed by the following experimental observations:

1. As much as 200 μmoles of Ca^{2+} may be transferred from the extracellular space to the myoplasm within minutes without causing contraction [7].
2. The magnitude of contraction depends on the rate of net cellular Ca^{2+} gain rather than on the magnitude of the gain [6].
3. Any Ca^{2+} released from the SR is always accompanied by stimulation of Ca^{2+} efflux into the extracellular space [8].
4. Tension development derived from activation of voltage-gated Ca^{2+} channels depends on the state of loading of the SR. A depleted SR will absorb all the stimulated Ca^{2+} entry until it is replete to near its physiological resting level [9].
5. Tonic aortic contractile tension depends on Ca^{2+} influx in a threshold-type fashion, and the Ca^{2+} influx threshold for tension development is decreased by opening SR Ca^{2+} channels and increased by stimulating the Ca^{2+}-ATPase which is responsible for Ca^{2+} accumulation [10].
6. SR Ca^{2+} depletion causes temporal dissociation of Ca^{2+}-induced luminescence of aequorin from tension and phosphorylation of the myosin light chains when Ca^{2+} influx is stimulated [11].
7. Ca^{2+}-sensitive K^+ channels of smooth muscle plasmalemma may be spontaneously activated without contractile activity being apparent [12].
8. Ryanodine enhances tonic depolarization-induced tension without affecting Ca^{2+} influx [13].
9. The average intracellular Ca^{2+} concentration for a certain level of tone is higher in a high-potassium-depolarized artery than for the same contracted by norepinephrine [14].

These data strongly suggest that (a) the superficial SR removes Ca^{2+} to reduce Ca^{2+} entry before it can activate myofilaments, (b) the state of the SR with respect to its permeability to Ca^{2+} and its rate of Ca^{2+}-ATPase activity partly determines the steady-state cellular Ca^{2+} concentration, and (c) that a variable Ca^{2+} gradient exists near the inner plasmalemmal surface with Ca^{2+} activity increasing nearer that surface. Figure 1 provides a plausible mechanism for the above. At rest or depolarization, which does not stimulate phospholipase C, part of the entering Ca^{2+} is taken up into the SR by way of its Ca^{2+}-ATPase. It is then preferentially released toward the plasmalemma to be extruded into the extracellular space. On the other hand, receptor activation by agonists shunts

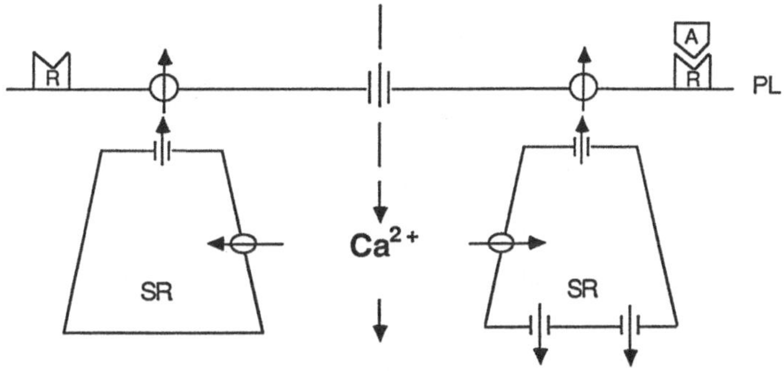

Fig. 1. Sarcoplasmic reticulum (*SR*) functioning as a superficial buffer barrier. At rest or depolarization, which does not stimulate phospholipase C, part of the entering Ca^{2+} through the leak and voltage-gated channels is taken up into the SR. It leaves the SR preferentially at the junctional surface, possibly due to backgroun IP_3 formation or Ca^{2+}-induced Ca^{2+} release. From there it is transported into the extracellular space by way of the plasmalemma (*PL*) Ca^{2+}-ATPase. Receptor (*R*) activation by agonist (*A*) shunts out this superficial Ca^{2+} cycle by the IP_3-mediated permeability increase of the SR membrane. Reproduced from [26]

out this superficial Ca^{2+} cycle by the inositol-1,4,5-trisphosphate (IP_3)-mediated permeability increase of the SR membrane.

Heterogeneity of Smooth Muscle SR

The volume of smooth muscle SR varies from 2%–7% and is related to the muscle capability to contract in Ca^{2+}-free experimental solution [4, 5]. In arterial smooth muscle the relative contribution made by the SR to Ca^{2+} delivery during agonist-induced stimulation decreases as the size of the lumen becomes smaller [15].

Smooth muscle SR variability also extends to its functional heterogeneity. The Ca^{2+} release from the SR can be indirectly observed as a stimulation of ^{45}Ca efflux by the agonists [16]. In the rabbit aorta, caffeine and norepinephrine stimulated ^{45}Ca efflux in a similar manner [17]. In the mesenteric artery, however, the caffeine-stimulated ^{45}Ca efflux is faster and bigger than the norepinephrine-stimulated ^{45}Ca efflux [8]. A similar difference between caffeine- and norephinephrine-induced contraction was observed in a Ca^{2+}-free experimental solution [18]. The loss and recovery of the caffeine-induced contraction upon Ca^{2+} removal and replenishment is slower than that of the rapid norephinephrine-induced contraction [18].

The development of the method of skinning smooth muscle cells with saponin has made it possible to study directly the Ca^{2+} transport mechanism of in situ SR [19]. Using saponin-treated skinned smooth muscle two mechanisms of Ca^{2+} release were clarified in the SR: Ca^{2+}-induced Ca^{2+} release [20, 21], and IP_3-induced Ca^{2+} release [22]. It is well-known that caffeine releases Ca^{2+} from the SR by facilitating the Ca^{2+}-induced Ca^{2+} release mechanism [23].

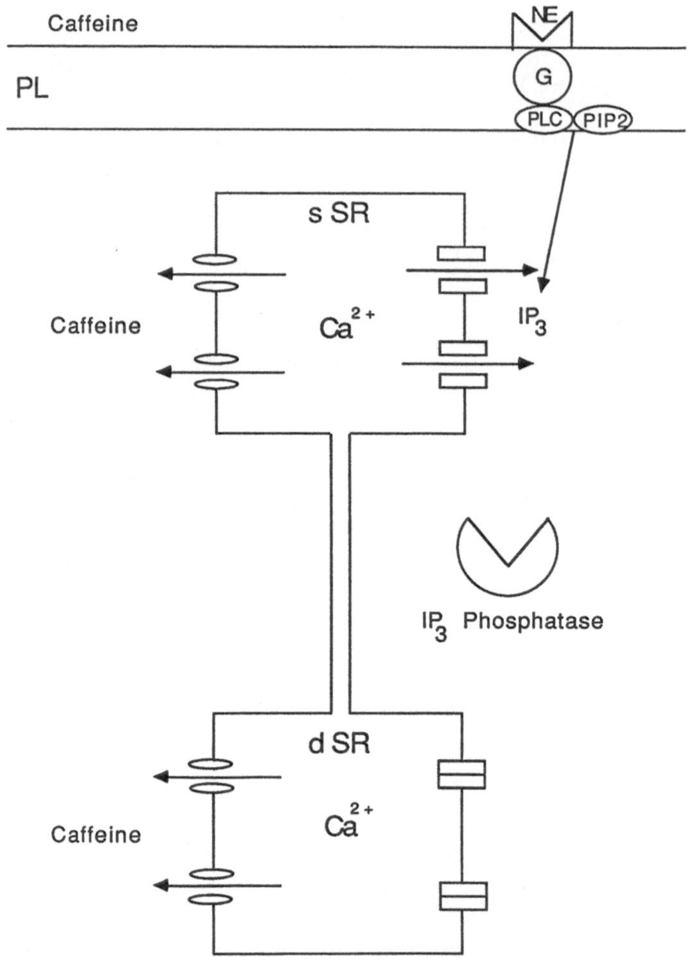

Fig. 2. A two compartment model for the sarcoplasmic reticulum (SR) of mesenteric artery. IP_3 generated by norepinephrine (*NE*) activation of the α-adrenergic receptor opens Ca^{2+} channels in the superficial SR (*sSR*) but it is hydrolyzed by its specific phosphatase before it reaches the deep SR (*dSR*). After release from the sSR Ca^{2+} is replenished from the dSR as well as from the extra-cellular space. Caffeine causes a rapid release of Ca^{2+} from both SR compartments. *Pl*, plasmalemma; *PLC*, phospholipase C; PIP_2, phosphatidylinositol 4,5-bisphosphate; *G*, GTP-binding protein. Reproduced from [26]

In the skinned mesenteric artery the caffeine-induced Ca^{2+} release from the SR is faster and bigger than the IP_3-induced Ca^{2+} release [24]. The IP_3-induced Ca^{2+} release requires GTP, whereas the caffeine-induced Ca^{2+} release does not [24]. The caffeine-induced Ca^{2+} release is insensitive to pertussis toxin and is more readily blocked by procaine than the IP_3-induced Ca^{2+} release [25]. Thus, it is clear that smooth muscle SR contains two separate types of Ca^{2+} release channels, namely, IP_3- and Ca^{2+} (caffeine)-activated ones.

The above data may be best explained by a functional model as depicted in Fig. 2. It should be noted that this model was deduced from the results observed in the rabbit mesenteric artery.

Concluding Remarks

In this paper, we briefly discussed how the SR functions as a superficial buffer barrier to Ca^{2+} entry from the extracellular space into the myoplasm in the mesenteric artery. We emphasize the importance of this Ca^{2+} cycle in the regulation of vascular tone and blood pressure.

References

1. Aoki K (1986) Calcium membrane theory of essential hypertension. In: Aoki K (ed) Essential hypertension. Springer, Tokyo Berlin Heidelberg New York, pp 223–242
2. Sugiyama T, Yoshizumi M, Takaku F, Urabe H, Tsukakoshi M, Kasuya T, Yazaki Y (1986) The elevation of the cytoplasmic calcium ions in vascular smooth muscle cells in SHR—measurement of the free calcium ions in single living cells by laser microfluorospectrometry. Biochem Biophys Res Commun 141: 340–345
3. Cauvin C, van Breemen C (1986) Membrane Ca^{2+} permeability and calcium antagonistic effects in resistance vessels of spontaneously hypertensive rats. In: Aoki K (ed) Essential hypertension. Springer, Tokyo Berlin Heidelberg New York, pp 27–33
4. Somlyo AP (1985) Excitation-contraction coupling and the ultrastructure of smooth muscle. Circ Res 57: 497–507
5. Devine CE, Somlyo AV, Somlyo AP (1972) Sarcoplasmic reticulum and excitation-contraction coupling in mammalian smooth muscles. J Cell Biol 52: 690–718
6. van Breemen C (1977) Ca^{2+} requirement for activation of intact aortic smooth muscle. J Physiol 272: 317–329
7. Aaronson P, van Breemen C (1981) Effects of Na gradient manipulation upon cellular Ca, ^{45}Ca fluxes and cellular Na in guinea pig taenia coli. J Physiol 319: 443–461
8. Leijten PAA, van Breemen C (1986) The relationship between noradrenaline-induced contraction and ^{45}Ca efflux stimulation in rabbit mesenteric artery. Br J Pharmacol 89: 739–749
9. Loutzenhiser R, van Breemen C (1983) The influence of receptor occupation on Ca^{2+} influx mediated vascular smooth muscle contraction. Circ Res 52: 97–103
10. van Breemen C, Leyten P, Yamamoto H, Aaronson P, Cauvin C (1986) Ca^{2+} activation of vascular smooth muscle. Hypertension 8: 1189–1195
11. Rembold C, Murphy RH (1988) Histamine induces mobilization of myoplasmic Ca from extracellular and intracellular stores in vascular smooth muscle. Biophys J 53: 594A
12. Benham CD, Bolton TB (1986) Spontaneous outward currents in single visceral and vascular smooth muscle cells of rabbit. J Physiol 381: 385–406
13. Hwang K, van Breemen C (1987) Ryanodine modulation of ^{45}Ca efflux and tension in rabbit aortic smooth muscle. Pflugers Arch Eur J Physiol 408: 343–350
14. Morgan JP, Morgan KG (1984) Stimulus specific intracellular calcium levels in smooth muscle of ferret portal vein. J Physiol 351: 155–167
15. Cauvin C, Loutzenhiser R, van Breemen C (1983) Mechanism of calcium antagonist-induced vasodilation. Ann Rev Pharmacol Toxicol 23: 373–396
16. van Breemen C, Deth R (1976) La^{3+} and excitation-contraction coupling in vascular smooth muscle. In: Betz E (ed) Symposium on the role of ions in transmission of

signals from tissue and blood to the vascular smooth muscle cells. Karger, Basel, pp 26–31
17. Leijten PAA, van Breemen C (1986) The relationship between noradrenaline-sensitive calcium store in rabbit aorta. J Physiol 357: 327–339
18. Saida K, van Breemen C (1984) Characteristics of the norepinephrine-sensitive calcium store in vascular smooth muscle. Blood vessels 21: 43–52
19. Saida K, Nonomura Y (1978) Characteristics of Ca^{2+}- and Mg^{2+}-induced tension development in chemically skinned smooth muscle fibers. J Gen Physiol 72: 1–14
20. Saida K (1981) Ca^{2+}- and "depolarization"-induced Ca^{2+} release in skinned smooth muscle fibers. Biomedical Res 2: 453–455
21. Saida K (1982) Intracellular Ca release in skinned smooth muscle. J Gen Physiol 80: 191–202
22. Somlyo AV, Bond M, Somlyo AP, Scarpa A (1985) Inositol trisphosphate-induced calcium release and contraction in vascular smooth muscle. Proc Natl Acad Sci USA 82: 5231–5235
23. Endo M (1977) Calcium release from the sarcoplasmic reticulum. Physiol Rev 57: 71–108
24. Saida K, van Breemen C (1987) GTP requirement for inositol-1,4,5-trisphosphate-induced Ca^{2+} release from sarcoplasmic reticulum in smooth muscle. Biochem Biophys Res Commun 144: 1313–1316
25. Saida K, Twort C, van Breemen C (1988) The specific GTP requirement for inositol-1,4,5-trisphosphate-induced Ca^{2+} release from skinned vascular smooth muscle. J Cardiovasc Pharmacol 12: S47–50
26. van Breemen C, Saida K (1989) Cellular mechaisms regulating $[Ca^{2+}]_i$ in smooth muscle. Ann Rev Physiol 51: 315–329

Contraction Induced by Ouabain and Potassium-free Solution in Human Umbilical Arteries

KOICHI SATO and KYUZO AOKI[1]

Summary. Ouabain-induced contraction is either of the myogenic or the neurogenic type depending on the variety of artery. In the present study, mechanisms of the ouabain and potassium-free solution induced contraction were investigated in the human umbilical arteries. The umbilical arteries were obtained from full-term, uncomplicated deliveries of normal infants. The arteries were cut helically into strips, which were mounted in the modified Krebs-bicarbonate solution. Ouabain (3×10^{-7} to $10^{-4} M$) induced concentration-dependent biphasic contractions which consisted of an early contraction—a transient rapid contraction—and a late contraction—a slow continuous contraction. Potassium-free solution also induced biphasic contractions. A calcium antagonist, nifedipine (5×10^{-9} and $1 \times 10^{-7} M$), inhibited the early contraction but did not inhibit the late contraction. In addition, Ca^{2+}-free solution containing 0.2 mM EGTA completely inhibited the early contraction but did not completely inhibit the late contraction. Neither the early nor late contractions altered phenoxybenzamine ($2 \times 10^{-6} M$). In conclusion, ouabain and potassium-free solution induced early contraction in umbilical arteries, which could have been caused by facilitation of Ca^{2+} entry into the vascular smooth muscle through voltage-sensitive Ca^{2+} channels.

Key words: Ouabain—Calcium channels—Sodium pump—Myogenic contraction—Biphasic contraction—Early contraction—Late contraction

Introduction

It is well known that the sodium-pump inhibitor, ouabain, potassium-free solution, and inhibition of the pump by cooling to 18°C induce contraction in a variety of arteries. The contraction is either of the myogenic or neurogenic type depending on the variety of artery. The myogenic mechanism for the contraction of the vascular smooth muscle has been shown in bovine facial vein [1], dog skeletal muscle [2, 3] and coronary vessels [4], dog cerebral artery [5], and guinea pig and rat aortas [6]; whereas the neurogenic mechanism (via norepinephrine release) has been demonstrated in dog mesenteric and rat tail arteries [7]. On the other hand, Hayashi and Park [8] have described an involvement of myogenic and

[1] Second Department of Internal Medicine, Nagoya City University Medical School, Mizuho-ku, Nagoya 467, Japan

neurogenic mechanisms for the contractile responses to both lowered potassium concentration and ouabain in the dog mesenteric arteries. The reduction in extracellular potassium causes an early contraction due to norepinephrine release from adrenergic nerves and a late contraction that appears to be of myogenic origin [8].

Ouabain might enhance both neurogenic and myogenic derived contractions in response to reduced extracellular potassium by exaggerating the inhibition of the pump for sodium-potassium ion exchange in membranes of adrenergic nerves and vascular smooth muscle. Marin et al. [9] stated that ouabain-induced contraction of cat cerebral arteries was due to a direct effect on vascular smooth muscle cells, while in femoral arteries it was due to norephinephrine release from adrenergic nerve terminals. In addition, nifedipine inhibited an increase in human forearm vascular resistance by ouabain infusion [10], and verapamil inhibited the potassium-induced contraction in human umbilical arteries [11].

Human umbilical arteries were devoid of sympathetic innervation, and thus it might be advantageous to study the direct action of drugs on vascular smooth muscle cells [12, 13]. In this study, the ouabain and potassium-free solution caused biphasic contractions, and the early contraction was nifedipine-sensitive in human umbilical arteries, which has never been described. It seems likely that Ca^{2+} available for the contraction induced by ouabain and potassium-free solution originates from (1) an extracellular Ca^{2+} pool, (2) a cell membrane-bound source, and (3) an intracellular source [14]. Therefore, the present experiments were designed to study the direct effects of nifedipine and Ca^{2+}-free solution containing 0.2 m*M* EGTA on the contraction induced by ouabain and potassium-free solution in isolated human umbilical arteries.

Materials and Methods

A total of 40 human umbilical cords from normal, full-term, spontaneous deliveries were utilized for this study. The human umbilical cords from mothers exhibiting hypertension, eclampsia, or other overt diseases were not included. The cords obtained by Caesarean sections were not utilized. Cords from mothers on medication were also excluded. In the delivery room, immediately after clamping, 10–15 cm segments were cut from the cords midway between the placentas and infants. The cords were nerve-free [12, 13]. The blood was allowed to drain, and the segment was placed in modified Krebs-bicarbonate solution at 4°C. The specimens were usually placed in the solution between 0 and 12 h after delivery. After removal of surrounding Wharton's jelly, the umbilical arteries were cut helically into strips 2 mm in width and 8 mm in length. The strips were mounted vertically between hooks in a water-jacketed muscle bath containing 20 ml of the modified Krebs-bicarbonate solution at 37°C bubbled with a mixture of 95% 0_2 and 5% CO_2. The upper end of the strips were connected to the lever of a force-displacement transducer (TB-612T, Nihon Koden Kogyo Co., Tokyo) for recording tension. The strips were stretched 200% of the original length. After the strips were allowed to equilibrate in the Krebs solution for approximately 4–5 h, the strips were adjusted for resting tension of about 2–3 g. This tension was maintained throughout the experiments.

Before the start of the experiments, the strips were allowed to equilibrate; and during this period, the solutions were replaced every 15 min. High-potassium (60 mM) solution was added to induce contraction which was considered as a reference response to 100%. Then, the high-potassium solution was replaced with Krebs solution and the resultant tension became the basal level. The dose response curves for the ouabain-induced contraction were determined by a single dose stimulation. Potassium-free solution-induced contraction was also studied. Nifedipine (5×10^{-9} and $1 \times 10^{-7} M$) or phenoxybenzamine ($2 \times 10^{-6} M$) were added to the bath 30 min before the administration of ouabain and potassium-free solution. In order to analyze the influence of extracellular Ca^{2+} on the contractile response elicited with ouabain and potassium-free solution, the arterial strips were exposed to Ca^{2+}-free solution containing 0.2 mM EGTA for 10 min before the ouabain and potassium-free solution.

The composition of the modified Krebs-bicarbonate solution (pH 7.2–7.4) was (mM): 115.0 NaCl, 4.7 KCl, 1.2 MgCl, 25.0 $NaHCO_3$, 1.2 KH_2PO_4, and 10.0 dextrose. Potassium-depolarizing solution (60 mM KCl) was made by equimolar substitution of KCl for NaCl. Potassium-free solution was made by omitting KCl and substituting NaH_2PO_4 for KH_2PO_4. Ca^{2+}-free solution was made by omitting $CaCl_2$ in the Krebs solution and adding 0.2 mM ethylene glycol bis- (beta-aminoethyl ether) N, N, N′, N′-tetraacetic acid (EGTA). Ouabain was dissolved in distilled water on the day of experiment. Stock solution of phenoxybenzamine ($2 \times 10^{-3} M$) was made in a saline (0.9% NaCl)-ascorbic acid (0.01%) solution with distilled water and 99.5% ethanol and kept at −20°C. Nifedipine was dissolved in ethanol to make a stock of 10^{-3} M and kept at −20°C and protected from light. Aliquots of these solutions were diluted before use with the distilled water to obtain the desired dose. The final dose of ethanol was not above 0.01% [15].

The drugs used were: ouabain octahydrate (Sigma), phenoxybenzamine HCl (Nakarai Chemical, Kyoto), and nifedipine (Bayer AG).

Results were expressed as mean ± SD. Statistical significance was assumed when $P < 0.05$. The Student's t-test was used.

Results

Ouabain induced biphasic contraction of the human umbilical arteries from 3×10^{-7} to 10^{-4} M. The contraction consisted of a transient rapid response as the early contraction and a slow continuous response as the late contraction in the umbilical arteries (Fig. 1). Both contractions were dose-dependent (Tables 1, 2).

The early contraction peaked rapidly; time to peak contraction was 10 ± 7 min for 10^{-4} M ouabain ($n = 8$). The magnitude of the early contraction was related to the dose of ouabain (Table 1). The early contractions were markedly inhibited by pretreatment with nifedipine (5×10^{-9} M) (Fig. 1, Table 1). The early contractions were not elicited in Ca^{2+}-free 0.2 mM EGTA solution (Fig. 2).

Table 1. The magnitude of the early contraction induced by ouabain ($10^{-8}-10^{-4}$ M) with or without nifedipine (5×10^{-9} M). The magnitude of ouabain displacement relative to 60 mM KCl contraction. Mean ± SD; n = 4–8

dose of ouabain (M)	control (%)	nifedipine (%)
10^{-8}	0	0
3×10^{-8}	0	0
10^{-7}	0	0
3×10^{-7}	16.8 ± 11.6	3.6 ± 2.9
10^{-6}	50.8 ± 13.9	5.6 ± 4.6
3×10^{-6}	91.4 ± 20.5	10.9 ± 8.2
10^{-5}	98.2 ± 19.7	10.1 ± 9.0
10^{-4}	97.2 ± 9.4	6.0 ± 4.5

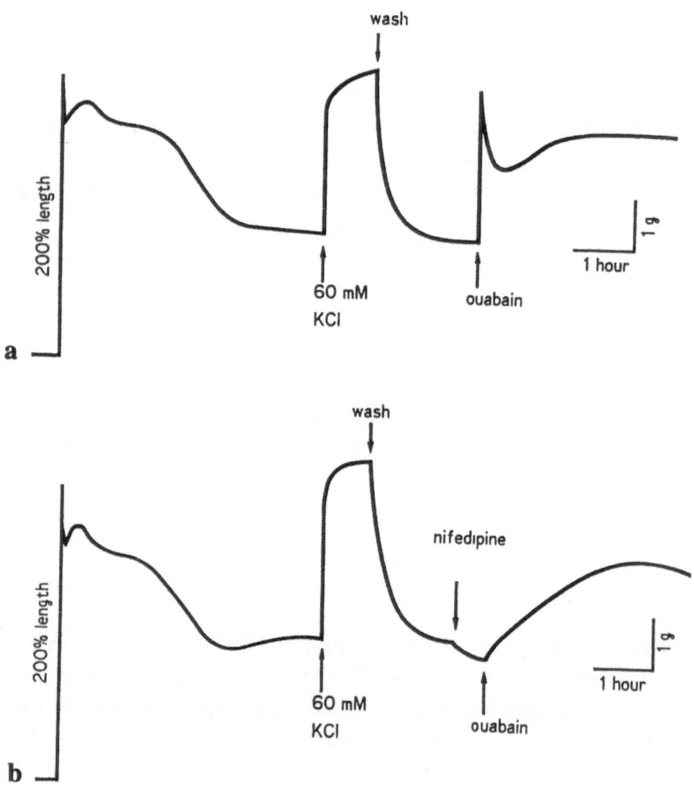

Fig. 1a,b. Contractile response of strips of human umbilical artery to ouabain ($10^{-5}M$) in the absence **a** and presence **b** of nifedipine ($5\times10^{-9}M$). Ouabain induced biphasic contractions which consisted of an early contraction as a transient rapid response and a late contraction as a slow, continuous response. Nifedipine inhibited ouabain-induced early contraction markedly

Table 2. The magnitude of the late contraction induced by ouabain ($10^{-8} - 10^{-4}$ *M*) with or without nifedipine (5×10^{-9} *M*). The magnitude of ouabain displacement relative to 60 m*M* KCl contraction. Mean ± SD; n = 4–8

dose of ouabain (M)	control (%)	nifedipine (%)
10^{-8}	0	0
3×10^{-8}	32.2 ± 15.4	39.7 ± 7.3
10^{-7}	65.9 ± 5.6	80.6 ± 7.5
3×10^{-7}	72.4 ± 6.7	88.0 ± 8.6
10^{-6}	80.2 ± 7.4	91.5 ± 8.2
3×10^{-6}	83.3 ± 9.2	93.6 ± 13.6
10^{-5}	98.2 ± 19.7	96.7 ± 8.6
10^{-4}	97.2 ± 9.4	102.9 ± 9.3

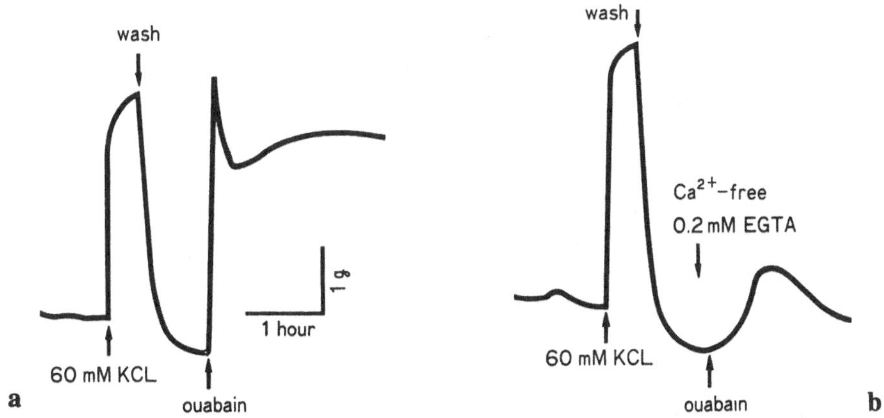

Fig. 2a,b. Contractile response to ouabain (10^{-4}*M*) in the absence **a** and presence **b** of Ca^{2+}-free solution. Ca^{2+}-free solution completely inhibited the early contraction, but not the late contraction

The late contraction reached maximum slowly, plateaued, and persisted over 4 h. The magnitude of the late contraction was related to the dose of ouabain (Table 2). The contractions were not inhibited by pretreatment with nifedipine (Fig. 1, Table 2). Yet, these contractions were not completely inhibited by pretreatment with Ca^{2+}-free solution containing 0.2 m*M* EGTA (Fig. 2). The biphasic contractions elicited with ouabain were not altered by phenoxybenzamine (2×10^{-6}*M*).

Potassium-free solution induced biphasic contractions which consisted of a transient rapid response (the early contraction) and a slow, continuous response (the late contraction) in the human umbilical arteries as with ouabain (Fig. 3). The magnitude of the early contraction was 65.6 ± 27.8% ($n = 6$) of to 60 m*M* potassium-induced contraction. The early contractions were markedly inhibited by the pretreatment with nifedipine (5×10^{-9}*M*) from 65.6 ± 27.8 to 6.3 ± 2.4%

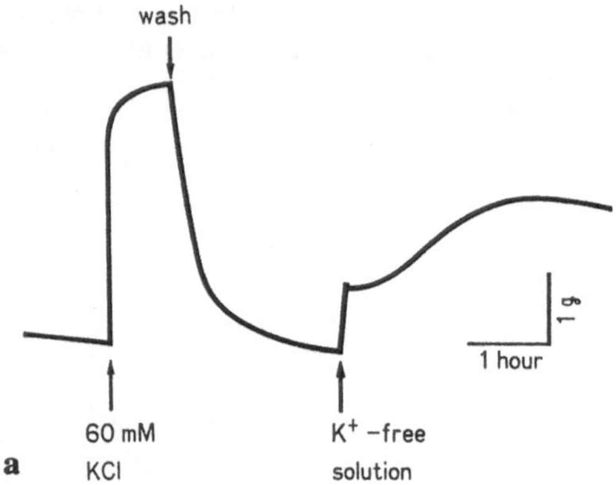

a

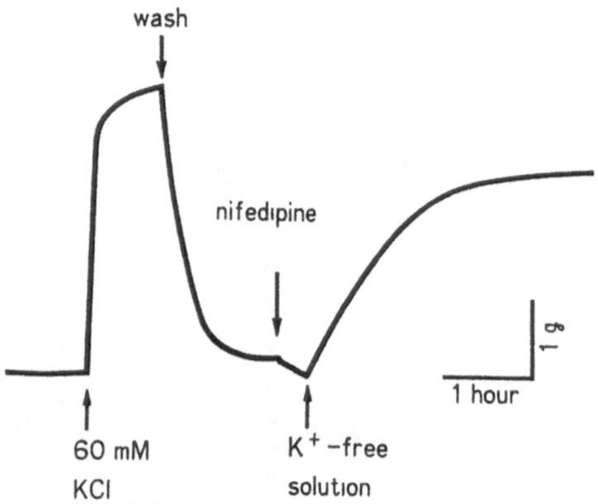

b

Fig. 3a,b. Contractile response to K^+-free solution in the absence **a** and presence **b** of nifedipine ($5 \times 10^{-9} M$). Nifedipine inhibited the early contraction markedly, but did not inhibit the late contraction

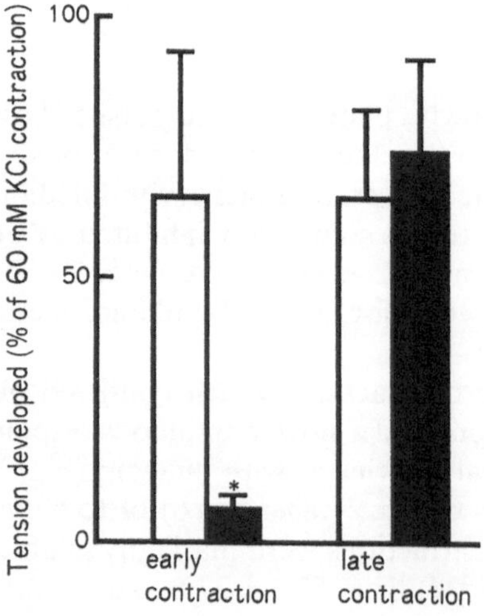

Fig. 4. Tension development of K^+-free solution in the early and the late contraction with (■) or without (□) nifedipine ($5 \times 10^{-9} M$). $P < 0.01$; $n = 6$

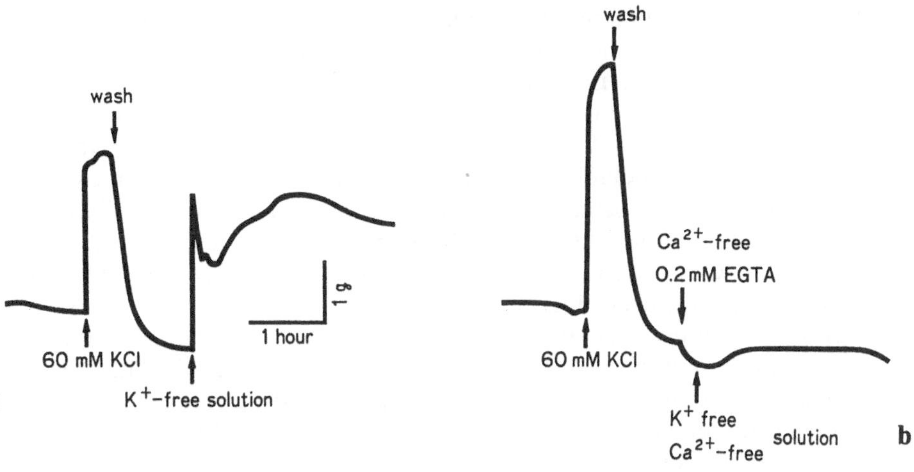

Fig 5a,b. Contractile response to K^+-free solution in the absence **a** and presence **b** of Ca^{2+}-free solution. Ca^{2+}-free solution completely inhibited the early and late contractions

(Fig. 4). The early contractions were not elicited in Ca^{2+}-free solution containing 0.2 m*M* EGTA (Fig. 5). The late contraction reached maximum slowly and formed a plateau which persisted over 4 h. The magnitude of the contraction was $65.3 \pm 16.7\%$ ($n = 6$) of 60 m*M* postassium-induced contraction. The late contractions were not inhibited by pretreatment with nifedipine ($5 \times 10^{-9} M$) from $65.3 \pm 16.7\%$ to $74.0 \pm 18.1\%$ (Fig. 4). However, these contractions were inhibited by pretreatment with Ca^{2+}-free solution containing 0.2 m*M* EGTA (Fig. 5). These biphasic contractions elicited with potassium-free solution were not altered by phenoxybenzamine ($2 \times 10^{-6} M$).

Discussion

It is not clear how increased intracellular sodium concentraction which is induced by sodium-pump inhibitor, ouabain, or potassium-free solution leads to contraction of vascular smooth muscle. There are at least two mechanisms that have been proposed to explain vascular contraction subsequent to sodium-pump inhibition. First, intracellular sodium concentration may increase after sodium-pump inhibition and cause a decrease in calcium extrusion via sodium-calcium exchange. Second, since the sodium pump is electrogenic, inhibition of the pump may cause depolarization and calcium influx through voltage-sensitive calcium channels [16, 17].

First, we demonstrated that in the human unbilical arterial strips, ouabain and potassium-free solution induced contractions consisted of the early and the late contractions; and the former was inhibited by a voltage-sensitive calcium channel blocker, nifedipine, and Ca^{2+}-free solution containing 0.2 m*M* EGTA, but the latter was not inhibited by nifedipine and not inhibited completely by Ca^{2+}-free solution containing 0.2 m*M* EGTA. Therefore, the early contraction of the human umbilical arteries was solely due to facilitated Ca^{2+} entry through voltage-sensitive calcium channels. In human umbilical arteries, sodium pump inhibition which is due to ouabain and potassium-free solution at the early contraction causes depolarization, open voltage-sensitive calcium channels, and Ca^{2+} entry through intracellular space and vascular smooth muscle contraction. The late contractions were not inhibited by nifedipine and were not completely inhibited by Ca^{2+}-free 0.2 m*M* EGTA solution. Therefore, the late contractions were induced by intracellular Ca^{2+} release.

It is a new observation that in human umbilical arteries, ouabain and potassium-free solution induced contraction has a biphasic nature. There is evidence that ouabain-induced contractions of central and peripheral arteries are due to enhanced influx of extracellular Ca^{2+}. For instance, Marin et al. [9] reported that ouabain-induced transient contraction of the cat cerebral arteries was reduced by a calcium antagonist (verapamil) and a Ca^{2+}-free medium. Ozaki et al. [18] suggested that the contraction induced by ouabain was mainly due to increased Ca^{2+} influx through voltage-sensitive Ca^{2+} channels. They did not report that the ouabain-induced contraction consisted of two components. However, Hayashi and Park [8] demonstrated that the contractions induced by ouabain in dog mesenteric arterial strips consisted of the following two portions. The first portion was neurogenic, since it was inhibited by phentolamine. The second portion was myogenic, since it was not inhibited by phentolamine, atropine, chlorpheniramine, cimetidine, saralasin, aspirin, or methysergide. Ouabain- and potassium-free solution-induced contraction is myogenic in human umbilical arteries, since the arteries in this area are nerve-free, and it is not inhibited by phenoxybenzamine. Furthermore, the early contraction (as a part of the biphasic contraction) was inhibited by calcium influx blocker and by Ca^{2+}-free solution containing 0.2 m*M* EGTA [19]. In conclusion, it is a new observation that the myogenic contraction in the human umbilical arteries is biphasic, and the early contraction is dependent on entry of extracellular Ca^{2+} [19].

References

1. Konold P, Gebert G, Brecht K (1968) The effect of potassium on the tone of isolated arteries. Pflugers Arch 301: 285–291
2. Anderson DK, Roth SA, Brace RA, Radawski D, Haddy FJ, Scott JB (1972) Effect of hypokalemia and hypomagnesemia produced by hemodialysis on vascular resistance in canine skeletal muscle: Role of potassium in active hyperemia. Circ Res 31: 165–173
3. Chen WT, Brace RA, Scott JB, Anderson DK, Haddy FJ (1972) The mechanism of the vasodilator action on potassium. Proc Soc Exp Biol Med 140: 820–824
4. Brace RA, Anderson DK, Chen WT, Scott JB, Haddy FJ (1974) Local effects of hypokalemia on coronary resistance and myocardial contractile force. Am J Physiol 227: 590–597
5. Toda N (1978) Mechanical response of isolated dog cerebral arteries to reduction of external K, Na and Cl. Am J Physiol 234: H404–411

6. Karaki H, Ozaki H, Urakawa N (1978) Effects of ouabain and potassium-free solution on the contraction of isolated blood vessels. Eur J Pharmacol 48: 439–443
7. Bonaccorsi A, Hermsmeyer K, Smith CB, Bohr DF (1977) Norepinephrine release in isolated arteries induced by K-free solution. Am J Physiol 232: H140–145
8. Hayashi S, Park MK (1984) Neurogenic and myogenic contractile responses of dog mesenteric arteries to reduced K^+ concentration and their interactions with ouabain. J Pharmacol Exp Ther 230: 527–533
9. Marin J, Sanchez-Ferrer CF, Salaices M (1988) Effects of ouabain on isolated cerebal and femoral arteries of the cat: a functional and biochemical study. Br J Pharmacol 93: 43–52
10. Schulte K-L, van Gemmeren D, Thiede H-M, Meyer-Sabellek W, Gotzen R, Distler A (1987) Ouabain-induced elevation in forearm vascular resistance, calcium entry and alpha-adrenoceptor blockade, and release and removal of noradrenaline. J Hypertens 5 (Suppl V): V-215–218
11. Ozaki H, Shibata S, Kitano H, Matsumoto P, Ishida Y (1981) A comparative study of relaxing effect of nitroprusside and verapamil on human umbilical vessels. Blood Vessels 18: 321–329
12. Davignon J, Lorenz RR, Shepherd JT (1965) Response of human umbilical artery to changes in transmural pressure. Am J Physiol 209: 51–59
13. Spivack M (1943) On the presence or absence of nerves in umbilical blood vessels of man and guinea pig. Anat Rec 85: 85–109
14. Cauvin C, Loutzenhiser R, van Breemen C (1983) Mechanism of calcium antagonist-induced vasodilation. Annu Rev Pharmacol Toxicol 23: 373–396
15. Yang H-Y, Shum Y-C, Ng H-T, Chen FC (1986) Effects of ethanol on human umbilical artery and vein in vitro. Gynecol Obstet Invest 21: 131–135
16. Mulvany MJ, Aalkajaer C (1985) Vasoconstrictor effects of ouabain: Mechanisms of action. Klin Wochenschr 63 (Suppl III): III-143–146
17. Cathy A, Bruner R, Webb RC, Bohr DF (1988) Vascular reactivity and membrane stabilizing effect of calcium in spontaneously hypertensive rats. In: Aoki K, Frohlich ED (eds) Calcium in essential hypertension. Academic Press, Tokyo, pp 275–306
18. Ozaki H, Kishimoto T, Karaki H, Urakawa N (1982) Effects of the Na ionophore monensin on the contractile response and monovalent cations in the vascular smooth muscle on rabbit aorta. Naunyn Schmiedegergs Arch Pharmacol 321: 140–144
19. Sato K, Aoki K (1989) The biphasic contraction induced by ouabain in human umbilical arteries. Eur J Pharmacol 158: 299–302

Pharmacology of Calcium Antagonists in Arterial Smooth Muscle

TÉOPHILE GODFRAIND, NICOLE MOREL, and MAURICE WIBO[1]

Summary. Interaction of calcium antagonists with calcium channels has been studied in vascular smooth muscle from normotensive and hypertensive (SHR) rats by measuring calcium fluxes, contraction and specific binding of dihydropyridines. Inhibition of KCl-evoked contraction by several dihydropyridines showed a pronounced time dependency, in which the inhibition increased slowly after depolarization to attain a steady-state value. The time course of the development of inhibition by (+)PN 200-110 paralleled the time course of binding to isolated membranes. The effect of the membrane potential on dihydropyridine binding was investigated in intact mesenteric arteries. Only one class of specific binding site was observed. Depolarization increased (+)[^{3}H]PN 200-110 binding by inducing a decrease in K_D without changing B_{max}. K_D or K_i values determined in depolarized arteries were similar to corresponding values measured from membrane preparations, and to IC_{50} values for steady-state inhibition of contractions. Thus, depolarization induces a conformation of calcium channels with enhanced affinity for dihydropyridines. Binding of these drugs to the high-affinity conformer induced by depolarization is responsible for inhibition of calcium influx and contraction. In SHR arteries the state of the calcium channels makes them more easily activated by agonist dihydropyridines, but the properties of the binding site in depolarized arteries appear unchanged.

Key words: Ca antagonists — Ca entry blockers — Arterial smooth muscle — Ca channels—Hypertension

Introduction

Calcium is the ultimate intracellular messenger of the vasoactive signals perceived by smooth muscle pericellular membrane which produce an increase in the active tone. Several transport mechanisms associated with the plasma membrane itself and with intracellular organelles maintain the calcium gradient existing across the plasmalemmal membrane. Vasoactive agents mobilize calcium from extracellular or intracellular pools according to the mechanism of activation of their receptors and to the intrinsic properties of the smooth muscle itself.

[1]Laboratoire de Pharmacodynamie Générale et de Pharmacologie, Université Catholique de Louvain, UCL 7350, Avenue Emmanuel Mounier, 73, B-1200 Brussels, Belgium

Table 1. Calcium modulators—agents affecting Ca^{2+} movements [1]

A. Inhibitors: calcium antagonists

1. Agents acting at the plasma membrane

    a. *Calcium entry blockers*

    Group I: Specific calcium entry blockers

    Subgroup 1: Agents selective for slow calcium channels in myocardium (slow channel blockers)
    - Phenylalkylamines: verapamil, gallopamil (D600); under investigation—anipamil, desmethoxyverapamil (D888), emopamil, falipamil (AQ-A-39), ronipamil
    - Dihydropyridines: nifedipine, nicardipine, niludipine, nimodipine, nisoldipine, nitrendipine, ryosidine; under investigation—amlodipine, azodipine, dazodipine (PY 108-068), felodipine, flordipine, FR 34235, iodipine, isradipine (PN 200-110), mesudipine, ni(l)vadipine, oxodipine, riodipine
    - Benzothiazepines: diltiazem; under investigation—fostedil (KB-944)

    Subgroup 2: Agents inactive on slow calcium channels in myocardium
    - Diphenylpiperazines: cinnarizine and flunarizine

    Group II: Non-specific calcium entry blockers

    Subgroup 1: Agents acting at similar concentrations on calcium channels and fast sodium channels
    - Bencyclane, bepridil, caroverine, etafenone, fendiline, lidoflazine, perhexiline, prenylamine, SKF 525A, terodiline, tiapamil

    Subgroup 2: Agents interacting with calcium channels while having another primary site of action
    - They include, among others: agents acting on sodium channels (local anesthetics, phenytoin); on catecholamine receptors (benextramine, nicergoline, phenoxybenzamine, phenothiazines, pimozide, propranolol, WB-4101, yohimbine derivatives); on benzodiazepine receptors (diazepam, flurazepam); on opiate receptors (loperamide, fluperamide); on cyclic nucleotide phosphodiesterases (amrinone, cromoglycate, papaverine); barbiturates; cyproheptadine; indomethacin; reserpine

    b. *Sodium-calcium exchange inhibitors*
    - Amiloride and derivatives

2. Agents acting within the cell

    a. Acting on sarcoplasmic reticulum
    - Dantrolene, TMB-8

    b. Acting on mitochondria
    - Ruthenium red

    c. Calmodulin antagonists
    - Phenotiazines: trifluoperazine, chlorpromazine
    - Naphthalene derivatives: W7
    - Local anesthetics: dibucaine
    - Dopamine antagonists: pimozide, haloperidol
    - Calmidazolium (R 24571)

(*Table 1 continued on following page*)

Table 1. *(continued)*

B. Facilitators
1. Agents acting at the plasma membrane
a. *Calcium agonists*
• Dihydropyridines: BAY K 8644, CGP 28392, YC 170
2. Agents acting on the sarcoplasmic reticulum
• Inositol 1,4,5-triphosphate
• Caffeine
3. Ionophores
• A 23187, ionomycin

Extracellular calcium is involved in tissue activation through the opening of membrane calcium channels operated by membrane depolarization or by receptors themselves. It is now recognized that some organic agents are able to block calcium entry through such calcium channels [1]. Those agents have been termed calcium antagonists, or calcium entry blockers and calcium channel blockers. The term calcium channel ligand is used not only to designate drugs that block calcium channels' opening but also those that increase the probability of their opening (also named calcium agonists). The more general term, calcium modulators, refers to all agents increasing or decreasing calcium bioavailability. These agents are listed in Table 1, according to their subcellular site of action, which is the most appropriate way to classify them [1].

The purpose of this report is to review the properties of calcium antagonists and calcium agonists in vascular smooth muscle with special consideration for the nature of the stimulus and for the influence of pathology, such as hypertension, in the tissue sensitivity to their action.

Calcium Movements and Contraction

Estimation of unidirectional calcium fluxes in smooth muscle requires appropriate methodologies. In vascular smooth muscle only one-tenth of the total calcium content is intracellular; therefore, when the tissue is bathed in ^{45}Ca-containing physiological solution, changes of its specific activity are obscured by the large fraction of extracellular calcium bound on the external surface of the smooth muscle cells and on connective tissue. Lanthanum allows us to delineate intra- and extracellular pools of calcium because it does not penetrate the cell and displaces calcium from its extracellular binding sites. We have developed a method to measure changes in the specific activity of the lanthanum-resistant calcium fraction in rat aorta [2]. Under resting conditions, the specific activity of the intracellular calcium defined as the lanthanum-resistant calcium fraction increases with the duration of incubation in radioactive solution and reaches a plateau after 1 or 2 h, when inward and outward movements become equal. However, the initial rate of uptake of ^{45}Ca may be taken as reflecting an inward flux. Similarly, when the intracellular fraction is completely loaded, the initial

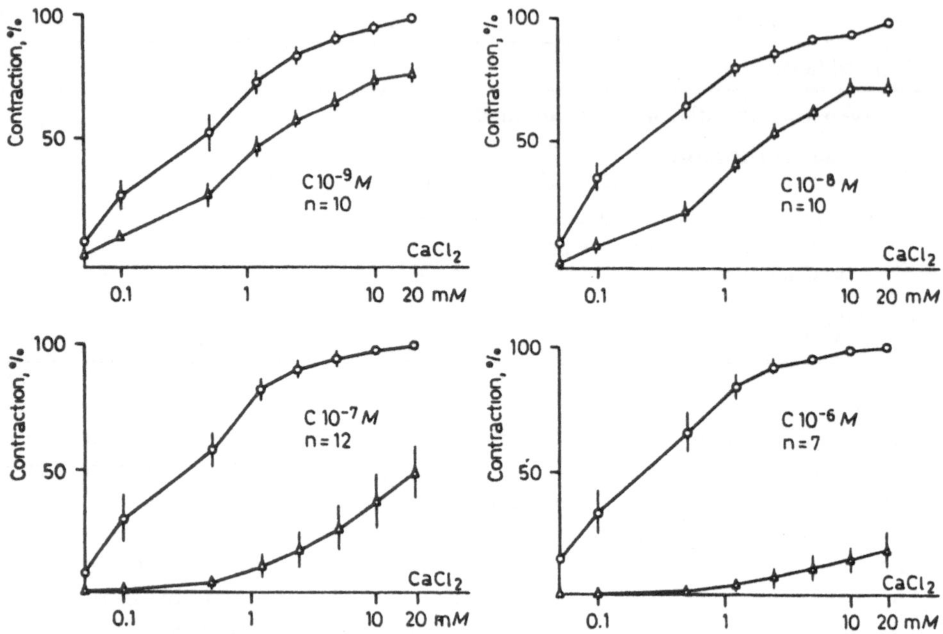

Fig. 1. Effect of cinnarizine on contractions evoked by Ca^{2+} in K^{+}-depolarized rabbit mesenteric arteries. Arterial preparations were preincubated in Ca^{2+}-free physiological solution, depolarized in Ca^{2+}-free, KCl-rich solution, and then further incubated with increasing Ca^{2+} concentrations. Cumulative concentration-effect curves were obtained before (*circles*) and after (*triangles*) addition of cinnarizine (*C*) at the concentrations indicated. Responses are expressed in percentage of maximal contraction evoked before addition of cinnarizine. The inhibitory effect of cinnarizine is observed at concentrations as low as 1 n*M* and resembles the action of antagonists in receptor studies. In view of this similarity, the term calcium antagonists in receptor studies. In view of this similarity, the term calcium antagonist was suggested to describe the action of cinnarizine. Reproduced from [5]

rate of ^{45}Ca loss after transfer to non radioactive solutions depends mainly on outward movement.

The calcium influx rate under resting conditions is about 0.014 and 0.038 pmol cm^{-2} s^{-1} in rat aorta [2] and in rabbit aorta [3], respectively. This calcium influx has been termed passive Ca leak, because it is thought to occur through pathways that differ from those involved in calcium influx evoked by stimulating agents. Indeed, most of this calcium influx is resistant to calcium entry blockers that block stimuli-dependent calcium influx [1].

Changes in the calcium concentration of the perfusing fluid do not usually affect the contractile state of smooth muscle at rest, except in arteries isolated from hypertensive rats [4]. The picture is quite different after stimulation by KCl depolarization or by agonists. For instance, in preparations immersed in solutions in which NaCl is partly replaced by KCl at an appropriate concentration, the cell membrane is depolarized, and a change in its permeability to Ca^{2+} may

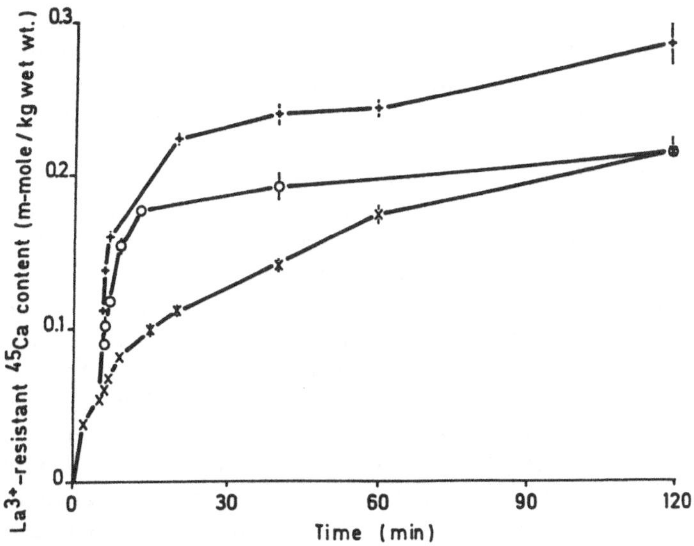

Fig. 2. Effect of norepinephrine (10^{-5} *M*) (*circles*) and of depolarization by 100 m*M* KCl (*crosses*) on ^{45}Ca influx in rat aorta.
Ordinate, ^{45}Ca content of the La^{3+}-resistant Ca^{2+} fraction expressed in mmol/kg wet wt. *Abscissa*, time in radioactive solution for controls (*x*) and treated arteries. *Vertical bars* represent SEM. Modified from [6]

be observed. This is demonstrated by a calcium-dependent increase in the tone of the muscle preparation; this increase is a function of the calcium concentration in the external medium [5] (Fig. 1). A more direct demonstration of such changes in calcium permeability may be obtained when ^{45}Ca fluxes are measured.

As Fig. 2 illustrates for rat aorta, norepinephrine stimulation and KCl depolarization evoke a nearly 20-fold increase in the rate of calcium influx [6]. There is a major difference between the two stimuli as shown by net ^{45}Ca uptake at steady state. Indeed, norepinephrine does not influence net ^{45}Ca uptake, whereas the net uptake is much higher in KCl-depolarized aortas. This observation is in agreement with the existence of receptor-operated calcium channels (ROCs) and of potential-operated calcium channels (POCs), as proposed in 1979 by Bolton [7] and by van Breemen et al. [8]. A third type of calcium channel has been postulated by Bevan and coworkers [9], the stretch-operated calcium channels (SOCs). These authors have supported their view by several functional observations. This functional recognition of three types of calcium channels is in agreement with the existence of three main categories of excitation: electrical, related to membrane depolarization by conducted potentials or by neurotransmitters; pharmacomechanical, related to activation by transmitters without cell membrane depolarization; and mechanical, induced by stretches causing or not causing depolarization.

Electrophysiological studies on Ca channels in vascular smooth muscle were initiated by Mras and Sperelakis [10], who reported Ca-dependent action potentials in cultured cells from rat aorta that had been treated with 10 m*M* tetraethylammonium; these spikes were blocked by calcium entry blockers. More recently, functional Ca channels have been studied in isolated cells using the patch-clamp technique. Tsien and his coworkers [11, 12] have reported distinct types of POCs in cultured sensory neurones of the chick dorsal root ganglion and in mammalian cardiac cells. In cardiac cells, they found two types of unitary conductance with different kinetic features, which can be attributed to the opening of two distinct calcium channels of T- and L-type. The first type is activated by a small depolarization and inactivates rapidly; the other requires stronger depolarization and inactivates slowly. Only the L-type is sensitive to dihydropyridines. In chick dorsal root ganglion cells, a third type of conductance (N-type) has been identified. Recent observations made on enzymatically isolated arterial smooth muscle cells using the patch-clamp technique [13–15] have established the existence of voltage-controlled calcium channels selective for divalent cations over monovalent cations. Worley et al. [13] have identified two conductance levels, 8 pS and 15 pS, which were both sensitive to nisoldipine. In contrast, Bean et al. [14] and Benham et al. [15] reported the existence of two current components much more similar to those found in other cell types (T and L currents); only the L-type current was influenced by dihydropyridines.

Calcium Movements and Relaxation

In the resting state, the cytosolic calcium concentration is probably not higher than 100 n*M*, and the ratio of extracellular to intracellular Ca^{2+} exceeds 10000. Upon stimulation, the cytosolic Ca^{2+} rises, partly as a result of an influx of extracellular Ca^{2+} and partly following the mobilization of internal Ca^{2+} stores, to reach about 1–10 μM. To restore the low resting cytosolic Ca^{2+} and to maintain Ca^{2+} homeostasis, the cell uses Ca^{2+} transport systems that are able to operate against large electrochemical gradients. Such systems have been demonstrated in the plasma membrane and in intracellular organelles, mainly the endoplasmic (sarcoplasmic) reticulum.

There are two main mechanisms allowing "uphill" Ca^{2+} transport. The first relies on the direct utilization of energy from ATP (Ca^{2+} transport ATPase), and the second uses the Na^+ electrochemical potential to drive the extrusion of Ca^{2+} in exchange for Na^+ entry (Na^+–Ca^{2+} exchange). ATP-dependent Ca^{2+} pumps with different properties are present in the plasma membrane and the endoplasmic reticulum. Their identification in smooth muscle has relied mainly on cell fractionation techniques, including the digitonin shift method [16]. The plasma membrane Ca^{2+} pump is activated by calmodulin, which acts mainly by enhancing the affinity of the pump for Ca^{2+}. In smooth muscle, the Ca^{2+} transport capacity of the Na^+–Ca^{2+} exchange system and its Ca^{2+} affinity seem to be distinctly lower than those of the plasmalemmal Ca^{2+} transport ATPase, in contrast to the heart sarcolemma where the transport capacity of the exchanger is much higher than that of the Ca^{2+} pump [16].

The Interaction of Calcium Antagonists with Potential-Operated Calcium Channels in Normotensive Vessels

Figure 1 shows the effect of cinnarizine on the contraction evoked by calcium in isolated rabbit mesenteric arteries. Isolated arteries were first depolarized by immersing them in a Ca^{2+}-free solution containing 100 m*M* KCl. Addition of Ca^{2+} to the depolarized solution evoked an increase in tension that is concentration dependent. In the presence of cinnarizine, concentration or dose-effect curves were displaced to the right, and their maximum was progressively depressed. The inhibitory effect is typical of that observed with other calcium entry blockers and resembles the action of antagonists in receptor studies. In view of this similarity, the term calcium antagonist was suggested in the 1960s to describe the action of cinnarizine [1, 5]. Like cinnarizine, nifedipine inhibits the entry of calcium in depolarized smooth muscle cell (Fig. 3). When the concentration of nifedipine is varied, there is a close correlation between inhibition of contraction and of ^{45}Ca entry [17]. That this action is related to specific receptor sites for nifedipine is indicated by a close similarity between IC_{50} values obtained in intact tissue and K_i values measured in binding studies performed on microsomal preparations, an observation confirmed with several other dihydropyridines (Table 2). Using the digitonin shift method, which allows us to characterize an activity associated with the plasma membrane, Godfraind and Wibo [18] have shown that the specific binding sites are located on the plasma membrane of the smooth muscle cell, in agreement with several lines of evidence showing that the specific receptor sites for dihydropyridines are located on plasma membrane Ca channels.

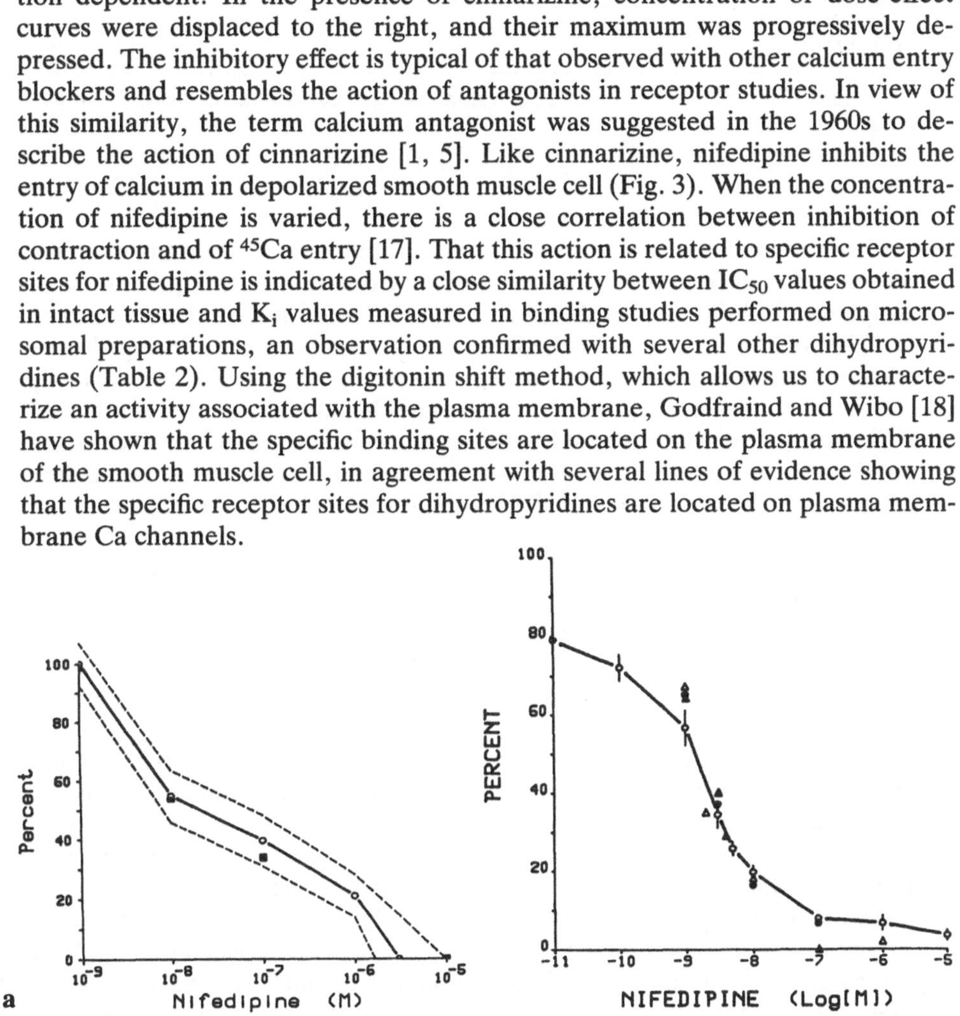

Fig. 3. **a** Depression by nifedipine of norepinephrine-evoked ^{45}Ca influx (*squares*) and contraction (*circles*) in rat aorta. *Ordinate*, nifedipine-sensitive component in percentage of the maximum. Each point is the mean calculated from at least five experiments. SEM are shown by the *dotted line*. **b** Nifedipine-induced depression of K-depolarization-evoked ^{45}Ca influx and contraction in rat aorta and mesenteric artery. Aorta contraction (*open circles*); aorta-corrected ^{45}Ca influx (*open triangles*); mesenteric artery contraction (*solid circles*); and mesenteric artery ^{45}Ca influx (*solid tiangles*). Each point is the mean of at least six determinations. SEM are shown for contractions. Modified from [17]

Table 2. K_i and IC_{50} values of various dihydropyridines in rat aorta[a]

Dihydropyridine	K_i (nM)	IC_{50} (nM)
(+)PN 200-100	0.085	0.092
Nisoldipine	0.148	0.071
(−)Nimodipine	0.546	0.234
Nitrendipine	0.958	0.321
(+)Nimodipine	2.89	1.37
Nifedipine	3.59	1.40
(−)PN 200-110	14.9	12.3
(+)BAY K 8644	36.3	35.3

[a] K_i values (binding studies) and IC_{50} values (KCl-evoked contractions, 35 min) were reported by Wibo et al. [20]

Figure 4 shows the time course of the contractions of rat aorta evoked by KCl before and after treatment with various dihydropyridines. It is quite obvious that the pattern of inhibition of the contraction depends on the calcium entry blocker. The inhibition of KCl-evoked contraction is unchanged during the 30-min stimulation in the presence of nifedipine. On the other hand, with nisoldipine, the inhibition develops gradually. This time-dependent pattern of inhibition of KCl-evoked contractions could be reproduced in successive contractions, separated by a resting period of 30 min in physiological solution in the presence of drug. As shown by Wibo et al. [20] with (+)PN 200-110, the kinetics of inhibition during depolarization is similar to the kinetics of binding to membranes isolated from rat aorta. This suggests that depolarization induces a conformation of calcium channels with increased affinity for dihydropyridines and that this high-affinity form is similar to that found in isolated membranes. This hypothesis has been examined in isolated rat mesenteric arteries by Morel and Godfraind [21], who have studied how membrane potential influences the specific binding of (+)PN 200-110 to intact arteries.

The time course of the K^+ contraction in the presence of (+)PN 200-110 was characteristic of a use-dependent action, with a time-dependent decrease in tension to a steady state inhibition as already shown in rat aorta (Fig. 5). Dose-effect curves for inhibition measured 2 and 30 min after initiation of depolarization were characterized by IC_{50} values equal to 270 ± 65 pM and 33 ± 3 pM, respectively. (−)PN 200-110 was less potent than (+)PN 200-110, and its inhibitory potency was also time dependent, IC_{50} values measured at 2 and 30 min in K^+ solution being equal to 16 ± 4 nM and 5 ± 3 nM, respectively. Thus, the effect of PN 200-110 was stereoselective, the (+) enantiomer being markedly more potent than the (−) enantiomer ($P < 0.01$).

(+)[^{3}H]PN 200-110 binding was determined in mesenteric arteries bathed in normal and high K^+ solutions. Non specific binding was not different in polarized and in depolarized arteries. On the other hand, as shown in Fig. 6, specific binding was markedly enhanced in depolarized arteries. Experimental estimates of specific binding were well fitted by one hyperbolic curve indicating a one-to-one binding of (+)PN 200-110 to a single class of sites. Computer least-squares

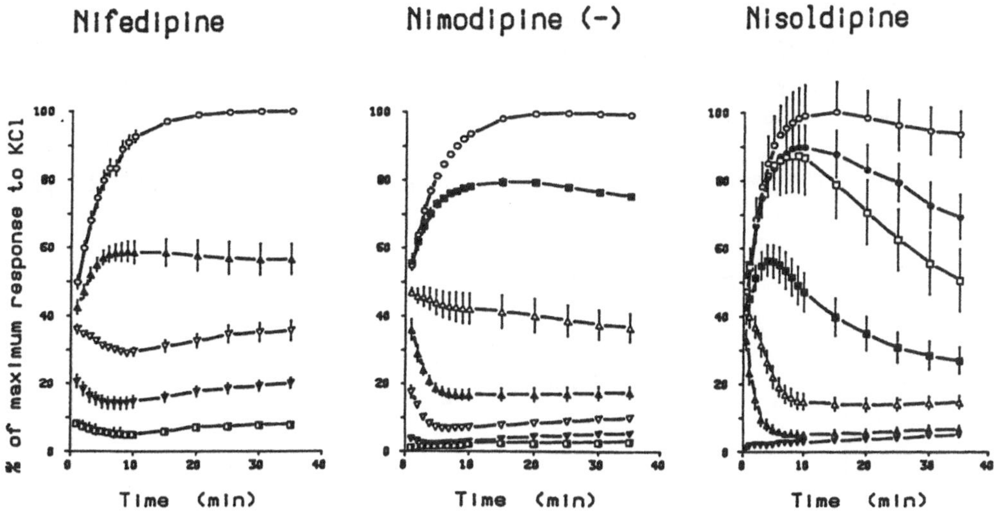

Fig. 4. Time course of the contraction of rat aorta evoked by K depolarization before (control) and after 90 min of pre-incubation with various concentrations of one of the dihydropyridines indicated on the graph at the following concentrations: (*solid circles*) 10 p*M*, (*open squares*) 30 p*M*, (*solid squares*) 100 p*M*, (*open triangles*) 300 p*M*, (*solid triangles*) 1 n*M*, (*open, downward triangles*) 3 n*M*, (*solid, downward triangles*) 10 n*M*, (*half-filled squares*) 100 n*M*. Note time-dependent inhibition mainly observable with nisoldipine. Modified from [19]

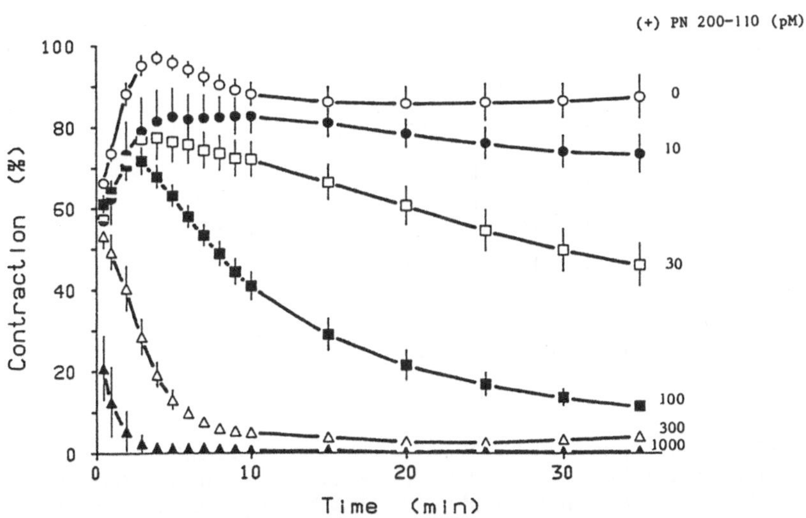

Fig. 5. Time course of the contraction of rat mesenteric artery evoked by K depolarization before (control) and after 90 min of pre-incubation with various concentrations of (+)PN 200-110 as indicated on the graph. Modified from [21]

fitting of the data reveals that membrane depolarization significantly lowered apparent K_D ($P<0.01$) while maximum binding capacity appeared not to be significantly affected ($P>0.1$). In the reciprocal plot of the same data it clearly appeared that both curves intersect at the same point on the 1/B axis, confirming that the B_{max} value is not different in arteries bathed in physiological and in K^+ depolarizing solutions. K_D value obtained in depolarized arteries (44 ± 2 p*M*) is

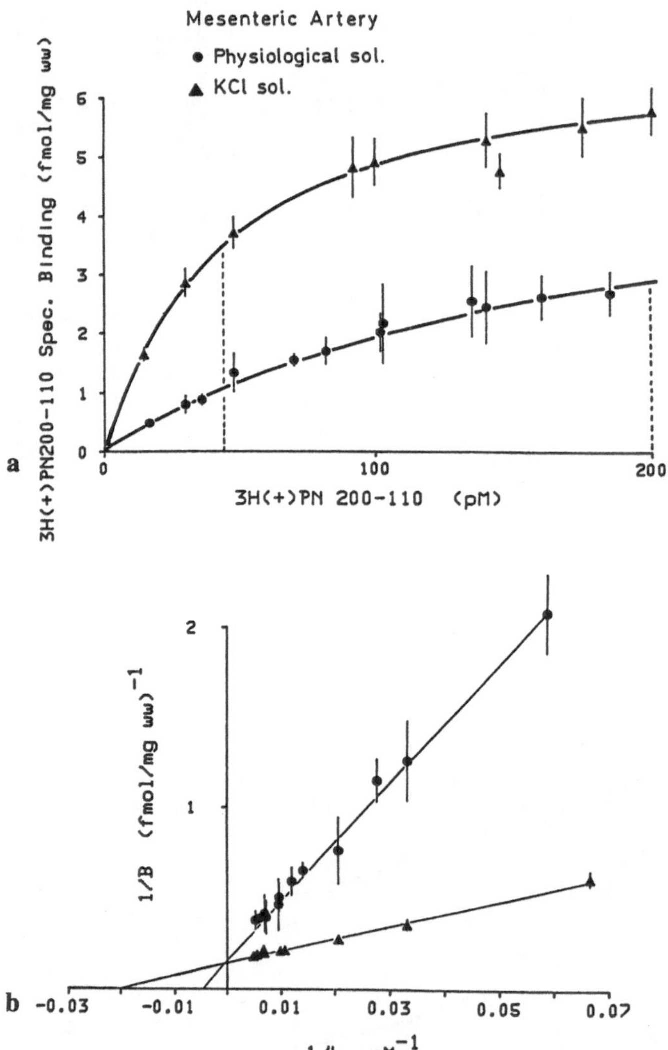

Fig. 6. Binding of (+)[^{3}H]PN 200-110 to isolated mesenteric arteries. **a** Saturation study of the specific binding of (+)[^{3}H]PN 200-110 in mesenteric arteries bathed in physiological solution (*solid circles*) and in KCl-depolarizing solution (*solid triangles*). Each point represents measurements from 6 to 18 rings ± SEM. *Dotted lines* indicate K_D values as calculated with a curve-fitting program for nonlinear models (physiological solution: B_{max} 5.8 ± 0.7 fmol/mg w.wt, K_D 200 ± 20 p*M*; KCl solution: B_{max} 7.0 ± 0.1 pmol/mg w.wt K_D 44 ± 2 p*M*). **b** Double reciprocal plot of the same data. Modified from [21]

similar to that reported for (+)PN 200-110 in membrane preparations from other tissues and is close to the IC_{50} measured after 30 min of depolarization (see Table 3). These data indicate that like its pharmacological effect, the binding of PN 200-110 to intact tissue is stereospecific and voltage-dependent.

Figure 7 compares the effects of nifedipine, (+)PN 200-110, and flunarizine on (+)[^{3}H]PN 200-110 specific binding in intact mesenteric arteries. Segments of mesenteric arteries were incubated either for 90 min in physiological solution in the presence of 100 p*M* (+)[^{3}H]PN 200-110 or for 60 min in physiological solution followed by 30 min in KCl-depolarizing solution, both containing 40 p*M* (+)[^{3}H]PN 200-110. These concentrations of radioligand labelled about 30% and 50% of the total number of binding sites in polarized and in depolarized arteries, respectively.

Nifedipine and (+)PN 200-110 inhibited the specific binding of (+)[^{3}H]PN 200-110 dose-dependently in polarized as well as in depolarized arteries. In arteries incubated in KCl medium, the curve describing the displacement of the tritiated ligand by unlabelled (+)PN 200-110 was shifted to the left compared with that obtained in NaCl medium. This is in agreement with saturation studies, which showed a significant increase of (+)[^{3}H]PN 200-110 binding affinity in depolarized arteries compared with arteries bathed in normal physiological solution [21]. K_i values are reported in Table 3. They are in good agreement with apparent K_D values estimated from saturation studies. On the contrary to what was observed with (+)PN 200-110, the effectiveness of nifedlipine was not signi-

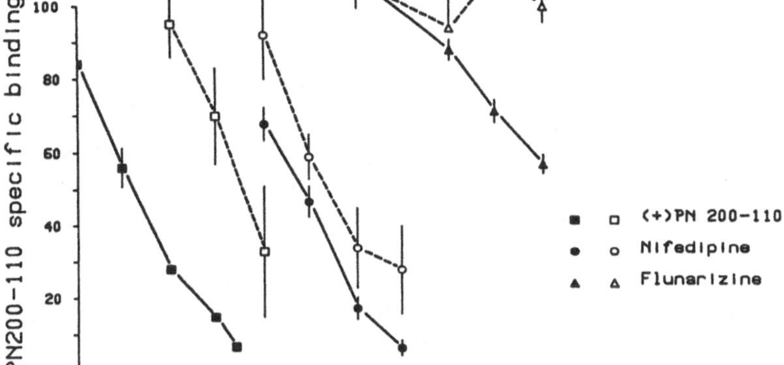

Fig. 7. Displacement of (+)[^{3}H]PN 200-110 specific binding by unlabelled (+)PN 200-110, nifedipine, and flunarizine in mesenteric artery rings bathed in physiological solution (*open symbols*) or in KCl-depolarizing solution (*solid symbols*). Mesenteric artery rings were incubated for 90 min in physiological solution containing 100 p*M* (+)[^{3}H]PN 200-110 and various concentrations of competitors or for 60 min followed by 30 min in KCl-depolarizing solution containing 40 p*M* (+)[^{3}H]PN 200-110 and various concentrations of competitors. Each value is the mean of 6 to 18 determinations. *Vertical bars* indicates SEM. Modified from [22]

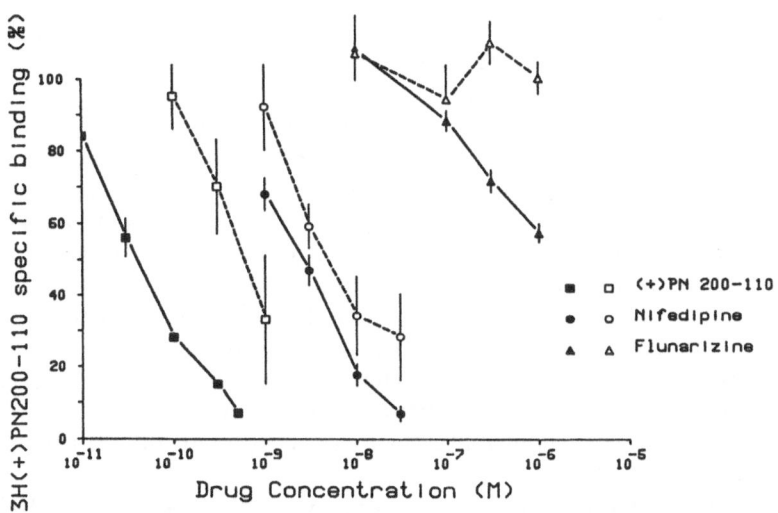

Table 3. Rat mesenteric artery: comparison of pharmacological and radioligand binding data

Drug	Intact artery				Membranes
	IC_{50} (nM)		K_i (nM)		K_i (nM)
	2 min	30 min	Physiological solution	K^+-rich solution	
(+)PN 200-110	0.270	0.033	0.410	0.025	0.055
(−)PN 200-110	16.0	5.0	23.0	1.0	–
Nifedipine	4.0	1.9	3.1	1.2	3.0
Flunarizine	37.0	2.0	unperceived	900.0	500.0

K_i values (binding studies) and IC_{50} values (KCl-evoked contractions, 2 and 30 min) were reported by Morel and Godfraind [21, 22]

ficantly modified by prolonged depolarization, and K_i values calculated from the displacement curves obtained in polarized and in depolarized arteries were not significantly different (Table 3). With both compounds K_i values are close to the concentrations inhibiting by 50% the K^+-induced contraction of rat mesenteric artery (Table 3).

Flunarizine up to 10^{-6} M was unable to displace (+)[^{3}H]PN 200-110-specific binding in arteries bathed in physiological solution, but it inhibited the binding in depolarized arteries. A 50% inhibition was obtained in the presence of about 10^{-6} M flunarizine. This concentration is much higher than the concentration required to block the K^+ contraction in the same tissue (Table 3) [22].

Binding experiments performed on intact mesenteric artery allow direct correlation of the binding and pharmacological properties of calcium entry-blocking drugs. The results reveal a good correlation between the existence of time dependency in the inhibition of KCl-induced contraction and the sensitivity of drug binding to depolarization. For all three compounds, binding data obtained with a membrane preparation from mesenteric arteries were similar to the data obtained with depolarized intact arteries (Table 3).

Depolarization of vascular smooth muscle is known to increase the cell membrane permeability to calcium by opening two types of membrane potential-dependent calcium channels (see above). Considering their kinetics and pharmacological properties, it seems that the slowly inactivating channel (L-type) mostly contributes to tonic contraction induced by 100 mM KCl depolarizing solution.

As shown previously, the inhibitory effect of (+)PN 200-110 on KCl contraction is markedly time dependent, and binding studies indicate that depolarization increases the binding affinity of arteries for this compound. In contrast to (+)PN 200-110, depolarization did not significantly affect the potency of nifedipine in displacement studies. Similarly, in contractility studies, the inhibition produced by nifedipine on the KCl contraction was little affected by the duration of the depolarization (the ratio of IC_{50} of nifedipine measured at 2 and 30 min of depolarization is equal to 2, while it reaches 8 for (+)PN 200-110; see Table 3).

These binding data indicate that, at pharmacologically relevant concentrations, (+)PN 200-110 binding to resting mesenteric arteries is very low while nifedipine is able to bind significantly to the calcium channel at rest. The low binding of (+)PN 200-110 in arteries incubated in physiological solution explains why inhibition of KCl contraction is very low during the first minutes of depolarization, despite a long period of equilibration of the tissue with the drug. Thus, this is in agreement with the observation that the time course of the binding of (+)PN 200-110 to isolated membranes can account for the typical time-dependent pattern of the inhibitory effect of this compound [20].

On the other hand, the absence of a marked time dependence in the effect of nifedipine can be accounted for by some degree of interaction of nifedipine with calcium channels in resting tissues. Binding of nifedipine to resting calcium channels has also been reported in atrial cells where nifedipine produces tonic block of the calcium current at holding potentials negative enough to ensure that calcium channels were not inactivated [23].

The voltage dependence of (+)[^{3}H]PN 200-110 binding may be analyzed according to the model proposed by Bean for cardiac cells [24], assuming that dihydropyridines bind tightly to the inactivated state (I) of the calcium channel and more weakly to the resting state (R). The fraction of channels in the R and I states is determined by the holding membrane potential, with more channels in I state at depolarized potential. Calcium channel binding of dihydropyridines is then expected to follow the form of a 1:1 binding with an apparent dissociation constant that depends on the fraction of channels that are in R and I states. Binding to open channels and the possible existence of different states of inactivated or resting channels are neglected in this model, which probably oversimplifies the actual situation but allows explanation of the results presented here. We found an apparent K_D value of 44 p*M* in arteries bathed in 100 m*M* KCl. Under these conditions, most channels are in I state, and the observed K_D corresponds to the actual dissociation constant for binding to the inactivated channel (K_{Inact}). This K_D is very close to the value found in membrane preparations from various tissues and in depolarized cardiac cells [25], suggesting that binding to the inactivated calcium channel is very similar in cells originating from different tissues. The apparent dissociation constant measured in preparations bathed in physiological solution is lower in mesenteric arteries than in cardiac cells (K_D: 730 p*M*) [25]. This difference could arise from different binding in resting channels in cardiac and vascular smooth muscle cells. However, the apparent dissociation constant observed in arteries bathed in physiological solution is likely to be lower than the true K_D for resting channels (K_{Rest}) as all the calcium channels are probably not in resting state in those conditions. Indeed from the two states model of Ca channels, it may be deduced [24] that

$$K_{App} = \frac{1}{[h/K_{Rest}] + [1-h]/K_{Inact}} \quad (1)$$

where h is the fraction of channels in the resting state. Assuming $K_{Inact} = 40$ p*M* and $K_{Rest}/K_{Inact} > 10$, the proportion of inactivated channels in arteries bathed in physiological solution can be estimated to be equal to 0.1–0.2. This proportion of channels inactivated in mesenteric artery at rest could be related to the level of the resting potential in this preparation; it could also be the result of the small depolarization induced by the destruction of vascular endothelium caused by experimental manipulations (about 10 mV) [26]. This relatively low proportion of inactivated channels can nevertheless induce an important decrease of the apparent K_D value. The difference between the actual dissociation constant of PN 200-110 binding to resting and inactivated channels in arteries could thus be much more important than that observed between arteries bathed in physiological and depolarizing solutions, and the higher value of the apparent dissociation constant found in polarized cardiac myocytes could be related to a lower proportion of inactivated channels in resting cardiac cells.

Another possible interpretation of our observations is that the time-dependent inhibition of the K^+ contraction is due to a selective sensitivity to dihydropyridines of different voltage-dependent calcium channel subtypes. If the increase in activity of PN 200-110 were due to the opening of a larger number of dihydropyridine-sensitive channels after prolonged depolarization, a marked increase of B_{max} (maximum binding capacity) should be observed. This is not consistent with our experimental data showing that maximum binding of (+)PN 200-110 was similar in polarized and depolarized arteries.

If we assume a one-to-one binding, the maximum binding capacity being equal to 7 fmol/mg wet weight and the cell surface area to 7300 cm^2/g tissue, there are about six dihydropyridine binding sites per μm^2 of cell membrane. In excised patches of rabbit mesenteric artery smooth muscle cells, Worley et al. [13] reported a minimal value of 0.2 channels per μm^2 membrane surface area, deduced from an average of one functional channel per patch. As the number of channels varied from 0 to 20 per patch, density of calcium channels can be estimated to be in the range 0.2–4 channels/μm^2, a value close to the estimation of the density of dihydropyridine binding sites in isolated arteries.

Calcium Antagonists and Potential-Operated Calcium Channels in Isolated Artery of Spontaneously Hypertensive Rats—Role of Endothelium

There are several reports in the literature describing differences in the contractile behaviour of arteries isolated from hypertensive rats and from normotensive ones. Hypertensive vessels show an increased reactivity to vasoconstrictor stimuli, namely to norepinephrine and to KCl [4, 27]. This increased responsiveness has been related to a difference in the resting membrane potential observed between arteries from hypertensive and normotensive animals [28]. As pointed out by Aoki and Asano [27], such a difference could account for the observation that BAY K 8644, a dihydropyridine that contracts vascular smooth muscle by increasing the probability of opening of POCs, evokes a contractile response more important in SHR than in WKY vessels. It has been shown that endothelium-dependent relaxation to acetylcholine and to the calcium ionophore A23187 are decreased in blood vessels of adult SHR [29–30]. We have reported that endothelium modulates the contractile response of isolated vessels to various stimuli, including clonidine, noradrenaline, KCl, and BAY K 8644 [26]. These observations have been interpreted assuming the existence of a direct control by endothelial factors on POCs and an indirect one through changes in membrane potential.

We have examined to what extent the difference in responsiveness between vessels of normotensive and hypertensive animals could be related to a different control exerted by the endothelium on the gating of POCs. Therefore, responses to KCl depolarization and to BAY K 8644 of aortas isolated from normotensive and from SHR have been examined under different experimental conditions.

It is known that the dihydropyridine BAY K 8644 increases the probability of opening of POCs. Therefore, the sensitivity of arteries to activating concentrations of KCl (between 6 and 20 m*M*) was studied in the presence of a submaximal (3×10^{-8} *M*) concentration of BAY K 8644. KCl dose-effect curves were different in SHR and WKY when they were established immediately after the initial incubation period of 60 min: ED_{50} of KCl was equal to 8 m*M* in SHR and to 12.5 m*M* in WKY, but the maximum responses were the same for the two strains. When this procedure was repeated after washout and resting incubation for 90 min in physiological solution, the contractile response was attenuated. This attenuation was more important in SHR than in WKY, in such a way that the difference between the two dose-effect curves became smaller. In endothelium-deprived arteries, the contraction developed was much higher than in the presence of endothelium, an observation in agreement with previous reports [26]. The inhibitory effect exerted by endothelium on the contractile response was more important in WKY than in SHR.

We have measured the amount of specific binding of (+)[^{3}H]PN 200-110 to intact aortas bathed in various solutions containing different concentrations of KCl. For KCl concentrations between 6 and 20 mmol/l, specific binding to SHR arteries was higher than to WKY arteries, but at 100 mmol/l KCl, the binding was similar in the two groups. Apparent K_D values were estimated for each KCl concentration. These values were lower in SHR than in WKY at low KCl concentrations (for KCl = 6 m*M*: WKY, 500p*M*; SHR, 200p*M*), but were similar at 100 m*M* KCl where a K_D value of 80 p*M* was found for both SHR and WKY. On the other hand, specific binding of (+)[^{3}H]PN 200-110 was measured in membrane preparations from aortas of SHR and WKY rats. K_D values were similar to those found in intact tissues bathed in 100 m*M* KCl.

The present experiments indicate that SHR arteries have a higher sensitivity to dihydropyridines because the state of their calcium channels differs from that of WKY arteries at physiological KCl concentration. A calculation based on Eqn. (1) shows that in aortas bathed in 6 m*M* KCl, 5% of the calcium channels are under the high affinity configuration in WKY. In SHR, at 6 m*M* KCl, 30% of the calcium channels are under this configuration. KCl concentrations required to trigger the contraction are lower in SHR than in WKY. This suggests an easier activation of calcium channels of SHR arteries, in line with their higher residual tone [19]. This may be due to differences in the resting membrane potential, but other factors cannot be ruled out.

Concluding Remarks

There is a large body of experimental information showing that the specific binding sites of calcium antagonists are involved in inhibition of stimulus-dependent Ca entry into arterial cells and in inhibition of the contractile response. These specific binding sites are located at the level of the plasma membrane and are very likely associated with calcium channels. The affinity of calcium channels for their ligands is related to the chemical structure of the blocker, but this affinity is also influenced by factors which may change the state of the channels. We have documented the role of the membrane potential showing that prolonged depolarization induces an increase of the binding affinity that could be related to a conformational change of the Ca channels, in agreement with the modulated receptor theory assuming that channels may be in a resting, an open, or an inactivated configuration.

In physiological solution, SHR arteries show a higher affinity than WKY arteries for the selective Ca channel ligand (+)PN 200-110. This explains why the endogenous tone of SHR arteries is highly sensitive to Ca agonists and Ca antagonists (dihydropyridines). Several factors may contribute to this enhanced affinity; their identification will allow a better understanding of the antihypertensive effect of Ca antagonists.

References

1. Godfraind T, Miller RC, Wibo M (1986) Calcium antagonism and calcium entry blockade. Pharmacol Rev 38: 321–416
2. Godfraind T (1976) Calcium exchange in vascular smooth muscle, action of noradrenaline and lanthanum. J Physiol (Lond) 260: 21–35
3. Khalil R, Lodge N, Saida K, van Breemen C (1987) Mechanism of calcium activation in vascular smooth muscle. J Hypertension 5 (Suppl 4): S5–S15
4. Bohr DF, Webb RC (1981) Membrane excitation in vascular smooth muscle, changes in hypertension. In: Godfraind T, Meyer P (eds) Cell membrane in function and dysfunction of vascular tissue. Elsevier/North Holland, Amsterdam, pp 168–192
5. Godfraind T, Kaba A (1969) Blockade or reversal of the contraction induced by calcium and adrenaline in depolarized arterial smooth muscle. Br J Pharmacol 36: 549–560
6. Godfraind T, Miller RC (1983) Specificity of action of Ca^{++} entry blockers. A comparison of their actions in rat arteries and in human coronary arteries. Circ Res 52 (Suppl I): 81–91
7. Bolton TB (1979) Mechanisms of action of transmitters and other substances on smooth muscle. Physiol Rev 59: 606–718
8. van Breemen C, Aaronson P, Loutzenhiser R (1979) Sodium-calcium interactions in mammalian smooth muscle. Pharmacol Rev 30: 167–208
9. Bevan JA, Bevan RD, Hwa JJ, Owen MP, Tayo FM, Winquist RJ (1982) Calcium intrinsic (myogenic) vascular tone. In: Godfraind T, Albertini A, Paoletti R (eds) Elsevier, Amsterdam, pp 125–132
10. Mras S, Sperelakis N (1981) Bepridil (CERM-1978) blockade of action potentials in cultured rat aortic smooth muscle cells. Eur J Pharmacol 71: 13–19
11. Nilius B, Hess P, Lansman JB, Tsien RW (1985) A novel type of calcium channel in ventricular heart cells. Nature 316: 443–446

12. Nowycky MC, Fox AP, Tsien RW (1985) Three types of neuronal calcium channel with different calcium agonist sensitivity. Nature 316: 440–443
13. Worley JF, Deitmer JW, Nelson MT (1986) Single nisoldipine-sensitive calcium channels in smooth muscle cells isolated from rabbit mesenteric artery. Proc Natl Acad Sci USA 83: 5746–5750
14. Bean BP, Sturek M, Puga A, Hermsmeyer K (1986) Calcium channels in muscle cells isolated from mesenteric arteries: modulation by dihydropyridine drugs. Circ Res 59: 229–235
15. Benham CD, Hess P, Tsien RW (1987) Two types of calcium channels in single smooth muscle cells from rabbit ear artery studied with whole-cell and single-channel recordings. Circ Res 61 (Suppl I): I10–I16
16. Godfraind T, Morel N, Wibo M (1986) The heterogeneity of calcium movements in cardiac and vascular smooth muscle cells. Scand J Clin Lab Invest 46 (Suppl 180): 29–40
17. Godfraind T (1983) Actions of nifedipine on calcium fluxes and contraction in isolated rat arteries. J Pharmacol Exp Ther 224: 443–450
18. Godfraind T, Wibo M (1985) Subcellular localization of (^{3}H)nitrendipine binding sites in guinea pig ileal smooth muscle. Br J Pharmacol 85: 335–340
19. Godfraind T, Eglème C, Wibo M (1985) Effects of dihydropyridines on human and animal isolated vessels. In: Fleckenstein A, van Breemen C (eds) Proceedings of Bayer symposia vol. IX. Springer, Berlin Heidelberg New York, pp 309–325
20. Wibo M, De Roth L, Godfraind T (1988) Pharmacological relevance of dihydropyridine binging sites in membranes from rat aorta: kinetic and equilibrium studies. Circ Res 62: 91–96
21. Morel N, Godfraind T (1987) Prolonged depolarization increases the pharmacological effect of dihydropyridines and their binding affinity in calcium channels of vascular smooth muscle. J Pharmacol Exp Ther 243: 711–715
22. Morel N, Godfraind T (1988) Selective modulation by membrane potential of the interaction of some calcium entry blockers with calcium channels in rat mesenteric artery. Br J Pharmacol H95: 252–258
23. Uehara A, Hume JR (1985) Interactions of organic calcium channel antagonists with calcium channels in single frog atrial cells. J Gen Physiol 85: 621–647
24. Bean BP (1984) Nitrendipine block of cardiac calcium channels: high-affinity binding to the inactivated state. Proc Natl Acad Sci USA 81: 6388–6392
25. Kokubun S, Prod'hom B, Becker H, Porzig H, Reuter H (1986) Studies on Ca channel in intact cardiac cells: voltage-dependent effects and cooperative interactions of dihydropyridine enantiomers. Molec Pharmacol 30: 571–584
26. Godfraind T, Eglème C, Alosachie I (1985) Role of endothelium in the contractile response to alpha adrenergic agonists. Clin Sci 68: 65S–71S
27. Aoki K, Asano M (1986) Effects of Bay K 8644 and nifedipine on femoral arteries of spontaneously hypertensive rats. Br J Pharmacol 88: 221–230
28. Cheung DW (1984) Membrane potential of vascular smooth muscle and hypertension in spontaneously hypertensive rats. Can J Physiol Pharmacol 62: 957–960
29. Konishi M, Su C (1983) Role of endothelium in dilator responses of spontaneously hypertensive rat arteries. Hypertension 5: 881–886
30. Winquist RJ, Bunting PB, Baskin EP, Wallace AA (1984) Decreased endothelium-dependent relaxation in New Zealand genetic hypertensive rats. J Hypertension 2: 541–546

III Membrane Calcium-Handling Abnormalities of Vascular Smooth Muscle in Spontaneously Hypertensive Rats

Increased Calcium Permeability of the Membrane of Vascular Smooth Muscle in Spontaneously Hypertensive Rats*

SIANGSHU CHAI, R. CLINTON WEBB, and DAVID F. BOHR[1]

Summary. An increased membrane permeability to calcium of vascular smooth muscle in spontaneously hypertensive rats (SHR) had been inferred from the observation that this muscle develops spontaneous tone in the presence of calcium, whereas that from the normotensive rat (WKY) does not. This functional evidence has special relevance to the mechanism that may be responsible for the increase in total peripheral resistance of hypertension. More recently, van Breemen et al. [12] reported that $^{45}Ca^{2+}$ influx in vascular smooth muscle from mesenteric resistance vessels was greater in SHR than in WKY at rest (leak channels), or during stimulation with norepinephrine (receptor-operated channels) or potassium chloride (potential-operated channels). Recent studies in our laboratory have demonstrated that the calcium-dependent spontaneous tone of the isolated vascular smooth muscle can be virtually abolished by removing the endothelium. Therefore, we believe that it may be the endothelium that is showing a greater calcium response in SHR than in WKY. It releases an "endothelium-derived contracting factor" in SHR. We have also observed a greater permeability of the lymphocyte membrane to calcium in SHR. These observations indicate that this greater calcium leak may be generalized cell membrane defect in the SHR.

Key words: Norepinephrine—Nifedipine—Calcium flux—Bay K 8644—Patch clamp

Introduction

Since 1934, when Goldblatt introduced the first experimental model of hypertension [1], the possible mechanisms responsible for the arterial pressure elevation have been extensively studied. It is clear that both structural and functional abnormalities of the wall of the resistance vessels are involved [2, 3]. The experimental model that has proven to be most satisfactory and has been most widely used in the past decade, is the spontaneously hypertensive rat (Aoki SHR) developed by Okomoto and Aoki [4]. In this model, the functional abnormality that appears to be most relevant to the development of hypertension resides in the membrane of the vascular smooth muscle cell. Several different types of evidence presented in this review support the conclusion that this membrane is excessively permeable to calcium.

*Work described in this manuscript was supported by grants from the National Institutes of Health (HL-27020 and HL-18575).
[1]Department of Physiology, University of Michigan, Ann Arbor, MI 48109–0622, USA

Functional Studies

Current functional studies support the postulate that calcium influx is greater in the vasculature of SHR than in that of normotensive rats (WKY). Noon et al. [5], Suzuki et al. [6], and Fitzpatrick and Szentivanyi [7] have demonstrated that isolated aortic strips from SHR develop spontaneous tone which is dependent on the calcium concentration of the physiological salt solution (PSS) in which they are bathed. Strips from WKY rats do not develop tone under these conditions. Similar observations have been made on helical strips of isolated basilar arteries from SHR (Fig. 1) as compared to those from WKY [8]. In this study, when the complete PSS in the muscle bath was replaced with a calcium-free PSS, the artery strip from the SHR relaxed, whereas that from the WKY displayed no change in tension. The latter was already completely relaxed. When calcium was returned to the bath it could again enter the leaky membrane of the cells in the strip from the SHR, so that its tension was gradually re-established. This did not happen in the normal vascular smooth muscle of the WKY, presumably because calcium could not enter the cell as fast. These studies have been interpreted as indicating that the membrane of the vascular smooth muscle is more permeable to calcium in the SHR than in the WKY.

Recent studies by Rinaldi and Bohr [9] have implicated an abnormal membrane of the endothelial cell in the hypertensive rat as contributing to its unusual spontaneous tone. Although these studies were carried out using a different type of experimental hypertension, they introduced another type of abnormal regulatory system that may be involved in the spontaneous tone of the isolated vascular

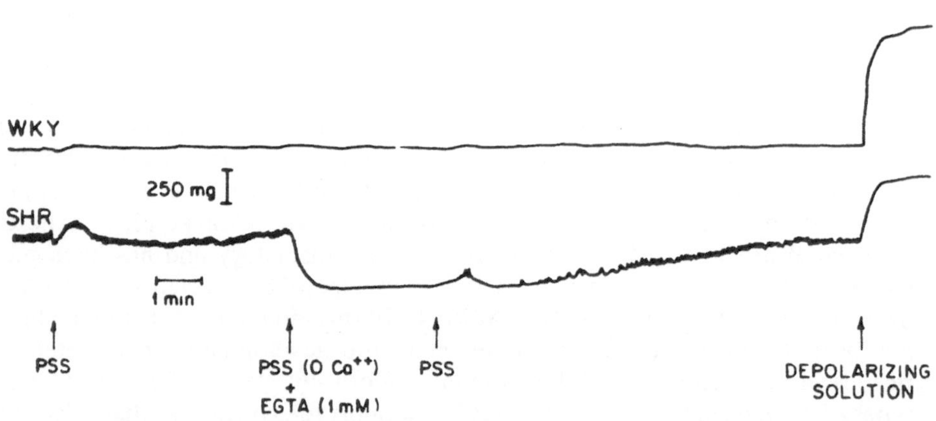

Fig. 1. Isometric force recordings of isolated basilar artery rings from a normotensive (*top*, *WKY*) and hypertensive (*bottom*, *SHR*) rat. Basilar arteries from SHR (but rarely those from WKY) were characterized by spontaneous activity which was inhibited following washing with physiological saline solution (*PSS*) containing no added Ca^{2+} but EGTA (1m*M*). The spontaneous activity, typically small amplitude oscillations superimposed on a tonic contraction, returned upon washing tissues in normal PSS. Rings from both arteries contract in response to the high potassium depolarizing solution. From [8], courtesy of The American Heart Association

smooth muscle from SHR. In Dr. Rinaldi's study, intact rings from the aorta of normotensive and DOCA-hypertensive Sprague-Dawley rats were mounted under tension in a calcium-free PSS. When calcium (2 m*M*) was added to this bath, the rings from the DOCA-hypertensive rats developed a large amount of tension, whereas those from the normotensive Sprague-Dawley rat developed only minimal tension. The involvement of the endothelium in these contractile responses was established by the observation that the responses could be virtually eliminated by gently rubbing the luminal surface of the rings to remove the endothelium. This procedure did not reduce the response of the ring to norepinephrine (NE). It was concluded that the spontaneous contraction resulted from the release of an endothelium-derived contracting factor, and that more of this factor was released in the ring from the DOCA-hypertensive rat than from that of the Sprague-Dawley normotensive rat. These observations were interpreted as indicating that the permeability of the endothelial membrane to calcium was greater in the DOCA-hypertensive animal, therefore generating a greater release of the contracting factor. The importance of this study on endothelium-derived contracting factor is emphasized by the observation of Yanagisawa et al. [10]. These investigators have recently isolated a potent 21-residue vasoconstrictor peptide from vascular endothelial cells. They named the peptide *endothelin*. Cloning and sequencing of pre-proendothelin complementary DNA showed that the peptide is generated through an unusual proteolytic processing, and regional homologies to several neurotoxins suggest that endothelin is a modulator of voltage-dependent ion channels.

In another recent study, Rinaldi (unpublished observations) made additional observations that support the hypothesis that the plasma membrane of vascular smooth muscle of stroke-prone SHR (SHRSP) is more permeable to calcium than is that of WKY. Tail artery rings were placed in a calcium-free PSS containing 2 m*M* EGTA for 3 min. They were then made to contract by stimulation with NE (10^{-6} *M*) while still in this medium. This contraction resulted from the release of intracellularly sequestered calcium and was transient. The sarcoplasmic reticulum was then reloaded with calcium by exposing the muscle to 1.6 m*M* calcium for time periods ranging from 0.5 to 20 min. The extent of reloading was tested by restimulation with NE (10^{-6} *M*) after a 3-min period in calcium-free PSS containing 2 m*M* EGTA. Following all time intervals of reloading, the rings from SHRSP had regained a greater percentage of their initial contractile activity than had those from WKY. These findings are compatible with the possibility that calcium entered the SHRSP cell more rapidly than it entered the WKY cell.

Flux Studies

Current concepts of the entrance of calcium into the cell depict it as occurring through four different types of channels [11]. This classification is based, on the one hand, on specific mechanisms of stimulation of the membrane and, on the other hand, on the specificity of the action of various calcium entry blockers on the activity of the channel. (a) One type of channel, referred to as the "calcium leak channel", appears to be present in all types of biological membranes com-

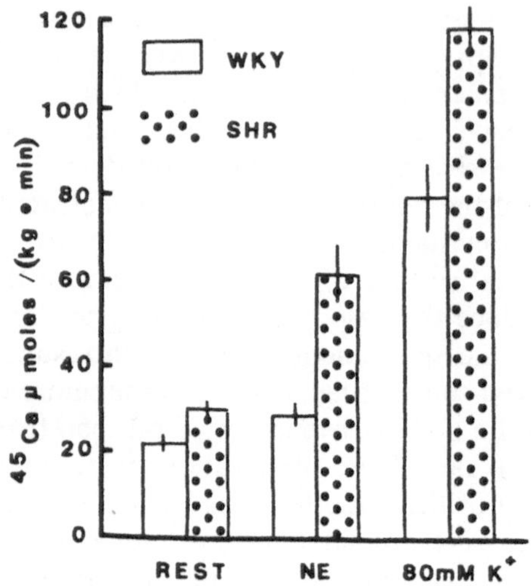

Fig. 2. $^{45}Ca^{2+}$ influx was measured over a 90-s period in vessels exposed to $^{45}Ca^{2+}$-labeled physiological saline solution (*PSS*) (*REST*) or $^{45}Ca^{2+}$-PSS $\times 10^{-4}$ *M* norepinephrine (*NE*) or 80 m*M* K^+-PSS, and then washed for 20 min in ice-cold 0 Ca^{2+}-PSS containing 2 m*M* EGTA. Resting $^{45}Ca^{2+}$ influx and in the presence of NE or 80 m*M* K^+ were significantly elevated in *SHR* vessels as compared to those of *WKY* ($P < 0.05$) (when expressed as Ca^{2+} flux per unit vessel weight). From [12], courtesy of *Federation Proceedings*

posed of phospholipids, cholesterol, and protein. It is active in the absence of any type membrane stimulation. (b) A potential-sensitive channel becomes increasingly available to calcium ions when the membrane is depolarized. (c) A third type of channel is activated by specific agonists such as serotonin, histamine, angiotensin II, and NE. This channel has a specific role in activating the phosphoinositide system. (d) It is now apparent that a fourth type of calcium channel may be activated by the stretch of vascular smooth muscle [12]. Although the "stretch-activated" calcium channels seem to be well-established for vascular smooth muscle, caution must be exercised in the interpretation of these studies because of the observation cited above indicating that the response to stretch may be initiated by the endothelium.

Van Breemen et al. [13] have compared the rate of calcium influx in vascular smooth muscle from resistance vessels of WKY and SHR through three of these channels (Fig. 2). Through the leak, the receptor-operated, and the potential-sensitive channels, the influx was greater in the smooth muscle from the SHR than in that from the WKY. The functional studies cited above demonstrating that the stretched muscle from the SHR contracts whereas that from the WKY remains relaxed suggest that the stretch-activated channel is also more sensitive in the hypertensive animal.

More detailed characterizations of the calcium channels of vascular smooth muscle have arisen from the use of a patch-clamp technique. In vascular smooth muscle, two types of voltage-operated calcium channels have been indeintified based on how quickly they are inactivated. Rusch and Hermsmeyer [14] have

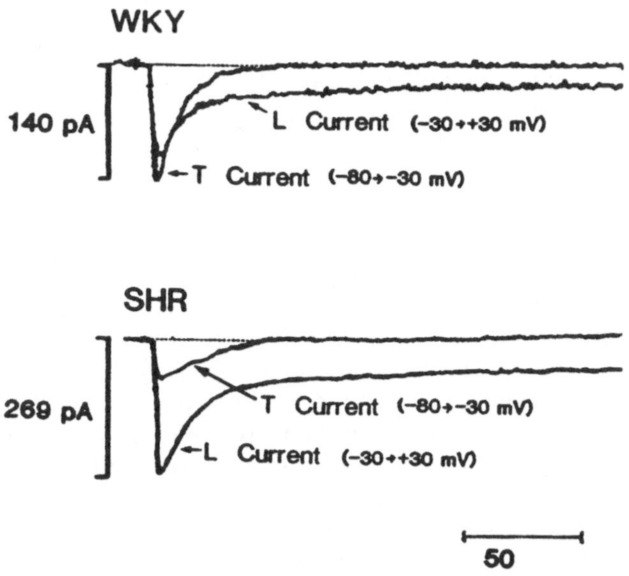

Fig. 3. Whole-cell calcium currents in *WKY* (*upper tracings*) and *SHR* (*lower tracings*) azygos venous mucle cells. The T current was measured during a 300-MS depolarizing pulse from −80 to −30 mV, while the L current was recorded during depolarization from −30 to +30 mV. The WKY cell showed predominantly T current (peaks of the T and L current were 140 and 112 pA, respectively), while the SHR cell showed more L current (peaks of the T and L current were 90 and 269 pA, respectively). Reproduced from [13], courtesy of *Circulation Research*

compared the whole-cell calcium currents of vascular smooth muscle from the azygos venous smooth muscle from both WKY and SHR (Fig. 3). They noted that in both types of cells two types of currents were observed: one transient and one long-lasting. Although the total calcium current did not differ between the cells from these two sources, the relative proportion of the two calcium currents was significantly different between the WKY and SHR cells. The transient current was greater in the WKY cells, whereas the long-lasting current was greater in the SHR cells. Since the long-lasting current delivers the calcium which is thought to regulate contraction, it is possible that this difference occurring in neonatal animals would explain a genetic component of increased peripheral resistance in the adults.

Calcium Channel Manipulations

Several types of studies have been carried out in which the functional effects of specific manipulations of calcium channels have been monitored. Studies which have compared the effects of specific manipulations on calcium channels from WKY and SHR have permitted an assessment of the similarities and differences between channels from these two sources.

Calcium Antagonists

Rinaldi et al. [15] compared the relaxing effect produced by nifedipine on aortic smooth muscle from SHR and WKY. The EC_{50} for this relaxation of SHR was (in $-\log$ M) 13.1 ± 0.4, whereas that for the WKY was 9.4 ± 0.2. Although other mechanisms may be involved, this marked difference in sensitivity to a calcium entry blocker suggests that the membrane receptor for nifedipine is far more sensitive in SHR than in WKY.

Lamb et al. [16] have described a marked difference in the responses to NE in tail arteries from WKY and from SHRSP. Whereas the response of the artery from WKY was a stable maintained contraction, that from SHRSP oscillated. In recent electrophysiological studies [17], these investigators observed that the oscillations were associated with action potentials (Fig. 4). These action potentials and oscillations in the response of the artery from SHRSP were eliminated by treatment with the calcium antagonist nifedipine. These findings suggest that the action potential reflects a calcium current. Apparently, the cell membrane from SHRSP has a calcium channel capable of generating action potentials whereas that from WKY does not.

In addition to blocking calcium channels in the cell membrane, the channel antagonists interfere with other membrane properties. For example, verapamil causes a small depolarization of the membrane in some vascular smooth muscle

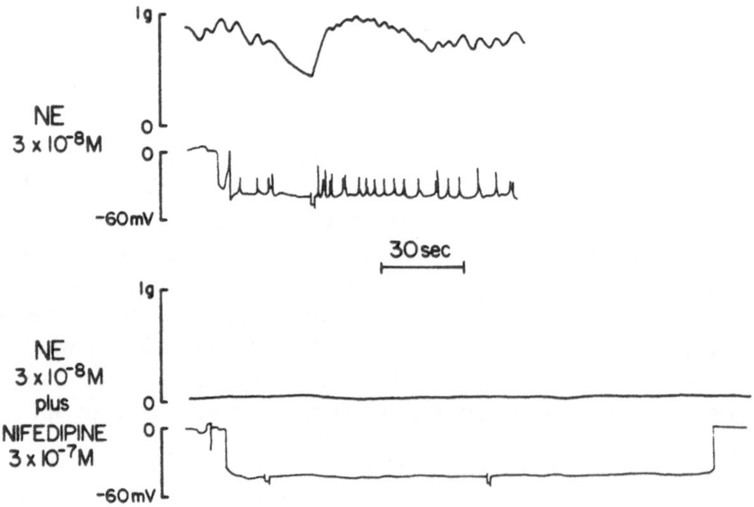

Fig. 4. Effect of nifedipine on membrane potential and force generation. Membrane potential was measured before and after treatment with nifedipine in two separate cells from the same SHRSP tail artery in which contractile force was being measured. Following treatment with norepinephrine (*NE*), tail arteries from SHRSP developed oscillatory contractions that were associated with action potentials (*top panel*). Nifedipine blocked force development and inhibited the action potential (*bottom tracing*). Nifedipine treatment is not associated with a change in membrane potential. From Lamb and Webb by permission, unpublished observation

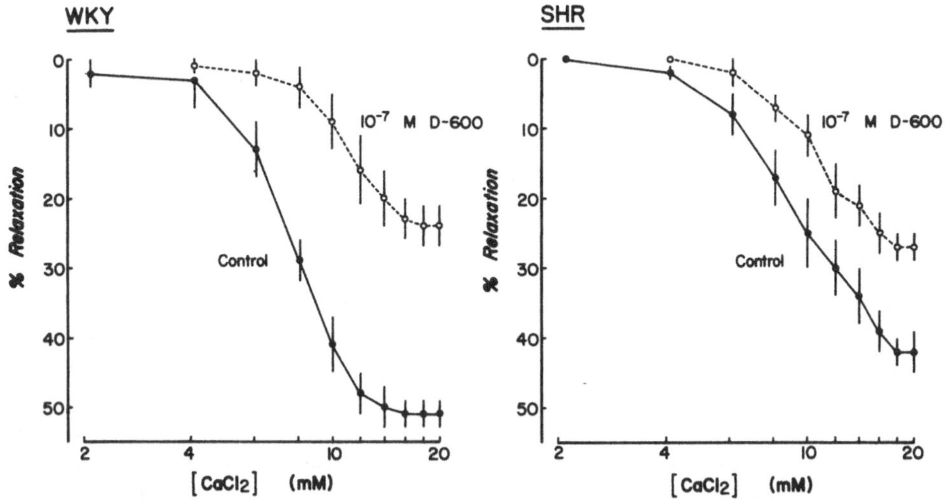

Fig. 5. D-600 and the membrane stabilizing effect of calcium. Tail artery strips from SHR and WKY were made to contract in response to 1.2×10^{-6} *M* methoxamine in the absence (*control*) and presence of 10^{-7} *M D-600*. After the contractile responses had reached a plateau, the calcium concentration of the PSS was raised from 1.6 m*M* to the concentrations indicated on the graph. The magnitude of relaxation in response to calcium was less in tail arteries of *SHR* (*right panel*) than in tail arteries from *WKY* (*left panel*). This relaxation response is caused by a membrane stabilizing effect of calcium, and it is inhibited by D-600. The magnitude of D-600 inhibition of the calcium-induced relaxation is less in tail arteries of SHR compared to WKY, as indicated by the smaller difference in relaxation between control and treated responses. From unpublished observations of Webb

preparations [18], and it is known that this antagonist has a direct action on phospholipid interactions in the membrane [19]. Presumably, these extra actions of the channel antagonists are related to their ability to bind at sites where calcium may exert a regulatory action.

We have examined the effects of D-600, an analogue of verapamil, on the ability of elevated calcium concentrations to relax tail arteries from SHR and WKY (membrane stabilizing effect of calcium [20]). Following contraction induced by methoxamine in normal PSS, tail arteries from SHR relaxed less in response to elevations in calcium concentration than tail arteries from WKY (compare control curves in Fig. 5). D-600 (10^{-7} *M*) inhibited the calcium-induced relaxation in arterial segments from both strains of rats, but the magnitude of the inhibition was less in SHR arteries than in WKY ones. This relationship is seen as a smaller shift to the right in the dose-response curve to calcium in the presence of D-600 in SHR arteries (with respect to control values) as compared to WKY arteries (Fig. 5). We attribute the lesser ability of calcium to cause relaxation in SHR to a decrease in the susceptibility of the membrane to the stabilizing actions of calcium. The reduced inhibiting action of D-600 in SHR arteries suggests that there may be a smaller number of binding sites that elicit membrane stabilization to calcium than in arteries from WKY.

Calcium Agonist

Recently, studies have compared the action of Bay K 8644—a dihydropyridine calcium agonist—on vascular smooth muscle from WKY and SHR. Asano et al. [21] observed that the vascular smooth muscle from SHR is more sensitive to this agonist than is that from WKY. Bruner and Webb (personal communication) have observed that carotid artery strips from SHRSP contracted in response to Bay K 8644 in PSS with normal potassium concentration, whereas those from WKY did not. When the concentration-response curve to Bay K 8644 was repeated in PSS containing 12 m*M* KCl, a small contraction developed in the WKY strip, and the magnitude of the response of the strip from SHRSP increased. When the KCl concentration was increased to 18 m*M*, both strips gave large and equal responses to Bay K 8644. It was concluded that in the resting vascular smooth muscle from SHRSP, more calcium channels are in a conformation that can be activated by Bay K 8644 than in the same muscle from WKY. This is a conformation that can be achieved by depolarization in vascular smooth muscle from either source.

A third manipulation which we have observed to influence the calcium channels in vascular smooth cuscle from SHRSP differentially is the mode of membrane depolarization [22]. Carotid artery strips from SHRSP are more sensitive to the contractile effects of elevated potassium level or of tetraethylammonium chloride than strips from WKY. Additionally, the rate of tension development in carotid arteries from SHRSP placed in potassium-free solution is faster than that in arterial strips from WKY. Each of these three experimental interventions causes contraction via a membrane-depolarizing event: (a) Elevations in extracellular potassium concentration cause depolarization by a reduction in the transmembrane gradient for the ion; (b) tetraethylammonium produces depolarization by inhibiting the potassium channels; and (c) potassium-free solution produces membrane depolarization by blocking the electrogenic sodium pump. In carotid arteries from both SHRSP and WKY, the contractile responses to all three experimental interventions were equal in terms of maximal effect, and they were all inhibited by calcium-free solution and by the calcium channel blocker verapamil. The striking difference between the various modes of depolarization was that the threshold sensitivity to tetraethylammonium is approximately 40 times lower in carotid arteries from SHRSP compared with that in arteries from WKY, whereas sensitivity to elevated potassium level and to potassium-free conditions increased by only a factor of 2 in SHRSP arteries. These results suggest that the calcium channels in SHRSP may be differentially regulated as compared to those in WKY.

Membrane Stabilization by Calcium

Several types of studies on different tissues have indicated that cell membrane function can be altered by changing the extracellular calcium concentration. Because the various effects of increasing this concentration represent a decrease in membrane function, it has seemed appropriate to refer to this action as a membrane stabilization by calcium. Some observations will be cited which indicate that a higher concentration of extracellular calcium is required to stabilize the membrane of tissues from SHR than from WKY.

Evidence of membrane stabilization by calcium in vascular smooth muscle is seen in its relaxation [23]. Figure 6 is a tracing of the effects of increasing calcium concentration on the contraction of femoral artery smooth muscle from normotensive and hypertensive rats. Relaxation was produced in the normotensive vascular smooth muscle by a concentration of 4.1 m*M* CaCl, whereas it required 13.1 m*M* CaCl to produce relaxation of vascular smooth muscle from the hypertensive animal. This finding is compatible with the hypothesis that there are fewer calcium-binding sites on the membrane of the hypertensive animal, and hence a higher concentration of calcium is needed in the extracellular fluid to achieve the same amount of bound calcium on the membrane to produce the stabilization.

Jones and Hart [24] have monitored potassium efflux as an index of membrane stability. They observed that as calcium concentration in the PSS was increased, the potassium efflux decreased. They also observed that it required a higher concentration of calcium to produce a given decrement in potassium efflux in aortic smooth muscle from DOCA hypertensive rats than in that from normotensive controls.

Furspan and Bohr [25] observed similar effects of increased calcium concentration on both potassium efflux and sodium influx in lymphocytes. In either of these fluxes, a higher concentration of calcium was required to stabilize the lymphocyte membrane from the hypertensive rat than was required to stabilize that from its normotensive control.

The interpretation of the mechanism responsible for this difference between SHR and WKY membranes in relation to calcium stabilization is substantiated by numerous direct measurements in which it is found that the calcium-binding capacity of the membranes from various tissues of SHR is less than that of similar tissues from WKY [26–28]. Recently, Kawarski et al. [29] have quantified the amount of calcium-binding portein, IMCAL, in various tissues from SHR and WKY, and find that it is reduced by approximately 30% in membranes from SHR.

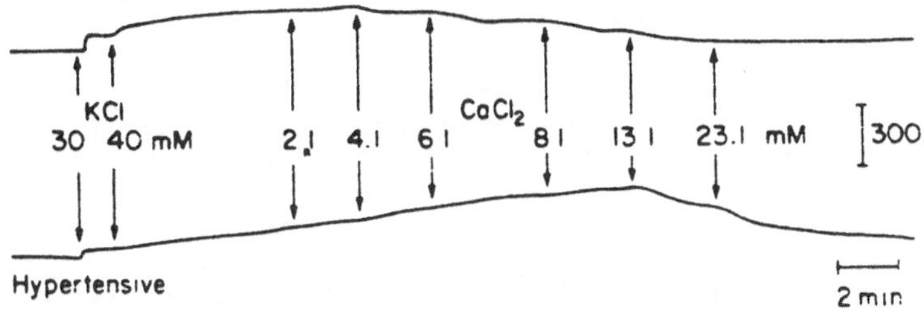

Fig. 6. Effect of Ca concentration on contractile response to 40 m*M* KCL of helical strips of femoral artery from WKY and SHR. Maximum contraction of smooth muscle from the normotensive rat occured at a Ca concentration of 2.1 m*M*, that from the hypertensive rat at 8.1 m*M*. From [23], courtesy of Springer-Verlag

Further evidence that supports the hypothesis of a less stable membrane contribution to increased vascular reactivity in SHR is apparent from studies on the pharmacological actions of ouabain and the calcium chelating agent, EGTA. Moreland et al. [30] observed that isolated aortic strips from SHR contract to ouabain (10^{-3} M) at a faster rate than do aortic strips from WKY. This contractile action of ouabain probably relates to the accumulation of intracellular sodium as a consequence of electrogenic sodium pump inhibition. As sodium accumulates in the cell, the transmembrane gradient of sodium is diminished, lessening the calcium extrusion by the sodium-calcium exchanger, and the smooth muscle contracts. Since the smooth muscle from SHR contracts at a faster rate than WKY aortic strips after 90 min of glycoside treatment yet still reaches the same magnitude, we interpreted these observations as evidence that in hypertension the membrane is more permeable to sodium. This interpretation was supported by the observation that monensin, a sodium ionophore, potentiates contractile responses to ouabain while amiloride, a sodium blocker, inhibits responses to the glycoside. Depending upon the concentration of these agents, contractions to ouabain in normal smooth muscle could be made to appear similar to those seen in aortic strips from SHR, and vice versa.

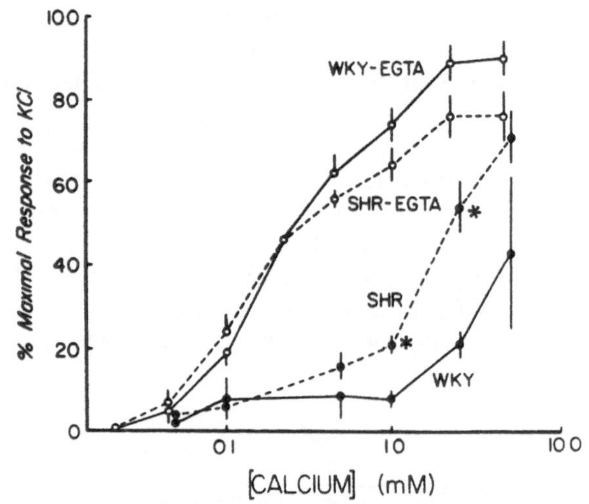

Fig. 7. Calcium and contractile responses to ouabain. *Closed symbols*: Aortic strips from *SHR* and *WKY* were made to contract in response to ouabain (1.0 mM) in the presence of different concentrations of calcium (0.05–5.0 mM). The 20-min responses of individual strips are shown here. At this interval, aortic strips from SHR developed greater force to ouabain than aortic strips from WKY. The number of rats tested at 1.0 and 5.0 mM calcium was five. All other points represent data from four animals. *Open symbols:* Aortic strips from SHR and WKY were made to contract in response to calcium (0.1–10.0 mM) in the presence of ouabain (1.0 mM) following incubation in EGTA. Dose-response curves to calcium in SHR aortic strips were not statistically different from those in aortic strips from WKY. Values are the mean ± SEM ($n = 5$ for both groups). *Asterisks* indicate a statistically significant difference between SHR and WKY (Student's t-test, $P < 0.05$). From Lamb et al. by permission, unpublished observation

The importance of calcium at the membrane and this same relationship of altered ouabain responsiveness in SHR were observed in another manner. When aortic strips from SHR and WKY were placed in PSS containing different concentrations of calcium (0.05–5.0 m*M*) and were then made to contract with 10^{-3} *M* ouabain, the dose-response relationship for 20-min contractions at different calcium concentrations in SHR aortic strips shifted to the left of that of WKY aortic strips (Fig. 7). The increased sensitivity to calcium in SHR strips treated with ouabain is probably related to altered membrane permeability to sodium, as mentioned above. Interestingly, when the aortic strips are first treated with 1.0 m*M* EGTA for 5 min and then placed in a calcium-free solution containing ouabain (10^{-3} *M*), the subsequent SHR dose-dependent contractions to added calcium at 20-min intervals are not statistically different from those in WKY (Fig. 7). Thus, incubation of the arteries with EGTA results in a faster contractile event to ouabain at all calcium concentrations and eliminates the difference between SHR and WKY arteries. Presumably, the brief incubation in EGTA removes calcium bound to stabilizing sites in the cell membrane, resulting in a greater, more excitable contraction to ouabain. Since the magnitude of the shift in the calcium dose-response curve following EGTA treatment was smaller in SHR, we conclude that deficient calcium binding to the cell membranes contributes to greater membrane lability and an increased sensitivity to vasoactive agents in this genetic model of hypertension.

In conclusion, these observations are compatible with the hypothesis that calcium channels in cell membranes from SHR permit a greater calcium influx than do those in membranes from WKY. The basis for this difference is that the membrane from the hypertensive animal is less readily stabilized by calcium.

References

1. Goldblatt H, Lynch J, Hanzal RF, Summerville WW (1934) Studies on experimental hypertension: production of persistent elevation of systolic pressure by means of renal ischemia. J Exp Med 59: 347
2. Follow B (1982) Physiological aspects of primary hypertension. Physiol Rev 62: 347–504
3. Webb RC, Bohr DF (1981) Recent advances in the pathogenesis of hypertension: consideration of structural, functional, and metabolic vascular abnormalities resulting in elevated arterial resistance. Am Heart J 102: 251–264
4. Okomoto K, Aoki K (1963) Development of a strain of spontaneously hypertensive rats. Jpn Circ J 27: 282–293
5. Noon JP, Rice PJ, Baldessarini RJ (1978) Calcium leakage as a cause of high resting tension in vascular smooth muscle from spontaneously hypertensive and normotensive rats. Proc Natl Acad Sci USA 75: 1605–1607
6. Suzuki A, Yanagawa T, Tajiri T (1979) Effects of some smooth muscle relaxants on the tonus and on the actions of contracile agents in isolated aorta of SHRSP. Jpn Heart J 20 (Suppl I): 219–221
7. Fitzpatrick DF, Szentivanyi A (1980) The relationship between increased myogenic tone and hyporesponsiveness in vascular smooth muscle of spontaneously hypertensive rats. Clin Exp Hypertension 2: 1023–1037
8. Winpuist RJ, Bohr DF (1983) Structural and functional changes in cerebral arteries from spontaneously hypertensive rats. Hypertension 3: 292–297

9. Rinaldi G, Bohr D (1989) Endothelium-mediated spontaneous response in aortic rings of deoxycorticosterone acetate-hypentensive rats. Hypertension 13: 259–261
10. Yanagisawa M, Kurihara H, Kimura S, Tomobe Y, Kobayashi M, Mitsui Y, Yazaki Y, Goto K, Masaki T (1988) A novel potent vasoconstrictor peptide produced by vascular endothelial cells. Nature 332: 411–415
11. Cauvin C, van Breemen C (1986) Membrane Ca^{2+} permeability and calcium antagonistic effects in resistance vessels of spontaneously hypertensive rats. In: K. Aoki (ed) Essential hypertension. Springer, Tokyo Berlin Heidelberg New York, pp 127–133
12. Hwa J, Bevan J (1986) Stretch-dependent (myogenic) tone in rabbit ear resistance arteries. Am J Physiol 250 (Heart Circ Physiol 19): H87–H95
13. van Breemen C, Cauvin C, Johns A, Leijten P, Yamamoto H (1986) Ca^{2+} regulation of vascular smooth muscle. Fed Proc 45: 2746–2751
14. Rusch NJ, Hermsmeyer K (1988) Calcium currents are altered in the vascular muscle cell membrane of spontaneously hypertensive rats. Circ Res (in press)
15. Rinaldi GJ, Cattaneo EA, Mattiazzi A, Cingolani HE (1987) Dissociation between calcium influx blockage and smooth muscle relaxation by nifedipine in spontaneously hypertensive rats. Circ Res 60: 367–374
16. Lamb FS, Myers JH, Hamlin MN, Webb RC (1985) Oscillatory contractions in tail arteries from genetically hypertensive rats. Hypertension 7 (Suppl I): 25–30
17. Lamb FS, Webb RC (1988) Bursting pacemaker activity associated with oscillatory contractions to norepinephrine (NE) in tail arteries from spontaneously hypertensive stroke-prone rats (SHR-SP). FASEB J 2: A787
18. Haeusler G (1972) Differential effect of verapamil on excitation-contraction coupling in smooth muscle and on excitation-secretion coupling in adrenergic nerve terminals. J Pharmac Exp Ther 180: 672–682
19. Mori T, Takai U, Minakuchi R, Yu B, Nishizuka Y (1980) Inhibitory action of chlorpromazine, dibucaine and other phospholipid-interacting drugs on calcium-activated, phospholipid-dependent protein kinase. J Biol Chem 255: 8378–8380
20. Webb RC, Bohr DF (1978) Mechanism of membane stabilization by calcium in vascular smooth muscle. Am J hysiol 235: C227–C232
21. Asano M, Aoki K, Matsuda T (1986) Action of calcium agonists and antagonists on femoral arteries of spontaneously hypertensive rats. In: K Aoki (ed) Essential hypertension: calcium mechanisms and treatment. Springer, Tokyo Berlin Heidelberg New York, pp 135–149
22. Thompson LP, Bruner CA, Lamb FD, King CM, Webb RC (1986) Calcium influx and vascular reactivity in hypertension. Am J Cardiol 59: 29A–34A
23. Holloway ET, Sitrin MD, Bohr DF (1972) Calcium dependence of vascular smooth muscle from normotensive and hypertensive rats. In: Genest J, Koiw E (eds) Hypertension 1972. Springer, Berlin Heidelberg New York Tokyo, pp 400–408
24. Jones AW, Hart RG (1975) Altered ion transport in aortic smooth muscle during deoxycorticosterone acetate hypertension in the rat. Circ Res 37: 333–341
25. Furspan PB, Bohr DF (1986) Calcium-related abnormalities in lymphocytes from genetically hypertensive rats. Hypertension 8 (Suppl II): 123–126
26. Devynck MA, Pernollet MG, Nunez AM, Aragon I, Montenary-Garestier T, Helene C, Meyer P (1982) Diffuse structural alterations in cell membranes of spontaneously hypertensive rats. Proc Natl Acad Sci USA 79: 5057–5060
27. Kwan CY, Grover AK, Sakai Y (1982) Abnormal biochemistry of subcellular membranes isolated from nonvascular smooth muscles of spontaneously hypertensive rats. Blood Vessels 19: 273–283
28. Postnov VY, Orlov NS (1984) Cell membrane alteration as a source of primary hypertension. J Hypertension 2: 1–6
29. Kawarski S, Cowen LA, Schachter D (1986) Decreased content of integral membrane calcium-binding protein (IMCAL) in tissues of the spontaneously hypertensive rat. Proc Natl Acad Sci USA 83: 1097–1100
30. Moreland RS, Major TC, Webb RC (1986) Contractile responses to ouabain and K^+-free solution in aorta from hypertensive rats. Am J Physiol 250: H612–H619

Membrane Potential and Calcium Influx in Vascular Muscle from Spontaneously Hypertensive Rats

KENT HERMSMEYER and NANCY J. RUSCH[1]

Summary. Comparison of calcium currents in voltage clamped vascular muscle cells shows differences in membrane excitation between WKY and Aoki SHR vascular muscle cells. Even in veins of new-born SHR, where blood pressure is normotensive, there is a significant shift towards a preponderance of the L-type, rather than the T-type, calcium channel. Changes in calcium influx in SHR would be expected to result in important changes in vascular muscle cell functions.

Key words: Excitation-contraction coupling—Ca^{2+}—Calcium currents—Hypertension

Introduction

Essential hypertension has been associated with an altered regulation of Ca^{2+} by the vascular muscle cell [1]. One aspect of this alteration is a change in the permeability and transport of Ca^{2+} by the cell membrane [2–6] in vascular muscle cells from spontaneously hypertensive rats (SHR). In this report, we will address the possibility that Ca^{2+} channels are altered in single vascular muscle cells from newborn SHR, suggesting that a genetic defect in the vascular muscle membrane proteins may account for changes in cell Ca^{2+} metabolism.

Methods

Primary cultures of vascular muscle cells from azygous veins of 1–4 day-old rats were prepared as previously described [7, 8]. Neonatal rats were from the Wistar-Kyoto (WKY) or spontaneously hypertensive (Aoki SHR) strains. Cells were grown on poly-*L*-lysine-coated cover slips in a 5% CO_2 incubator at 95% humidity at 37°C and used after 3–7 days in culture.

[1] University of Iowa and Cardiovascular Research Laboratory, Chiles Research Institute, Providence Medical Center, and Department of Medicine and Cell Biology, Oregon Health Sciences University, Portland, OR 97213, USA

For measurement of whole-cell Ca^{2+} currents, cover slips of cells were placed in a perfusion chamber containing an external solution consisting of (m*M*): 20 $CaCl_2$, 135 tetraethylammonium (TEA) chloride, 1 $MgCl_2$, 10 glucose, and 10 N-2-Hydroxyethylpiperagine-N′-2-ethanesulphonic acid (HEPES) (pH adjusted to 7.4 with TEA OH, 22°C). Patch pipettes were filled with a cesium glutamate solution to eliminate outward K^+ currents. The pipette solution was composed of (m*M*): 150 cesium glutamate, 0.1 ethyleneglycol-bis-(β aminoethyl ether) N,N,N′,N′-tetraacetic acid (EGTA), 1 $MgCl_2$, and 10 HEPES (pH adjusted to 7.4 with CsOH). Whole-cell Ca^{2+} currents were recorded after suction break-through into the cell and measured during 300 ms depolarizing pulses. The voltage-clamp currents were amplified by a List EPC-7 patch clamp amplifier with a 0.5 Gohm Feedback resistor, and filtered at 500 Hz. All data were digitized (sampling rate, 5000/s) and stored on floppy disks to permit analyses at a later time. Leak and capacitative currents were subtracted from each record by summation of currents during depolarizing pulses with linearly scaled current obtained during 10 mV hyperpolarizing pulses. Figures were traced from corrected currents printed from digitized data.

Statistical comparisons were made between vascular muscle cells from SHR and WKY by unpaired *t*-tests. Comparisons of currents within a sample of SHR or WKY were analyzed by paired *t*-test with a P value of <0.05 always accepted as significant.

Results

Two types of Ca^{2+} channels can be distinguished during whole-cell voltage-clamp recording from azygous venous cells, as previously described [9]. Current through these two channels can be isolated by the use of different holding potentials in this preparation. Transient (T) current, which is activated and inactivated at negative membrane potentials, is measured during a test pulse from −80 mV to −30 mV. The long-lasting (L) current is elicited by depolarizing from −30 mV to +30 mV, since it is activated and inactivated in a more positive range of membrane potential.

Using this method, Fig. 1 shows whole-cell T and L Ca^{2+} currents in a WKY azygous venous cell. We initially identified T and L types of Ca^{2+} currents in these vascular muscle cells from normotensive rats and characterized them according to their voltage-dependent and pharmacological properties [9]. In a later study on a larger sample of WKY azygous venous cells, we showed that approximately two-thirds of WKY vascular cells have predominantly the T type of Ca^{2+} channel [10].

When we recorded whole-cell T and L Ca^{2+} currents in similar azygous cells from SHR, the predominant type of Ca^{2+} current measured was the L current [10]. Eighty percent of SHR venous cells studied showed more L than T Ca^{2+} current, as demonstrated in the recording in Fig. 2.

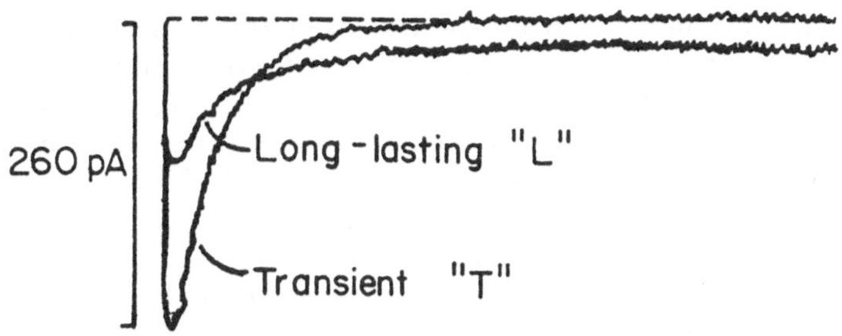

Fig. 1. Whole-cell calcium currents in a Wistar-Kyoto rat (WKY) azygous venous muscle cell. The *T* current was measured during a 300 ms depolarizing pulse from −80 to −30 mV, while the *L* current was recorded during depolarization from −30 to +30 mV. The WKY cell showed predominantly *T* current. Reproduced with permission of American Heart Association

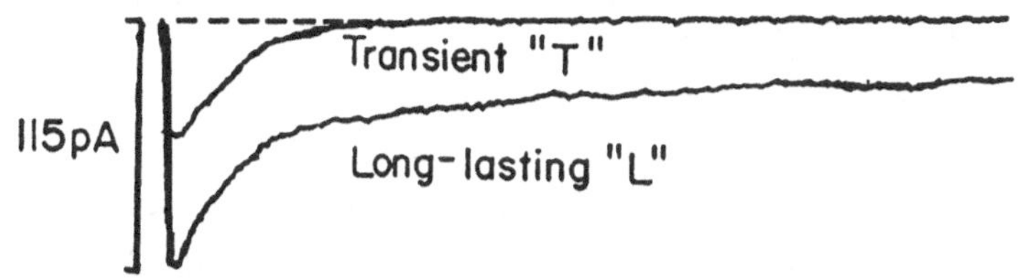

Fig. 2. Whole-cell calcium currents in a spontaneously hypertensive rat (SHR) azygous venous muscle cell. The *T* current was measured during a depolarizing pulse from −80 to −30 mV, while the *L* current was recorded during depolarization from −30 to +30 mV (300 ms voltage clamp duration). The SHR cell showed predominantly *L* current. Reproduced with permission of American Heart Association

Discussion

Two types of Ca^{2+} current have been measured in vascular muscle cells by this and other laboratories [9, 11–13]. The T and L Ca^{2+} currents differ in their thresholds for activation, decay rates during maintained depolarization, and sensitivity to permeant charge carriers and organic Ca^{2+} antagonists. We believe that the T current, which is activated at negative membrane potentials and inactivates rapidly, may play a role in pacemaker activity and initiating contraction. In contrast, the L current, which activates at more positive membrane potentials and is sustained during depolarization, may provide extracellular Ca^{2+} to maintain contraction.

Evidence previously presented shows that the proportion of T and L Ca^{2+} current is altered in vascular muscle cells from SHR [10]. This suggests that voltage-dependent Ca^{2+} channels are altered in the vascular muscle membrane, and concurs with earlier reports that genetic hypertension may be related to a basic membrane defect [1]. Whether the alteration in Ca^{2+} currents represents changes in Ca^{2+} channels strictly of genetic origin or is due to the influence of neuronal or humoral factors is unknown. However, since we found altered Ca^{2+}

channels in vascular muscle from 1–4 day-old rats, it is unlikely that the membrane properties were altered by neurogenic influences. Cells from these animals are not effectively innervated until 1–3 weeks after birth [14], making a trophic function of the sympathetic nervous system on the vascular muscle membrane unlikely. It is also unlikely that a greater distending pressure altered Ca^{2+} channels in the vascular cell membrane of SHR, as the azygous cells studied were isolated from venous tissue.

Calcium channels in vascular muscle not only are important in membrane excitation, but also supply activator Ca^{2+} to initiate and maintain contraction. Thus, alteration of these channels may profoundly affect vascular muscle contractility and ultimately peripheral vascular tone. Further studies are needed to elucidate the functional roles of the different Ca^{2+} channels and their modulation by physiological and pathophysiological factors.

Acknowledments. This research was supported by Grants HL14388, HL16328, HL38537, and HL38645 from the National Institutes of Health and the Postdoctoral Fellowship HL06907 to N.J. Rusch.

References

1. Kwan CY (1985) Dysfunction of calcium handling by smooth muscle in hypertension. Can J Physiol Pharmacol 63: 366–374
2. Bhalla RC, Webb RC, Singh D, Ashley T, Brock T (1978) Calcium fluxes, calcium binding and adenosine cyclic 3′5′-monophosphate-dependent protein kinase activity in the aorta of spontaneously hypertensive and Kyoto-Wistar normotensive rats. Mol Pharmacol 14: 468–477
3. Somlyo AP, Somlyo AV (1968) Vascular smooth muscle. I. Normal structure, pathology, biochemistry, and biophysics. Pharmacol Rev 20: 197–272
4. Moore L, Hurwitz L, Davenport GR, Landon EJ (1975) Energy-dependent calcium uptake activity of microsomes from the aorta of normal and hypertensive rats. Biochim Biophys Acta 413: 432–443
5. Fitzpatrick DF, Szentivanyi A (1980) The relationship between increased myogenic tone and hyperresponsiveness in vascular smooth muscle of spontaneously hypertensive rats. Clin Exp Hypertens [A] 2: 1023–1037
6. Hermsmeyer K (1984) Altered arterial muscle ion transport mechanism in the spontaneously hypertensive rat. J Cardiovasc Pharm 6: S10–S15
7. Marvin WJ, Robinson RB, Hermsmeyer K (1979) Correlation of function and morphology of neonatal rat and embryonic chick cultured cardiac and vascular muscle cells. Cir Res 45: 528–540
8. Hermsmeyer K, Mason R (1982) Norepinephrine sensitivity and desensitization of cultured single vascular muscle cells. Circ Res 50: 627–632
9. Sturek M, Hermsmeyer K (1986) Calcium and sodium channels in spontaneously contracting vascular muscle cells. Science 233: 475–478
10. Rusch NJ, Hermsmeyer K (1988) Calcium currents are altered in the vascular muscle cell membrane of spontaneously hypertensive rats. Circ Res 63: 977–1002
11. Bean BP, Sturek M, Puga A, Hermsmeyer K (1986) Calcium channels in muscle cells isolated from rat mesenteric arteries: modulation by dihydropyridine drugs. Circ Res 59: 229–235

12. Loirand G, Pacaud P, Mironneau C, Mironneau J (1986) Evidence for two distinct calcium channels in rat vascular smooth muscle cells in short-term primary culture. Pflugers Arch 407: 566–568
13. Yatani A, Seidel CL, Allen J, Brown AM (1987) Whole-cell and single-channel calcium currents of isolated smooth muscle cells from saphenous vein. Circ Res 60: 523–533
14. Ljung B, Stage D, Carlsson C (1975) Postnatal ontogenetic development of neurogenic and myogenic control in the rat portal vein. Acta Physiol Scand 94: 112–127

Cell Membrane Properties of the Arterial Smooth Muscle from Spontaneously Hypertensive Rats*

RAMESH C. BHALLA[1], LUSIANE M. BENDHACK[1,2], and RAM V. SHARMA[1]

Summary. We have measured the cytoplasmic free calcium concentration ($[Ca^{2+}]_i$) and cell contraction simultaneously using freshly isolated single vascular smooth muscle (VSM) cells from bovine carotid artery. In K^+-stimulated cells the increase in $[Ca^{2+}]_i$ preceded the initiation of cell contraction and remained elevated after the maximum contraction was achieved. In contrast, in hormone-stimulated cells, the rise in $[Ca^{2+}]_i$ was transient, lasting less than 30 s, and it then declined gradually to near resting levels while the cells continued to contract. Thus, during hormonal stimulation low levels of Ca^{2+} may act synergistically with protein kinase C to produce sustained VSM contraction. The hypothesis that increased Ca^{2+} sensitivity of norepinephrine (NE)-stimulated SHR caudal artery is due to an increased Ca^{2+} influx, an increased Ca^{2+} release, and/or an increased activation of protein kinase C was tested. NE-stimulated Ca^{2+} influx as well as Ca^{2+} release from SR was significantly increased ($P < 0.05$) in SHR caudal artery as compared with WKY. Furthermore, phorbol esters, potent activators of protein kinase C, produced significantly greater ($P < 0.05$) potentiation of NE-stimulated contraction in SHR caudal artery rings as compared with WKY. Thus, a synergistic action between increased $[Ca^{2+}]_i$ and protein kinase C could account for the increased Ca^{2+} sensitivity observed in the SHR VSM cell.

Key words: Bovine carotid artery—Caudal artery—Fura-2—Phorbol esters—Serotonin

Introduction

It is now well documented that the vascular smooth muscle (VSM) of the spontaneously hypertensive rat (Aoki SHR) is hyperreactive to a variety of vasoconstrictive hormones [1–3]. Vasoconstrictive hormones induce VSM contraction through the receptor-stimulated hydrolysis of phosphatidylinositol-4,5-bisphosphate (PIP_2) resulting in the production of inositol-1,4,5-trisphosphate

*This work was supported by the National Institute of Health Grants HL 35682 and HL 14388.

[1] Department of anatomy, The University of Iowa, College of Medicine, Iowa City, IA 52242, USA

[2] Present address: Faculdade de Medicina de Ribeirao Preto, USP, Department de Fisiologia, Ribeirao Preto, SP 14049 Brasil

(IP_3) and diacylglycerol (DAG) in VSM cells [4–6]. The water-soluble IP_3 then triggers calcium release from the sarcoplasmic reticulum (SR) to initiate muscle contraction [7, 8]. The increase in IP_3 and $[Ca^{2+}]_i$ in response to hormonal simulation is transient, only lasting approximately 30–90 s [6, 9–12]. This would imply that sustained stress maintenance in response to hormone stimulation may depend on mechanisms that can function at or just above the resting $[Ca^{2+}]_i$. Recently, it has been suggested that protein kinase C, which is widely distributed in mammalian tissues including VSM [13], may be involved in the control of sustained VSM contraction [14, 15]. Under physiological conditions, protein kinase C is activated by DAG [16]. Recent evidence suggests that the DAG production may be sustained in smooth muscle during hormonal stimulation for longer periods than IP_3 [6, 17]. Thus, it would appear that hormone-stimulated phosphatidylinositol hydrolysis may be increased in the VSM of SHR leading to increased $[Ca^{2+}]_i$ in the cell and/or increased activation of protein kinase C. Previous studies from our laboratory [3, 18] as well as other laboratories [19, 20] have demonstrated that the Ca^{2+} sensitivity of SHR arterial smooth muscle is increased during hormonal stimulation. Hormone-mediated vasoconstriction is a complex process that may involve several different mechanisms, any one or more of which may be altered in SHR arteries. Therefore, it is possible that the increased Ca^{2+} sensitivity of SHR caudal arteries stimulated with agonists may be due to: (a) increased Ca^{2+} permeability through receptor-operated calcium channels; (b) increased intracellular Ca^{2+} release; or (c) increased activation of protein kinase C. Since phorbol esters have been shown to elicit VSM contraction, probably via the activation of Protein kinase C (see [14]), we have studied the effect of the low-affinity phorbol ester 12-deoxyphorbol 13-isobutyrate 20-acetate (DPBA) on the contraction of caudal artery rings from WKY and SHR in order to test whether protein kinase C-mediated mechanisms are increased in SHR.

Considerable progress has been made in understanding the molecular mechanisms by which the VSM cell membrane regulates the $[Ca^{2+}]_i$; however, the direct measurement of $[Ca^{2+}]_i$ transients during VSM cell contraction has been difficult. The availability of a highly specific and sensitive Ca^{2+} indicator, fura-2, and isolated VSM cells which contract in response to hormonal stimulation and membrane depolarization, has made it possible to study simultaneously changes in $[Ca^{2+}]_i$ and contraction of single VSM cells. We have measured changes in $[Ca^{2+}]_i$ in hormone- and K^+-stimulated single cells and correlated the increase in $[Ca^{2+}]_i$ with cell contraction.

Materials and Methods

Isolation of Vascular Smooth Muscle Cells

VSM cells were prepared from bovine carotid artery according to the procedure of Warshaw et al. [21] with slight modifications given below. Calcium was omitted from the enzyme solution and was decreased to 50 μM in the incubation salt solution. The elastase concentration was decreased to 20 mU/ml, and ATP was omitted. The cell suspension was centrifuged at a very low speed (66 × *g*) for 5 min to remove contaminating enzymes and cell debris. The cell pellet was gently dispersed in ice-cold Ca^{2+}-free Physiological salt solution (PSS) and kept on ice. Cells were made calcium tolerant by gradually increasing the calcium concentration in several steps (1 μM, 10 μM, 100 μM, 0.5 mM, and finally 1.6 mM) at 30-min intervals. After about 3 h from initial isolation, cells were used to measure $[Ca^{2+}]_i$ transients and cell contraction. The cells remained relaxed and responsive for at least 24 h when stored on ice in the presence of 50 μg/ml gentamycin to retard bacterial growth. Cells isolated by this procedure did not consistantly respond to hormones.

Measurement of $[Ca^{2+}]_i$ by Digital Image Analysis of Fura-2-Loaded Cells

Briefly, $[Ca^{2+}]_i$ was measured in individual VSM cells using fluorescent videomicroscopy and digital image analysis of fura-2 loaded cells as described by Sharma and Bhalla [12]. 0.2 ml of calcium-tolerant cells (2–2.8 × 10^5 cells/ml) were transferred to a glass tube and slowly brought to room temperature and were loaded with 10 μM fura-2 for 45 min at 37°C. Cells were then diluted 10-fold with PSS and incubated for an additional 30 min at 37°C. Cells were washed once by contrifugation at 66 × *g* for 5 min and dispersed in normal Ca^{2+} PSS. The fura-2-loaded cells were stored on ice and used within 4–6 h to measure $[Ca^{2+}]_i$. This procedure allowed loading of a sufficient quantity of fura-2 into the cell for detection of fluorescence signals. Then, 30 μl of the cell suspension were pipetted onto a no. 1 glass coverslip mounted in a temperature-controlled chamber (36° ± 0.5°C), and the cells were allowed to attach to the coverslip for 30 min. This time period was sufficient for the attachment of cells to the glass. All experiments were carried out at 36°C unless otherwise indicated. The extent of fura-2 fluorescence varied considerably from cell to cell even in the same batch of cells loaded together. The reasons for these differences are not clear. High potassium (70 mM K^+ final concentration) in the bathing medium was achieved by adding an equal quantity of PSS in which NaCl was substituted with KCl.

Fluorescence images were obtained using a silicon-intensified target (SIT) video camera (Hamamatsu, C2400) attached to a Zeiss IM 35 inverted microscope equipped with epifluorescence, 75W Xenon lamp, and a Zeiss ultrafluor 100 ×/1.25 quartz glycerine immersion objective. The SIT camera output was fed to a monitor and to the Gould FD 5000 image processing system. Excitation wavelengths were obtained by passing the light through interference filters centered at 340 nm and 380 nm (12-nm half width). A variable speed, computer-controlled shutter/filter changer was used to open the shutter when measurements were made, and to switch automatically between 340 nm and 380 nm excitation filters (Fig. 1). Experiments were conducted with a 500-nm band pass filter (half width 40 nm) to maximize the light transmittance. The fluorescence images obtained at 340 nm and 380 nm excitations were corrected for background fluorescence and camera dark current by subtracting a blank frame of the same excitation wave-length made at the end of each experiment by using an area where there were no cells.

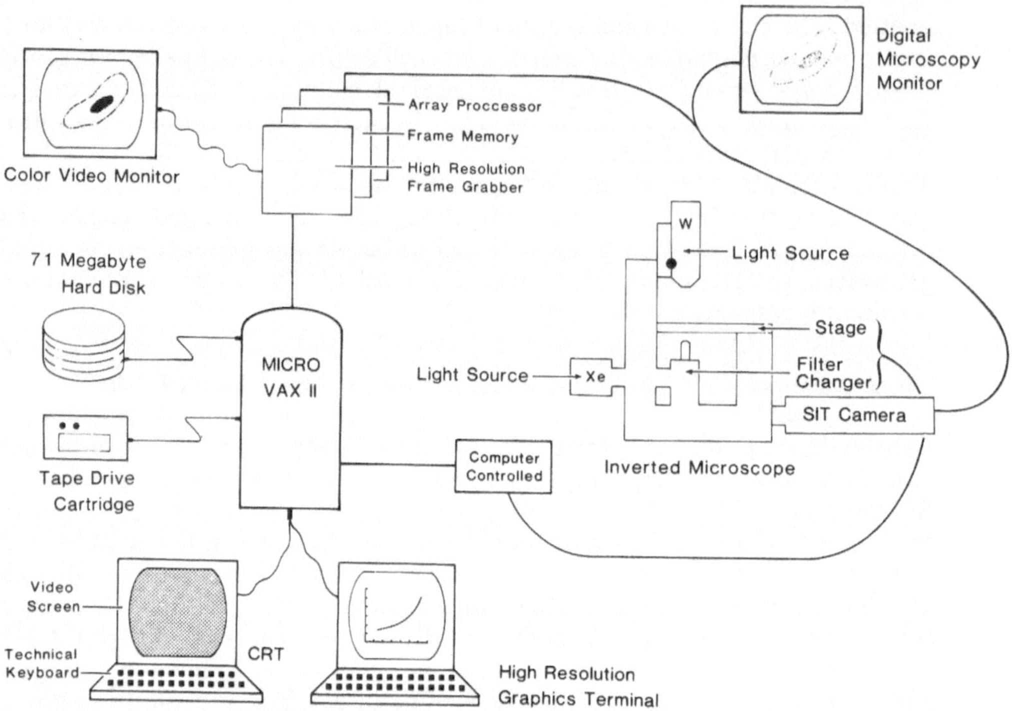

Fig. 1. Schematic representation of microscopic digital image analysis system. The SIT camera output is fed to a monitor and to the Gould FD5000 image processing system. The FD5000 digitizes the video signal to an eight-bit resolution (256 gray levels) and sums successive frames, typically 8, at each of the two excitation wavelengths. All operations of FD5000 are controlled by a Micro Vax II (Digital Equipment Corporation) super mini-computer

For standardization of fluorescence ratios (340/380) at zero calcium (<10 n*M*) and a saturating calcium (>1 m*M*), acid-etched glass coverslips were used. Intermediate ratios (R) were measured using Ca-EGTA buffers with known free calcium concentration. In these experiments Mops buffered medium was used containing: K^+ 100 m*M*, $MgCl_2$ 1 m*M*, EGTA 4 m*M*, Mops 10 m*M* buffered to pH 7.2 with KOH. R_{max}, R_{min}, and Fo/Fs were determined at three separate times to compute $[Ca^{2+}]_i$, accourding to the equation given below:

$$[Ca^{2+}] = K_D \left[\frac{R\text{-}R_{min}}{R_{max}\text{-}R}\right](Fo/Fs) \qquad (1)$$

The minimum fluorescence ratio ($R_{min} = F340/F380$) determined in buffer with no added calcium and 4 m*M* EGTA was between 0.78 and 0.83. The maximum ratio ($R_{max} = F340/F380$) determined in the presence of 1.6 m*M* calcium in buffer was between 8.9 and 9.2. The ratio of fluorescence at 380 nm excitation at zero and saturating levels of calcium (Fo/Fs) was between 5.0 and 5.2. The 340/380 ratio images were converted to an actual $[Ca^{2+}]_i$ in a given cell by substituting in equation (1): $R_{max} = 9.05$, $R_{min} = 0.82$, $Fo/Fs = 5.1$, and $K_D = 224$.

Fluorescence ratio images were formed by dividing the corrected 340 nm image by the corrected 380-nm image on a pixel by pixel basis. The thresholding operation was performed prior to the division so that the background or noncellular regions of the image appear as a constant grey level. These grey levels are then preserved in the pixel by pixel division and yield a crisper and less mottled ratio image.

Measurement of Caudal Artery Contractions

Contractions of caudal artery rings obtained from 20–22-week-old, male, spontaneously hypertensive rats (SHR; BP = 202 ± 5 mmHg, $n = 15$) and control Wistar Kyoto normotensive rats (WKY; BP = 136 ± 7 mmHg, $n = 14$) were measured by the procedures described earlier [18]. Rats were killed by direct heart puncture under ether anesthesia, and the ventral caudal artery was immediately removed. The proximal part of the caudal artery was cut into approximately 3-mm long pieces under a dissecting microscope and mounted by inserting two stainless steel hooks in the lumen of the artery to record tension in the circumferential axis of the vessel wall. Resting tension was adjusted to 2000 mg. Arteries were denervated with 6-hydroxydopamine [18] and then allowed to recover for 3 h in a well-oxygenated PSS containing NaCl 130 m*M*, KCl 4.7 m*M*, $NaHCO_3$ 14.9 m*M*, NaH_2 PO_4 1.18 m*M*, $MgSO_4$ 1.17 m*M*, dextrose 5.5 m*M*, $CaCl_2$ 1.6 m*M*, EDTA 0.027 m*M*, and HEPES 13 m*M*. PSS was aerated with 95% O_2/5% CO_2, and pH was maintained at 7.4.

Prior to the experiments, the arteries were repeatedly contracted with a given contractile agent and only the arteries evoking comparable contractions were used. The effects of DPBA (10^{-6} *M*) alone and/or DPBA plus contractile agents were measured on the same arterial rings. Verapamil, when used, was added to the muscle bath along with NE and DPBA at the time indicated. When the experiments were performed in Ca^{2+}-free PSS, calcium was omitted from the normal PSS. High potassium solution was prepared by equimolar substitution of NaCl with KCl.

DPBA (LC Services Corp, Woburn, MA) was dissolved in dimethylsulfoxide (DMSO) at a final concentrationn of 10 m*M*. Verapamil (Calbiochem Corp., San Diego, CA) was dissolved in ethanol at a concentration of 10 m*M*. The final concentration of DMSO and ethanol were less than 0.1% in the muscle bath, and these concentrations did not affect caudal artery contraction. Results are expressed as maximum isometric tension ($N \times 10^5/m^2$). Surface area was calculated as described earlier [18]. Statistical analysis was done using Student's *t*-test, and differences were considered significant at *P* values less than 0.05.

Results

Basal $[Ca^{2+}]_i$ in the Freshly Isolated Cells

Enzymatically isolated cells from bovine carotid artery appeared relaxed (Fig. 2). Individual cells varied in length from 55 μm to 262 μm. The mean cell length

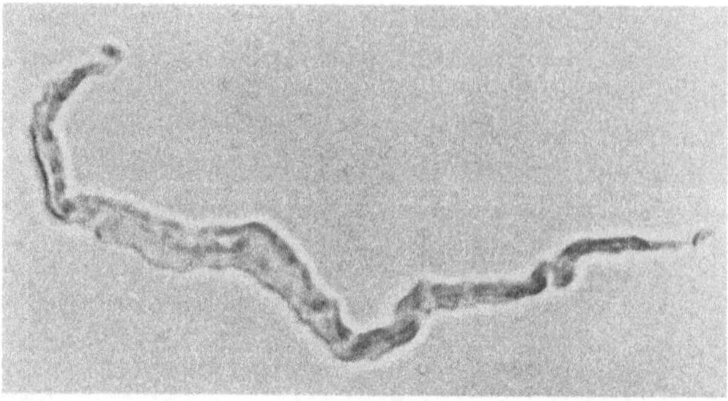

Fig. 2. Light microscopic image of a single vascular smooth muscle cell isolated from bovine carotid artery by collagenase/elastase digestion. The cells were dispersed in physiological salt solution containing 1.6 m*M* Ca^{2+}. Cell length is 246 μm, measured from an image of a live cell attached to a glass covership using an interactive computer program which allows the use of multiple line segments passing through the midsection of the cell. $\times 330$

in a typical preparation was 146 ± 5 μm ($n = 75$). Viability of cells tested by trypan blue exclusion was greater than 85%. The majority of these cells attached to glass coverslips.

In the relaxed cells the basal intracellular calcium concentration ($[Ca^{2+}{}_i)$ was 163 ± 23 n*M* (mean $\pm$ SEM, $n = 55$). Occasionally, fully relaxed and viable cells showed a mean basal $[Ca^{2+}]_i$ up to 350 n*M*. Distribution of $[Ca^{2+}]_i$ was almost homogenous throughout the cell except for a few hot spots. The identity of these hot spots is not known at the present time. No significant differences were observed between the cytosolic and nuclear regions, when the comparisons were made by tracing the nuclear region, which could be identified in raw images (either 340 or 380) due to greater dye uptake in the nucleus in our loading protocols, and tracing that region back in the ratio images. Resting $[Ca^{2+}]_i$ was dependent on extracellular Ca^{2+}, since removal of Ca^{2+} from the extracellular medium significantly decreased basal $[Ca^{2+}]_i$ (88 ± 13 n*M*, $n = 9$).

Effect of Potassium Depolarization on $[Ca^{2+}]_i$

Exposure of cells to 70 m*M* K^+ resulted in a progressive increase in fluorescence at 340 nm excitation and decrease in fluorescence at 380 nm excitation (data not shown). Within 10 s, addition of high potassium near the cell produced a significant increase in $[Ca^{2+}]_i$. The $[Ca^{2+}]_i$ continued to increase gradually up to 120 s and then remained elevated for at least 5 min in the continued presence of potassium (Fig. 3b). The mean peak $[Ca^{2+}]_i$ achieved in these cells was 933 ± 41 n*M* (mean $\pm$ SEM, $n = 36$). The increase in $[Ca^{2+}]_i$ was almost uniform throughout the cell at all time interval after stimulation with high potassium. Although most cells responded to potassium stimulation, we observed significant cell to cell variations in the time course and extent of increase in $[Ca^{2+}]_i$. Thus, in some cells K^+ produced maximum increase in 60 s, while in others it took up to 4–5 min. Similarly, the maximum increase in $[Ca^{2+}]_i$ varied from just above 500 n*M*

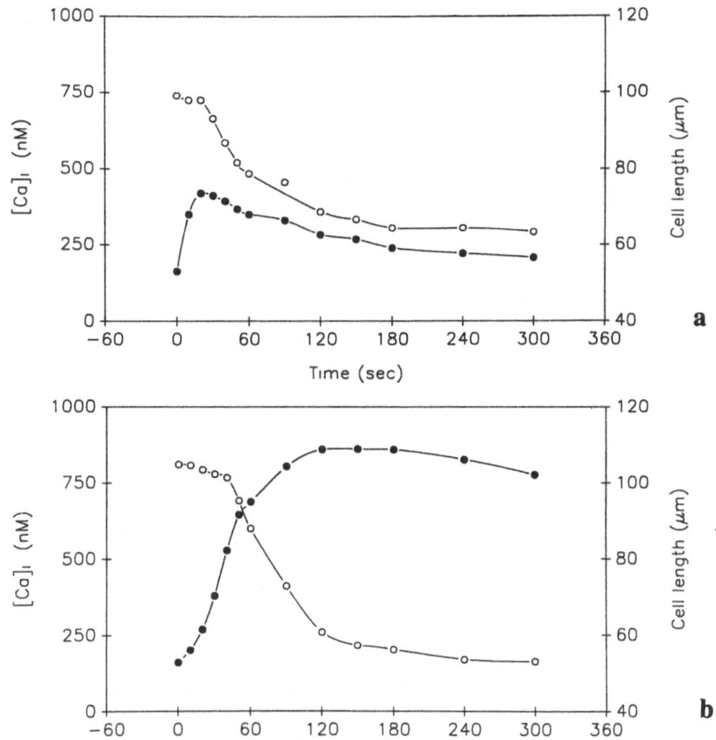

Fig. 3a,b. Changes in $[Ca^{2+}]_i$ (*closed circles*) and cell length (*open circles*) in response to 10 μM serotonin (**a**) and 70 mM K^+ (**b**). The cells were bathed in a physiological salt solution containing NaCl 120 mM, KCL 5.0 mM, $CaCl_2$ 1.6 mM, $MgCl_2$ 1.5 mM, Na_2HEPES 25 mM, glucose 10 mM; BSA 1mg/ml, pH 7.4. All cell images were obtained, processed, and displayed at the same camera gain settings. The time interval between dual wavelength image pairs was defined at the start of each experiment. The intensities of all pixels within the cellular region were summed and converted to ionized Ca^{2+} concentrations

to 1800 nM in different cells. For this reason only the results of a representative cell are plotted.

Potassium-induced increases in $[Ca^{2+}]_i$ were dependent on the extracellular calcium. Prior exposure of cells to calcium channel antagonists verapamil (100 μM) almost completely abolished the increase in $[Ca^{2+}]_i$ in response to high potassium. These results would suggest that membrane depolarization by high potassium increases the opening probability of cell membrane calcium channels, resulting in increased calcium influx.

Effect of Receptor-Stimulation on $[Ca^{2+}]_i$

Addition of 10 μM serotonin resulted in a transient rise in $[Ca^{2+}]_i$ which reached its maximum in 20 s (Fig. 3a). The mean peak $[Ca^{2+}]_i$ was 320 ± 36 nM

(mean ± SEM, $n = 5$). After the initial rise, $[Ca^{2+}]_i$ started to decrease in the continued presence of hormone and returned to near basal values in about 4 min (Fig. 3a). All cells did not respond to serotonin, and among the cells that responded, cell to cell variability was observed both with respect to the time course and extent of increase in $[Ca^{2+}]_i$. For this reason, again, data for a representative cell are plotted. The cell to cell variability could be an inherent property of cells or could be due to an alteration in the hormone receptors during the cell isolation procedure.

Hormone- and K^+-Stimulated Cell Contraction

An increase in $[Ca^{2+}]_i$ in response to high potassium and serotonin resulted in contraction of vascular smooth muscle cells (Fig. 3). Cell contraction was measured from unprocessed fluorescence images of the cell obtained at 340 nm excitation since it was not possible to grab light images during the fluorescence ratioing protocol. During potassium stimulation, the onset of cell shortening was observed approximately 40 s after the first detectable increase in $[Ca^{2+}]_i$. $[Ca^{2+}]_i$ reached around 500 n*M* in this particular cell at the time when the cell started to contract, and the maximum cell contraction was obtained almost 60 s after $[Ca^{2+}]_i$ reached its peak value (Fig. 3b). During maximum cell contraction, $[Ca^{2+}]_i$ remained elevated at or near peak levels in all potassium-stimulated cells. In those cells stimulated with 10 μM serotonin, the oneset of cell contraction started at 30 s, by which time $[Ca^{2+}]_i$ had passed peak values and had started to decline (Fig. 3a). The maximum cell shortening occurred when $[Ca^{2+}]_i$ was declining towards resting levels.

Altered Ca^{2+} Metabolism in SHR Caudal Artery Rings

We have observed that the maximum isometric tension developed by SHR caudal artery rings was significantly more compared with WKY when arteries were stimulated by NE but not when stimulated by K^+. The Ca^{2+} sensitivity of NE-stimulated arterial rings was about twofold higher compared with WKY. This increase in Ca^{2+} sensitivity was specifically due to changes in the α_1-adrenergic receptor-mediated mechanisms in SHR. However, the Ca^{2+} sensitivity of K^+-stimulated arterial rings was comparable between WKY and SHR. These results indicate that the mechanisms involved in K^+-depolarization-dependent contraction are not altered in SHR. However, the mechanisms involved in the coupling of α_1-adrenergic receptors and smooth muscle contractions may be altered in SHR caudal artery [18].

Hormonal stimulation has been shown to increase $[Ca^{2+}]_i$ in vascular smooth muscle cells by increasing transmembrane Ca^{2+} influx and Ca^{2+} release from the sarcoplasmic reticulum (SR) [2]. Thus, it is possible that the increased Ca^{2+} sensitivity of SHR arterial rings could be due to increased Ca^{2+} influx and/or increased Ca^{2+} release from the SR. ^{45}Ca influx in response to NE- and methoxamine-stimulation was significantly increased ($P < 0.05$) in SHR caudal artery rings compared to WKY. On the other hand, ^{45}Ca influx in response to K^+-depolarization was comparable between the two strains [22].

We have measured intracellular Ca^{2+} release from the SR by studying the NE- and methoxamine-induced contraction of SHR and WKY caudal artery rings in the absence of extracellular Ca^{2+} [23]. The magnitude of contraction produced by different concentrations of NE was significantly greater ($P<0.05$) in SHR compared to WKY in Ca^{2+}-free PSS. From these observations it can be concluded that the increased Ca^{2+} sensitivity of hormone-stimulated SHR arterial rings appears to be due to both an increased Ca^{2+} influx through the receptor-operated calcium channels and an increased Ca^{2+} release from the SR. These alterations do not appear to be due to changes in the affinity or density of α_1-adrenoceptors in SHR caudal arteries for the following reasons: (a) The threshold dose of NE for the initiation of contraction was comparable between WKY and SHR [18, 22]; (b) the dose-response curves for NE- and methoxamine-stimulated contraction were parallel in WKY and SHR caudal artery [18, 22, 23]; (c) the affinity and density of α_1-adrenoceptors estimated by [^{3}H]prazosin binding and displacement with agonists was comparable bwtween WKY and SHR caudal artery rings [22].

Effect of DPBA on Contraction of WKY and SHR Caudal Artery

Since our observations with single cells indicated that in response to serotonin cell activation continued during the time $[Ca^{2+}]_i$ was declining, it is possible that protein kinase C activation may play an important role during sustained cell activation. Moreover, several recent observations suggest that activation of protein kinase C by active tumor-promoting phorbol esters produce slowly developing sustained contraction in VSM [14, 15]. The hypothesis that protein kinase-C-mediated mechanisms may be increased in SHR VSM cell was tested by using DPBA as an exogenously added protein kinase C activator. DPBA alone (1 μM) did not produce any contraction in the caudal artery rings of WKY and SHR for up to 2 h; however, 30-min preincubation of both WKY and SHR arterial rings with DPBA significantly augmented NE-induced contraction. The effect of DPBA could be completely and promptly reversed by washing.

DPBA-mediated potentiation of NE-stimulated contraction was dependent on extracellular Ca^{2+}. Removal of Ca^{2+} from the bathing medium completely and promptly inhibited NE plus DPDA-mediated sustained contraction. Similarly, addition of 10 μM verapamil to the bathing medium completely inhibited NE plus DPBA-induced contractions (Fig. 4). These results demonstrate that phorbol ester-mediated protein kinase C activation augments caudal artery contraction produced by neurotransmitters and that this augmentation is reversible and requires the continued presence of extracellular Ca^{2+}.

Using these protocols, we have examined the role of protein kinase C in SHR VSM contraction. The results presented in Table 1 show that DPBA augmented caudal artery contractions evoked by norepinephrine, vasopressin, and potassium, and the extent of the increase was almost twofold higher in SHR as compared with WKY. The extent of DPBA-mediated augmentation was not dependent on the excitatory stimuli.

To test whether the DPBA-mediated augmentation of hormone- or K-stimulated artery contraction is due to increased Ca^{2+} influx, we studied the

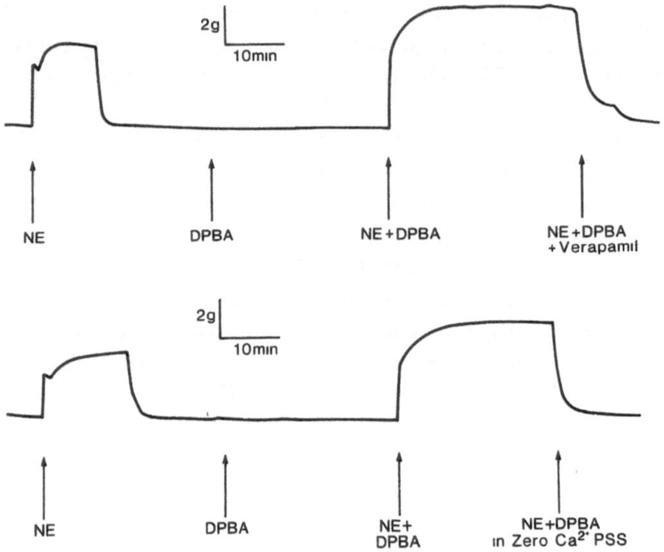

Fig. 4. Dependence of norepinephrine (*NE*) plus *DPBA*-mediated contraction on extracellular calcium. *Upper tracing* shows the effect of 10 μM verapamil on the contractile response to 1 μM NE plus 1 μM DPBA. *Lower tracing* shows the effect of zero calcium physiological salt solution (no EGTA) on the NE plus DPBA-mediated contractile response. In both cases the contractile response was completely and promptly reversed

effect of DPBA on Ca^{2+} influx in the presence as well as the absence of hormones or K^+. DPBA did not augment either basal, NE-, vasopressin-, or K^+-stimulated ^{45}Ca influx in the caudal artery rings from WKY and SHR. Moreover, pretreatment of arterial rings for 30 min with DPBA had no effect on either basal, NE-, vasopressin-, or K^+-stimulated ^{45}Ca influx in both WKY and SHR.

Discussion

We have demonstrated that the Ca^{2+} sensitivity of the caudal artery rings stimulated with NE and the α_1-agonist, methoxamine, is about twofold higher in SHR as compared with WKY [18]. Similarly, the maximum tension developed by NE and methoxamine stimulation at all concentrations of extracellular Ca^{2+} was significantly greater in SHR as compared with WKY. On the other hand, neither the Ca^{2+} sensitivity nor the maximum tension developed in response to K^+ depolarization was different between WKY and SHR caudal arteries. Therefore, our results suggest an increase in the α_1-adrenoceptor-mediated responses in the caudal artery of SHR. These results are consistent with earlier findings of increased Ca^{2+} sensitivity of vascular beds [24], isolated portal vein [20], and isolated mesenteric resistance vessels [19] of SHR stimulated with NE.

Table 1. Effect of DPBA on WKY and SHR caudal artery contractions

Drug	Maximum isometric tension ($N \times 10^5/m^2$)			
	Hormones/protein kinase C activator		Net effect of DPBA	
	WKY	SHR	WKY	SHR
l-Norepinephrine (10^{-6} *M*)	1.5 ± 0.15 (18)	2.66 ± 0.21 (10)[a]	–	–
l-Norepinephrine (10^{-6} *M*) plus DPBA (10^{-6} *M*)	2.12 ± 0.19 (18)[b]	3.88 ± 0.25 (10)[a,b]	0.62	1.22
Vasopressin (2 mU/ml)	0.99 ± 0.05 (4)	1.82 ± 0.11 (4)[a]	–	–
Vasopressin (2 mU/ml) plus DPBA (10^{-6} *M*)	1.59 ± 0.14 (4)[b]	3.48 ± 0.14 (4)[a,b]	0.60	1.66
	Potassium/protein kinase C activator			
Potassium (60 m*M*)	0.79 ± 0.14 (13)	1.0 ± 0.17 (13)	–	–
Potassium (60 m*M*) plus DPBA (10^{-6} *M*)	1.65 ± 0.17 (13)[b]	2.57 ± 0.19 (13)[a,b]	0.86	1.57

Experiments were carried out according to protocols given in legend to Fig. 4. Values are mean ± SEM of *n* given in parentheses
[a] Significant differences ($P < 0.05$) between SHR and WKY
[b] Significant differences ($P < 0.01$) between durgs alone and drugs plus DPBA

In VSM cells, it has been shown that upon activation by hormones or K^+, the cytosolic free calcium concentration ($[Ca^{2+}]_i$) rises, and it is believed that this rise is the major determinant of the contractile state of the VSM [25]. On the basis of the apparent direct relationship between $[Ca^{2+}]_i$ and the contractile state of the VSM cell, and in view of the increased peripheral resistance in SHR, it is logical to assume that $[Ca^{2+}]_i$ may be increased in the VSM cell of SHR. Since the increase in Ca^{2+} sensitivity of SHR caudal artery was observed in response to hormone stimulation but not in response to membrane depolarization, these data prompted us to study $[Ca^{2+}]_i$ profiles in response to hormone and potassium stimulation. The results demonstrate dramatic differences in the kinetics of increase in $[Ca^{2+}]_i$ in serotonin- and K^+-stimulated bovine carotid artery VSM cells. In the case of serotonin stimulation the $[Ca^{2+}]_i$ was declining while the cell contraction was increasing. On the other hand, during K^+ stimulation the increase in $[Ca^{2+}]_i$ preceded cell length shortening and during maximum cell contraction the $[Ca^{2+}]_i$ remained at or near peak levels. Similar observations have been made previously in intact vascular muscle strips using aequorin [9, 11], and in freshly isolated coronary artery cells using quin-2 [10]. From these results, it would appear that an increase in $[Ca^{2+}]_i$ is important in initiating the VSM contraction in response to hormones. However, unlike K^+-stimulated cells, the sustained increase in $[Ca^{2+}]_i$ is not necessary for the sustained contraction during prolonged hormone exposure.

The cellular mechanisms regulating sustained cellular contraction in the face of falling $[Ca^{2+}]_i$ in response to hormonal stimulation are not fully understood. Recently, it has been shown that tumor-promoting phorbol esters produce sustained VSM contraction [14, 15]. Since phorbol esters produce activation of protein kinase C by interacting at the DAG binding site and increasing its calcium sensitivity [26, 27], it is possible that DAG-mediated activation of protein kinase C may be responsible for the sustained contraction of VSM cells in response to hormone stimulation at very low levels of $[Ca^{2+}]_i$. This concept is further strengthened by the finding that hormones produce a sustained rise in DAG [6, 17]. The sustained increase in DAG production thus can maintain sustained activation of protein kinase C which may be playing an important role in hormone-mediated sustained contraction of the VSM cell at calcium concentrations that are not significantly higher than the resting $[Ca^{2+}]_i$.

On the basis of these observations it would appear that receptor-mediated contraction of smooth muscle involves two different mechanisms, namely: (a) The initial increasing $[Ca^{2+}]_i$ due to release of Ca^{2+} from SR may be involved in myosin light chain (MLC) phosphorylation and initiation of contraction; (b) the sustained transmembrane Ca^{2+} cycling dependent on receptor-mediated mechanisms may be involved in the activation of protein kinase C which may be responsible for the maintenance of contraction [15]. Either or both of these mechanisms may be increased in the SHR caudal artery during hormonal stimulation to produce greater Ca^{2+} sensitivity and greater contraction.

Measurement of ^{45}Ca influx in response to NE, methoxamine, and K^+ indicated that ^{45}Ca influx through receptor-operated Ca^{2+} channels (ROCs), but not potential-operated Ca^{2+} channels, is increased in the SHR caudal artery as compared with WKY [22]. Similarly, Cauvin et al. [28] have shown that NE-stimulated ^{45}Ca influx through ROCs is significantly increased in SHR mesenteric arteries as compared with WKY. In addition to increased ^{45}Ca influx, the Ca^{2+} release from intracellular organelles, probably SR, is also increased in SHR caudal artery rings. The contractile response of SHR caudal artery rings to NE, methoxamine, and vasopressin was significantly increased in Ca^{2+}-free solution [2, 23]. Similarly, it has been shown that SHR mesenteric artery contraction in response to NE is significantly greater in the absence of extracellular Ca^{2+} as compared with WKY [28]. In addition to an increase in hormone-stimulated $[Ca^{2+}]_i$, it appears that protein kinase C activity or protein kinase C-mediated mechanisms are also increased in SHR VSM. We have shown that potent tumor-promoting phorbol esters produced greater potentiation of agonist- and K^+-stimulated contractions in SHR caudal artery as compared with WKY (Table 1; also see [29]). Similarly, it was found that phorbol esters produce greater contraction in SHR mesenteric arteries as compared with WKY [30, 31].

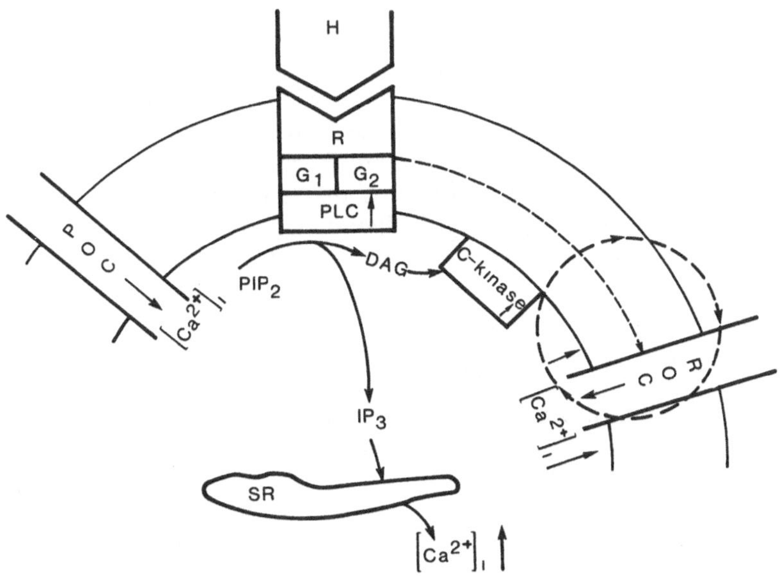

Fig. 5. Model for the hormone-dependent phosphatidylinositol metabolism and the bifurcating signal transduction pathways in vascular smooth muscle. In response to hormones (*H*) the receptor (*R*) initiates activation of phospholipase C (*PLC*) probably via guanine nucleotide binding proteins (G_1, G_2), resulting in the hydrolysis of phosphatidylinositol 4,5-bisphosphate (*PIP_2*) and the generation of two second messengers, inositol 1,4,5-triphosphate (*IP_3*) and diacylglycerol (*DAG*). The water soluble IP_3 causes a transient release of Ca^{2+} from the sarco-plasmic reticulum (*SR*). DAG causes activation of protein kinase C (*C-kinase*) by synergistic action with $[Ca^{2+}]_i$. Activation of receptor-operated Ca^{2+} channels (*ROC*) occurs probably via G-proteins; however, the specificity of the G-proteins involved and the mechanism of their action on phospholipase C activation and activation of ROC are not clear. The activation of ROC results in an increase in $[Ca^{2+}]_i$ and/or cycling of Ca^{2+} across the plasma membrane to produce full activation of protein kinase C. (↑) denotes possible sites of alteration in SHR

A model summarizing possible sites of defect in hormone-mediated signal transduction in the SHR VSM cell is given in Fig. 5. It appears that the increased Ca^{2+} sensitivity of hormone-stimulated arterial rings in SHR is due to alterations in both Ca^{2+} metabolism and protein kinase C activity. We have demonstrated that (a) The $^{45}Ca^{2+}$ influx in response to hormone stimulation is increased in SHR; (b) the Ca^{2+} release from the SR in response to hormone stimulation is increased in SHR; (c) protein kinase C mediated contraction is increased in SHR.

All the observed lesions can result from a single alteration in the plasma membrane function at the level of hormone-receptor coupling to phospholipase C activation, probably via guanine nucleotide protein(s) [32]. The specificity of the G-proteins involved and their mechanism of action in the activation of phospholipase C are yet to be worked out in VSM cells. These changes can be produced by increased PIP_2 degradation in response to hormone stimulation in SHR VSM. The increased PIP_2 metabolism will result in increased production of IP_3 and DAG. The increased IP_3 production can account for the increased Ca^{2+} release from the SR observed during hormonal stimulation. Furthermore, a synergistic interaction between Ca^{2+} and DAG will result in increased activation of protein kinase C and in turn increased sustained contractile response in SHR VSM cell. The increased PIP_2 metabolism does not appear to be due to an alteration in receptor number or affinity [23]. Therefore, it is possible that either the phospholipase C activity per se is increased or the activation of phospholipase C via G-proteins is increased. Recent evidence has linked the G-proteins with regulation of several types of ion channels (see [32] for review). Thus, it is conceivable that G-protein-mediated activation of ROC is augmented in SHR leading to an increased Ca^{2+} influx and/or increased Ca^{2+} cycling across the plasma membrane. Increased production of DAG and increased Ca^{2+} cycling across the plasma membrane may be synergistically producing greater activation of protein kinase C and increased contraction of SHR VSM. In summary, we believe that an alteration in the VSM plasma membranes of SHR has occurred at the site of hormone-receptor coupling. At this time it is not known whether this change is due to an increase in phospholipase C activity and/or an alteration in the G-proteins in SHR VSM.

References

1. Berecek KH, Schwertschlag U, Gross L (1980) Alterations in renal vascular resistance and reactivity in spontaneous hypertension of rats. Am J Physiol 238: H287–H293.
2. Bhalla RC, Sharma RV, Bendhack LM (1989) Increased Ca^{2+} sensitivity of α_1-adrenoceptor-mediated contraction in arterial smooth muscle from spontaneously hypertensive rat. In: Aoki K, Frohlich ED (eds) Calcium in essential hypertension: Academic Press, San Diego, pp 334–358
3. Webb RC (1984) Vascular changes in hypertension. In: Antonaccio M (ed) Cardiovascular pharmacology. Raven, New York, pp 215–255

4. Baron CB, Cunningham M, Strauss JF, Coburn RF (1984) Pharmacomechanical coupling in smooth muscle may involve phosphatidylinositol metabolism. Proc Natl Acad Sci USA 81: 6899–6903
5. Cambell MD, Deth RC, Payne RA, Honeyman TW (1985) Phosphoinositide hydrolysis is correlated with agonist-induced calcium flux and contraction in the rabbit aorta. Eur J Pharmacol 116: 129–136
6. Griendling KK, Rittenhouse SE, Brock TA, Ekstein LS, Gimbrone MA Jr, Alexander RW (1986) Sustained diacylglycerol formation and inositol phosholipids in angiotensin II-stimulated vascular smooth muscle cells. J Biol Chem 251: 5901–5906
7. Somlyo AV, Bond M, Somlyo AP, Scarpa, A. (1985) Inositol trisphosphate-induced calcium release and contraction in vascular smooth muscle. Proc Natl Acad Sci USA 82: 5231–5235
8. Hashimoto T, Hirata M, Itoh T, Kanamura Y, Kuriyama H (1986) Inositol 1,4,5-trisphosphate activates pharmacomechanical coupling in smooth muscle of rabbit mesenteric artery. J Physiol (Lond.) 370: 605–618
9. Morgan JP, Morgan KG (1984) Neuroeffector function of isolated portal vein from spontaneously hypertensive and Wistar-Kyoto rats: dependence on external calcium concentration. Blood Vessels 18: 89–99
10. Sumimoto K, Kuriyama H (1986) Mobilization of free Ca^{2+} measured during contraction relaxation cycles in smooth muscle cells of the porcine coronary artery using Quin-2. Pflugers Arch 406: 173–180
11. Takuwa Y, Takuwa N, Rasmussen H (1987) Measurement of cytoplasmic free Ca^{2+} concentration in bovine tracheal smooth muscle using aequorin. Am J Physiol 253: C817–827
12. Sharma RV, Bhalla RC (1989) Regulation of cytosolic free Ca^{2+} concentration in vascular smooth muscle cells by A- and C-kinases. Hypertension 13: 845–850
13. Yu B (1981) Calcium-activated phospholipid-dependent protein kinase in smooth muscle and its possible relation to phosphatidylinositol turnover. Kobe J Med Sci 27: 225–237
14. Abdel-Latif AA (1986) Calcium mobilizing receptors, polyphosphoinositides and the generation of second messengers. Pharmacol Rev 38: 227–271
15. Rasmussen H, Takuwa Y, Park S (1987) Protein kinase-C in the regulation of smooth muscle contraction. FASEB J 1: 177–185
16. Kishimoto A, Takai Y, Mori T, Kikkawa U, Nishizuka Y (1980) Activation of calcium and phospholipid-dependent protein kinase by diacylglycerol, its possible relation to phosphatidylinositol turnover. J Biol Chem 255: 2273–2276
17. Takuwa Y, Takuwa N, Rasmussen H (1986) Carbochol induces a rapid and sustained hydrolysis of polyphosphoinositides in bovine tracheal muscle. J Biol Chem 261: 14670–14675
18. Aqel MB, Sharma RV, Bhalla RC (1986) Increased Ca^{2+} sensitivity of α_1-adrenoceptor-stimulated contraction in SHR caudal artery. Am J Physiol 250: (Cell Physiol 19): C275–C282
19. Mulvany MJ (1984) Resistance vessel abnormalities in spontaneously hypertensive rats. J Cardiovasc Pharmacol 6: S656–S665
20. Pegram BL, Ljung B (1981) Neuroeffector function of isolated portal vein from spontaneously hypertensive and Wistar-Kyoto rats: dependence on external calcium concentration. Blood Vessels 18: 89–99
21. Warshaw DM, Szarek JL, Hubbard MS, Evans JN (1986) Pharmacology and force development of single freshly isolated bovine carotid artery smooth muscle cells. Circ Res 58: 399–406
22. Aqel MB, Sharma RV, Bhalla RC (1987) Increased ^{45}Ca influx in response to α_1-adrenoceptor-stimulation in spontaneously hypertensive rat caudal artery. J Cardiovasc Pharmacol 10: 205–212
23. Aqel MB, Sharma RV, Bhalla RC (1987) Increased norepinephrine-sensitive intracellular Ca^{2+} pool in the caudal artery of spontaneously hypertensive rats. J Hypertension 5: 249–253

24. Folkow B, Hallback M, Jones JV, Sutter M (1977) Dependence on external calcium for noradrenaline contractility of the resistance vessels in spontaneously hypertensive and renal hypertensive rats, as compared with normotensive controls. Acta Physiol Scand 101: 84–97
25. Bolton TB (1979) Mechanism of action of transmitters and other substances on smooth muscle. Physiol Rev 58: 606–718
26. Blumberg PM, Jahen S, Koning B, Sharkey NA, Leach KL, Jeng SY, Yeh E (1984) Mechanism of action of the phorbol ester tumor promoters: specific receptors for lipophilic ligands. Biochem Pharmacol 33: 933–940
27. Nishizuka Y (1986) Studies and perspectives of protein kinase-C. Science 233: 305–312
28. Cauvin C, Johns A, Yamamoto M, Hwang O, Gelband C, van Breemen C (1989) Ca^{2+} movements in vascular smooth muscle and their alterations in hypertension. In: Kwan CY (ed) Membrane abnormalities in hypertension, vol 1. CRC Press, Boca Raton, pp 146–179
29. Bendhack LM, Sharma RV, Bhalla RC (1988) Contractile response of spontaneously hypertensive rat caudal artery to phorbol esters. Hypertension 11 (Suppl I): I–112–I–116
30. MacKay MJ, Cheung DW (1987) Increased reactivity in the mesenteric artery of spontaneously hypertensive rats to phorbol esters. Biochem Biophys Res Commun 145: 1105–1111
31. Turla MB, Webb RC (1987) Enhanced vascular reactivity to protein kinase-C in genetically hypertensive rats. Hypertension 9 (Suppl III): III150–III154
32. Allende JE (1988) GTP-mediated macromolecular interactions: the common features of different systmes. FASEB J 2: 2356–2367

Decreased α_1-Adrenoceptor Reserve in Arteries from Spontaneously Hypertensive Rats

Masayoshi Kojima[1], Kyuzo Aoki[2], Masahisa Asano[1], Yasuaki Dohi[1], and Tomohiro Matsuda[1]

Summary. Pharmacological characteristics of α-adrenoceptor-mediated contractile responses were examined in mesenteric arterial strips from spontaneously hypertensive rats (Aoki SHR, 13 weeks old, male) and age- and sex-matched normotensive Wistar-Kyoto rats (WKY). The α_1-adrenoceptor agonist, phenylephrine, caused dose-dependent contractions of the strips from both strains. The SHR strips developed a greater tension with a smaller pD_2 value as compared with WKY. The contractile-responses to phenylephrine were induced by the activation of α_1-adrenoceptors in the arteries from both strains. The affinity of α_1-adrenoceptors did not differ between SHR and WKY. The α_1-adrenoceptor occupancy-response relationship for phenylephrine was a hyperbolic curve in both strains. The curve was steeper in WKY than in SHR. α_1-Adrenoceptor occupancy at a half-maximum response to phenylephrine was significantly greater in SHR. The calcium antagonist, nifedipine, inhibited the phenylephrine-induced contraction to a greater extent in SHR artery. The contractions were more susceptible to inhibition by nifedipine in arteries from which α_1-adrenoceptor reserves were removed by phenoxybenzamine. In the phenoxybenzamine-treated arteries, the extent of the inhibitory effect of nifedipine became the same between SHR and WKY. These results suggest that a receptor reserve for the α_1-adrenoceptor-mediated contraction decreased in SHR mesenteric artery as compared with WKY, which may lead to the greater inhibitory effect of nifedipine on phenylephrine-induced contractions and may contribute to the enhancement of vasoconstrictor responsiveness in SHR.

Key words: α-Adrenoceptor—Mesenteric artery—Nifedipine—Phenylephrine—Receptor reserve

Introduction

The elevated arterial pressure of essential hypertension is the result of an increased total peripheral vascular resistance [1]. There is much evidence that α-adrenoceptor-mediated contractile responses are enhanced in arterial smooth muscle from spontaneously hypertensive rats (SHR) [2–6]. The enhanced responses may contribute to an increased arterial vasoconstriction in hypertension. Evidence has been reported in alterations of α-adrenoceptors [7–9], receptor-

Department of Pharmacology[1] and 2nd Department of Internal Medicine[2], Nagoya City University Medical School, Mizuho-ku, Nagoya 467, Japan operated calcium channels [10–11], and interactions of α-adrenoceptors and receptor-operated calcium channels [4].

It has been demonstrated that contractile responses to α-adrenoceptor stimulation were inhibited by organic calcium channel blockers (calcium antagonists) to a greater extent in SHR arteries than in normotensive Wistar-Kyoto rats (WKY) [4, 12–14]. van Breemen et al. [15] suggested that voltage-dependent and receptor-operated calcium channels exist in vascular smooth muscle. Therefore, the greater inhibitory effects of calcium antagonists on the contractile responses of SHR arteries have been explained by abnormalities of either voltage-dependent calcium channels [16] or receptor-operated calcium channels [4, 12].

It has been suggested that the inhibitory effect of calcium antagonists on α-adrenergic contractions is associated with the magnitude of a receptor reserve [17–20]. Recently, it has been shown that a smaller α_1-adrenoceptor reserve could explain a greater sensitivity of noradrenaline-induced contractions to calcium antagonist in SHR aorta [21]. Therefore, it is possible that a relationship between α-adrenoceptor occupancy and contractile response is altered in SHR arterial smooth muscle.

The present study was designed to clarify pharmacological characteristics of α-adrenoceptor-mediated contractile responses in mesenteric arteries from SHR and WKY. We found that the α_1-adrenoceptor reserve for phenylephrine was significantly smaller in SHR than in WKY and that the smaller receptor reserve may be responsible for the abnormal α_1-adrenoceptor-mediated contractile response in SHR.

Methods

Blood Pressures of Rats

SHR (Aoki strain, 13 weeks old, male) [22] and age- and sex-matched WKY were used. Systolic blood pressures of the rats, measured by tail-cuff plethysmography [22], were 189 ± 9 mmHg (SHR, $n = 29$) and 119 ± 4 mmHg (WKY, $n = 29$; $P < 0.001$), respectively. Body weights at that age were not significantly different between SHR (271 ± 4 g) and WKY (270 ± 4 g).

Preparation of Arterial Strips

The rats were stunned and exsanguinated. The distal portion of mesenteric artery (0.7–0.9 mm outside diameter) was dissected. Excess fat and adherent tissue were removed; then the artery was cut into a helical strip of 0.8 mm in width and 7 mm in length. The endothelium of the strip was removed by rubbing the endothelial surface with cotton pellets.

The strips were mounted vertically between hooks in water-jacketed muscle baths. The upper end of the strip was connected to a force-displacement transducer (TB-612T, Nihon Kohden Kogyo Co., Tokyo). The strips were stretched passively by imposing a resting tension of 0.5 g and were equilibrated for 90 min.

Then, a maximum contraction in the strip was obtained by 60 m*M* KCl. The contraction was repeated at 40-min intervals until the response was reproducible. Loss of endothelial function was confirmed by the disappearance of relaxation by acetylcholine (10^{-6} *M*) [23]. The muscle baths contained 20 ml of the oxygenated Krebs-bicarbonate solution (composition in m*M*: NaCl, 115.0; KCl, 4.7; $CaCl_2$, 2.5; $MgCl_2$, 1.2; $NaHCO_3$, 25.0; KH_2PO_4, 1.2; dextrose, 10.0), which was aerated with a mixture of 95% O_2 and 5% CO_2 at 37.0° ± 0.5°C.

Contractile Response to Phenylephrine

Cumulative dose-response curves of the strips for phenylephrine (*l*-phenylephrine hydrochloride) were determined by adding increasing concentrations of phenylephrine ($10^{-9} - 10^{-3}$ *M*) to the solution in the presence of the β-adrenoceptor antagonist, timolol (5×10^{-7} *M*, timolol maleate).

The dose-response curves for phenylephrine were determined in the presence of the α_1-adrenoceptor antagonist, prazosin (prazosin hydrochloride), or the α_2-adrenoceptor antagonist, yohimbine (yohimbine hydrochloride). The first dose-response curve in the absence of these antagonists was taken as a control, and the second, third, and fourth dose-response curves were determined in the presence of three concentrations of these antagonists. Dose-ratio for phenylephrine, i.e., ED_{50} in the presence of prazosin or yohimbine divided by the ED_{50} in the absence of the antagonist, was obtained at the concentrations of the antagonist. The data were subjected to a Schild plot analysis by the method of Arunlakshana and Schild [24], and affinity constant (pA_2) values and slopes of the regression lines were obtained for prazosin and yohimbine using phenylephrine as the agonist. The dose-response curves for phenylephrine in the presence of the calcium antagonist, nifedipine, were determined in a similar fashion and were also determined after receptor reserves were removed by irreversible inactivation of partial α-adrenoceptors using phenoxybenzamine (phenoxybenzamine hydrochloride).

Determination of Dissociation Constant, Receptor Occupancy and Receptor Reserve for Phenylephrine

Acid ionization dissociation constants (K_A) were determined according to the receptor theory [25]. The first dose-response curve for phenylephrine was obtained as a control. The second curve was determined after the treatment with phenoxybenzamine (3×10^{-9} *M*) for 30 min. Equieffective agonist concentrations before (A) and after phenoxybenzamine treatment (A′) were obtained, and a plot of 1/A against 1/A′ was made. The slope of the regression line and the *y*-intercept were used to calculate K_A value, i.e., $K_A = (\text{slope} - 1)/\text{intercept}$ [25]. The fraction of receptors occupied by phenylehrine (receptor occupancy) was calculated using the formula $A/(K_A + A)$ [26, 27]. The control response to phenylephrine was replotted as a function of the fractional receptor occupancy. In addition, receptor occupancy at a half-maximum response ($ED_{50}/(K_A + ED_{50})$) was used to estimate the magnitude of receptor reserve. The receptor reserve was calculated using the formula antilog ($pD_2 - pK_A$), where pD_2 and pK_A are negative log ED_{50} and negative log K_A value, respectively [28].

Statistical Analysis

Data are expressed as means ± SE (n = number of preparations) and statistically analyzed by Student's t-test for paired or unpaired data. The difference was considered significant when $P < 0.05$.

Results

α-Adrenoceptors in Mesenteric Arteries

Dose-response curves of mesenteric arteries for the α_1-adrenoceptor agonist, phenylephrine, were obtained in the presence of the β-adrenoceptor antagonist, timolol. A maximum response to phenylephrine was significantly greater in SHR than in WKY (Fig. 1). The pD_2 value for phenylephrine was significantly smaller in SHR arteries than in WKY arteries (Table 1).

The response to phenylephrine in the presence of either the α_1-adrenoceptor antagonist, prazosin, or the α_2-adrenoceptor antagonist, yohimbine, was determined to classify the subtype of α-adrenoceptors. Prazosin and yohimbine shifted the dose-response curve for phenylephrine to the right with no reduction of the maximum response in the arteries from SHR and WKY. The pA_2 values for prazosin against phenylephrine were approximately 1000-fold higher than those for yohimbine, indicating an existence of α_1-adrenoceptors in the mesenteric arteries (Table 1). Moreover, the obtained pA_2 values were in good agreement with the reported values for the antagonists for α_1-adrenoceptors [29–31]. Prazosin or yohimbine exhibited similar pA_2 values between both strains. The

Table 1. Maximum responses, pD_2 values (-log ED_{50}), pK_A values (-log K_A), and receptor reserves for phenylephrine, and pA_2 values for prazosin and yohimbine antagonism against phenylephrine in mesenteric arterial strips from SHR and WKY [37]

	SHR (n)	WKY (n)
Maximum response	162.2 ± 4.7* (10)	138.1 ± 2.1 (10)
pD_2 value	6.93 ± 0.06* (10)	7.35 ± 0.05 (10)
pK_A value	6.48 ± 0.18 (7)	6.45 ± 0.08 (7)
Receptor reserve	2.69 ± 0.84* (7)	7.92 ± 1.77 (7)
pA_2 value		
Prazosin	9.41 ± 0.17 (5)	9.40 ± 0.20 (6)
Yohimbine	6.39 ± 0.09 (4)	6.62 ± 0.14 (5)
Slope		
Prazosin	1.05 ± 0.21 (5)	1.03 ± 0.21 (6)
Yohimbine	1.01 ± 0.10 (4)	0.98 ± 0.11 (6)

Maximum response to phenylephrine (10^{-3} M) is expressed as a percentage of the maximum contraction by 60 mM KCl

Slopes are expressed as means ± 95% confidence limits and the others are expressed as means ± SE

*Significantly different from WKY ($P < 0.05$)

PE, phenylephrine; receptor reserve, antilog ($pD_2 - pK_A$) [28]

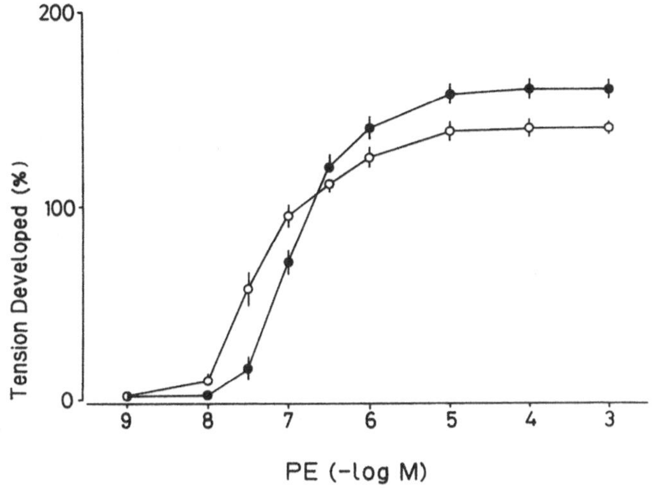

Fig. 1. Dose-response curves for phenylephrine (*PE*) in mesenteric arterial strips from SHR (●, $n = 10$) and WKY (○, $n = 10$). Contractions are expressed as a percentage of the response to 60 m*M* KCl. *Vertical bars* represent SE [37]

slopes of lines for these antagonists did not differ significantly from unity. These results suggest the existence of α_1- but not α_2-adrenoceptors in the arteries from both strains.

Affinity of α_1-Adrenoceptors and Receptor Occupancy-Response Relationship for Phenylephrine

The phenoxybenzamine pretreatment shifted the dose-response curves for phenylephrine to the right with a significant reduction of the maximum response. The reduction of the response was significantly greater in SHR (46% ± 5%, $n = 7$) than in WKY (31% ± 4%, $n = 7$). The difference in pD_2 values for phenylephrine before and after the phenoxybenzamine treatment was significantly smaller in SHR (0.38 ± 0.08) than in WKY (0.67 ± 0.10) (Fig. 2).

Dissociation constants (K_A) for phenylephrine in the arteries did not differ significantly between the two strains (Table 1). Receptor occupancy-response relationship curves constructed by using the K_A values, which represent the contractile responses as a function of receptor occupation, were hyperbolic curves. The hyperbolic curves indicate an existence of α_1-adrenoceptor reserves. In addition, the curve was steeper in WKY than in SHR. The receptor occupancy required to produce a half-maximum response was significantly greater in SHR (32% ± 4%, $n = 7$) than in WKY (14% ± 3%, $n = 7$) (Fig. 3).

Effects of Nifedipine on the Dose-Response Curves for Phenylephrine

Nifedipine caused a noncompetitive inhibition of the contractile response to phenylephrine (Fig. 4). The reduction of the maximum responses by nifedipine was significantly greater in SHR (48% ± 4%, 10^{-8} *M* and 75% ± 3%, 10^{-7} *M*;

$n = 6$) than in WKY (31% ± 3%, 10^{-8} *M* and 60% ± 2%, 10^{-7} *M*; $n = 6$). The α_1-adrenoceptor reserves were removed from the arteries by phenoxybenzamine treatment. Contractions by phenylephrine in the phenoxybenzamine-treated arteries were more susceptible to inhibition by nifedipine. Moreover, the inhibitory effect of nifedipine on the contractions of the arteries did not differ significantly between SHR and WKY (Fig. 5).

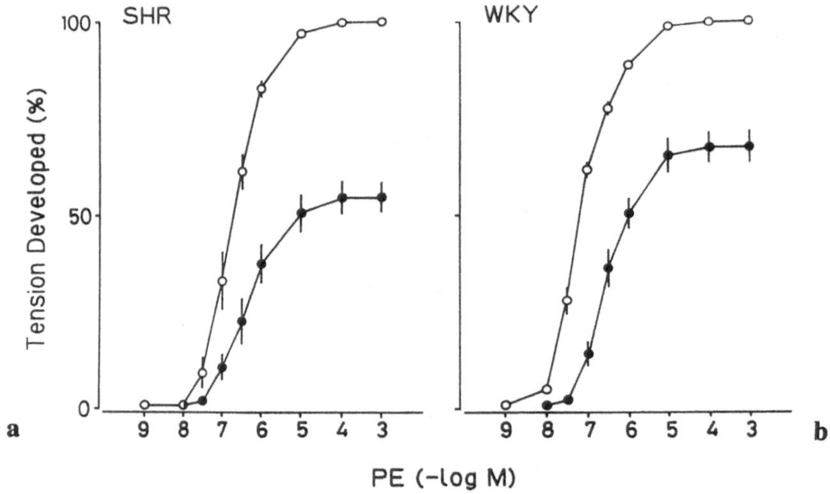

Fig. 2a,b. Dose-response curves for phenylephrine (*PE*) determined before (○) and after (●) irreversible inactivation of partial α-adrenoceptors by phenoxybenzamine (3×10^{-9} *M*) in mesenteric arterial strips from **a** *SHR* ($n = 7$) and **b** *WKY* ($n = 7$). The first curve (○) was taken as a control and the second curve (●) was determined after phenoxybenzamine treatment. The maximum contraction by 10^{-3} *M* phenylephrine in the control curve was taken as 100%. *Vertical bars* represent SE [37]

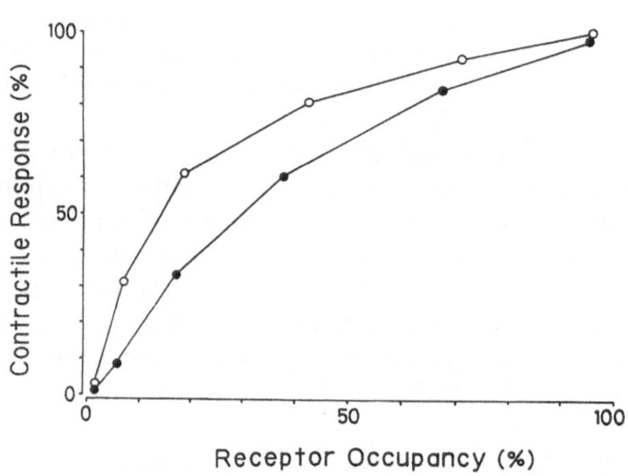

Fig. 3. Occupancy-response relationship for phenylephrine in mesenteric arterial strips from SHR (●, $n = 7$) and WKY (○, $n = 7$) plotted as a linear occupancy axis. Contractile responses are expressed as a percentage of the maximum response to phenylephrine. The data are replotted from Fig. 2

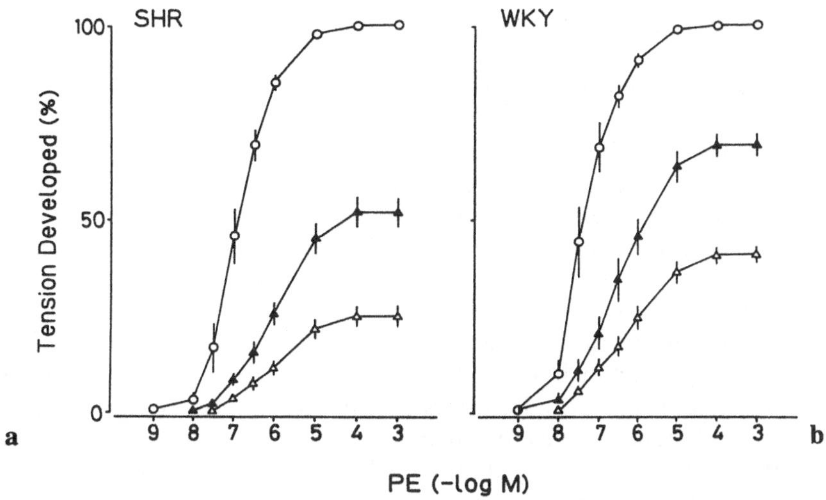

Fig. 4a,b. Effects of nifedipine on the dose-response curves for phenylephrine (*PE*) in mesenteric arterial strips from **a** SHR ($n = 6$) and **b** WKY ($n = 6$). The curves were determined in the absence (○) and presence of nifedipine (1×10^{-8} *M*, ▲; 1×10^{-7} *M*, △). The maximum contraction by 10^{-3} *M* phenylephrine in the absence of nifedipine was taken as 100%. *Vertical bars* represent SE [37]

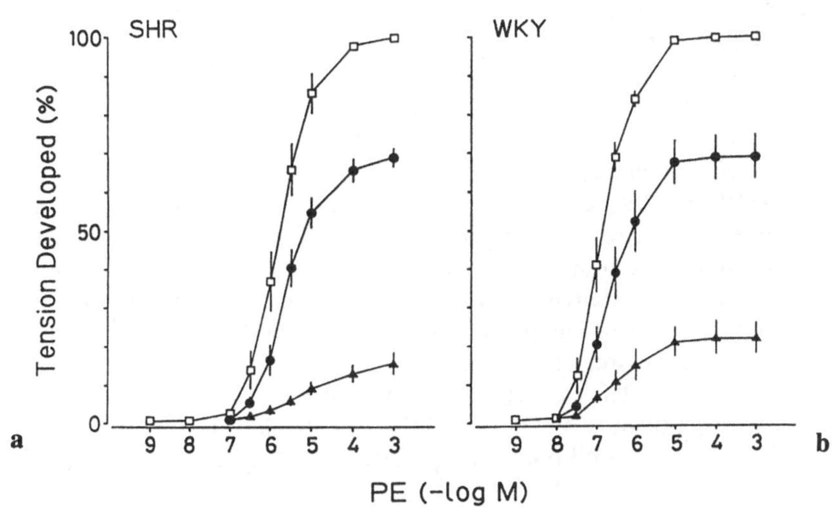

Fig. 5a,b. Effects of nifedipine on the dose-response curves for phenylephrine (*PE*) in mesenteric arteries from **a** SHR ($n = 6$) and **b** WKY ($n = 6$) after irreversible inactivation of partial α-adrenoceptors by phenoxybenzamine. The control curve (□) was determined after phenoxybenzamine treatment, and the curves were then determined in the presence of nifedipine (1×10^{-9} *M*, ●; 1×10^{-8} *M*, ▲). The maximum contraction by 10^{-3} *M* phenylephrine in the control curve was taken as 100%. *Vertical bars* represent SE [37]

Discussion

The maximum contractile response induced by the α_1-adrenoceptor agonist, phenylephrine, was significantly greater in SHR mesenteric artery than in WKY, and the sensitivity to phenylephrine was significantly less in SHR. The greater contractile response in SHR arteries has been reported previously [2–6]. The potency of the α_1-adrenoceptor antagonist, prazosin, in antagonizing the phenylephrine-induced contractions was approximately 1000-fold greater than that of the α_2-adrenoceptor antagonist, yohimbine. In addition, the pA_2 values obtained agreed with those of prazosin and yohimbine reported for α_1-adrenoceptors in rat mesenteric artery [5, 30]. The pA_2 values for the antagonists did not differ between SHR and WKY, suggesting that phenylephrine acts on α_1-but not α_2-adrenoceptors in mesenteric arteries from both strains. Moreover, K_A values for phenylephrine did not differ significantly between the two strains, suggesting that the affinity of α_1-adrenoceptors was not altered in SHR. It has been reported that the dissociation constant for norepinephrine did not differ between SHR and normotensive rats [32].

It has been shown that an α-adrenoceptor reserve modulates a responsiveness of α-adrenoceptor agonists in vascular smooth muscle [33, 34]. Thus, we determined the magnitude of α_1-adrenoceptor reserves for the contractile responses to phenylephrine in mesenteric arteries from SHR and WKY. Dose-response curves for full agonists which have a large receptor reserve are greatly shifted to the right by phenoxybenzamine. On the other hand, phenoxybenzamine causes great reductions in maximum response in the case of partial agonists which have a small receptor reserve [26]. In the present study, phenoxybenzamine reduced the maximum response to phenylephrine to a greater extent in SHR mesenteric artery than in WKY. The difference in pD_2 values for phenylephrine before and after the phenoxybenzamine treatment was significantly smaller in SHR (0.38 ± 0.08, $n = 7$) than in WKY (0.67 ± 0.10, $n = 7$). In addition, antilog of the difference between the pD_2 value for phenylephrine and the corresponding pK_A value, which estimates the magnitude of receptor reserve [28], was 2.69 ± 0.84 in SHR and 7.92 ± 1.77 in WKY ($P < 0.05$) (Table 1). These findings suggest that the receptor reserve for phenylephrine is smaller in SHR mesenteric artery than in WKY. A relationship between α_1-adrenoceptor occupancy and contractile response to phenylephrine was a hyperbolic function in both strains. The hyperbolic curve was steeper in WKY than in SHR. Phenylephrine produced a half-maximum response by occupying $32\% \pm 4\%$ ($n = 7$) of α_1-adrenoceptors in SHR and $14\% \pm 3\%$ ($n = 7$; $P < 0.01$) in WKY. Thus, the α_1-adrenoceptor reserve was significantly smaller in SHR mesenteric artery than in WKY, which might lead to an altered relationship between α_1-adrenoceptor occupancy and contractile response in SHR.

Recently, it has been demonstrated that the inhibition induced by calcium antagonists of α-adrenoceptor-mediated contractions is dependent on the size of the α-adrenoceptor reserve [17–20]. The present study showed that the nifedipine-induced inhibition of the response to phenylephrine was significantly greater in SHR than in WKY. We examined whether the greater inhibitory effect of nifedipine results from the smaller α_1-adrenoceptor reserve for phenylephrine in SHR artery. The phenylephrine-induced contractions were more susceptible to inhibition by nifedipine in mesenteric arteries from which α_1-adrenoceptor reserves were removed by phenoxybenzamine. In addition, the inhibitory effect of nifedipine on the contractions of the phenoxybenzamine-treated arteries reached the same extent for SHR and WKY. These findings suggest that the smaller α_1-adrenoceptor reserve in SHR is responsible for the greater antagonism by nifedipine in the response to phenylephrine. Our results are supported by the observation that a small α_1-adrenoceptor reserve explains the greater sensitivity of noradrenaline-induced contractions to calcium antagonist in SHR aortae [21].

It was shown that α-adrenoceptor reserves for norepinephrine were smaller in aorta from old rats than in those from young rats, and that a greater inhibition of norepinephrine contractions by diltiazem was due to a reduction in the receptor reserve [34]. Therefore, the decreased α_1-adrenoceptor reserve in SHR artery may imply that age-related changes of α-adrenoceptors appear at the earlier stage of development in SHR as compared with WKY. Purdy and Weber [35] showed that angiotensin II enhances a vasoconstrictor response to α-adrenergic stimulation in rabbit femoral artery and that the magnitude of enhancement is greater under conditions of reduced α-adrenoceptor reserve. It is likely that in vivo enhancement of responses to norepinephrine by angiotensin II is greater in SHR than in WKY, since α_1-adrenoceptor reserves decrease in SHR.

In view of the present study, it would be interesting to speculate whether a receptor reserve influences α_1-adrenoceptor-mediated contractions via transmembrane calcium movement. Contractile responses to full α_1-adrenoceptor agonists are caused by both an influx of extracellular calcium and a release of intracellular calcium, while the responses to partial α_1-adrenoceptor agonists are caused by the influx of extracellular calcium. The component of the responses which is coupled to the translocation of extracellular calcium has a high efficiency of coupling, indicating that a large α_1-adrenoceptor reserve exists for this component of calcium utilization. On the other hand, the component of the mobilization of intracellular calcium has a low efficiency of coupling, leading to a small α_1-adrenoceptor reserve [36]. Therefore, a decrease in α_1-adrenoceptor reserve for the total response may be associated with a fall in the ability to mobilize intracellular calcium to induce contractions. The decreased α_1-adrenoceptor reserve in SHR may imply the abnormality of calcium mobilization [37].

In conclusion, the findings of the present study demonstrate that the α_1-adrenoceptor reserve for phenylephrine decreases in SHR mesenteric artery. The decrease in α_1-adrenoceptor reserve in SHR artery may be associated with the enhancement of arterial vasoconstrictions and may contribute to elevated vascular resistance, which may result in high blood pressure in SHR.

References

1. Aoki K, Sato K (1986) Decrease in blood pressure and increase in total peripheral vascular resistance in supine resting subjects with normotension or essential hypertension. Jpn Heart J 27: 467–474

2. Webb RC, Bohr DF (1981) Recent advances in the pathogenesis of hypertension: Consideration of structural, functional, and metabolic vascular abnormalities resulting in elevated arterial resistance. Am Heart J 102: 251–264
3. Mulvany MJ, Nyborg N (1980) An increased calcium sensitivity of mesenteric resistance vessels in young and adult spontaneously hypertensive rats. Br J Pharmacol 71: 585–596
4. Aqel MB, Sharma RV, Bhalla RC (1986) Increased Ca^{2+} sensitivity of α_1-adrenoceptor stimulated contraction in SHR caudal artery. Am J Physiol 250: C275–C282
5. Aoki K, Asano M, Matsuda T (1988) Antagonism of α_1-adrenoceptor-mediated vascular contraction by urapidil in isolated arterial strips of spontaneously hypertensive rats. J Cardiovasc Pharmacol 12: 167–178
6. Aoki K, Mochizuki A, Hotta K (1981) Noradrenaline and calcium-induced tension in aortic strips of normotensive and spontaneously hypertensive rats. Jpn Circ J 45: 547–551
7. Kobayashi H, Wada A, Izumi F, Magnoni MS, Trabucchi M (1985) α-Adrenergic receptors in cerebral microvessels of normotensive and spontaneously hypertensive rats. Circ Res 56: 402–409
8. Weiss RJ, Webb RC, Smith CB (1984) Comparison of alpha$_2$-adrenoceptors on arterial smooth muscle and brain homogenates from spontaneously hypertensive and Wistar-Kyoto normotensive rats. J Hypertens 2: 249–255
9. Hicks PE, Tierney C, Langer SZ (1985) Preferential antagonism by diltiazem of α_2-adrenoceptor mediated vasoconstrictor responses in perfused tail arteries of spontaneous hypertensive rats. Naunyn Schmiedebergs Arch Pharmacol 328: 388–395
10. Bhalla RC, Aqel MB, Sharma RV (1986) α_1-Adrenoceptor mediated responses in the vascular smooth muscle of spontaneously hypertensive rats. J Hypertens 4 (Suppl III): III-S65–S67
11. Cauvin C, Hwang O, Yamamoto M, van Breemen C (1987) Effects of dihydropyridines on tension and calcium-45 influx in isolated mesenteric resistance vessels from spontaneously hypertensive and normotensive rats. Am J Cardiol 59: B116–B122
12. Lederballe Pedersen O, Mickelsen E, Anderson KE (1978) Effects of extracellular calcium on potassium-and noradrenaline-induced contractions in the aorta of spontaneously hypertensive rats: increased sensitivity to nifedipine. Acta Pharmacol Toxicol 43: 137–144
13. Mochizuki A, Aoki K, Kondo S, Mizuno T, Hotta K (1979) Specificity of tension development and calcium flux of the arterial smooth muscle in SHR. Jpn Heart J 20 (Suppl I): I-225–227
14. Aoki K, Kawaguchi Y, Sato K, Kondo S, Yamamoto M (1982) Clinical and pharmacological properties of calcium antagonists in essential hypertension in humans and spontaneously hypertensive rats. J Cardiovasc Pharmacol 4: S298–S302
15. van Breemen C, Aaronson P, Loutzenhiser R (1979) Na-Ca interactions in mammalian smooth muscle. Pharmacol Rev 30: 167–208
16. Aoki K, Asano M (1986) Effects of Bay K 8644 and nifedipine on femoral arteries of spontaneously hypertensive rats. Br J Pharmacol 88: 221–230
17. Hamilton CA, Reid JL, Sumner DJ (1983) Acute effects of phenoxybenzamine on α-adrenoceptor responses in vivo and in vitro: relation of in vivo pressor responses to the number of specific adrenoceptor binding sites. J Cardiovasc Pharmacol 5: 868–873
18. Ruffolo RR Jr, Morgan EL, Messick K (1984) Possible relationship between receptor reserve and the differential antagonism of alpha-1 and alpha-2 adrenoceptor-mediated pressor responses by calcium channel antagonists in the pithed rat. J Pharmacol Exp Ther 230: 587–594
19. Jim KF, Macia RA, Mattews WD (1986) Role of receptor reserve in the inhibition of alpha-1 adrenoceptor-mediated pressor responses by calcium antagonists in the pithed rat. J Pharmacol Exp Ther 238: 89–94

20. Pedrinelli R, Tarazi RC (1985) Calcium entry blockade by nitrendipine and alpha adrenergic responsiveness in vitro: comparison with noncalcium entry blocker vasodilators in absence and presence of phenoxybenzamine pretreatment. J Pharmacol Exp Ther 233: 636–642
21. Holck MI (1988) α_1-Adrenoceptor reserve and effects of a Ca^{2+} entry blocker (Ro 18–3981) on aorta of spontaneously hypertensive rats. Eur J Pharmacol 148: 9–15
22. Okamoto K, Aoki K (1963) Development of a strain of spontaneously hypertensive rats. Jpn Circ J 27: 282–293
23. Furchgott RF, Zawadzki JV (1980) The obligatory role of endothelial cells in the relaxation of arterial smooth muscle by acetylcholine. Nature (Lond) 228: 373–376
24. Arunlakshana O, Schild HO (1959) Some quantitative uses of drug antagonists. Br J Pharmacol 14: 48–58
25. Furchgott RF (1966) The use of β-haloalkylamines in the differentiation of receptors and in the determination of dissociation constants of receptor-agonist complexes. In: Harper A, Simmonds B (eds) Advances in drug research, vol 3. Academic Press, London, pp 21–55
26. Furchgott RF, Bursztyn P (1967) Comparison of dissociation constants and of relative efficacies of selected agonists acting on parasympathetic receptors. Ann NY Acad Sci 144: 882–898
27. Ruffolo RR Jr (1982) Important concepts of receptor theory. J Auton Pharmacol 2: 277–295
28. Kenakin TP (1984) The classification of drugs and drug receptors in isolated tissues Pharmacol Rev 36: 165–222
29. Ruffolo RR Jr, Waddell JE, Yaden EL (1981) Postsynaptic alpha adrenergic receptor subtypes differentiated by yohimbine in tissues from the rat. Existence of alpha-2 adrenergic receptors in rat aorta. J Pharmacol Exp Ther 217: 235–240
30. Agrawal DK, Triggle CR, Daniel EE (1984) Pharmacological characterization of the postsynaptic alpha adrenoceptors in vascular smooth muscle from canine and rat mesenteric vascular beds. J Pharmacol Exp Ther 229: 831–838
31. Drew GM (1985) What do antagonists tell us about α-adrenoceptors? Clin Sci 68(Suppl X): X-S15–S19
32. Strecker RB, Hubbard WC, Michelakis AM (1975) Dissociation constant of the norepinephrine-receptor complex in normotensive and hypertensive rats. Circ Res 37: 658–663
33. Purdy RE, Stupecky GL (1984) Characterization of the alpha adrenergic receptor properties of rabbit ear artery and thoracic aorta. J Pharmacol Exp Ther 229: 459–465
34. Wanstall JC, O'Donnell SR (1988) Inhibition of norepinephrine contractions by diltiazem on aorta and pulmonary artery from young and aged rats: influence of alpha-adrenoceptor reserve. J Pharmacol Exp Ther 245: 1016–1020
35. Purdy RE, Weber MA (1988) Angiotensin II amplification of α-adrenergic vasoconstriction: role of receptor reserve. Circ Res 63: 748–757
36. Nichols AJ, Ruffolo RR Jr (1988) The relationship of α-adrenoceptor reserve and agonist intrinsic efficacy to calcium utilization in the vasculature. Trend Pharmacol Sci 9: 236–241
37. Kojima M, Asano M, Aoki K, Yamamoto M, Matsuda T (in press) Decrease in α_1-adrenoceptor reserve in mesenteric arteries isolated from spontaneously hypertensive rats. J Hypertens 7

Diminished β-Adrenoceptor-Mediated Relaxation of Arteries from Spontaneously Hypertensive Rats

SEIGO FUJIMOTO[1], YASUAKI DOHI[1], KYUZO AOKI[2], and TOMOHIRO MATSUDA[1]

Summary. Beta-adrenoceptor-mediated relaxation was studied with vascular preparations from spontaneously hypertensive rats (SHR) and rats of the Wistar-Kyoto strain (WKY). Femoral arteries from SHR were less sensitive to isoproterenol (ISO) and mesenteric arteries were less sensitive to ISO and norepinephrine (NE) than the similar arteries from WKY. The abnormality in relaxation was found in young SHR before the development of hypertension. The diminished response of the arteries to ISO was not accounted for by the decreased sensitivity of the arteries to cyclic AMP (cAMP); the defect may be due to a decreased number of functional beta-adrenoceptors and/or to abnormal coupling of the receptor to adenylate cyclase. Relative potency ratios between ISO, fenoterol, and NE and Schild plot data for atenolol and butoxamine indicated that the relaxation responses of the femoral and mesenteric arteries to ISO were mediated through beta-1 and beta-2 adrenoceptors, respectively. The diminished relaxation was not explained by an altered subtype of beta-adrenoceptor. The subtype was not changed during early stages of hypertension. It was suggested that beta-1 adrenoceptor-mediated relaxation was diminished in the resistant vessels of prehypertensive SHR (PHSHR).

Key words: Beta-adrenoceptor—Isoproterenol—Schild plot—Spontaneously hypertensive rat—Vasorelaxation

Introduction

Since Aoki [1] succeeded in separating spontaneously hypertensive rats (SHR) from a group of rats of the Wistar strain (Wistar-Kyoto strain; WKY), SHR have been used as an animal model of essential hypertension. In almost all cases, an increase in total peripheral resistance (TPR) is characteristic of essential [2] and spontaneous hypertension [3]. Cardiac output becomes normal once the elevation of systemic blood pressure (BP) is sustained in essential hypertension. Hypernoradrenergic innervation has been suggested in certain blood vessels of SHR prior to the onset of hypertension [4]. Thus, it seems likely that vascular smooth muscle cells (VSMC) of the immature SHR are under increased noradrenergic discharge.

Department of Pharmacology[1] and 2nd Department of Internal Medicine[2], Nagoya City University Medical School, Mizuho-ku, Nagoya 467, Japan

Norepinephrine (NE) binds alpha- and beta-adrenoceptors in VSMC to increase and decrease vascular tones, respectively. In other words, TPR reflects summation of alpha- and beta-adrenoceptor activities in resistance vessels. In fact, catecholamines dilate blood vessels in which alpha-adrenoceptors are blocked, and vascular contractile response to NE is potentiated with beta-adrenoceptor antagonists [5]. It is known that beta-adrenoceptor-mediated relaxation of peripheral arteries is diminished in prehypertensive SHR (PHSHR) [6]. Although it is not known whether the diminished relaxation is related to an initiation and development of hypertension or not, the reduced beta-adrenoceptor activity could account for an enhanced vasocontractile response to NE, which might be a critical factor for the development of hypertension [7]. The defect seems to be explained by abnormalities in beta-adrenoceptor density and receptor-adenylate cyclase coupling [8]. In addition, abnormal vasorelaxing pathways distal to cAMP formation have been proposed for the diminished relaxation response to beta-adrenoceptor agonists; cAMP-dependent protein kinase activity and Ca^{2+} efflux or sequestration were reported to be reduced in SHR arteries compared with WKY tissues [9].

Response of contractile elements to Ca^{2+} in skinned arteries from SHR was reported to be similar to that in WKY arteries [10], suggesting that there are abnormal foci in the membrane of VSMC in SHR.

TPR and muscle blood flow are effectively changed with beta-2 adrenoceptor agonists [11], suggesting that beta-2 subtypes are more responsive to vascular dilatation caused by beta-2 adrenoceptor agonists than beta-1 subtypes. By radioligand binding technique [12], autoradiographical study [13], and pharmacological approach, it was demonstrated that the aorta and the mesenteric and pulmonary arteries contained predominantly beta-2 adrenoceptors. Coronary arteries contain only beta-1 subtypes [14]. We also demonstrated that the femoral artery relaxed through beta-1 but not beta-2 adrenoceptors [15]. Some veins contain beta-1 and/or beta-2 subtypes of beta-adrenoceptors [16, 17]. Thus, in this chapter, it was described which subtype of beta-adrenoceptors was involved in the diminution of vascular relaxation response to beta-agonists in PHSHR.

Methods and Materials

Animals and BP Determination

Male WKY and SHR (2–10 weeks old) were used. Systolic BP was determined in conscious, preheated rats by the tail-cuff method. Measurements were repeated at least 3 times and mean values were determined in each rat.

Vascular Prepartations

Femoral, superior mesenteric and caudal arteries and aortae were isolated and then cut into helical strips. The strips were mounted in muscle chambers filled with Krebs-Henseleit bicarbonate buffer (KHB). The composition of the KHB and the optimal resting tensions applied to each strip were described previously

[6]. The strips were connected to a force-displacement transducer (TB-612T, Nihon Kohden Kogyo Co., Tokyo) coupled to a pen recorder. In order to observe relaxation response to beta-adrenoceptor agonists, the strips were incubated for 1 h with 2×10^{-6} *M* phenoxybenzamine (POB).

Time Course of Changes in the Sensitivity of Arteries to Isoproterenol (ISO) and Salbutamol (SAL)

The arterial strips were contracted with 30 m*M* KCl. After the contraction had reached a steady state, cumulative dose-response curves for ISO- and SAL-induced relaxations were obtained. Maximum relaxation caused by papaverine (10^{-4} *M*) was expressed as 100% in the present study. ED_{50} values are the molar concentration which produces 50% of the maximum relaxation in the dose-response curve of the agonist.

Difference between WKY and SHR in Dose-Response Curves for beta-Agonists

The first set of experiments was carried out with normal KHB. The arterial strips were contracted with appropriate concentrations of KCl by 85% of the maximal KCl contraction. Then, cumulative dose-response curves for ISO-, fenoterol (FEN)-, and NE-induced relaxations were obtained. In the second set of experiments, the arterial strips were incubated in Ca^{2+} free, high K^+ (30 m*M*) buffer. Cumulative dose-response curves for ISO-, dibutyryl (DB) cAMP-, and forskolin-induced relaxations were obtained in the arterial strips which had been contracted with appropriate concentrations of $CaCl_2$ by 75% of the maximal $CaCl_2$ contraction.

Beta-Adrenoceptor Subtypes Mediating Relaxation

In the arterial strips contracted with 30 m*M* KCl, cumulative dose-response curves for ISO, FEN, and NE were determined in each strip. The first was carried out in the absence of beta-adrenoceptor antagonists. The others were carried out in the presence of increasing dose of atenolol or butoxamine. Schild plots, i.e., plots of log (CR-1) vs log molar dose of the antagonist (log [B]), were obtained, where CR was the ratio of the agonist ED_{50} value in the presence of the antagonist to that in the first dose-response curve. If the Schild plots had a slope which did not differ significantly from unity, the affinity constant (pA_2) values were calculated for each dose of the antagonist from the equation $pA_2 = \log(CR\text{-}1) - \log[B]$.

Response of Skinned Fiber to Ca^{2+}

The caudal arteries isolated from 4-week-old rats were cut into circular strips. The strips were 300 μm long, 80 μm wide, and 30–40 μm thick. The strips were mounted in muscle chambers filled with normal KHB for recording isometric force and stretched passively to 120% of the original length. The muscle chambers were then filled with relaxing solution (130 m*M* potassium propionate, 20 m*M* tris-malate, 4 m*M* $MgCl_2$, 4 m*M* Na_2ATP, 2 m*M* EGTA; pH 6.8) maintained at 25°C. The tissue preparation was then treated with saponin (30–50 μg/ml of the relaxing solution) for 20 min. The skinned fiber was quickly washed by replacing the relaxing solution.

Drugs and Solutions

The following drugs were dissolved in distilled water and diluted with normal KHB or 0.9% NaCl to obtain the desired concentrations: (±)atenolol (ICI, Cheshire, UK), (±)butoxamine HCl (Japan Wellcome, Osaka), DB cAMP (Sigma, St. Louis, USA), (±)FEN HCl (Boehringer Ingelheim Japan, Hyogo, Japan), forskolin (Sigma), (±)ISO HCl (Sigma), (−)NE bitartrate (Sigma), papaverine HCl (Wako, Osaka, Japan), POB HCl (Nakarai, Kyoto, Japan), and SAL (Sankyo, Tokyo, Japan). Saponin (ICN Biochem. Inc., N.Y., USA) was dissolved with relaxing solution.

Results

Four-week-old but not 2-week-old SHR were smaller than the respective control WKY (Table 1). The BP of 4-week-old SHR was similar to that of age-matched WKY. SAL elicited a dose-dependent relaxation in the arterial strips from WKY and SHR (Fig. 1). The dose-response curves in SHR were shifted to the right and upward of those in WKY. ISO also elicited a dose-dependent vasorelaxation; sensitivities to ISO of the femoral and mesenteric arteries from SHR at ages from 4 to 10 weeks were less than those from age-matched WKY (Fig. 2). The sensitivity to ISO remained unchanged during developing stages of hypertension.

To differentiate carefully agonist-induced relaxation of the arteries between WKY and SHR, both groups of the arteries were contracted either with KCl by 85% of the maximal KCl contraction or with $CaCl_2$ by 75% of the maximal $CaCl_2$ contraction. The dose-response curves for ISO in the femoral arteries and for NE in the mesenteric arteries from SHR were shifted to the right of those in the WKY tissues (Fig. 3). The dose-response curves for ISO in the SHR mesenteric arteries were shifted to the right and upward of those in the WKY tissues. The dose-response curves for FEN in the SHR mesenteric arteries were quite similar to those in the WKY tissues. Table 2 shows the ED_{50} values of ISO- and NE-induced relaxations in the arterial strips.

ISO-induced relaxations of the femoral and mesenteric arteries precontracted with $CaCl_2$ were diminished in SHR compared to those in WKY (Fig. 4). DB cAMP-induced relaxation of the femoral but not mesenteric arteries from SHR was diminished as compared to those from WKY. Response of the SHR mesenteric arteries to forskolin was similar to that of the WKY tissue. The mean ED_{50} values for ISO-, FEN-, and NE-induced relaxations of the arteries precontracted with 30 m*M* KCl are summarized in Table 3. Relative potency ratios of ISO/FEN/NE were determined with their ED_{50} values. Relative potency ratios on the femoral artery were markedly different from those on the mesenteric artery

Table 1. Body weight and blood pressure in WKY and SHR

Age (weeks)	WKY		SHR	
	2	4	2	4
Body weight (g)	31 ± 1	84 ± 4	31 ± 2	74 ± 3*
Blood pressure (mmHg)	ND	100 ± 2	ND	104 ± 1

Two-week-old animals were too small for blood pressure to be determined by tail-cuff method
*Significantly different from age-matched WKY ($P < 0.05$)

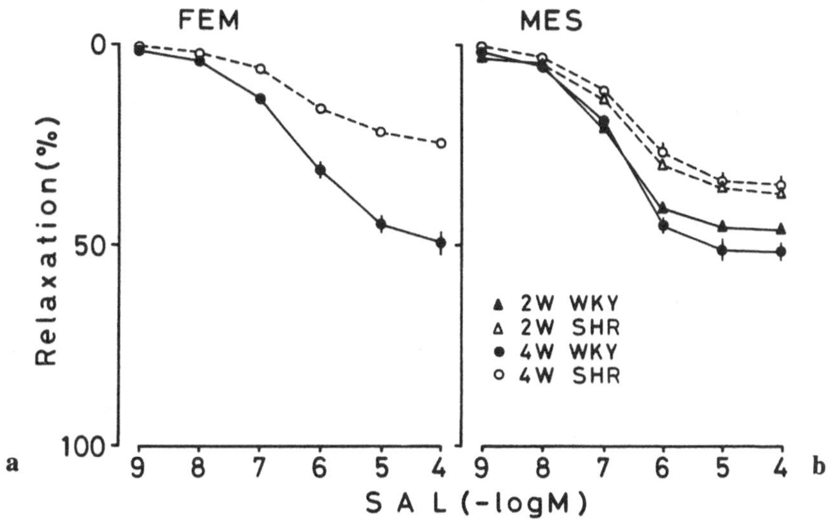

Fig. 1a,b. Dose-response curves for salbutamol (*SAL*)-induced relaxation of **a** femoral and **b** mesenteric arteries from 2-week-old (▲, △) and 4-week-old (●, ○) WKY and SHR, respectively. Relaxation elicited with 10^{-4} *M* papaverine is expressed as 100%. *Vertical bars* represent SE of means ($n = 8$)

of the same animal. There were no marked differences in the relative potency ratios between either WKY and SHR or between 4-week-old SHR and 7-week-old SHR.

Atenolol and butoxamine competitively inhibited NE- and FEN induced relaxations of the femoral and mesenteric arteries (Fig. 5). Atenolol was more effective than butoxamine in inhibiting relaxation responses of the femoral artery to either NE or FEN; the reverse was true for the mesenteric artery. For both antagonists on the WKY femoral arteries, the Schild plot obtained when using NE as the agonist was superimposed on that obtained with FEN as the agonist (Fig. 6). The slopes of the Schild plots did not differ from unity. Yet, for the antagonists used on the WKY mesenteric arteries and aortae, the Schild plots when using NE and FEN as the agonists were not superimposable. While the slopes of the Schild plots were not necessarily 1.0, there were no marked differences in the Schild plots either between WKY and SHR or between 4-

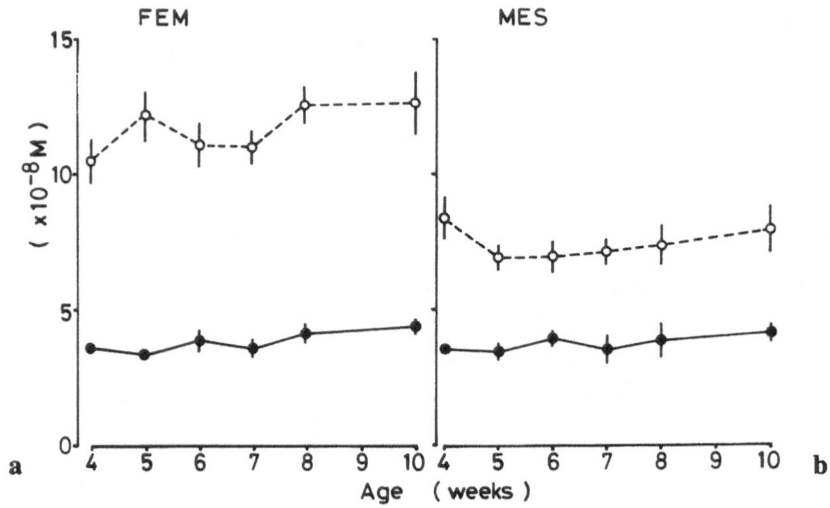

Fig. 2a, b. Time course of changes in ED_{50} values of isoproterenol (ISO)-induced relaxation of **a** femoral and **b** mesenteric arteries from WKY (●) and SHR (○). To determine the dose-response curves of ISO, the arteries were contracted with 30 m*M* KCl

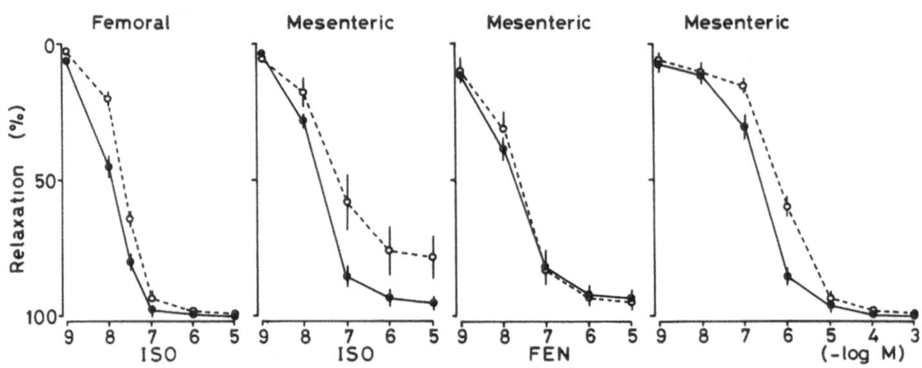

Fig. 3. Dose-response curves for isoproterenol (*ISO*)-, fenoterol (*FEN*)-, and norepinephrine (*NE*)-iduced relaxation of femoral and mesenteric arteries from 4-week-old WKY (●) and SHR (○). The arteries were contracted with KCl to 85% of the maximal KCl contraction. Papaverine-induced relaxation is expressed as 100%. *Vertical bars* represent SE of means ($n = 7$–18)

week-old SHR and 7-week-old SHR. The pA_2 values for atenolol and butoxamine on the SHR blood vessels were essentially the same as those on the tissues from WKY or conventional Wistar rats (Tables 4, 5). On the other hand, the pA_2 values for these antagonists on the femoral artery were significantly different from those on either the mesenteric artery or the aorta from the same animal.

Table 2. ED_{50} values of isoproterenol- and norepinephine-induced relaxations in femoral and mesenteric arteries

Artery	Agonist	ED_{50} ($\times 10^{-8}$ *M*)	
		WKY	SHR
Femoral	Isoproterenol	1.19 ± 0.14	2.48 ± 0.20*
Mesenteric	Isoproterenol	2.30 ± 0.23	4.19 ± 0.79*
Mesenteric	Norepinephrine	22.6 ± 2.7	66.1 ± 14.0*

The arterial strips (pretreated with phenoxybenzamine) were contracted with KCl, which was added to normal Krebs-Henseleit bicarbonate buffer to produce 85% of maximal KCl-contractions
* Significantly different from WKY ($P < 0.05$)

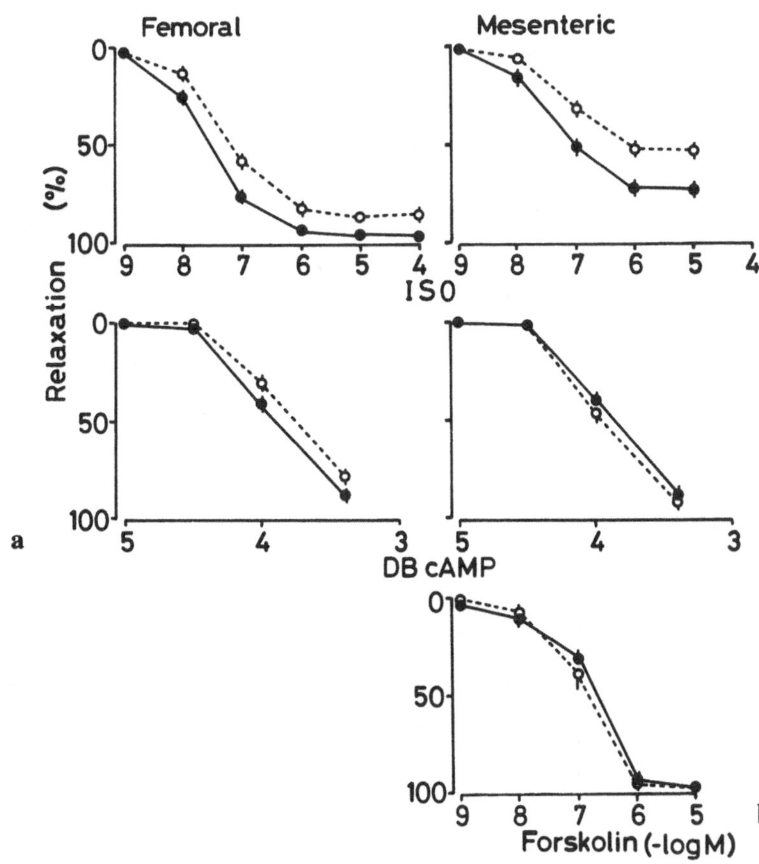

Fig. 4a,b. Dose-response curves for isoproterenol (*ISO*)-, dibutyryl cyclic AMP (*DB cAMP*)-, and forskolin-induced relaxation of **a** femoral and **b** mesenteric arteries from 4-week-old WKY (●) and SHR (○). The arteries, incubated with Ca^{2+} free, high K^+ (30 m*M*) medium, were contracted with $CaCl_2$ to 75% of the maximal $CaCl_2$ contraction. Papaverine-induced relaxation is expressed as 100%

Table 3. ED_{50} values of isoproterenol (ISO)-, fenoterol (FEN)-, and norepinephrine (NE)-induced relaxations in femoral and mesenteric arteries[a]

Agonist	ED_{50} value ($\times 10^{-8}$ *M*)			Relative potency (as dl isomers)[b]		
	Wistar	WKY	SHR	Wistar	WKY	SHR
Femoral (4 weeks)						
(±) ISO	ND	3.63 ± 0.14	10.50 ± 0.80	ND	100	100
(±) FEN	ND	59.4 ± 7.3	133.4 ± 13.5	ND	6	8
(−) NE	ND	26.1 ± 1.9	68.2 ± 6.3	ND	7	7.5
Femoral (7 weeks)						
(±) ISO	6.11 ± 0.44	3.64 ± 0.28	10.90 ± 0.60	100	100	100
(±) FEN	128 ± 16	67.1 ± 10.8	212.6 ± 8.6	5	5	5
(−) NE	22.2 ± 1.2	20.1 ± 1.0	59.4 ± 3.5	4	9	9
Mesenteric (4 weeks)						
(±) ISO	ND	3.55 ± 0.06	8.40 ± 1.26	ND	100	100
(±) FEN	ND	4.45 ± 0.64	5.54 ± 0.52	ND	80	152
(−) NE	ND	342 ± 47	914 ± 129	ND	0.5	0.5
Mesenteric (7 weeks)						
(±) ISO	4.50 ± 0.33	3.53 ± 0.68	5.90 ± 1.00	100	100	100
(±) FEN	4.20 ± 0.69	4.38 ± 0.68	5.33 ± 0.72	107	81	111
(−) NE	456 ± 67	313 ± 38	511 ± 45	0.5	0.5	0.5

[a] The phenoxybenzamine-treated arterial strips were contracted with 30 m*M* KCl
[b] Relative potency (as dl isomers) = (mean ED_{50} value ISO/mean ED_{50} value agonist) × 100

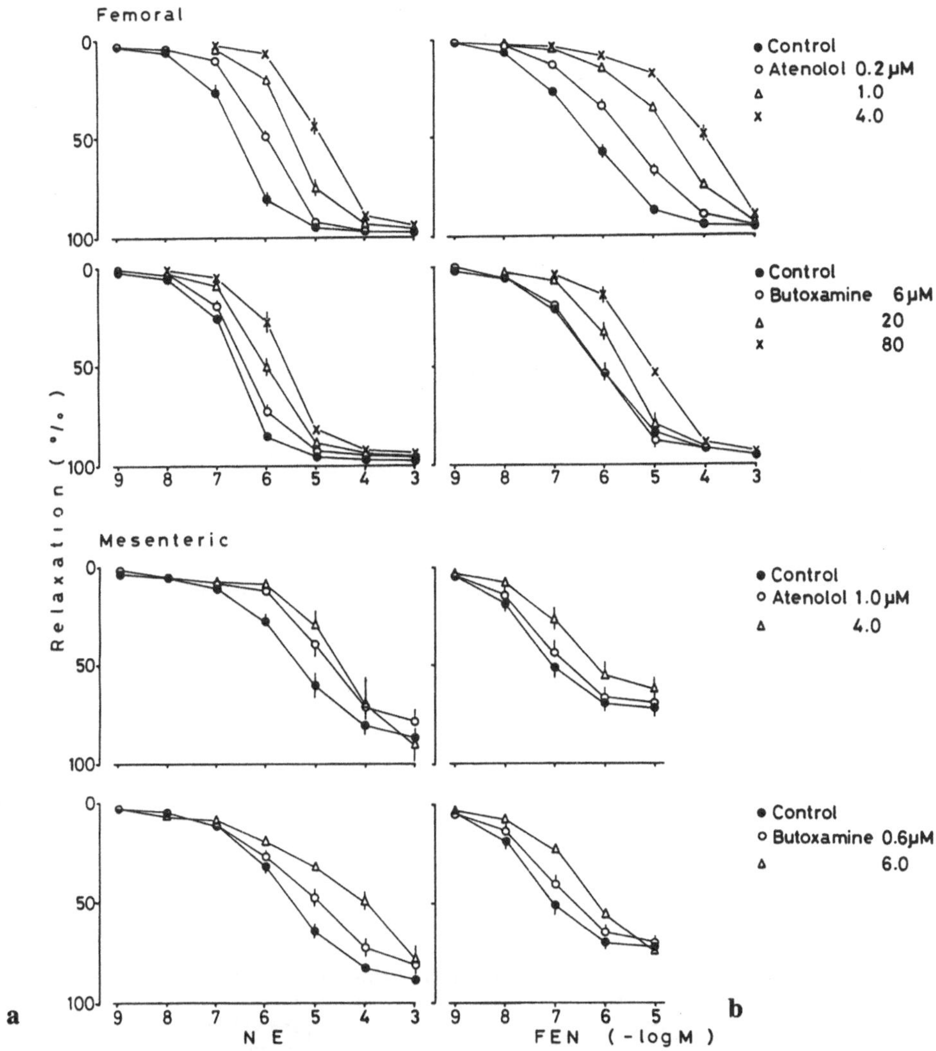

Fig. 5a, b. Dose-response curves for **a** norepinephrine (*NE*)- and **b** fenoterol (*FEN*)-induced relaxation of femoral and mesenteric arteries from 4-week-old WKY. The arteries were treated with atenolol or butoxamine for 60 min before determination of the dose-response curve. Relaxation in response to 10^{-4} *M* papaverine is expressed as 100%

It was then determined whether the diminished relaxation was due to an enhanced noradrenergic outflow in the resistance vessels. SHR were treated i.p. with reserpine (1 mg/kg daily) for 3 weeks or with 6-hydroxydopamine (OHDA) (100 mg/kg twice a week) for 3–4 weeks. These studies were started with immature SHR at the age of just 3 weeks. The treatments delayed the initiation of hypertension but did not change the potency and efficacy of ISO in the arteries tested (Table 6). Subcutaneous infusion of nifedipine at a dose of 1 mg/kg daily for 1 week (Alzet osmotic minipump) inhibited the development of hypertension

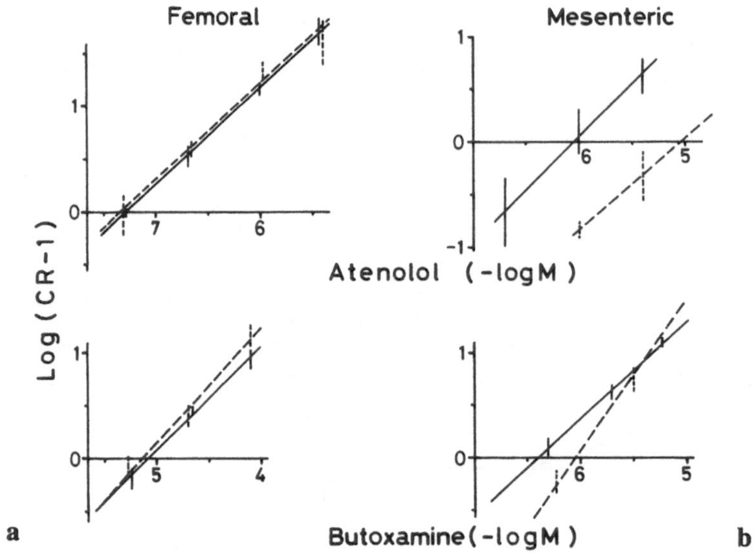

Fig. 6a,b. Schild plots for atenolol and butoxamine in the **a** femoral and **b** mesenteric arteries of 4-week-old WKY. Agonists used were norepinephrine (beta-1 selective) (*solid lines*) and fenoterol (beta-2 selective) (*dashed lines*). *CR*, concentration ratio

Table 4. The pA_2 values for atenolol and butoxamine on the femoral arteries[a]

Antagonist	Agonist	pA_2 value[b]		
		Wistar	WKY	SHR
Four weeks				
Atenolol	NE	ND	7.12 ± 0.08	7.19 ± 0.10
	FEN	ND	7.23 ± 0.05	7.90 ± 0.05
Butoxamine	NE	ND	4.97 ± 0.02	5.11 ± 0.06
	FEN	ND	4.99 ± 0.03	5.00 ± 0.14
Seven weeks				
Atenolol	NE	7.25 ± 0.09	7.18 ± 0.10	7.17 ± 0.07
	FEN	7.43 ± 0.13	7.44 ± 0.09	7.31 ± 0.06
Butoxamine	NE	5.04 ± 0.03	4.81 ± 0.03	5.32 ± 0.19
	FEN	5.27 ± 0.10	5.06 ± 0.03	4.91 ± 0.05

pA_2, affinity constant; NE, norepinephrine; FEN, fenoterol
[a] The arterial strips were treated with phenoxybenzamine and contracted with 30 m*M* KCl
[b] For pA_2 value = log (CR-1) − log [B]. The slopes of the Schild plots were not different from 1.0

Table 5. The pA_2 values and slopes of the Schild plots for atenolol and butoxamine on the mesenteric arteries and thoracic aortae[a]

Antagonist	Agonist	Wistar		WKY		SHR	
		pA_2 value	Slope	pA_2 value	Slope	pA_2 value	Slope
Mesenteric (4 weeks)							
Atenolol	NE	ND	ND	6.38*	0.87	6.10*	0.62
	FEN	ND	ND	5.63*	1.39	5.85*	1.02
Butoxamine	NE	ND	ND	6.37 ± 0.05	1.11 ± 0.12	6.14*	0.68
	FEN	ND	ND	6.17*	0.91	6.59*	0.94
Mesenteric (7 weeks)							
Atenolol	NE	6.39 ± 0.21	1.11 ± 0.11	7.00*	0.82	6.60 ± 0.09	0.91 ± 0.12
	FEN	ND	ND	6.05*	0.82	5.69*	1.61
Butoxamine	NE	6.44*	0.83	6.27 ± 0.13	0.94 ± 0.13	6.35*	0.72
	FEN	ND	ND	6.05*	1.11	6.35*	0.80
Thoracic (7 weeks)							
Atenolol	NE	6.57*	0.73	ND	ND	ND	ND
Butoxamine	NE	6.58*	0.89	ND	ND	ND	ND

pA_2, affinity constant; NE, norepinephrine; FEN, fenoterol
*Values obtained by extrapolation of Schild plot to log (CR-1) = 0, since the slopes were less than 1.0
[a]The phenoxybenzamine-treated vascular strips were contracted with 30 m*M* KCl

Table 6. Effects of reserpine, 6-hydroxydopamine (OHDA), and nifedipine on the blood pressure (BP) and the isoproterenol-induced relaxation of the arteries[a]

	BP (mmHg)	Femoral		Mesenteric	
		ED_{50} (× 10^{-8} *M*)	Relaxation (%)[b]	ED_{50} (× 10^{-8} *M*)	Relaxation (%)[b]
SHR	135 ± 2	11.0 ± 0.8	77 ± 1	7.1 ± 0.7	28 ± 4
SHR + reserpine	107 ± 3*	9.9 ± 1.0	80 ± 2	5.9 ± 0.7	35 ± 1
SHR	153 ± 8	10.6 ± 0.6	80 ± 1	7.1 ± 0.7	28 ± 4
SHR + 6-OHDA	102 ± 11*	10.8 ± 0.5	75 ± 5	5.4 ± 1.0	25 ± 4
SHR	131 ± 8	12.2 ± 1.2	79 ± 2	6.8 ± 0.6	33 ± 5
SHR + nifedipine	111 ± 3*	8.7 ± 0.9*	85 ± 2*	5.8 ± 0.7	66 ± 5*
WKY	108 ± 7	3.4 ± 0.2	96 ± 1	3.3 ± 0.3	75 ± 2
WKY + nifedipine	93 ± 6*	2.7 ± 0.1*	98 ± 1	3.3 ± 0.6	83 ± 5

*Significantly different from untreated animals ($P < 0.05$)
[a]The arterial strips were contracted with 30 m*M* KCl
[b]Maximum response to papaverine (10^{-4} *M*) was expressed as 100%

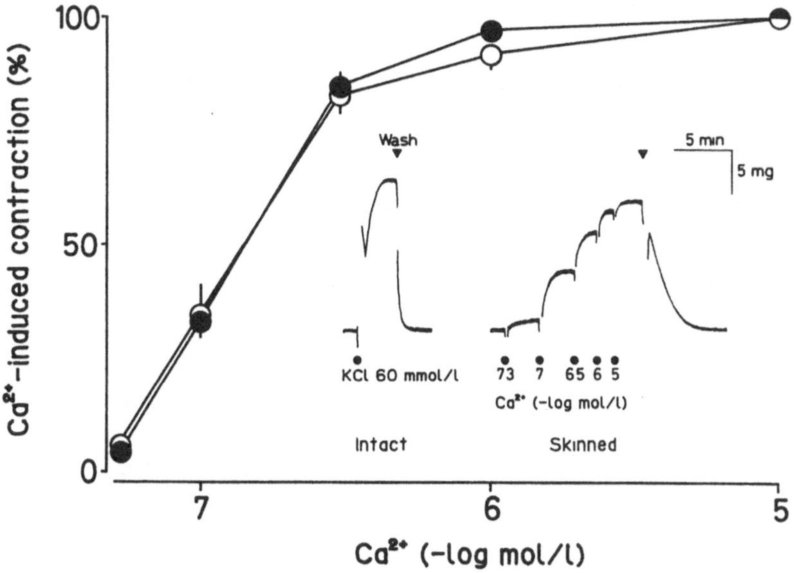

Fig. 7. Responses of skinned muscle fibers from the rat caudal arteries to Ca^{2+}. Ca^{2+} concentration-tension relation of skinned fibers from WKY (○) and SHR (●), and typical recording of Ca^{2+}-induced contraction of skinned fibers from WKY. Vertical bars represent SE of means (n = 9). [29]

and did increase the potency and efficacy of ISO in the SHR femoral artery. Nifedipine also increased the potency of ISO in the WKY femoral artery.

Skinned fibers from the caudal arteries contracted well with the addition of Ca^{2+} (Fig. 7). There was not a difference in the contractile responses of the skinned fibers to Ca^{2+} between WKY and SHR.

Discussion

It is reasonable to speculate that abnormal abilities of resistance vessels to contract or relax account for spontaneous hypertension. It is known that vascular preparations from SHR are less responsive to the beta-adrenoceptor agonist, ISO, compared to those from WKY (Table 7) [6–8, 18–22]. These studies, however, were carried out with aortae prepared from adult SHR, which had already developed hypertension. Cheng and Shibata [23] first suggested that the diminished response of the aorta to ISO was not due to an elevated BP. Using ISO and other beta-adrenoceptor agonists (dobutamine, fenoterol, orciprenaline, and salbutamol), we thereafter demonstrated that beta-adrenoceptor-mediated relaxation of the resistance vessels was diminished in immature SHR at ages as early as 2–4 weeks [6, 8, 24]. In our experiments, the vascular preparations were contracted with KCl or $PGF_{2\alpha}$ at equimolar concentrations in WKY and SHR. In another series of experiments, the preparations were contracted

Table 7. Relaxation responses of blood vessels from SHR to isoproterenol (ISO) in different studies

Age (weeks)	Vascular preparation	Contractile agent used	α-blocker added	Dose of ISO (μM)	Response to ISO	References
6–8[a]	Aorta, helical	5-HT (0.5–1.0 μM)		0.01–1.0	Decrease	[18]
20–24	Aorta, circular	KCl (100 mM)[b] CaCl$_2$ (WKY; 1 mM) (SHR; 3 mM)	PTA	0.01–3.0	Decrease	[19]
10–12	Aorta, helical	5-HT (ED$_{50}$)[c]	PTA	0.01–1.0	Decrease	[20]
10–12	Femoral, ring	KCl (ED$_{50}$)[c]	PTA	0.01–1.0	Decrease	[21]
7–10	Aorta, helical	5-HT (8 μM)		0.4	Increase	[22]
6, 14	Femoral, helical, mesenteric	KCl (30 mM)	POB	0.01–100.0	Decrease	[7]
3–35	Aorta, helical	5-HT (10 μM) NE (5 μM)		0.005–1.0 0.1–1.0	Decrease	[23]
2–10	Femoral, helical, mesenteric	KCl (30 mM) PGF$_{2\alpha}$ (1 μM)	POB	0.01–100.0	Decrease	[6]
4	Femoral, helical, mesenteric	KCl (30 μM) CaCl$_2$ (WKY; 0.9–1.0 mM) (SHR; 0.8–2.1 mM)	POB	0.01–100.0	Decrease	[8]

NE, norepinephrine; PGF$_{2\alpha}$, prostaglandin F$_{2\alpha}$ PTA, phentolamine; POB, phenoxybenzamine
[a] SHR 6–8 months old
[b] High K^+ medium was used; contractions of the tissues from WKY with 1 mM Ca^{2+} were similar to those from SHR with 3 mM Ca^{2+}
[c] The tissue was contracted with a given agent by 50% of the maximum responses to the agent. Actual doses used were not given

with $CaCl_2$ by 75% or with KCl by 85% of the maximal $CaCl_2$ (KCl)-contractions.

Several lines of evidence suggested that there was hypernoradrenergic innervation in peripheral vasculatures of prehypertensive SHR [4, 25]. In addition, increased levels of NE caused desensitization of beta-adrenoceptors. We found that a decrease in ISO-induced vasorelaxation was associated with deoxycorticosterone-salt hypertension in rats [15]. This type of hypertension was accompanied with an increased noradrenergic activity. It was found that reserpine and 6-OHDA delayed the onset of spontaneous hypertension but did not potentiate the response of the SHR arteries to ISO, suggesting that the diminished relaxation was not accounted for by elevated BP nor by the increased noradrenergic outflow. The potency of ISO in the arteries of SHR remained almost unchanged during 10 long weeks after birth, while hypertension became marked [6]. We are of the opinion that the defect is genetic in origin and that mild increases in BP and/or adrenergic activity do not cause a further decrease in beta-adrenergic vasorelaxation in SHR.

The number of beta-adrenoceptors was decreased in blood vessels of young SHR before the development of hypertension as compared to that in the WKY tissues [26]. The number of beta-adrenoceptors in SHR remained unchanged up to the age of 25 weeks. We challenged the general concept that unlike beta-1 adrenoceptors, beta-2 adrenoceptors were responsible for relaxation of the peripheral blood vessels in response to beta-adrenoceptor agonists. We tried to subclassify beta-adrenoceptors in the femoral artery of PHSHR. The results were compared to those obtained with the mesenteric artery and aorta of the rat in which beta-2 subtypes had been reported to be predominant [5, 12]. Relative potency ratios of ISO/FEN/NE on the WKY femoral arteries suggested that the artery contained predominantly beta-1 subtypes of beta-adrenoceptors. Schild plots for atenolol were superimposed whenever NE or FEN was used as the agonist. Schild plots for butoxamine were also superimposed. The pA_2 values obtained for atenolol and butoxamine were very similar to those obtained previously on beta-1 adrenoceptors. The slopes of Schild plots for the antagonists did not differ from unity. These results, altogether, suggested that beta-adrenoceptors in the rat femoral artery were classified as a beta-1 subtype. We are the first to demonstrate that there is a peripheral artery other than the coronary artery which contains only beta-1 subtypes of beta-adrenoceptors. The femoral arteries from PHSHR and SHR at early stages of hypertension contained a homogenous population of beta-1 adrenoceptors. Evidence obtained suggested for the first time that the diminished relaxation response to the beta-agonists was not explained with changes in the beta-adrenoceptor subtypes. We also found that deoxycorticosterone-salt hypertension did not alter the population of the subtype of beta-adrenoceptors [15].

Relative potency ratios of ISO/FEN/NE on the mesenteric artery and thoracic aorta were quite different from those on the femoral artery and suggested that beta-adrenoceptors on these blood vessels were of beta-2 subtypes. This concept was supported by the pA_2 values for atenolol and butoxamine. On the other hand, Schild plots for butoxamine (and atenolol to a lesser extent) were separated when using NE and FEN as agoists. The slopes of the Schild plots which differed from unity suggested that these blood vessels contained a major population of beta-2 subtypes and a minor population of beta-1 subtypes. The present study also suggested that the subtype of beta-adrenoceptors on the mesenteric artery and aorta was not altered by the early stages of hypertension. Thus, we concluded that beta-1 adrenoceptor-mediated relaxation of the femoral and mesenteric arteries was already diminished in PHSHR (Figs. 3, 4).

DB cAMP was less potent as a relaxant in the femoral artery of PHSHR than in the WKY artery. The decreased response to DB cAMP has also been found in aortae from SHR at early stages of hypertension [20]. The response of the PHSHR mesenteric artery to forskolin was similar to that of the WKY artery. These results suggested that the diminished relaxation is related to events distal and proximal to activation of adenylate cyclase (or cAMP formation) in the femoral and mesenteric arteries, respectively. It was reported that cAMP-dependent protein kinase activity was decreased [27] or not changed [28] in the PHSHR aorta compared with the WKY aorta. There was no difference in cAMP-dependent protein kinase response to forskolin between WKY and SHR aortae. Thus, this result suggested that the diminished relaxation response of the aorta to forskolin is due to an abnormality in events distal to activation of cAMP-dependent protein kinase [19], although the defect in the SHR aorta may not be generalized to other vasculatures.

Phosphorylation of membranous protein by protein kinase was followed by transmembrane efflux of Ca^{2+}, Ca^{2+} uptake into the sarcoplasmic reticulum (SR), and thus decreased the amounts of Ca^{2+} available for maintenance of tonicity of VSMC. The cAMP-stimulated phosphorylation of caudal arterial SR membranes and Ca^{2+} uptake by SR are decreased in PHSHR [9, 27].

The present study demonstrated that the response of the skinned muscle fiber from the PHSHR caudal artery to Ca^{2+} was quite similar to that from the WKY artery. A result similar to ours was reported previously [10]. Thus, the diminished relaxation may not be explained by abnormal sensitivity of contractile elements to Ca^{2+}. In other words, increased amounts of Ca^{2+} available for the arterial contraction are important for initiation of hypertension. We found by chance that in the PHSHR caudal artery, optimal concentrations of saponin in terms of a good dose-response curve of Ca^{2+}-induced contraction were different from those in the WKY arteries, suggesting an alteration in VSMC membrane of PHSHR [29]. Nifedipine, a highly lipophilic compound, affects membranous voltage-dependent Ca^{2+} channels to inhibit Ca^{2+} influx and then to reduce arterial contractility (and thus to normalize hypertension). We found that nifedipine increased beta-adrenoceptor-mediated relaxation of the femoral artery. The effect of nifedipine in PHSHR was more marked than that in WKY. It has been reported that in man, nifedipine increased the density of atrial beta-adrenoceptors [30]. Nifedipine has been shown to inhibit NE release from adrenergic nerve endings in rabbit heart [31]. Consequently, chronic administration of the drug via osmotic minipumps could lead to the "up-regulation" of beta-adrenoceptors.

In conclusion, this study was first to demonstrate that beta-1 adrenoceptor-mediated relaxation of resistance vessels was diminished in SHR before the development of hypertension.

References

1. Aoki K (1986) Discovery of the spontaneously hypertensive rat. In: Aoki K (ed) Essential hypertension. Springer-Verlag, Tokyo, pp 3–7
2. Frohlic ED, Tarazi RC, Dustan HP (1969) Re-examination of the hemodynamics of hypertension. Am J Med Sci 257: 9–23
3. Tobia AJ, Walsh GM, Lee JY (1974) Hemodynamic alterations in the young spontaneously hypertensive rat: elevated total systemic and hindquarter vascular resistance. Proc Soc Exp Biol Med 146: 670–673
4. Donohue SJ, Stitzel RE, Head RJ (1988) Time course of changes in the norepinephrine content of tissues from spontaneously hypertensive and Wistar-Kyoto rats. J Pharmacol Exp Ther 245: 24–31
5. O'Donnell SR, Wanstall JC (1984) Beta-1 and beta-2 adrenoceptor-mediated responses in preparations of pulmonary artery and aorta from young and aged rats. J Pharmacol Exp Ther 228: 733–738
6. Fujimoto S, Dohi Y, Aoki K, Asano M, Matsuda T (1987) Diminished β-adrenoceptor-mediated relaxation of arteries from spontaneously hypertensive rats before and during development of hypertension. Eur J Pharmacol 136: 178–187
7. Asano M, Aoki K, Matsuda T (1982) Reduced beta-adrenoceptor interactions of norepinephrine enhance contraction in the femoral artery from spontaneously hypertensive rats. J Pharmacol Exp Ther 223: 207–214
8. Fujimoto S, Dohi Y, Aoki K (1988) Diminished β-adrenergic relaxation in arterial smooth muscle from spontaneously hypertensive rats. In: Aoki K (ed) Calcium in essential hypertension. Academic Press, Tokyo, pp 359–380
9. Bhalla RC, Sharma RV, Ramanathan S (1980) Possible role of phosphorylation-dephosphorylation in the regulation of calcium metabolism in cardiovascular tissues of SHR. Hypertension 2: 207–214
10. Mrwa U, Güth K, Haist C, Troschka M, Herrmann R, Wojciechowski R, Gagelmann M (1986) Calcium-requirement for activation on skinned vascular smooth muscle from spontaneously hypertensive (SHRSP) and normotensive control rats. Life Sci 38: 191–196
11. Rothwell NJ, Stock MJ, Sudera DK (1987) Changes in tissue blood flow and β-receptor density of skeletal muscle in rats treated with the β_2-adrenoceptor agonist clenbuterol. Br J Pharmacol 90: 601–607
12. Tsujimoto G, Hoffman BB (1985) Desensitization of β-adrenergic receptor-mediated vascular smooth muscle relaxation. Mol Pharmacol 27: 210–217
13. Summers RJ, Molenaar P, Stephenson JA (1987) Autoradiographic localization of receptors in the cardiovascular system. Trend Pharmacol Sci 8: 272–276
14. Purdy RE, Stupecky GL, Coulombe PR (1988) Further evidence for a homogenous population of beta-1 adrenoceptors in bovine coronary artery. J Pharmacol Exp Ther 245: 67–71
15. Fujimoto S, Dohi Y, Aoki K, Matsuda T (1988) Altered vascular beta-adrenoceptor-mediated relaxation in deoxycorticosterone-salt hypertensive rats. J Pharmacol Exp Ther 244: 716–723
16. Tokudome T, Taira N (1981) Characterization of β-adrenoceptors in the dog saphenous vein. Jpn J Pharmacol 31: 731–736
17. Tsuru H, Negita S (1987) Effects of catecholamines on isolated canine facial veins. Jpn J Pharmacol 44: 489–492
18. Cohen ML, Berkowitz BA (1976) Decreased vascular relaxation in hypertension. J Pharmacol Exp Ther 196: 396–406
19. Silver PJ, Michalak RJ, Kocmund SM (1985) Role of cyclic AMP protein kinase in decreased arterial cyclic AMP responsiveness in hypertension. J Pharmacol Exp Ther 232: 595–601
20. Triner L, Vulliemoz Y, Verosky M, Manger WM (1975) Cyclic adenosine monophosphate and vascular reactivity in spontaneously hypertensive rats. Biochem Pharmacol 24: 743–745

21. Field FP, Soltis EE (1985) Vascular reactivity in the spontaneously hypertensive rat: effect of high pressure stress and extracellular calcium. Hypertension 7: 228–235
22. Spector S, Fleisch JH, Maling HM, Brodie BB (1969) Vascular smooth muscle reactivity in normotensive and hypertensive rats. Science 166: 1300–1301
23. Cheng JB, Shibata S (1981) Vascular relaxation in the spontaneously hypertensive rat. J Cardiovasc Pharmacol 3: 1126–1140
24. Fujimoto S, Dohi Y, Aoki K, Matsuda T (1988) Beta-1 and beta-2 adrenoceptor-mediated relaxation responses in peripheral arteries from spontaneously hypertensive rats at prehypertensive and early hypertensive stages. J Hypertens 6: 543–550
25. Scott TM, Pang SC (1983) The correlation between the development of sympathetic innervation and the development of medial hypertrophy in jejunal arteries in normotensive and spontaneously hypertensive rats. J Auton Nerv Syst 18: 25–32
26. Limas CJ, Limas C (1979) Decreased number of β-adrenergic receptors in hypertensive vessels. Biochim Biophys Acta 582: 533–536
27. Bhalla RC, Webb RC, Singh D, Ashley T, Brock T (1978) Calcium fluxes, calcium binding, and adenosine cyclic 3', 5'-monophosphate-dependent protein kinase activity in the aorta of spontaneously hypertensive and Kyoto Wistar normotensive rats. Mol Pharmacol 14: 468–477
28. Coquil JF, Hamet P (1980) Activity of cyclic AMP-dependent protein kinase in heart and aorta of spontaneously hypertensive rat. Proc Soc Exp Biol Med 164: 569–575
29. Dohi Y, Aoki K, Fujimoto S, Kojima M, Matsuda T (in press) Alteration in sarcoplasmic dependent contraction of tail arteries to caffeine and noradrenaline in spontaneously hypertensive rats. J Hypertens 7
30. Hedberg A, Kempf F Jr, Josephson ME, Molinoff PB (1985) Coexistence of beta-1 and beta-2 adrenergic receptors in the human heart: effects of treatment with receptor antagonists or calcium entry blockers. J Pharmacol Exp Ther 234: 561–568
31. Stark K, Schnemann JH (1973) Wirkung von Nifedipine auf die Funktion der Sympathischen Nerven der Herzens. Arzneim ittelforschung 23: 193–197

Changes of Vascular β-Adrenoceptors in Spontaneously Hypertensive Rats: Characterization and Technical Problems*

CHIU-YIN KWAN[1], ROBERT M.K.W. LEE[2], and EDWIN E. DANIEL[1]

Summary. Derangement of Ca handling in vascular smooth muscle (VSM) and accelerated VSM cell growth have been related to the etiology of spontaneous hypertension. Alteration of the β-adrenoceptor has also been implicated in the functional (impaired relaxation) as well as structural (increased proliferation) changes of VSM cells in hypertension. In this study, we utilized subcellular membranes isolated from aortic muscle and cultured aortic smooth muscle cells (CASMC) from SHR and WKY to investigate the properties of vascular β-adrenoceptors by $[^{125}I]$-ICYP binding. We observed (a) ICYP binding sites (B_{max}) are located primarily on the plasma membranes of VSM and occur more frequently in small arteries than in large arteries; (b) ICYP binding to crude tissue homogenate often yielded a nonlinear Scatchard plot, whereas ICYP binding to relatively purified membranes yielded a linear Scatchard plot (nonlinearity was also observed in microsomal fractions isolated from smaller branches of mesenteric arteries); (c) microsomes from SHR aorta failed to show reduced β-adrenoceptor density compared with those from WKY or normotensive Wistar rats; (d) cultured aortic muscle cells from SHR showed substantially higher B_{max} for ICYP binding than those from WKY without a change in K_d. We conclude that the use of a purified membrane fraction is advantageous in receptor binding studies. In a comparative study of the receptor density, knowledge of the relative purity of the membranes for the control and test groups is essential. Increased B_{max} of ICYP binding to SHR aortic cells may be related to the increased cell size grown in culture and, if present in vivo, may reflect impaired relaxation.

Key words: Adrenoceptor—Hypertension—Vascular smooth muscle—Cell culture—Subcellular membranes

Introduction

Effectors of vascular contractile function, whether as neurotransmitters or as humoral substances, interact with vascular smooth muscle (VSM) cells by first binding to their specific receptors, presumably located in the cell membranes. This is followed by post-membrane mechanisms which ultimately lead to

*This work was supported by the Heart and Stroke Foundation of Ontario and the Medical Research Council of Canada.
Smooth Muscle Research Program, Departments of Biomedical Sciences[1] and Anaesthesia[2], Faculty of Health Sciences, McMaster University, Hamilton, Ontario, Canada L8N 3Z5

changes in the cytoplasmic level of free calcium concentration, resulting in contraction and relaxation of blood vessels [2, 41]. Abnormalities in the ligand-receptor interactions can conceivably result in altered contractile function (responsiveness and/or sensitivity) upon stimulation by these vasoactive hormones. This has been suggested to be the case for the changes in vascular responses to β-adrenergic stimulation in hypertension [1, 5, 7, 9, 11, 39], diabetes mellitus [8, 15, 34], and aging processes [10, 17].

A widely employed technique to investigate directly the binding characteristics of these hormones to their membrane receptors involves the study of binding characteristics of radiolabelled synthetic ligands to isolated vascular muscle membranes. In such ligand-binding studies, crude particulate fractions usually of ill-defined or unknown nature were commonly used. The design of the experimental protocol in such studies often focused on the saturability, reversibility, and specificity of the ligand binding to its receptors. Recently, increased attention has been placed upon the technical advantage as well as the importance of using well-defined and better purified vascular membrane fractions for the characterization of adrenoceptors, particularly in comparative studies of ligand binding parameters of vascular tissues in health and disease [20, 23].

Results on β-adrenoceptor changes in hypertension, i.e., the affinity (K_d) and the maximum number of binding sites (B_{max}) for the radioligands derived from binding studies, have not been very consistent. They also raise questions about the cause of the large variations in B_{max} values reported in earlier studies. For example, in cardiac muscle membranes from spontaneously hypertensive rats (Aoki SHR), some investigators [28] reported that B_{max} was decreased with unchanged K_d, whereas others found an increased B_{max} with unchanged K_d [32]. Similar variations in findings have been reported in renal hypertension, e.g., decreased B_{max} with a change in K_d [44] and decreased K_d without a change in B_{max} [12]. Admittedly, variables due to age, sex, regional difference in vascular segments, or duration of hypertension could be some of the contributing factors, but at the moment differences which might have arisen from the use of crude and poorly defined membrane fractions, instead of any other variables, cannot be excluded and have not been carefully evaluated.

In studies using VSM, the B_{max} for β-adrenoceptor binding of rat mesenteric arterial membranes varied from 10 fmol/mg [44] to greater than 250 fmol/mg [23] under similar conditions, including the same ligand and the same source of smooth muscle. Furthermore, using a crude membrane fraction from rat mesenteric artery, Tsujimoto and Hoffman [40] reported a K_d value of 82 p*M* and a B_{max} value of 28 fmol/mg. This is in sharp contrast to the corresponding values ($K_d = 14$ p*M* and $B_{max} = 280$ fmol/mg) recently reported in the same vascular tissue using highly purified plasma membranes [23].

Another way of studying the vascular β-adrenoceptor changes in hypertension is the use of radioligand binding to cultured VSM cells obtained from hypertensive and normotensive animals. This approach, if employed under properly controlled conditions, provides a number of technical advantages, including a purer source and larger amount of smooth muscle cells. Furthermore, since vascular muscle cells from normotensive and hypertensive animals are grown in the same culture medium under experimental conditions free of differential extrinsic factors imposed upon them (e.g., transmural pressure, humoral substances, and neurotransmitters), the differences observed can be regarded as intrinsic changes associated with the VSM cells independent of blood pressure. We have previously adopted such an approach to study the Ca transport [18] and elastase-like activities [21a] of aortic smooth muscle cell membranes from SHR. In this chapter, we detail our recent findings concerning the properties of vascular β-adrenoceptors using the above two approaches employing isolated subcellular vascular muscle membranes and cultured aortic smooth muscle cells from SHR and Wistar-Kyoto (WKY) normotensive rats. Special attention is given to the identification and the resolution of the technical pitfalls we have encountered.

Methods

Fractionation of Vascular Smooth Muscles

Animals and vascular tissues. Thoracic aortae and mesenteric vascular trees were removed from male or female dogs weighing 10–20 kg and were trimmed to remove the surrounding nonarterial tissues as previously described in detail [24, 25]. The mesenteric nerves removed during this careful dissection were also used as a source of membranes. The tissue-trimming procedures we used removed almost completely the endothelial and adventitial layers from the aortae and mesenteric arteries. Special care was taken to obtain a pure source of mesenteric arterial muscle and the nerves [24]. For comparative studies of membrane fractions isolated from aortae or mesenteric arteries of rats of different strains, tissues from SHR, WKY, and Wistar rats were always processed at the same time under paired conditions using our previously described method [22].

Membrane fractionation. Differential centrifugation and a discontinuous sucrose density gradient were employed to isolate various subcellular membranes from the above-mentioned vascular tissues as previously described [24, 25] with a minor modification for the ligand-binding studies [23]. Microsomal fractions from mesenteric nerve fibers were prepared by exactly the same procedures used to prepare microsomes from the arteries. The scheme of differential centrifugation adopted for the isolation of microsomal membrane fractions from all the arterial tissues employed is the same as that reported previously [25]. Subfractionatin of the microsomal fraction on a discontinuous sucrose density gradient was also carried out in some experiments to obtain the highly enriched plasma membrnes from dog aortae [25] and mesenteric arteries [24].

Characterization of subcellular membranes. Morphological features of the membranes used and biochemical characterization with membrane marker enzymes routinely used in this laboratory have been described elsewhere (for a recent review, see [23]). Generally, 5′-nucleotidase, NADPH-cytochrome c reductase, and cytochrome c oxidase are used as representative marker enzymes for plasma membrane, endoplasmic reticulum, and mitochondria, respectively.

Aortic Smooth Muscle Culture Cells

Animals and culture conditions. We employed the secondary cultures of smooth muscle cells derived from the thoracic aortae of SHR and WKY at 3–4 weeks old. At that age, we did not detect any significant difference in blood pressure between SHR and WKY. Cells from passages 3–7 were used. In each experiment, cells of SHR and WKY from the same passage were used. Cells from each SHR and WKY were maintained separately, so that we could compare cells derived from individual SHR and WKY, instead of pooled cells from several SHR and WKY, as is the case in some studies. Cells were cultured in Dulbecco's modified Eagle medium with 10% fetal calf serum at 37°C in an atmosphere of 95% O_2 and 5% CO_2 until confluency. Cells were harvested using 0.25% trypsin at 37°C.

Measurement of cell size, number, viability, and multinuclearity. The number and the size of the smooth muscle cells were determined using a Coulter counter connected to a P-64 model size distribution analyzer. Viability of the cells was determined using the trypan blue dye exclusion method. Cell viability obtained was 90% or better. Multiculture well trays were used to compare the rate of proliferation of the cells, at an initial seeding density of 2×10^4 cells/well. To determine the incidence of multinucleated smooth muscle cells in the culture, cells at confluency were fixed with 10% neutral formalin, and subsequently stained with hematoxylin. Under the high-power phase contrast microscope, the number of smooth muscle cells containing one or more nuclei was counted, until the total number of cells counted exceeded 500. The percentage of cells containing one nucleus, and those containing more than one nucleus were calculated. Student's *t*-test and linear regression analyses were carried out to compare SHR with WKY. The *n* values indicate the number of different cell lots derived from different SHR and WKY, and not repeats of pooled cells from SHR and WKY. *P* values less than 0.05 were considered significant.

Ligand Binding Studies

Binding assays were performed in triplicate using [^{125}I]-monoiodinated cyanopindolol (2200 μCi/nmol), hereafter abbreviated as ICYP (New England Nuclear). The total binding incubation volume was 500 μl and consisted of 100 μl ICYP (dilution of desired final concentrations). 100–200 μl of either cultured aortic smooth muscle cells ($2–4 \times 10^5$ cells) or isolated subcellular membrane fractions (15–50 μg protein), and 200–300 μl of Tris buffer solution containing final concentrations of 150 m*M* NaCl, 1 m*M* $MgCl_2$, and 0.1 m*M* ascorbic acid in 10 m*M* Tris buffered at pH 7.4. Nonspecific binding was determined by parallel binding experiments containing 1 μ*M* DL-propranolol in the incubation medium. Binding experiments were started by addition of membrane fractions or cell suspensions, and the incubation mixture was maintained at 37°C with constant shaking for 60 min. The reaction was stopped with 2.5 ml of incubation buffer and filtered through Millipore HAWP filters on vacuum manifolds followed by four additional rinses of 2.5 ml each of incubation buffer. Filters were counted for 2 min in a Beckman 5500 gamma counter with a 75% counting efficiency. Nonspecific binding to the filters in the absence of added cells as blanks and 100 μl each of ICYP dilution as standards were counted in each run under similar conditions. Any deviation from the above standard conditions will be specified in the descriptions of thr corresponding ligands.

Results

Properties of ICYP Binding to Subcellular Membranes of Vascular Smooth Muscles

Subcellular distribution of β-adrenoceptors. One of the powerful advantages of the subcellular membrane approach lies in the acquisition of information on the subcellular localization of the biochemical parameters in question. This technical advantage has not been effectively utilized in the β-adrenoceptor studies of VSM in spite of repeated demonstration of the existence of β-adrenoceptors in isolated VSM membranes for the past decade. Table 1 summarizes the previous work on ligand binding studies of β-adrenoceptors using membrane fractions isolated from VSM. We have recently reported for the first time in three different vascular tissues on the subcellular localization of the β-adrenoceptors based upon the subcellular distribution of ICYP binding sites. This is reflected in the correlation studies, in which the distribution profiles of ICYP binding were compared with those of several marker enzymes in all subcellular fractions. Figure 1 shows the excellent positive correlation between the ICYP binding sites and the plasma membrane marker enzyme activities. This is further confirmed by membrane density perturbation studies using digitonin which binds to plasma membrane cholesterol molecules and increases the buoyant density of the plasma membrane in a sucrose density gradient [23, 24]. That vascular muscle β-adrenoceptor is of plasmalemmal origin has now been experimentally demonstrated. Figure 1 also indicates that the large artery has an intrinsically lower receptor density than the small artery.

Crude versus more purified membrane fractions. Table 1 indicates a considerable variation in the K_d values for ICYP binding. To investigate whether membrane fractions of different purity in the plasma membrane content contribute to the variability in K_d values, we studied the saturation profiles of ICYP binding to the crude homogenate, microsomal, and highly purified plasma membrane fractions of dog mesenteric arteries [23]. A linear Scatchard relationship was obtained for ICYP binding to the microsomal and plasma membrane fractions. whereas the crude homogenate fraction showed multiple binding sites. It seemed that the contaminating materials in the crude homogenate or total particulate fraction may either interfere with the ICYP binding to the smooth muscle receptors or themselves contain β-adrenoceptors of different affinity for ICYP. Whatever the nature of the interference, it could be eliminated by further purifying the membrane fractions. However, we also observed a nonlinear Scatchard rela-

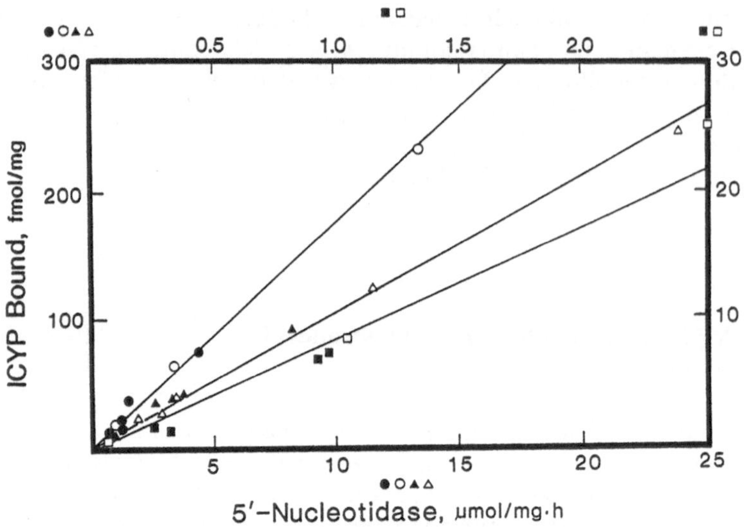

Fig. 1. Correlation between specific ICYP binding and 5′-nucleotidase activities in subcellular fractions isolated by differential centrifugation (*solid symbols*) and sucrose density gradient (*open symbols*) from dog mesenteric arteries (*circles*) and aortae (*squares*). Fractions of rat mesenteric arteries were isolated by differential centrifugation with (*open triangles*) and without (*solid triangles*) Mg-induced membrane aggregation [19]. ICYP concentration is 15 p*M* and 25 p*M* for fractions from rat and dog vascular tissues, respectively

Table 1. Some examples of reported binding parameters using vascular muscle membranes and cultured smooth muscle cells

Vascular muscle	Ligand[a]	K_d(T)	B_{max} (fmol/mg)	References
Rat aorta	DHA	10 n*M* (37°C)	105	[29]
Rat vena cava	DHA	14 n*M* (37°C)	135	[29]
Rat brain vessels[b]	IHYP	95 p*M* (37°C)	132	[30]
Rat mesenteric artery	IHYP	120 p*M* (37°C)	10	[44]
	ICYP	82 p*M* (37°C)	30	[40]
Rat mesenteric artery[c]	ICYP	14 p*M* (37°C)	280	[23]
Dog mesenteric artery[c]	ICYP	14 p*M* (37°C)	360	[23]
Dog aorta[c]	ICYP	20 p*M* (37°C)	50	[23]
Pig coronary artery	ICYP	90 p*M* (37°C)	47	[37]
Cultured cell lines[d]				
BC, 3H-1	ICYP	105 p*M* (20°C)	9000	[14]
DDT_1, MF-2	ICYP	17 p*M* (37°C)	21337	[35]
$A7_r5$	ICYP	70 p*M* (37°C)	130000	[13]
A10	ICYP	100 p*M* (37°C)	48000	[13]

[a] DHA, [^{3}H]-dihydroalprenolol; IHYP, [^{125}I]-iodohydroxybenzylpindolol; ICYP, [^{125}I]-iodocyanopindolol
[b] Binding was carried out using intact blood vessels
[c] Plasma membrane-enriched fractions were employed
[d] Binding was carried out using intact cultured cells, and B_{max} values are expressed as binding sites per cell

Table 2. ICYP binding parameters of microsomal fractions prepared from different levels of arterial branches of dog mesenteric artery

Parameters	1°	2°	3°
B_{max} (fmol/mg)	80.7	102.1	85.9
K_d (p*M*)	16.9	18.2	16.3
Hill coefficient	0.9	0.61	0.65

Data represent the average from two separate membrane preparations using tissues pooled from 5–6 dogs in each experiment

tionship of ICYP binding to the microsomal membranes isolated from the smaller vascular branches of the dog mesenteric arterial bed. Table 2 shows the ICYP binding parameters obtained from the microsomal fractions isolated from the primary (1°), secondary (°), and tertiary (3°) branches of the dog mesenteric arteries. Although there was no marked difference in K_d and B_{max} values among these three levels of arterial branches, the Hill coefficient, n, for the binding of ICYP to microsomes isolated from the smaller arterial branches was consistently lower than that from the major branch of the superior mesenteric artery. This agrees with the notion that the relative contamination by a nonvascular smooth muscle component (e.g., the nerve fibers and nerve endings) is greater in smaller arterial branches (due to greater innervation) in the mesenteric vasculature. In spite of these findings, the use of a highly purified plasma membrane fraction (thus higher receptor density) is technically advantageous for the characterization of β-adrenoceptors in tissues with low receptor density or radioligands with low-specific activity.

Smooth muscle versus non-smooth muscle membrances. One of the possible causes for the nonlinear Scatchard relationship for ICYP binding to the vascular muscle homogenate of dog mesenteric arteries or the microsomal fraction of the smaller arterial branches could be the contaminating membranes from the periarterial nerves. We have therefore compared the properties of ICYP binding to the microsomal membranes isolated from mesenteric vascular muscle and nerve fibers under similar conditions. Although the B_{max} of ICYP binding was significantly higher in the nerve membranes than in the muscle membranes, both membrane types showed a single homogenous ICYP binding site with similar K_d and n values [23]. Therefore, the nonlinearity of the Scatchard plot for ICYP binding to the homogenate fraction of this highly innervated vascular tissue cannot be interpreted solely on the basis of nerve contamination.

However, we did observe a major difference between the muscle and nerve microsomal membranes in the effect of GTP on ICYP binding to the β-adrenoceptors. Figure 2 shows that in the presence of a nonhydrolysable GTP analogue, the binding of a β-adrenoceptor agonist, isoproterenol, to the β-adrenoceptor changed from a high to a low affinity state in the muscle membranes. In the nerve membranes, which seemed to lack high affinity binding sites for the agonist, the GTP analogue had no effect on the binding of isoproterenol to the β-adrenoceptor. All the interaction represented a binding to a low affinity

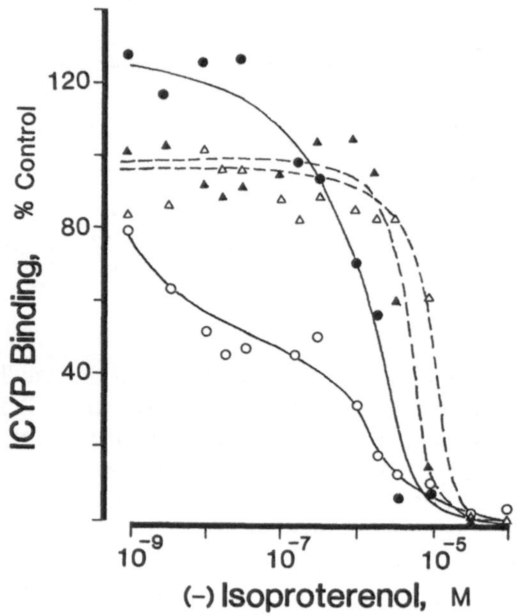

Fig. 2. Competition of ICYP binding by (−) isoproterenol in the presence (*solid symbols*) and absence (*open symbols*) of the nonhydrolysable GTP analogue, GDP(NH)P, in the microsomal fraction of dog mesenteric artery (*solid lines*) and nerves (*dotted lines*). ICYP concentration is 15 p*M*

state. The nature of such a difference remains to be determined, but it raises the possibility that β-receptors from nerve differ in their interaction with agonists and/or in the presence of G-proteins. The ICYP binding sites in the microsomal fraction of dog mesenteric nerves may represent the vascular presynaptic β-adrenoceptors. This is supported by the demonstration of a functional presynaptic β_2-adrenoceptor, which was stimulated by epinephrine, resulting in the facilitation of field-stimulated vasoconstriction of the isolated superior dog mesenteric artery [21]. We also showed the presence of β_2-adrenoceptors in dog mesenteric nerve microsomes by displacement of ICYP binding with the selective β_2-adrenoceptor antagonist, ICI 118,551 [21].

Normotension versus hypertension. Table 3 shows a comparison of the ICYP binding parameters obtained from a crude microsomal membrane fraction isolated from SHR, WKY, and normotensive Wistar rats (NWR). No further purification was attempted due to the limited amount of vascular material and the high cost of the rats. To take into account the possible differential purity of the fractions obtained from these three groups, the enrichment (MIC/PNS activity ratio) of two plasma membrane marker enzymes in the microsomal fractions was also compared with ICYP binding. It is clear that there were some statistically significant differences in the number of specific binding sites between the NWR and SHR groups but not between the WKY and SHR groups, ICYP bind-

Table 3. ICYP binding parameters and enzyme activities in crude microsomal fraction (MIC) isolated from aortae of NWR, WKY, and SHR

	NWR (n = 4)	WKY (n = 3)	SHR (n = 4)
Binding at 25 pM ICYP			
PNS	10.0 ± 2.1	20.3 ± 6.2	18.4 ± 7.5[a]
MIC	20.1 ± 4.5	35.8 ± 13.4	35.8 ± 7.9[a]
	(0.62 ± 0.10)[b]	(1.10 ± 0.42)[b]	(1.00)[b]
MIC/PNS	2.2 ± 0.3	1.9 ± 0.2	2.0 ± 0.4
Scatchard analysis			
K_d (p*M*)	13.3 ± 2.4	22.7 ± 5.5	14.5 ± 3.1[a]
B_{max} (fmol/mg)	22.8 ± 5.0	65.7 ± 12.8[a]	41.0 ± 9.4
Hill coefficient	0.93 ± 0.02	0.94 ± 0.01	0.96 ± 0.01
Plasma Membrane Marker Enzymes			
5′-nucleotidase			
in MIC (μmol/mg.h)	3.57 ± 0.75	5.14 ± 2.26	5.97 ± 0.90[a]
MIC/PNS	2.1 ± 0.3	1.7 ± 0.2	1.8 ± 0.4
Phosphodiesterase I			
in MIC (μmol/mg.h)	1.34 ± 0.12	1.24 ± 0.20	1.21 ± 0.10
MIC/PNS	2.0 ± 0.1	2.6 ± 0.5	2.1 ± 0.3

NWR, normotensive Wistar rats; SHR, spontaneously hypertensive rats. Six rats per group were used for each of the four separate experiments. Experiments on all three groups were carried out under paired conditions except for one experiment, in which WKY rats were not available. The activity ratio MIC/PNS represents the enrichment factor in MIC over the postnuclear supernatant (PNS).
[a] Significantly different from NWR group ($P < 0.05$, paired t-test)
[b] Numbers in parentheses indicate the mean ± SEM of the relative ICYP binding activities taking the specific binding of ICYP to SHR microsomes as unity.

ing site density being greater in SHR than in NWR. The values for the WKY group were more variable but appeared to be very similar to those of SHR. Since the enrichment of ICYP binding and the two plasma membrane marker enzymes in the crude microsomal fraction were similar in all three groups, the difference in ICYP binding between the SHR and NWR groups cannot be due to the difference in membrane purity among the groups. Conceivably, the higher density of β-adrenoceptors in resistance arteries of SHR (and WKY) compared to NWR may reflect different genetic control of these receptors in the inbred rat strains, compared with NWR.

Properties of ICYP Binding to Cultured Aortic Smooth Muscle Cells

In the first series of experiments, after the cultured aortic muscle cells from SHR and WKY reached confluency, the cells were trypsinized, centrifuged to remove the culture media, and resuspended in an isotonic medium (ISOTON from Coulter Electronics Ltd. of Canada: an azide-free, balanced electrolyte solution, diluent for blood cell counting and sizing for the Coulter Counter). After washing the cells an additional three times in the isotonic medium, the cells were resuspended in the same isotonic medium at a density of $1-2 \times 10^6$ cells/ml and

Table 4. ICYP binding parameters of lysed culture aortic muscle cells from SHR and WKY

	WKY	SHR
B_{max}	1.00	4.29 ± 0.9
K_d, p*M*	12.8 ± 4.1	12.0 ± 3.3

Data are expressed as mean ± SEM from five separate paired experiments. Aortic muscle cells were derived from different pairs of SHR and WKY in each experiment. Due to the large variation in B_{max} values obtained from one experiment to another (500–3000 sites/cell for WKY, and 3000–12000 sites/cell for SHR), the B_{max} data were normalized by taking the value for WKY cells as unity. The binding sites in SHR cells were three to seven fold in excess of those in WKY cells

were used directly in the ICYP binding assay for their β-adrenoceptors. This series of experiments consistently indicates a higher density of β-adrenoceptor sites (at 30 p*M* ICYP) in the aortic muscle cells cultured from SHR compared with those from WKY. Howecer, we also found that repeated washing of the cells caused as many as 90% of the cells to take up the trypan blue dye to various degrees, indicating cell damage.

In order to determine the effect of cell integrity on the β-adrenoceptor binding parameters, we performed the binding experiments using unwashed intact cells (suspended in the culture media) and lysed cells (suspended in water to cause osmotic damage). Other various ways of damaging the cells were also evaluated for their effect on β-adrenoceptor density. It was found that intact aortic muscle cells from both SHR and WKY took up such a high level of ICYP, which behaved like nonspecific binding, that it masked the specific binding sites. On the other hand, cells lysed by osmotic shock, by freeze-and-thaw or by gentle homogenization with an all-glass homogenizer showed a consistently higher level of specific binding similar to that obtained using isolated membranes as previously discussed. The cells lysed by osmotic shock by dilution in distilled water followed by one freeze-and-thaw cycle gave the most satisfactory results. Using such an experimental protocol, we found that the lysed cells from SHR showed a variable but consistent increase of the specific ICYP binding sites compared with those from WKY in nine separate paired experiments. Scatchard analysis of the data from five separate experiments is summarized in Table 4. It is clear that the major difference in ICYP binding between SHR and WKY cells is in the quantity (B_{max}), not the affinity (K_d), of the β-adrenoceptor.

Growth and Morphology of Aortic Smooth Muscle Cells from SHR and WKY

When the number of cells per well at different days after seeding were compared between SHR and WKY, the number of smooth muscle cells from WKY was always higher than SHR, suggesting a faster growth rate. Similarly, the rate of percentage cell number increase was also higher in WKY than SHR, and the difference was significant ($P < 0.005$, ANOVA). Figure 3 indicates that the average cell diameter was larger in SHR (14.1 ± 0.6 μm, mean ± SEM, $n = 6$) than

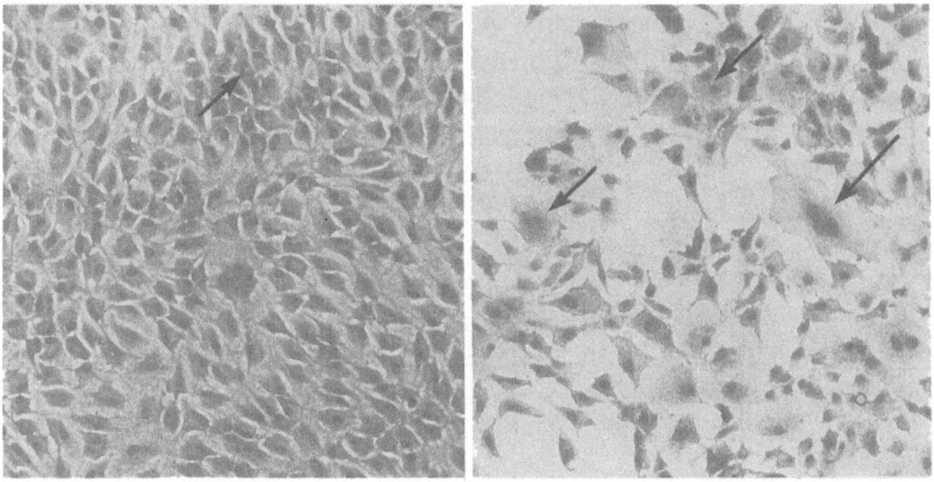

a b

Fig. 3a,b **a** Smooth muscle cells from WKY at confluency. Note the cells are usually of similar size. *Arrow* points to a binucleated cell. ×90. **b** Smooth muscle cells from SHR at confluency. Note the presence of large cells with multiple nuclei (*arrows*). ×90

WKY (12.1 ± 0.3, $n = 6$, $P < 0.0005$). Figure 3 also shows that the percentage of cells with multiple nuclei was higher in SHR (2.8% ± 0.5%, mean ± SEM, $n = 6$) than WKY (0.6% ± 0.1%, $n = 6$, $P < 0.0005$). The observation of the increased proportion of multinuclear aortic muscle cells from SHR is consistent with previous reports on increased nuclear polyploidy and vascular hypertrophy in the VSM cells from hypertensive rats [6, 16, 33]

Discussion

Major Novel Findings

In this work, we have presented direct evidence that β-adrenoceptors are located in the vascular muscle cell membranes isolated from different blood vessels from different animal species. We have also demonstrated that cultures enriched in the plasma membranes isolated from aorta and mesenteric arteries of adult (4–5 months old) SHR and WKY did not show any marked difference in the β-adrenoceptor density as reflected by ICYP binding sites. However, cultured aortic smooth muscle cells from young (4-week-old) SHR and WKY showed a contrasting difference in their β-adrenoceptor density. SHR arotic muscle cells contained many more ICYP binding sites than the WKY aortic muscle cells without alteration in K_d values in membranes of comparable purity. These findings concerning changes of vascular β-adrenoceptors in hypertension are in sharp contrast to some earlier studies that showed β-adrenoceptor density was reduced in vascular membranes isolated from hypertensive rats compared with those from the corresponding normotensive controls. Table 5 shows a sum-

Table 5. Changes of β-adrenoceptor density in tissues from hypertensive rats based on radioligand binding studies

References	Tissues or membranes	Ligand	Changes	Hyper-tension	Comments
Cardiac muscle					
[12]	MIC	DHA	⟷	RH	↑ K_d (2 vs 1 nM)
[43]	MIC	IHYP	↓	DH	K_d not reported
[42]	MIC	IHYP	↓	RH	⟷K_d (~ 2 nM)
[38]	MIC	DHA	↓	RH	⟷K_d (~ 0.5 nM)
[45]	MIC	DHA	↓	SH	⟷K_d (80–100 pM)
[32]	MIC	ICYP	↑	SH	K_d not reported
[31]	MIC	ICYP	⟷	SH	Beta-2 ↑ ; Beta-1 ↓
Vascular smooth muscle					
[29]	Aorta MIC	DHA	↓	SH	⟷K_d (~ 10 nM)
[29]	Inferior vena cava MIC	DHA	↓	SH	⟷K_d (~ 15 nM)
[44]	Mesenteric arteries MIC	ICYP	↓	DH	⟷K_d (100 pM)
[30]	Unfractionated brain microvessels	IHYP	↓	SH	⟷K_d (80–100 pM)
[30]	Unfractionated brain microvessels	IHYP	↓	DH	⟷K_d (80–100 pM)
Present Study	Mesenteric arteries	ICYP	⟷	SH	⟷K_d (10–20 pM) B_{max} ⟷ compared to WKY; B_{max} ↑ compared to Wistar
Other tissues or cells					
[42]	Kidney	DHA	⟷	RH	
[42]	Lung	DHA	⟷	RH	
[31]	Kidney	ICYP	↑	SH	Beta-2 ↑ ; Beta-1 ↑
[31]	Lung	ICYP	↑	SH	Beta-2 ↑ ; Beta-1⟷
[4]	Human lymphocytes	ICYP	↑	EH	
[4]	Rat lymphocytes	ICYP	↑	SH	
[3]	Human lymphocytes	ICYP	↑	EH	

MIC, crude postmitochondrial or microsomal fraction; SH, spontaneous hypertension; DH, deoxycorticosterone-salt hypertension; RH, renovascular hypertension; EH, essential hypertension; abbreviations for ligands were the same as listed in Table 1; ↑, increase; ↓, decrease; ⟷, unchanged mary of the results from a collection of comparative studies of β-adrenoceptor changes in hypertension using membranes isolated from vascular smooth muscle and other tissues. We have also described some methodological advantages as well as the associated technical problems in studying vascular β-adrenoceptors using isolated subcellular membranes and culture cells. Some of these problems may provide a reasonable basis for the different findings obtained in this and earlier studies by other investigators.

Technical Problems

We have pointed out a number of technical pitfalls that may influence the interpretation of ICYP binding data. This includes the use of very crude membrane fractions. In our hands, the use of a total particulate fraction from dog mesenteric arteries for the ICYP binding study resulted in a nonlinear Scatchard binding profile. When the membranes were partially purified by making a microsomal fraction, such nonlinearity was eliminated as reflected by the Hill coefficient close to unity. However, this is not necessarily true for smaller branches of dog mesenteric arteries. The reason for the nonlinear Scatchard plot obtained from using microsomal fractions isolated from small mesenteric arterial branches is still not known. The presence of β-adrenoceptors in nerve plexus or in endothelium, which are more difficult to removed completely in smaller arteries, is a possible explanation. Due to technical limitations, we have been able only to study the ICYP binding to the mesenteric nerve membranes derived primarily from axon. We found a homogenous single binding site as reflected by a linear Scatchard with a K_d similar to that obtained from the vascular smooth muscle membranes [23]. It is possible that β-adrenoceptors in nerve endings and in axonal membrane have different functions and different binding characteristics. Although ICYP binding to both the smooth muscle and nerve membranes was of a predominantly β-adrenoceptor subtype [21], modulation by GTP of β-adrenergic agonist binding differed in the two membranes, as shown in Fig. 2. Whether such a difference is the result of different susceptibilities of the G-protein to the various membrane preparation procedures remains to be clarified.

We have previously shown that the regional difference in the β-adrenoceptor density in different vascular muscles was not due to the difference in membrane purity [23]. In this work, we have also demonstrated that the difference in the number of ICYP binding sites of the aortic microsomes between the NWR and SHR groups was not due to the difference in the plasma membrane content. Neither was the difference in ICYP binding related to blood pressure, since the ICYP binding was not different between the WKY and SHR groups, whereas there were more ICYP binding sites in WKY than in NWR in spite of their similar blood pressures in the normotension range. It is likely that the changes in aortic muscle β-adrenoceptor seen in SHR compared with that in NWR merely reflect genetic differences. This may receive further support from the results obtained using cultured aortic muscle cells derived from young (3–4 weeks old) WKY and SHR when their blood pressure levels were similar and were both within the normotensive range.

To prevent the high level of nonspecific cellular uptake of the lipophilic ligand, ICYP, that masked the specific binding of ICYP in studies using aortic culture cells, we intentionally disrupted the cells by diluting them in deionized and distilled water. These damaged cells, after one freeze-and-thaw cycle, showed levels of total versus nonspecific ICYP binding that could be expected of isolated membranes. The difference between results obtained with microsomal fractions derived from intact aortic muscle and those with lysed cells in SHR and WKY groups was probably due to the age of the rats used. However, we cannot rule out an intrinsic difference in β-adrenoceptor properties due to the sample preparation (e.g., intact tissue vs culture cells). The intrinsic difference in Na^+ dependence of the β-adrenergic responses between the intact vascular tissue and enzymatically isolated single cells has recently been reported [36].

Changes of β-Adrenoceptors in Vascular Muscle in Hypertension

Our results on the changes of β-adrenoceptors in VSM membranes from SHR, obtained either with intact tissue or with cultured cells, were in variance with those reported previously as clearly shown in Table 5. We found no changes of β-adrenoceptor density in the isolated aortic microsomes and an increase in β-adrenoceptor density in the aortic culture cells from SHR as compared with those from WKY. Leaving aside the contributions due to previously mentioned technical problems, decreased vascular β-adrenoceptor density in SHR observed by others as listed in Table 5 has been interpreted to account for the decreased contractile responses of vascular muslce to β-adrenergic stimulation [29, 30, 44]. There is still no general consensus as to whether the decrease in vascular β-adrenergic response is a primary or a secondary event of hypertension [1, 5, 7, 9, 11]. On the other hand, the increased β-adrenoceptor activity associated with hypertensive disease has also been reported in a number of tissues, as shown in Table 5. At least in the VSM, an increase in β-adrenoceptors has been implicated [27, 46] in the cellular growth associated with the vascular structural changes in hypertension, e.g., hypertrophy of aortic smooth muscle cells [16, 26, 33]. Since we have shown that some vascular structural changes were also genetically determined in spontaneous hypertension in rats [26] it is not surprising to find higher vascular β-adrenoceptor density in culture aortic cells from young SHR than in those from young WKY. We have indeed found a greater number of large cells, frequently multinuclear in appearance, in aortic culture cells from SHR than in those from WKY. It would be of interest to determine the growth pattern and density of β-adrenoceptors in cultures of mesenteric artery for WKY and SHR since these may respond differently compared with the large, elastic arteries. It β-adrenoceptor chages in intact vascular tissue are involved with both functional and structural changes in hypertension, and if the regulation of β-adrenoceptor density changes corresponding to the functional change (e.g., reduced contractile response to β-adrenergic stimulation) or structural change (e.g., vascular hypertrophy) in hypertension operates in an opposite manner, the net change of β-adrenoceptor density in the intact vascular tissue in hypertension may then depend upon the relative dominance of the functional or structural changes. These changes could be quite variable depending on the duration and severity of hypertension and the maintenance of the rats, thus contributing to different findings in the literature.

References

1. Asano M, Aoki K, Matsuda T (1982) Reduced beta-adrenoceptor interactions of norepinephrine enhanced contraction in the femoral artery from spontaneously hypertensive rats. J Pharmacol Exp Ther 223: 207–214
2. Bolton TB (1976) Mechanisms of action of transmitters and other substances on smooth muscle. Physiol Rev 3: 606–718
3. Brodde OE, Prywarra A, Daul A, Anlauf M, Bock KD (1984) Correlation between lymphocyte β_2-adrenoceptor density and mean arterial blood pressure: elevated β-adrenoceptors in essential hypertension. J Cardiovasc Pharmacol 6: 678–682
4. Bruschi G, Orlandini G, Pavarani C, Spaggiari M, Tacinelli L, Cavatorta A (1984) Lymphocytic beta-adrenoceptor abnormality in primary hypertension. IRCS Med Sci 12: 461–462
5. Chang JB, Shibata S (1981) Vascular relaxation in the spontaneously hypertensive rat. J Cardiovasc Pharmacol 3: 1126–1140
6. Chobanian A, Lichtenstein AH, Schwartz JH, Hanspal J, Brecher P (1987) Effects of deoxycorticosterone/salt hypertension on cell ploidy in rat aortic smooth muscle cells. Circulation 75 (Suppl I): 102–106
7. Cohen ML, Berkowitz BA (1976) Decreased vascular relaxation in hypertension. J Pharmacol Exp Ther 196: 396–406
8. Dowell RT, Atkins FL, Love S (1986) Integrative nature and time course of cardiovascular alterations in diabetic rat. J Cardiovasc Pharmacol 8: 406–413
9. Field FP, Soltis EE (1985) Vascular reactivity in the spontaneously hypertensive rat: effect of high pressure stress and extracellular calcium. Hypertension 7: 228–235
10. Fleisch JH, Maling HM, Brodie BB (1970) Beta-receptor activity in aorta. Variations with age and species. Circ Res 26: 15;–16
11. Fujimoto S, Dohi Y, Aoki K, Asano M, Matsuda T (1987) Diminished β-adrenoceptor-mediated relaxation of arteries from spontaneously hypertensive rats before and during development of hypertension. Eur J Pharmacol 135: 179–187
12. Giachetti AL, Clark TL, Berti F (1979) Subsensitivity of cardiac beta-adrenoceptors in renal hypertensive rats. J Cardiovasc Pharmacol 1: 467–471
13. Hirata Y, Tomita M, Ikeda M (1985) Characterization of alpha- and beta-adrenergic and angiotensin receptors in cultured vascular smooth muscle cells of rat aorta. Jpn Circ J 49: 1043–1051
14. Hughes RJ, Boyle MR, Brown RD, Taylor P, Insel PA (1982) Characterization of coexisting $alpha_1$- and $beta_2$-adrenergic receptors on a cloned muscle cell line, BC3H-1. Mol Pharmacol 22: 258–266
15. Ingebretsen CG, Hawelu-Johnson C, Ingebretsen WR (1983) Alloxan-induced diabetes reduces beta-adrenergic receptor number without affecting adenylate cyclase in rat ventricular membranes. J Cardiovasc Pharmacol 5: 454–461
16. Kanbe T, Nara T, Tagami M, Yamori Y (1983) Studies of hypertension-induced vascular hypertrophy in cultured smooth muscle cells from spontaneously hypertensive rats. Hypertension 5: 887–892
17. Kobayashi H, Maoret T, Spano PF, Trabucchi M (1982) Effect of age on β-adrenergic receptors on cerebral microvessels. Brain Res 244: 374–377
18. Kwan CY (1985) Dysfunction of calcium handling by smooth muscle in hypertension. Can J Physiol Pharmacol 63: 366–374
19. Kwan CY (1986) Aggregation of smooth muscle membranes and its use in the preparation of plasma membrane enriched fraction from gastric fundus smooth muscle. Biochem Cell Biol 64: 535–542

20. Kwan CY(1987) Preparation of smooth muscle plasma membranes: a critical evaluation. In: Kidwai AM (ed) Sarcolemmal biochemistry. CRC Florida, pp 59–97
21. Kwan CY, Borkowki K, Daniel EE (1988) Presynaptic vascular beta-adrenoceptors in canine mesenteric arteries. In: Halpern W, Brayden J, McLaughlin M, Osol G, Pegram B, MacKey K (eds) Proceedings of the 2nd international symposium on resistance arteries. Perinatology Press, New York, pp 428–434
21a. Kwan CY, Lee RMKW, Ito H (1987) Elastase-like activity is elevated in aortic muscle extract of intact tissue but not in the cultured aortic cells of spontaneously hypertensive rats. Biochem Arch 3: 217–221
22. Kwan CY, Lee RMKW, Daniel EE (1981) Isolation of plasma membranes from mesenteric veins: a comparison of their physical and biochemical properties with arterial membranes. Blood Vessels 18: 171–186
23. Kwan CY, Sipos S, Osterroth A, Daniel EE (1987) Beta-adrenoceptors in vascular smooth muscle with special reference to subcellular localization and number of binding sites. J Pharmacol Exp Ther 243: 1074–1081
24. Kwan CY, Triggle CR, Grover AK, Lee RMKW, Daniel EE (1983) An analytical approach to the preparation and characterization of subcellular membranes from canine mesenteric arteries. Preparative Biochem 13: 275–315
25. Kwan CY, Triggle CR, Grover AK, Lee RMKW, Daniel EE (1984) Membrane fractionation of canine aortic smooth muscle: subcellular distribution of calcium transport activity. J Mol Cell Cardiol 16: 747–764
26. Lee RMKW (1987) Structural alterations of blood vessels in hypertensive rats. Can J Physiol Pharmacol 65: 1528–1535
27. Leitschuh M, Chobanian AV (1987) Inhibition of nuclear polyploidy by propranolol in aortic smooth muscle cells of hypertensive rats. Hypertension 9(Suppl III): 106–109
28. Limas C, Limas CJ (1978) Reduced number of beta-adrenergic receptors in the myocardium of spontaneously hypertensive rats. Biochem Biophys Res Comm 83: 710–714
29. Limas CJ, Limas C (1979) Decreased number of beta-adrenergic receptors in hypertensive vessels. Biochim Biophys Acta 582: 533–536
30. Magnoni MS, Kobayashi H, Cazzaniga A, Izumi F, Spano PF, Trabucchi M (1983) Hypertension reduces the number of beta-adrenergic receptors in rat brain microvessels. Circulation 67: 610–613
31. Michel MC, Wang XL, Schlicker E, Gothert M, Beckeringh JJ, Brodde OE (1987) Increased β_2-adrenoceptor density in heart, kidney and lung of spontaneously hypertensive rats. J Auton Pharmacol 7: 41–51
32. Mochizuki M, Ogawa K (1984) Increase of cardiac beta-adrenergic receptors in young spontaneously hypertensive rats. Jpn Heart J 25: 411–423
33. Owens GK (1987) Influence of blood pressure on development of aortic medial smooth muscle hypertrophy in spontaneously hyertensive rats. Hypertension 9: 178–187
34. Ramanadham S, Tenner TE (1987) Alterations in the myocardial β-adrenoceptor system of streptozotocin-diabetic rats. Eur J Pharmacol 136: 377–390
35. Scarpace PJ, Baresi LA, Sanford DA, Abrass IB (1985) Desensitization and resensitization of β-adrenergic receptors in a smooth muscle cell line. Mol Pharmacol 28: 495–501
36. Scheid CR (1987) β-Adrenergic relaxation of smooth muscle: differences between cells and tissues. Am J Physiol 253: C369–C374
37. Schwartz J, Velly J (1983) The beta-adrenoceptor of pig coronary arteries: determination of beta-1 and beta-2 subtypes by radioligand binding. Br J Pharmacol 79: 409–414
38. Sharma RV, Kemp DB, Gupta RC, Bhalla RC (1982) Properties of adenylate cyclase in the cardiovascular tissues of renal hypertensive rats. J Cardiovasc Pharmacol 4: 622–628

39. Spector S, Fleisch JH, Maling HM, Brodie BB (1969) Vascular smooth muscle reactivity in normotensive and hypertensive rats. Science 166: 1300–1301
40. Tsujimoto G, Hoffman BB (1985) Desensitization of beta-adrenergic receptor-mediated vascular smooth muscle relaxation. Mol Pharmacol 27: 210–217
41. van Breemen C, Leijten P, Yamamoto H, Aaronsin P, Cauvin C (1986) Calcium activation of vascular smooth muscle. State of the art lecture. Hypertension 8 (Suppl II): 89–95
42. Woodcock EA, Johnston CI (1980) Cardiovascular adrenergic receptors in experimental hypertension in the rat. Circ Res 46: 145–146
43. Woodcock EA, Funder JW, Johnson CI (1979) Decreased cardiac β-adrenergic receptors in deoxycorticosterone-salt and renal hypertensive rats. Circ Res 45: 560–565
44. Woodcock EA, Olsson CA, Johnson CI (1980) Reduced vascular beta-adrenergic receptors in deoxycorticosterone-salt hypertensive rats. Biochem Pharmacol 29: 1465–1468
45. Yamada S, Ishima T, Tomita T, Hayashi M, Okada T, Hayashi E (1984) Alterations in cardiac alpha and beta adrenoceptors during the development of spontaneous hypertension. J Pharmacol Exp Ther 228: 454–460
46. Yamori Y, Mano M, Nara Y, Horie R (1987) Catecholamine-induced polyploidization in vascular smooth muscle cells. Circulation 75 (Suppl I): 92–95

Vascular Mechanisms in Development and Maintenance of Hypertension in Spontaneously Hypertensive Rats

MICHAEL J. MULVANY[1]

Summary. This paper reviews previous work from our laboratory concerning the role of mesenteric resistance vessel abnormalities in the pathogenesis of high blood pressure in the spontaneously hypertensive rat (SHR). A number of resistance vessel characteristics differ between those from SHRS and those from normotensive control Wistar-Kyoto rats (WKYs), including vascular structure (expressed as media: lumen ratio), calcium-sensitivity, sensitivity of the noradrenaline concentration-response curve to cocaine (expressed as a "cocaine-shift," and possibly being a marker for amount of sympathetic innervation), and oscillatory activity (rhythmic variations in tone when activated with noradrenaline). However, when these characteristics were examined in SHR/WKY F_2-populations, only the oscillatory activity correlated clearly with the blood pressure, correlations of structure and "cocaine-shift" with blood pressure being at best weak, and there was no indication of correlation concerning calcium sensitivity. These suggestions that resistance vessels structure may not be a prime determinant of blood pressure were supported by experiments in which rats were treated from age 4 weeks to 24 weeks. When treatment was then withdrawn, the rate of rise of blood pressure did not correlate with the effect of the treatment on the mesenteric resistance vessel structure. Thus, although in hypertension certain resistance vessel characteristics, in particular structure, are clearly altered, the evidence is not strong for their being related to the prime cause of the increased blood pressure.

Key words: Spontaneously hypertensive rat—Resistance vessels—Vascular structure—Oscillatory activity—Antihypertensive treatment—Treatment withdrawal—Recovery of blood pressure

Introduction

Over the past decade, much interest has been directed towards trying to understand the extent to which abnormalities of the vasculature are responsible for the high blood pressure of SHRs. That the vasculature is involved in producing the high blood pressure is of course clear: the peripheral resistance is increased, indicating that there must be an effective narrowing of the resistance vasculature [1, 2]. This narrowing could, though, just be due to increased levels of activation, without any alteration in the vasculature itself. The most clear-cut way of

[1]Biophysics Institute, Aarhus University, 8000 Aarhus C, Denmark

solving this problem is to take the vessels concerned out of the animal and examine them in a controlled environment. With this in view, we have used our technique for investigating small vessels [3], and the following points have been established as regards mesenteric resistance vessels taken from SHRs and control normotensive WKYs. In these vessels there is:

1. An increased oscillatory (i.e., rhythmic or phasic) activity, that is, the SHR vessels show rhythmic variations in force when stimulated by, for example, noradrenaline [4]
2. An increased sensitivity to calcium [5]
3. An increased "cocaine-shift," that is, cocaine causes a greater increase in noradrenaline sensitivity (interpreted as indicating that the nerve terminals in the SHR vessels have a greater uptake of noradrenaline, this abnormality thus possibly being a marker for an increased degree of innervation) [6]
4. An increased structure, as typified by their media: lumen ratio [4]

The question has been, however, whether any of these changes are associated with the factors which cause the increased blood pressure. Since the SHR and WKY are 2 different breeds of rat, separated by many generations, they are likely to differ in more respects than just their difference in pressure. Thus, any of these abnormalities could be quite incidental to the increased blood pressure. To answer this question, we have made 2 types of experiments. First, we made a genetic experiment in which we crossed SHRs and WKYs and then crossed their progeny to obtain the F_2-generation [7]. The idea was that if an abnormality was associated with hypertension, then the abnormality should be increased in those F_2-hybrids which had the higher blood pressures. Second, rats have been treated with various antihypertensive drugs to determine if treatment causes regression of the abnormalities, and if so, whether such regression reduces the rate of increase of blood pressure if treatment is then withdrawn [8]. This paper will review the results of these studies.

Crossbreeding Experiments

Methods

In these experiments [8], a total of 130 male F_2 rats were bred, and from this population, we chose 32 rats with lower blood pressures and 32 rats with higher blood pressures. Up to 4 mesenteric resistance vessels were taken from each animal, and the vessels were tested in pairs on a double myograph so that vessels from high- and low-pressure rats could be directly compared. Using this technique, we then tested whether the 4 abnormalites mentioned above could be associated with hypertension. Small arteries were taken from the mesenteric bed of adult SHRs and control WKYs. The vessels had internal diameters of 100–200 μm and were thus small enough to be considered resistance vessels. The vessels were threaded onto fine wires which were then attached to supports mounted, respectively, on a force transducer and a micrometer [3]. The supports were suspended in a chamber containing bicarbonate buffered saline solution held at

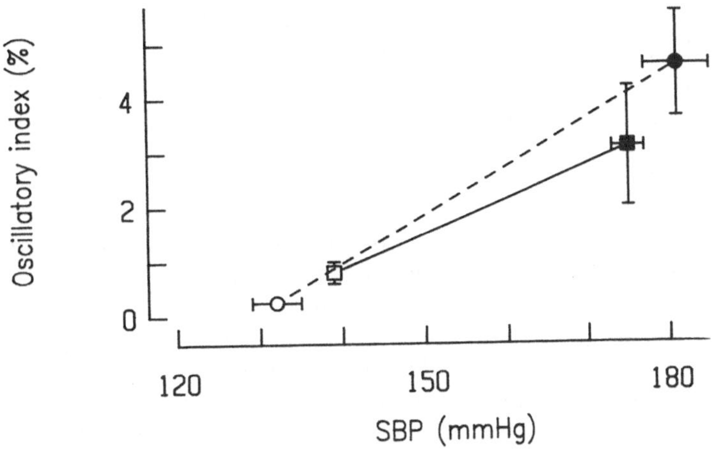

Fig. 1. Relation between oscillatory activity and systolic blood pressure (*SBP*) of SHRs (*filled circle*), WKYs (*open circle*), high-SBP F_2-SHR/WKYs (*filled square*), and low-SBP F_2-SHR/WKYs (*open square*). Values show mean ± SE. For clarity, *stippled lines* join points for pure strains; *solid lines* join points for high- and low-SBP F_2-SHR/WKYs. *Oscillatory index* is the oscillation amplitude (peak-to-peak) expressed as a percentage of the maximum response of the vessel. From data reported in [7] and reviewed in [15]

37°C. Using the micrometer the vessels could be stretched out to an extent corresponding to a transmural pressure of 100 mmHg, while the force transducer could measure the force response to agonists and antagonists which were added to the bath. The myograph was mounted on the stage of a microscope allowing the media thickness and corresponding lumen diameter to be measured. From this the media: lumen ratio was estimated with the vessel internal circumference being 90% of the internal circumference when the vessel was subjected to a transmural pressure of 100 mmHg [4].

Oscillatory Activity and Calcium Sensitivity

It has been a common finding that vessels taken from SHRs show marked oscillatory activity [9, 10, 11], and this characteristic was examined in the F_2 rats. Figure 1 shows a measure of the oscillatory activity of the mesenteric resistance vessels plotted against the blood pressure of the rats concerned. As indicated, taking the pure strains first, the SHRs had a high degree of oscillatory activity, corresponding to their high blood pressure, while the WKYs with their low blood pressure had virtually no phasic activity. In the F_2 rats, those having the higher blood pressure also had an increased degree of oscillatory activity, compared to the F_2 rats with the lower blood pressure. The difference between the oscillatory activity of the vessels from the low- and high-pressure F_2 rats was statistically significant ($P<0.05$). This result corresponds precisely to recent findings [12] concerning the oscillatory activity of tail artery of stroke-prone SHRs (SHRSPs), where in the F_2 generation of SHRSPs and WKYs, the oscillatory activity was increased in those F_2 rats having the greater pressure. Furthermore, in experiments we have made in which WKYs were made hypertensive by the Goldblatt technique, we observed no oscillatory activity in their mesenteric resistance vessels, suggesting that it is not just a result of hypertension [13]. Thus, there seem to be grounds to believe that the oscillatory activity which we have observed may be associated with the cause of hypertension in these animals.

According to a number of investigators, oscillatory activity may be due to the interaction between potential-dependent calcium channels and calcium-dependent potassium channels [12, 14]. According to this model, a depolarization of the membrane leads first to opening of potential-dependent calcium channels and the influx of calcium. Second, the rise in intracellular calcium then activates the calcium-dependent potassium channels, increasing the potassium permeability, so that the membrane potential then goes towards the potassium diffusion potential, thus becoming more negative. Third, with the resulting reduction in the potassium permeability, the membrane depolarizes again and the process repeats itself [15]. Support for this model comes from experiments which have shown that inhibitors of the calcium channels, e.g., dihydropyridines, inhibit oscillatory activity, while other workers [16] have shown that activation of the channels by Bay k 8644 enhances the activity. Furthermore, inhibitors of the calcium activated potassium channel, such as quinidine and spartein, inhibit the oscillatory activity [11]; while we have preliminary evidence that pinacidil, which appears to activate potassium channels, can initiate oscillatory activity [Videbæk L and Mulvany MJ, unpublished observations]. On this basis, we were interested to explore whether some defect in this mechanism could be responsible for the greater oscillatory activity seen in the SHR vessels. A prime candidate would be the calcium channel, and we do indeed have some evidence that this is altered in the SHR.

A number of years ago we showed that the "calcium sensitivity" of SHR mesenteric resistance vessels was greater than that of corresponding WKY vessels [5]. Furthermore, we have shown that the increased calcium sensitivity can be eliminated by blocking the potential-dependent calcium channels with the calcium antagonist feldoipine [17]. Thus, this supported the possibility that the increased oscillatory activity could be due to alterations in the calcium sensitivity. However, the results from the F_2 rats did not support this. Figure 2 shows the relation between calcium sensitivity of the mesenteric resistance vessels and the blood pressure of the rats concerned. As indicated, compared to the pure WKYs, the pure SHRs have a high calcium sensitivity corresponding to their high pressure. However, the F_2 rats with the higher pressure do not have a greater calcium sensitivity than those with the lower pressure. We can therefore draw 2 conclusions. First, calcium sensitivity is not associated with increased blood pressure. Second, the increased calcium sensitivity does not seem to be a cause of the oscillatory activity. Thus, the increased oscillatory activity is apparently not due to abnormalities in the potential-dependent calcium channel.

If the increased oscillatory activity of the vessels is not due to altered calcium channels, this suggests that the reason may lie in the calcium-activated potassium channels. That is, the increased oscillatory activity of the vessels from hybrids with the higher blood pressure could be due to an increased activity of the

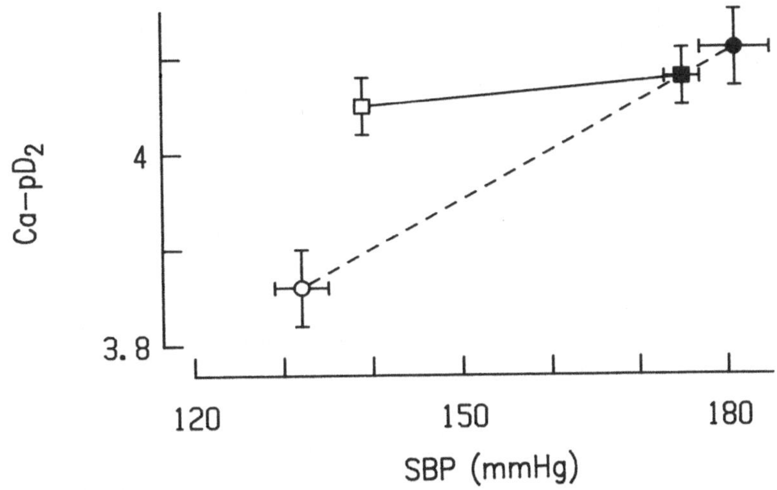

Fig. 2. Relation between calcium sensitivity and systolic blood pressure (*SBP*) of SHRs, WKYs, high-SBP F_2-SHR/WAYs, and low-SBP F_2-SHR/WAYs using same nomenclature as in Fig. 1. Calcium-sensitivity is defined [5] as the *Ca-pD*$_2$, i.e., the negative logarithm of the calcium concentration (in mol/l) required to give half-maximal response of a vessel when activated with noradrenaline. From data reported in [7] and reviewed in [15]

calcium-activated potassium channels. The literature supports this possibility, for it has been a consistent finding that the potassium permeability of vascular smooth muscle from hypertensive individuals is increased [18], and Jones and colleagues have shown that this may be associated with calcium-activated potassium channels [14]. It seems unlikely that such a defect could be a cause of hypertension—potassium channel activators causes vasodilation and blood pressure reduction. It seems more likely to be a reaction to some other defect in the vascular smooth muscle cells of these animals—a defect for example which causes a slight increase in the cytoplasmic calcium level. Thus, the increased oscillatory activity may be just a result of slightly increased cytoplasmic calcium concentration.

Structure

I have previously shown that the media: lumen ratio of vessels taken from a variety of vascular beds is increased [19]. Thus, not only in the mesenteric vasculature, but also in cerebral and renal vasculature, the media: lumen ratio of SHR vessels is increased, although the effect is most pronounced in the mesenteric bed. The results of measurements of this characteristic in mesenteric vessels of the F_2 rats are shown in Fig. 3. As found previously [7], the SHRs with the high blood pressure have a higher resistance vessel media: lumen ratio than the vessels from the WKYs, with their lower media: lumen ratio. In the F_2 rats with the higher blood pressure, the media: lumen ratio was a little greater, although the effect was not as marked as might be expected. Figure 3b shows the difference between the high- and low-pressure rats, the difference on a paired basis,

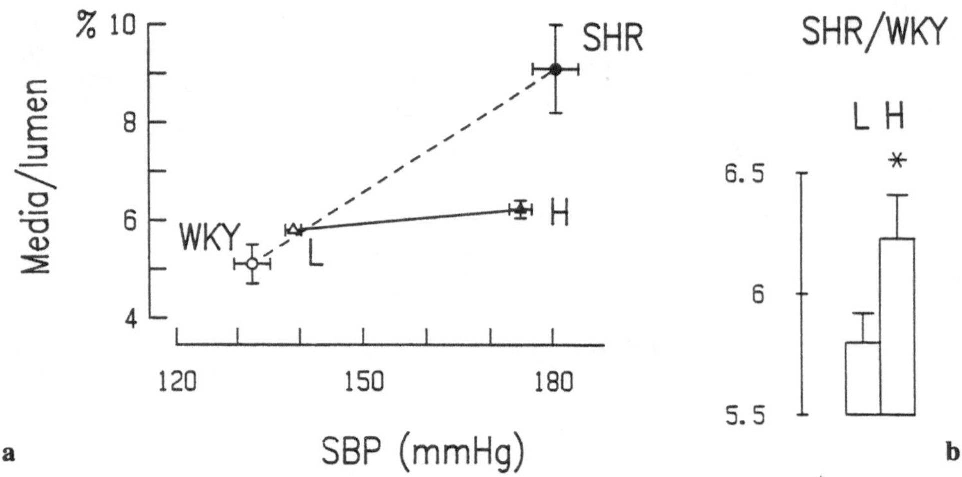

Fig. 3. a Relation between media : lumen ratio and systolic blood pressure (*SBP*) of *SHR*s, *WKY*s, high-SBP F_2-SHR/WKYs (*H*), and low-SBP F_2-SHR/WKYs (*L*) using same nomenclature as in Fig. 1 ($n = 32$). Media : lumen ratio determined for vessels being held with a transmural pressure of 100 mmHg [6]. **b** The media : lumen ratios of the high- and low-SBP F_2-SHR/WKYs on an expanded scale, the difference being significant ($P < 0.05$, paired *t*-test). From data reported in [7] and reviewed in [15]

just reaching the level of significance. However, the modesty of the difference suggests that the difference in structure is not an actual cause of the hypertension.

Sympathetic Innervation

Previous investigation of the effect of cocaine on the noradrenaline concentration-response relation of mesenteric resistance vessels from SHRs and from WKYs indicated that under control conditions there was no difference in sensitivity [4]. However, when cocaine (3 μmol/l) was added, in both cases the curves were shifted to the left, and the shift was greatest for the SHR vessels. The increase in sensitivity is explained in terms of cocaine inhibiting the uptake of noradrenaline into the nerve terminals, so that the concentration of noradrenaline reaching the smooth muscle cells is closer to that of the bathing solution. This interpretation is supported by the findings of Whall et al. [20], who found essentially the same result if uptake was inhibited by destroying the nerves with 6-hydroxydopamine (OHDA). Thus, although the greater cocaine shift seen in the SHR vessels could be explained in terms of a greater activity of the neuronal amine pump, a more likely explanation is that the SHR vessels contain a greater degree of sympathetic innervation.

Since increased sympathetic nerve activity has been shown to correlate with blood pressure in SHR/WKY F_2-populations and backcrosses [21], it might be thought that the cocaine-shift should also correlate with blood pressure. However, as indicated in Fig. 4, the results were negative. As indicated, the SHRs with

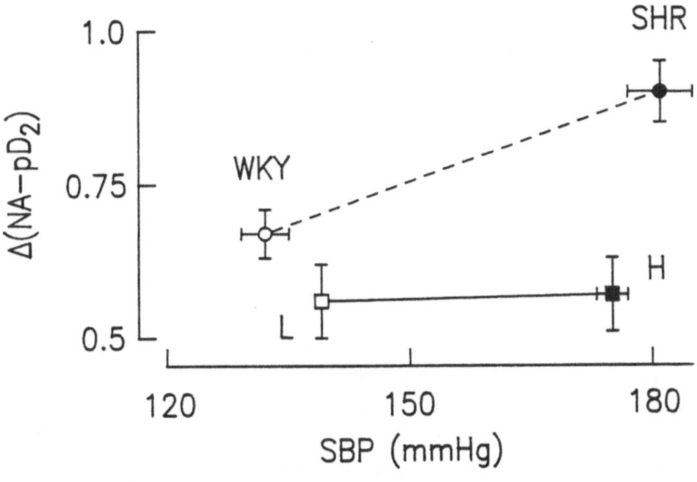

Fig. 4. Relation between "cocaine-shift" (Δ(NA-pD_2) pressure (*SBP*) of *SHR*s, *WKY*s, high-SBP F_2-SHR/WKYs (*H*), and low-SBP F_2-SHR/WKYs (*L*) using same nomenclature as in Fig. 1 ($n = 22$). Cocaine-shift is the logarithm of the ratio of the noradrenaline-EC_{50} in the presence and in the absence of cocaine [4]. From data reported in [7] and reviewed in [15]

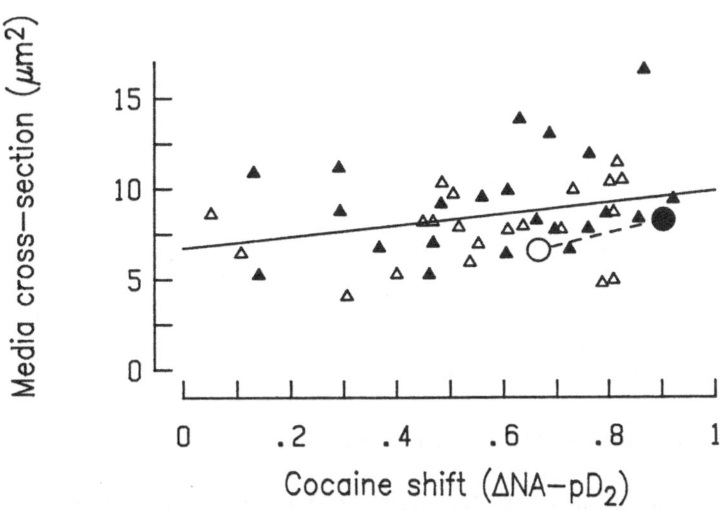

Fig. 5. Relation between "cocaine-shift" and media : lumen ratio for those rats comprising the data shown in Figs. 3 and 4 in which the "cocaine-shift" was determined (22 high-SBP F_2-SHR/WKY, *filled triangles*; 22 low-SBP F_2-SHR/WKY, *open triangles*; 6 SHR, *filled circles*; 6 WKY, *open circles*). The full line is a regression line through the F_2-SHR/WKY points, the slope being almost significantly different to zero, $P = 0.06$

the high blood pressure had vessels in which the cocaine-shift was greater than the cocaine-shift of the vessels from the WKYs, but there was no difference in the cocaine-shift of the vessels from the high- and low-pressure F_2-rats. There was, however, a weak correlation ($P = 0.06$) between the cocaine-shift and structure (Fig. 5), thus supporting the possibility that there was in fact a slight increase in the content of both smooth muscle and sympathetic innervation in the F_2-rats with the higher blood pressures. However, the modesty of the increase suggests that neither were a direct cause of the higher blood pressure seen in these animals.

Antihypertensive Experiments

The rather suprising findings concerning the lack of a direct relation between vascular structure and blood pressure in the F_2-rats has been investigated in another way. In these experiments, the aim was to determine the relation between vascular structure during antihypertensive treatment, and the rate of rise of blood pressure when treatment is withdrawn. If vascular structure is an important determinant of blood pressure, then those treatments which have the most pronounced effect on vascular structure might be expected to be treatments which would have the most prolonged effects on blood pressure when treatment was withdrawn.

Methods

Rats were treated from the age of 4 weeks–24 weeks with one of the following treatments: captopril (angiotensin-converting enzyme inhibitor [ACEI], Squibb, 60 mg/kg daily, perindopril (ACEI, Servier, 1.5 mg/kg daily), isradipine (calcium antagonist, Sandoz, 42 mg/kg daily), metoprolol (beta$_1$-blocker, Hässle, 130 mg/kg daily), and hydralazine (25 mg/kg daily). In each case, the dose was titrated so that blood pressure was approximately normalized to the untreated WKY level. Blood pressure was measured regularly using the tail-cuff technique. At the age of 24 weeks, measurement of the 24-h blood pressure was made. Also, a biopsy containing 2 mesenteric resistance vessels was taken, the measurements being made using an intra-arterial catheter, with the rat able to move freely in its cage. Treatment was then stopped, and the rats were observed for up to an additional 12 weeks.

Vascular Structure and Blood Pressure at 24 weeks

Figure 6 shows the blood pressure and vascular structure of the rats at the age of 24 weeks, just before treatment was withdrawn. As shown by Fig. 6a, the blood pressure was reduced from the untreated SHR level to a level close to the untreated WKY level, to the largest extent by perindopril. Figure 6b illustrates the media: lumen ratio at that time. As indicated, the media: lumen ratio was normalized only in the perindopril group. In the captopril, isradipine, and hydralazine groups, the media: lumen ratio was only partially normalized, while in the metoprolol group there was no effect on the media: lumen ratio.

Vascular Structure and Blood Pressure on Withdrawal of Treatment

The next stage of the project was to determine the rate of rise of blood pressure when treatment was withdrawn. We had expected that the rate of rise of blood

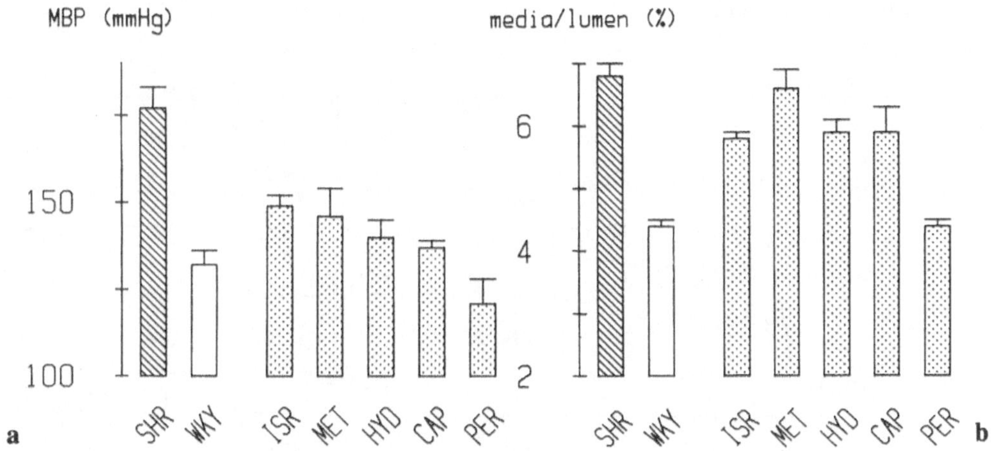

Fig. 6a,b. Effect of treatment of SHRs from 4 weeks to 24 weeks on **a** 24-h mean blood presssure (*MBP*) and **b** vascular structure (*media:lumen* ratio). Treatments were: *ISR*, isradipine; *MET*, metoprolol; *HYD*, hydralazine; *CAP*, captopril; *PER*, perindopril. *SHR* and *WKY* denote untreated SHRs and WKYs, respectively. Values show mean ± SE, respectively; $n = 6–23$. From data published in [8]

pressure would be related to the degree of regression of vascular structure. If this were the case, then the perindopril group (with the lowest media: lumen ratio) would increase slowest; the captopril, isradipine, and hydralazine groups (all of which had a similar, partial regression of vascular structure) would have an intermediate rate of rise; and the metoprolol group (where media: lumen ratio was unaffected) would rise fastest. The results are shown in Fig. 7, which shows the redevelopment of blood pressure when treatment was stopped for the rats grouped by the media: lumen ratio at 24 weeks. In the perindopril group (Fig. 7a), the blood pressure remained low for the entire 3-month period, while in the metoprolol group (Fig. 7c) with the unaffected media: lumen ratio, the blood pressure rose rapidly. However, in the intermediate group (Fig. 7b), there was no correlation between vascular structure and the rate of rise of blood pressure. Although all 3 groups had the same media: lumen ratio at age 24 weeks, blood pressure remained low in the captopril group (Fig. 7b), while blood pressure rose rapidly in the other 2 groups. Thus, vascular structure does not seem to be a direct determinant of blood pressure, supporting the findings obtained with the F_2 rats. The prolonged effects of blood pressure treatment on blood pressure appeared to be more related to the type of drug used, similar prolonged effects having been reported earlier for ACEIs [22, 23].

Conclusions

The results of these experiments have suggested that although it is clear that in the SHR the peripheral vasculature plays a key role in the sequence of events which produces the increased peripheral resistance, in general, abnormalities of

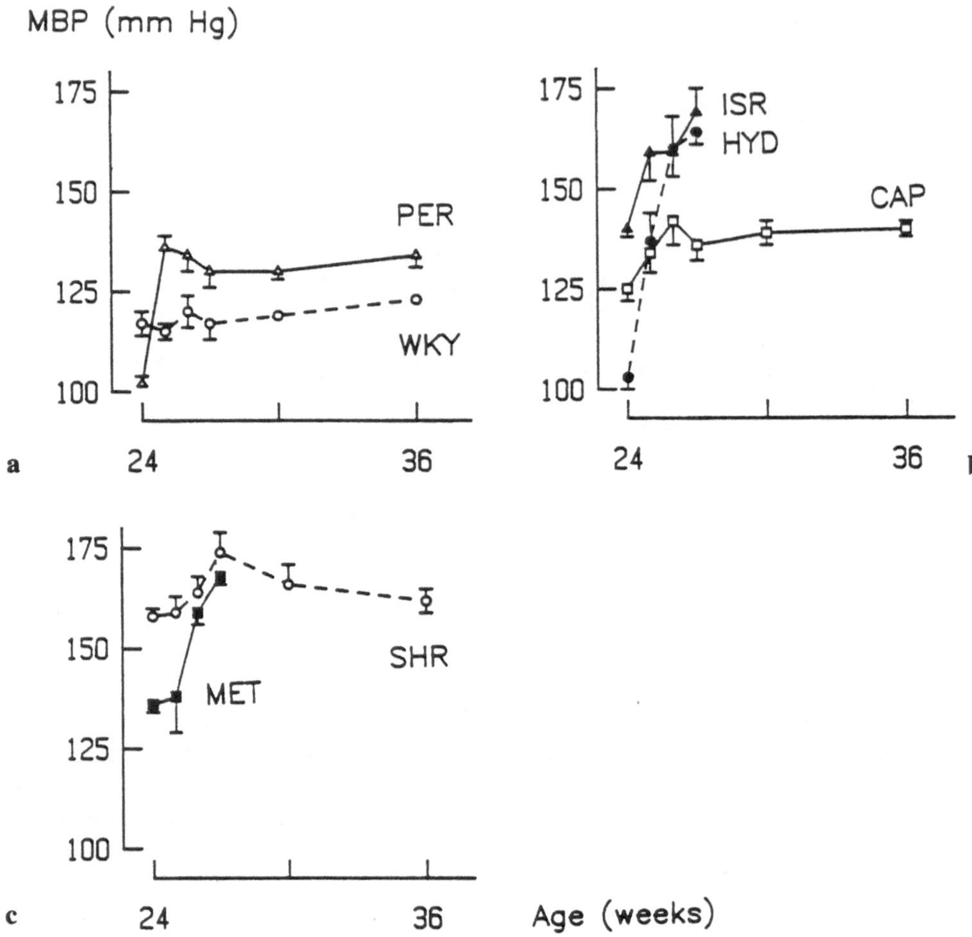

Fig. 7a–c. Redevelopment of mean arterial blood pressure (*MBP*) following withdrawal of treatment at age 24 weeks. Treatments (see legend to Fig. 6) grouped according to the effect of treatment on media:lumen ratio. **a** *PER* group and untreated *WKY*, where media:lumen at 24 weeks was about 4.4%. *b ISR*, *HYD*, and *CAP* groups, where media:lumen ratio at 24 weeks was about 5.8%. **c** *MET* group and untreated *SHR*, where media:lumen ratio at 24 wk was about 6.7%. From data published in [8]

the peripheral vasculature are not the primary cause of the high blood pressure. Both the crossbreeding experiments and the antihypertensive treatment experiments have suggested that this is the case for vascular structure, while the crossbreeding experiments have suggested that this is also the case for the calcium sensitivity and the degree of sympathetic innervation (as expressed by the cocaine-shift). The only vascular abnormality which does seem to be associated with the cause of hypertension is the oscillatory activity of the resistance vessels, although even here the mechanism for this abnormality is still uncertain.

These results are unfortunately rather negative but point to the following two things. First, they emphasize the importance of not accepting any difference between SHR and WKY as being necessarily related to the cause of hypertension in the SHR (let alone being the cause of hypertension in man) without further evidence. Second, they suggest that abnormalities of the vasculature, with the possible exception here of the "oscillatory activity," are not in themselves important determinants of blood pressure. The vasculature is altered in hypertension, yes, particularly as regards the structure. But this alteration does not appear to be the *cause* of the hypertension. It seems that the cause must be sought elsewhere.

References

1. Ferrone RA, Walsh GM, Tsuchiya M, Frohlich ED (1979) Comparison of hemodynamics in conscious spontaneous and renal hypertensive rats. Am J Physiol 236: H403–H408
2. Mulvany MJ (1983) Do resistance vessel abnormalities contribute to the elevated blood pressure of spontaneously hypertensive rats? A review of some of the evidence. Blood Vessels 20: 1–22
3. Mulvany MJ, Halpern W (1977) Contractile properties of small arterial resistance vessels in spontaneously hypertensive and normotensive rats. Circ Res 41: 19–26
4. Mulvany MJ, Aalkjaer C, Christensen J (1980) Changes in noradrenaline sensitivity and morphology of arterial resistance vessels during development of high blood pressure in spontaneously hypertensive rats. Hypertension 2: 664–671
5. Mulvany MJ, Nyborg N (1980) An increased calcium sensitivity of mesenteric resistance vessels in young and adult spontaneously hypertensive rats. Br J Pharmacol 71: 585–596
6. Mulvany MJ, Aalkjaer C, Hansen PK (1979) Evidence that the greater vascular reactivity in spontaneously hypertensive rats may in part be due to structural differences in the vasculature. In: Bevan JA (ed) Vascular neuroeffector mechanisms. Raven Press, New York, pp 370–372
7. Mulvany MJ (1988) Role of resistance vessel structure and function in the etiology of hypertension studied in F_2-generation hypertensive-normotensive rats. J Hypertens 6: 655–663
8. Christensen KL, Jespersen LT, Mulvany MJ (1989) Development of blood pressure in spontaneously hypertensive rats after withdrawal of long-term treatment related to resistance vessel structure. J Hypertens 7: 83–90
9. Bandick NR, Sparks HV (1970) Contractile response of vascular smooth muscle of renal hypertensive rats. Am Physiol 219: 340–344
10. Holloway ET, Bohr DF (1973) Reactivity of vascular smooth muscle in hypertensive rats. Circ Res 33: 678–685
11. Lamb FS, Myers JH, Hamlin MN, Webb RC (1985) Oscillatory contractions in tail arteries from genetically hypertensive rats. Hypertension 7 (Suppl I): I–25–30
12. Bruner CA, Myers JH, Sing CF, Jokelainen PT, Webb RC (1986) Genetic association of hypertension and vascular changes in stroke-prone spontaneously hypertensive rats. Hypertension 8: 904–910
13. Mulvany MJ, Korsgaard N (1983) Correlations and otherwise between blood pressure, cardiac mass and resistance vessel characteristics in hypertensive, normotensive and hypertensive/normotensive hybrid rats. J Hypertens 1: 235–244
14. Magliola L, McMahon EG, Jones AW (1986) Alterations in active Na-K transport during mineralocorticoid-salt hypertension in the rat. Am J Physiol 250: C540–C546
15. Mulvany MJ (1988) Possible role of vascular oscillatory activity in the development of high blood pressure in the spontaneously hypertensive rat. J Cardiovasc Pharmacol 12 (Suppl 6): S16–S20

16. De Mey JGR, Boonen HCM, Struyker-Boudier HAJ (1988) Rhythmic contractile activity in resistance-sized arteries of spontaneously hypertensive rats. In: Halpern W, Pegram B, Brayden J, Mackey K, McLaughlin MK, Osol G (eds) Resistance arteries. Perinatology Press, Ithaca NY, pp 336–341
17. Nyborg NCB, Byg-Hansen J, Mulvany MJ (1985) Effect of felodipine on resistance vessels from spontaneously hypertensive and normotensive rats. J Cardiovasc Pharmacol 7 (Suppl VI): VI–S43–S46.
18. Aalkjaer C, Mulvany MJ (1989) Sodium and calcium metabolism of resistance vessels in hypertension. In: Lee C-Y (ed) Membrane abnormalities in hypertension. CRC Press, Boca Raton (in press)
19. Mulvany MJ (1986) Role of vascular structure in blood pressure development of the spontaneously hypertensive rat. J Hypertens 4 (Suppl III): III–S61–S63.
20. Whall CW, Myers MM, Halpern W (1980) Norepinephrine sensitivity, tension development and neuronal uptake in resistance arteries from spontaneously hypertensive and normotensive rats. Blood Vessels 17: 1–15
21. Judy WV, Watanabe AM, Murphy WR, Aprison BS, Yu PL (1979) Sympathetic nerve activity and blood pressure in normotensive backcross rats genetically related to the spontaneously hypertensive rat. Hypertension 1: 598–604
22. Guidicelli JF, Freslon JL, Glasson S, Richer C (1980) Captopril and hypertension development in the SHR. Clin Exp Hypertens [A] 2: 1083–1096
23. Harrap SB, Nicolaci JA, Doyle AE (1986) Persistent effect on blood pressure and renal haemodynamics following chronic angiotensin-converting enzyme inhibition with perindopril. Clin Exp Pharmacol Physiol 13: 753–765

Effects of Hypertension on Endothelial-Smooth Muscle Cell Interactions in Pressurized Cerebral Arteries*

GEORGE OSOL, SONJA KNUTSON, and WILLIAM HALPERN[1]

Summary. The purpose of this study was to evaluate the effects of hypertension on cerebral artery tone and reactivity in small pial arteries from normotensive (WKY) and hypertensive (SHRSP) rats. Vessels (100–200 μm) were studied in vitro, under conditions in which an arterial segment is cannulated, pressurized, and able to constrict or dilate. Changes in lumen diameter were measured in response to pharmacological probes of smooth muscle and endothelial cell function. The principal findings were: (a) Although the vascular wall of cerebral arteries was clearly hypertrophied in hypertensives, the average lumen diameters of vessels in a passive state, or in a state of partial activation due to an intrinsic (vasogenic) tone were not significantly different. (b) Arteries from hypertensive rats did not dilate after application of an endothelium-dependent dilator (acetylcholine, ACh), yet dilation to sodium nitroprusside (NP), which activates smooth muscle cell guanylate cyclase, was significantly augmented. (c) The application of serotonin (5-HT), an indoleamine having smooth muscle and endothelial effects, induced a much greater narrowing of SHRSP arteries. In summary, ACh and 5-HT responses were altered in favor of hypercontractility and may be the result of specific and possibly deleterious changes in endothelial function. Conversely, similarities in the level of vasogenic tone and the nitroprusside data may reflect adaptive changes in vascular smooth muscle which counteract the diminished inhibitory influence of the endothelium.

Key words: Cerebral artery—Myogenic—Endothelium—Hypertension—Serotonin

Introduction

An association between hypertension and cerebrovascular disease has been recognized for many years, and characterization of hypertension-induced changes in blood vessel function has been a subject of many studies in and ex vivo. In the past decade, however, our perception of the vascular wall has been expanded to include a previously unrecognized and active interaction between smooth muscle and endothelial cells in determining vascular reactivity. Judging from the most recent work, this interaction is complex, and the endothelium appears to be

* Supported by National Institutes of Health Grants Numbers HL 38231, HL 17335, and Grant 526738 from the Vermont Chapter of the American Heart Association.
[1] Departments of Obstetrics and Gynecology and Physiology and Biophysics University of Vermont College of Medicine, Burlington, VT 05405, USA

capable of secreting a number of factors, some of which cause constriction, others dilation [1]. The precise nature of endothelial influence varies with the size, type, and anatomic location of the blood vessel studied [2].

Systemic hypertension is thought to alter cerebral endothelial function in several ways. For example, recent studies suggest that although blood vessels from normotensive animals dilate in response to acetylcholine, this may be completely abolished in hypertensive rats [3]. This effect, observed in vivo through a pial window preparation, could result from (a) loss of the endothelium-derived relaxing factor (EDRF) secretion; (b) an increase in the secretion of other, as yet unidentified, constricting factors; or (c) an insensitivity of vascular smooth muscle to the action of EDRF. There is other evidence for hypertension-induced alterations in endothelial surface charge [4] and blood-brain barrier functions [5].

Other studies describe hypertension-induced changes in the vascular smooth muscle of cerebral arteries, such as reductions in actin and myosin content [6]. Changes in distensibility have also been recorded in pial vessels [6], a property which is dependent, in large part, on the connective tissue composition of a vessel. Thus, it is clear that hypertension induces a profound remodelling of the vascular wall, with specific changes in both the cellular and noncellular components.

The results described in this paper are from a set of preliminary experiments which were designed to test the hypothesis that hypertension-induced changes in tone and reactivity occur in the cerebral vasculature, and that these changes are due to specific alterations in both endothelial and smooth muscle cell function. Relatively small arteries were studied because of their importance in cerebral blood flow regulation and recognized involvement in the pathogenesis of ischemic stroke. Measurements of wall thickness and lumen diameter were made on vessels fully relaxed with papaverine, constricted with a depolarizing potassium solution, and in normal physiological saline solution (PSS) containing 1.6 m*M* Ca^{2+}. In the PSS, cerebral arteries develop a stable tone which appears to be an intrinsic response of the cells of the vascular wall to transmural pressure and is not the product of neurogenic or metabolic stimuli [7]. The effects of endothelium-dependent (acetylcholine) and -independent (sodium nitroprusside) vasodilators on basal tone were compared in arteries from normotensive (WKY) and hypertensive (SHRSP) rats. Vascular reactivity to serotonin, a naturally occurring indoleamine which has been implicated in the pathogenesis of cerebral vasospasm [8], was examined as well. The actions of serotonin on the cerebral vasculature are complex and are thought to involve an interplay between excitatory smooth muscle and inhibitory endothelial effects [9, 10].

Materials and Methods

Experiments were carried out in vitro, using a method in which isolated vessels are mounted on glass microcannulae and pressurized, and the lumen diameter and wall thickness are measured by a video-electronic system [7, 11]. This approach offers the advantages of studying an artery under conditions which may be selectively controlled, and in which the cells of the wall assume their natural geometry. This becomes important when one considers that endothelial secretory activities are affected profoundly by physical forces such as stretch, pulsatile pressure, and shear stress [12, 13].

Measurements were made on vessels from adult (16–30-week-old) male WKY ($n = 9$) and SHRSP ($n = 10$) rats. The animals were lightly anesthetized with either and decapitated. The cranium was opened with rongeur forceps, the entire brain removed and immediately immersed in oxygenated PSS. By following the posterior cerebral artery from the Circle of Willis and over the occipital cortex, a 1–2 mm tertiary branch was removed, gently cleaned of connective tissue, and transferred into the chamber of a specially designed arteriograph. The proximal end of the vessel segment was grasped with forceps and slipped onto a PSS-filled glass microcannula with an outer tip diameter of approximately 100 μm. Once it was secured with two single strands of nylon teased from braided surgical suture, the distal end of the artery was slipped onto a second microcannula and secured in a similar manner.

Polyethylene tubing was used to couple the outlet cannula to a miniature, 2-way, solenoid poppet valve. The inlet cannula was connected to a 2-ml syringe connected, in turn, to the threaded shaft of a linear motor. Its movement induced a displacement of the fluid, thereby altering the pressure in the tubing, which was sensed with an in-line transducer.

Once a vessel was mounted on the cannulae, the transmural pressure was set to be half of the animal's systolic blood pressure (measured by tail cuff plethysmography prior to the experiment), and then allowed to equilibrate for 60–90 min at 37°C. The artery was superfused with saline from a reservoir which was warmed, re-circulated, and continually oxygenated with a gas mixture of 10% O_2, 85% N_2 and 5% CO_2. The composition of the PSS was: NaCl 119 mM, $NaHCO_3$ 24 mM, KCl 4.7 mM, KH_2PO_4 1.18 mM, $MgSO_4$-$7H_20$ 1.17 mM, $CaCl_2$ 1.6 mM, glucose 5.5 mM, Na-EDTA 0.026 mM.

The arteriograph containing the pressurized artery was positioned on the stage of an inverted microscope with a side-attached video camera. A 500 μm length of vessel, visible on a television monitor at a final magnification of 350x, was made to intersect perpendicularly the television scan lines by rotating the camera about the optical axis of the microscope. When the video signal was processed by an electronic dimension analyzer, analogue voltages proportional to the lumen diameter and thickness of the vessel wall were generated, displayed in microns on digital voltmeters, and recorded on a multi-channel strip chart recorder.

Spontaneous tone developed during the equilibration period, or shortly thereafter following a change in transmural pressure. This tone is insensitive to TTX and conventional adrenergic or cholinergic blockade, and is maintained with or without luminal flow [7]. Arterial dimensions were measured at a stable transmural pressure of 75 mmHg, and the differences evaluated for statistical signifcance by using an unpaired Student's t-test. All pharmacological solutions were made fresh on a daily basis. These included stock solutions of: acetylcholine chloride, papaverine, serotonin (5-hydroxytryptamine, creatinine sulfate complex), and sodium nitroprusside. High potassium (125 mM) depolarizing solution consisted of PSS containing an equimolar substitution of KCl for NaCl and

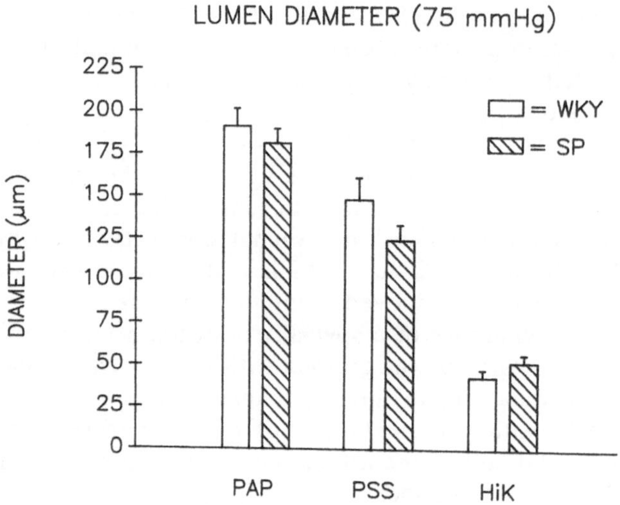

Fig. 1. Lumen diameter of cerebral arteries from normotensive (*WKY*) and hypertensive (*SP*) rats at a transmural pressure of 75 mmHg. Dimensions were measured in 8×10^{-5} *M* papaverine (*PAP*, fully relaxed), in physiological saline (*PSS*, intrinsic tone), and in 125 m*M* potassium (*HiK*, fully constricted). WKY vs SHRSP differences were not statistically significant

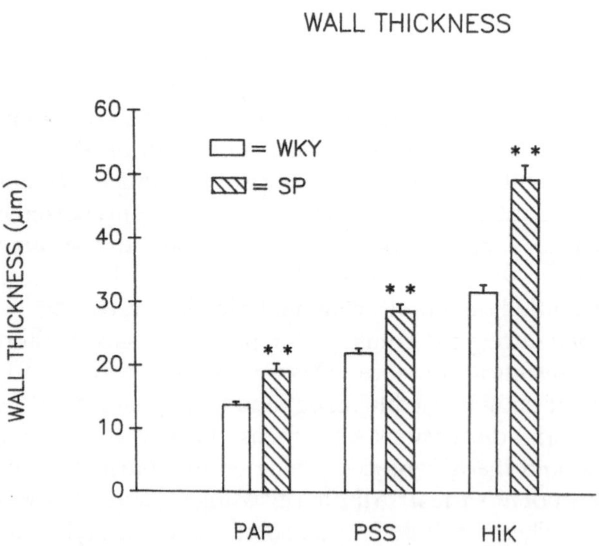

Fig. 2. The thickness of the vascular wall as measured on the video monitor under the same conditions as those described for Fig. 1. Significant differences were observed between hypertensive and normotensive vessels as indicated by asterisks (**, $P < 0.01$).

a $CaCl_2$ concentration of 5 m*M*. Drugs were added to the reservoir and the diameter response of each vessel allowed to stabilize prior to making the final measurement, usually 10–15 min.

Results

Body weights of the rats used in this study averaged: WKY = 360 ± 9 g; SHRSP = 299 ± 8 g ($P < 0.01$). Systolic blood pressures, measured by tail cuff plethysmography were: 130 ± 3 mmHg and 184 ± 5 mmHg for WKY and SHRSP, respectively ($P < 0.01$).

At a transmural pressure of 75 mmHg, lumen diameters of WKY and SHRSP vessels did not significantly differ in 8×10^{-5} *M* papaverine (fully relaxed), 1.6 m*M* Ca^{2+} PSS (intrinsic tone), or in 125 m*M* potassium (maximally constricted) (see Fig. 1). The thickness of the vascular wall, however, was significantly increased in each case (Fig. 2).

The addition of acetylcholine (10^{-5} *M*) induced a significant dilation in WKY vessels; in SHRSP arteries, the observed response was either no change in diameter of a slight constriction (Fig. 3).

Sodium nitroprusside (5×10^{-8} *M*) produced a notable dilation in both WKY and SHRSP vessels, although the degree of dilation was significantly ($P < 0.01$) increased in the latter (Fig. 4).

Serotonin (5×10^{-7} *M*) effected a constriction in every vessel studied, and the degree of constriction was significantly [$P < 0.01$] greater in SHRSP arteries (Fig. 5).

If the distribution of vascular resistance as a function of arterial size remains the same in hypertension, arteries in same anatomical location would operate at proportionately higher transmural pressures in hypertensive arteries. One concern was that the observed differences in the degree of constriction with serotonin might be related to a reduction of wall stress due to the hypertrophy of the vascular wall (approximately 30%–35%); i.e., at the same transmural pressure, smooth muscle cell stress is also likely to be lower and may contribute towards an enhanced constriction. To eliminate this possibility, some vessels ($n = 6$) were given serotonin (5×10^{-7} *M*) at 75 or at 110 mmHg, and the responses compared. As shown in Fig. 6, the constrictions were of equal magnitude, although the amplitude was somewhat diminished, possibly due to tachyphylaxis.

Discussion

Following equilibration in PSS having an ionic composition similar to that of cerebrospinal fluid, isolated rat pial arteries were constricted by approximately 30%. This "vasogenic" tone remains quite stable for hours, positions the vessel at a point from which further dilation or constriction is possible, and occurs in the absence of innervation or surrounding parenchymal tissues which could act as a source of vasoactive metabolites. Hence, it appears to be a direct response of the vascular wall to transmural pressure.

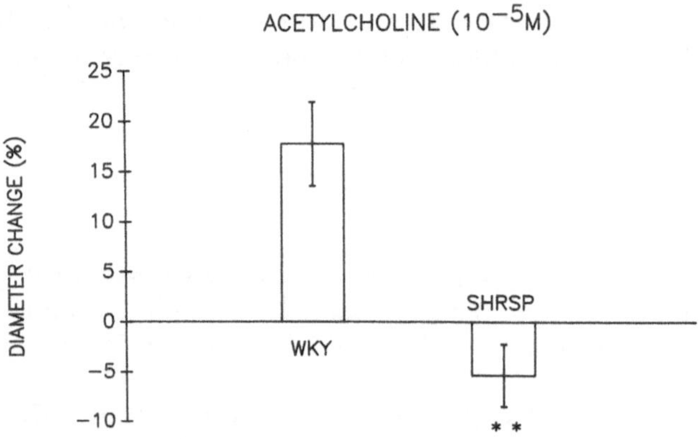

Fig. 3. Effect of acetylcholine on the intrinsic tone in normotensive (*WKY*) and hypertensive (*SHRSP*) vessels. **, $P<0.01$

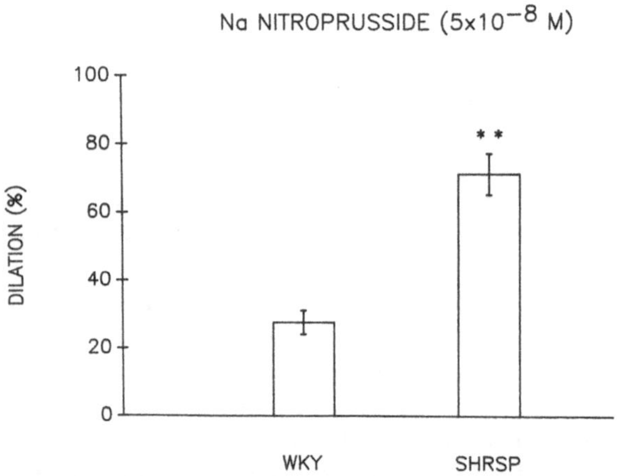

Fig. 4. Effect of sodium nitroprusside on intrinsic tone in normotensive (*WKY*) and hypertensive (*SHRSP*) vessels. **, $P<0.01$

There were no significant differences between WKY and SHRSP vessels in resting lumen diameter (in papaverine) or in the level of intrinsic tone although the thickness of the vascular wall was increased by 30%–35% in hypertensives. Hypertension-induced hypertrophy of the cerebral vasculature has been reported previously and appears to be a direct consequence of increased systemic pressures [14, 15].

Since the initial observations [16], many studies have demonstrated that vascular relaxation induced by acetylcholine is mediated by the release of an endothelium-derived relaxing factor (EDRF). In the cerebral circulation, the dilation is not nearly as dramatic as in vessels from other vascular beds. For example, when acetylcholine is added to cannulated rat mesenteric arteries comparable in size to those used for this study and preconstricted with norepinephrine, rapid and complete dilation can be induced repeatedly, and at lower

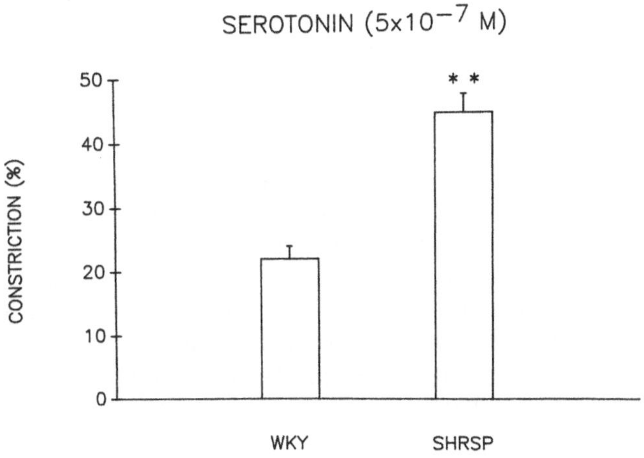

Fig. 5. Arterial constriction in response to serotonin. **, $P<0.01$; *WKY*, normotensive; *SHRSP*, hypertensive

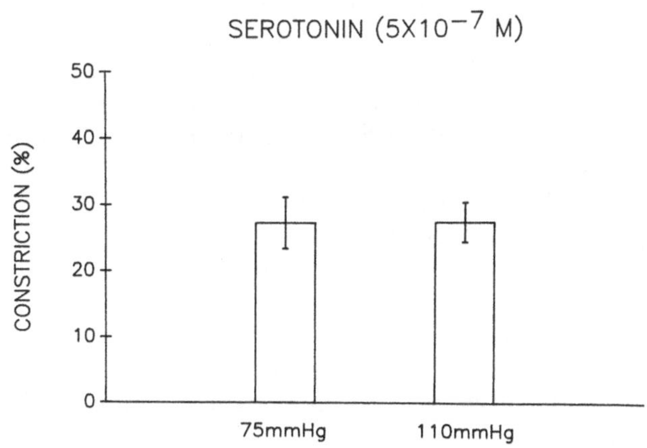

Fig. 6. Constriction subsequent to repeated doses of serotonin given to the same vessels at transmural pressures of 75 mmHg and 110 mmHg. The degree of constriction was nearly identical

concentrations [17]. In the brain, under the same experimental conditions, WKY cerebral arteries dilate only approximately 17%, and vessels from the SHRSP rats were either completely unresponsive or constricted slightly to the same concentration of acetylcholine. A similar degree of dilation, and its attenuation in SHRSP pial arteries, have recently been described in vivo [3] and are consistent with the observations made here under in vitro conditions.

The mechanism underlying the acetylcholine dilation is thought to be via the release of EDRF which then binds to vascular smooth muscle, activates guanylate cyclase, and elevates the cytoplasmic concentration of soluble cGMP [18]. When sodium nitroprusside was used to provoke the smooth muscle guanylate cyclase system directly, heightened responses were observed in arteries from the hypertensive strain. Thus, it appears that the smooth muscle not only retains but augments its ability to dilate via a cGMP mechanism. At this time, we are unable to determine whether the secretion of EDRF is decreased, or if another, constricting factor is released, as appears to be the case in the rat aorta.

There is some precedent for an increased constrictor effect of serotonin in the cerebral vasculature of SHR. Winquist and Bohr [19] report a significant shift to the left in the EC_{50} of serotonin and a greater maximal response to this agonist in basilar arteries from SHR compared with those from WKY rats. In their study, the effect of a different wall stess was not evaluated since a wire-mounted, isometric preparation was used, and the degree of stretch was selected on the basis of maximal responsiveness to a depolarizing KCl solution. We found that sensitivity was not different in SHRSP vessels (data not given), but that the maximal response was approximately doubled. This difference was not due to wall stress since the degree of constriction was virtually identical at 75 or 110 mmHg.

Serotonin is thought to effect the release of endothelial-derived substances which modulate its direct constrictor effects on vascular smooth muscle [9, 10]. Although we did not test the effects of endothelial removal in this series of experiments, in an earlier study of WKY cerebral arteries [20], constriction to serotonin was increased following endothelial denudation, while the constrictor effects of PGF_{2a}—a fairly selective smooth muscle agonist—were not. Therefore, it is possible that the increased constriction with serotonin and a loss of dilation with acetylcholine both occur as a result of specific and potentially deleterious hypertension-induced changes in endothelial function.

In the cerebral circulation, serotonin is present in high concentrations within circulating blood platelets, in periarterial nerves, and in mast cells, which are often closely associated with blood vessels [8, 21]. In view of its alleged involvement in the pathogenesis of cerebrovascular disorders such as ischemic brain disease, subarachnoid hemorrhage, and cerebral vasospasm, an amplification of serotonin constrictor responses in hypertension—a major risk factor cerebrovascular disease—may have clinical implications which warrant further investigation.

References

1. Vanhoutte PM, Rubanyi GM, Miller VM, Houston DS (1986) Modulation of vascular smooth muscle contraction by the endothelium. Ann Rev Physiol 48: 307–320
2. DeMey JG, Vanhoutte PM (1982) Heterogenous behavior of the canine arterial and venous wall; importance of the endothelium. Circ Res 51: 439–447
3. Mayhan WG, Faraci FM, Heistad DD (1987) Endothelium-dependent responses of cerebral arterioles in stroke-prone spontaneously hypertensive rats. Fed Proc 46: 642
4. Nag S (1984) Cerebral endothelial surface charge in hypertension. Acta Neuropathol 63: 276–281
5. Werber AH, Heistad DD (1984) Effects of chronic hypertension and sympathetic nerves on the cerebral microvasculature of stroke-prone spontaneously hypertensive rats. Circ Res 55: 286–293
6. Brayden JE, Halpern W, Brann LR (1983) Biochemical and mechanical properties of resistance arteries from normotensive and hypertensive rats. Hypertension 5: 17–25

7. Osol G, Halpern W (1985) Myogenic properties of cerebral blood vessels from normotensive and hypertensive rats. Am J Physiol 249: H914–H921
8. Edvinsson L, Hogestatt E, MacKenzie ET (1983) 5-Hydroxytryptamine and the cerebral circulation. In: Bevan JA (ed) Vascular neuroeffector mechanisms: 4th international symposium. Raven, New York, pp 251–258
9. Cohen RA, Shepherd JT, Vanhoutte PM (1983) 5-Hydroxytryptamine can mediate endothelium-dependent relaxation of coronary arteries. Am J Physiol 245: H1077–H1080
10. Angus JA, Cocks TM (1984) Role of endothelium in vascular responses to norepinephrine, serotonin and acetylcholine. In: Proc 3rd Int Symp Mechanisms of Vasodilatation. Karger, Basel, pp 43–52
11. Halpern W, Osol G, Coy GS (1984) Mechanical behavior of pressurized in vitro pre-arteriolar vessels determined with a video system. Ann Biomed Eng 12: 463–479
12. Skarlatos SI, Hollis TM (1987) Cultured bovine aortic endothelial cells show increased histamine metabolism when exposed to oscillatory shear stress. Atherosclerosis 64: 55–61
13. Pohl U, Forstermann U, Busse R, Bassenge E (1985) Endothelium-mediated modulation of arterial smooth muscle tone and PGI_2-release: pulsatile versus steady flow. In: Schor S (ed) Prostaglandins and other eicosanoids in the cardiovascular system. Karger, Basel, pp 553–558
14. Nordborg C, Fredriksson K, Johansson BB (1985) The morphometry of consecutive segments in cerebral arteries of normotensive and spontaneously hypertensive rats. Stroke 16: 313–320
15. Osol G, Halpern W (1986) Effect of antihypertensive treatment on myogenic properties of brain arteries from the stroke-prone rat. J Hypertension 4 (Suppl 3): S517–S518
16. Furchgott RF, Zawadski JV (1980) The obligatory role of endothelial cells in the relaxation of arterial smooth muscle by acetylcholine. Nature (Lond) 288: 373–376
17. Tesfamariam B, Halpern W, Osol G (1985) Effects of perfusion and endothelium on the reactivity of isolated resistance arteries. Blood Vessels 22: 301–305
18. Rapoport RM, Muard F (1983) Agonist-induced endothelium-dependent relaxation in rat thoracic aorta may be mediated through cGMP. Circ Res 52: 352–357
19. Winquist RJ, Bohr DF (1983) Structural and functional changes in cerebral arteries from spontaneously hypertensive rats. Hypertension 5: 292–297
20. Osol G, Knutson S, Osol R, Halpern W (in press) Endothelial influence on cerebral artery tone and reactivity to transmural pressure change. In: Halpern W et al. (eds) Proceedings of international symposium on resistance arteries. Perinatology Press, Ithaca
21. Steinbusch HWM, Verhofstad AAJ, Joosten HWJ (1978) Localization of serotonin in the central nervous system by immunohistochemistry: description of a specific and sensitive technique and some applications. Neuroscience 3: 811–819

Protective Effects of Calcium Antagonists on Hypertensive Diseases in Heart, Brain, and Kidney of Hypertensive Rats

ANDREAS KNORR[1], BERNWARD GARTHOFF[1], CLAUDIA HIRTH[1], STANISLAV KAZDA[1], CARLOS LAGUNA[1], GEORG LUCKHAUS[2], JOHANNES-PETER STASCH[1]

Summary. The most evident result of calcium antagonist action in hypertension is the lowering of blood pressure. However, the protective effects of this type of agents on tissue structure and function may be largely independent of their vasorelaxant action. The natriuretic effect of dihydropyridine calcium antagonists is found in their antihypertensive dose range and was demonstrated to prevent volume expansion in (1-kidney, 1-clip)-hypertension. The functional recovery of the post-ischemic kidney was improved by treatment with nifedipine. With the help of biopsies of small mesenteric arteries of salt-loaded Dahl-S-rats with preexistent hypertension before and after 6 weeks of nifedipine, the formation of new arterial endothelium and internal elastic lamina was observed. Morbidity, mortality, and incidence of organ lesions were largely reduced or prevented by the treatment of adult stroke-prone SHR with blood-pressure-neutral doses of nitrendipine or nimodipine. Nimodipine was shown largely to normalize brain and kidney calcium concentrations without altering serum ionized calcium concentration or blood pressure. It is concluded that the direct protective effects of calcium antagonists on cell structure and function might be therapeutically more important than their depressor effect.

Key words: Calcium — Dihydropyridine — Nifedipine — Nitrendipine — Nimodipine — Parathyroid — Renal action

Introduction

Calcium antagonists have long been proven to be useful antihypertensive drugs [1]. Their prominent cellular action is the inhibition of transmembrane calcium influx, causing a decrease in total peripheral resistance. This mechanism appears to be well understood. However, calcium antagonists have actions over and above pure blood pressure reduction. Thus, dihydropyridine and other calcium antagonists have been demonstrated to improve or preserve both the structure and function of various organs in hypertensive rats [2–8]. They also have a diuretic effect [9]. Further protective actions of dihydropyridine calcium antagonists on the structure and function of some organs of hypertensive rats are reported here.

Institute of Pharmacology[1] and Institute of Toxicology[2], Bayer AG, 5600 Wuppertal 1, Federal Republic of Germany

Influence on Renal Function

Normal Kidney

In contrast with conventional vasodilators, which cause Na and water retention [10], calcium antagonists increase renal electrolyte and water excretion in man [11] and rats [9]. In stroke-prone spontaneously hypertensive rats (SHRSP) sodium and volume excretion were dose-dependently increased in the oral dose range between 1 and 10 mg/kg. At 10 mg/kg both were approximately doubled in comparison with control (Fig. 1). There was no significant rise in potassium excretion. Whereas these effects were observed in the antihypertensive dose range of nisoldipine—that is, at reduced blood pressure—depressor doses of minoxidil dose-dependently depressed sodium, potassium, and volume excretion (Fig. 1).

The mechanism of the natriuretic effect of calcium antagonists is not definitely explainable at present, but it appears that glomerular filtration rate (GFR) does not necessarily have to be increased [12]. Possibly, selective preglomerular vasodilation plays a role [13]. However, evidence for an inhibition of proximal tubu-

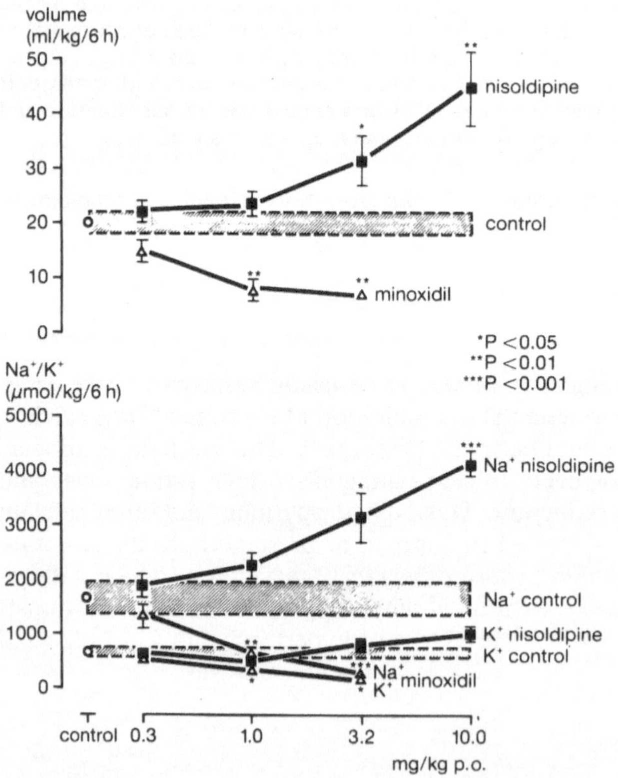

Fig. 1. Influence of nisoldipine and minoxidil on volume, sodium, and potassium excretion in male stroke-prone spontaneously hypertensive rats. Animals were challenged by oral administration of 30 ml/kg of saline. The period of collection was 6 h. Both agents were administered in the dose range of their antihypertensive activity

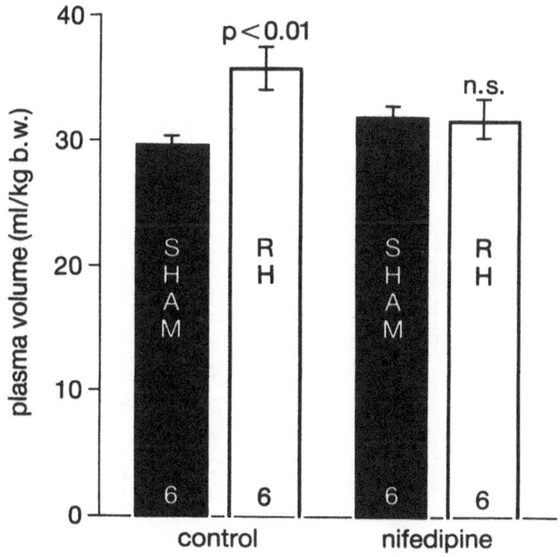

Fig. 2. Influence of nifedipine, administered for 2 weeks at 300 ppm in the solid feed, on hypervolemia of rats with (1-kidney,1-clip)-renal hypertension (*RH*), studied 2 weeks after surgery. *SHAM*, Sham-operated control

lar Na^+ reabsorption has accumulated [14–18] as well as that for distal inhibition of tubular Na^+ reabsorption [19, 20].

In spite of conflicting results with regard to its precise localisation, the tubular action of dihydropyridine calcium antagonists appears to play a predominant role in their natriuretic action.

In rats with (1-kidney, 1-clip)-renal hypertension, at 2 weeks after surgery an increase in plasma volume was observed. In animals treated with nifedipine after surgery, the plasma volume was similar in sham-operated rats and in those with renal hypertension (Fig. 2). Nifedipine-treated rats with renal hypertension were almost normotensive with a mean systolic blood pressure of 139 ± 11 mmHg ($n = 6$). These data indicate that prolonged treatment with nifedipine produces a steady-state normalisation of plasma volume in hypervolemic hypertension.

Ischemic Kidney

Since calcium is a primary mediator of irreversible hypoxic cell damage, possible protective effects of nifedipine against ischemic renal damage were investigated. In uninephrectomized Wistar rats the vascular pedicle was occluded in situ with a vascular clamp for 60 min. Sham-operated rats were prepared identically, with the exception of renal artery occlusion. One group was treated with nifedipine, 300 ppm in the solid feed, beginning 4 days before renal artery clipping. Additionally, immediately beforehand a single oral dose of 10 mg/kg was administered.

Serum creatinine and urea were excessively increased on the first day after temporary ischemia in control over sham-operated animals. On the third day, a

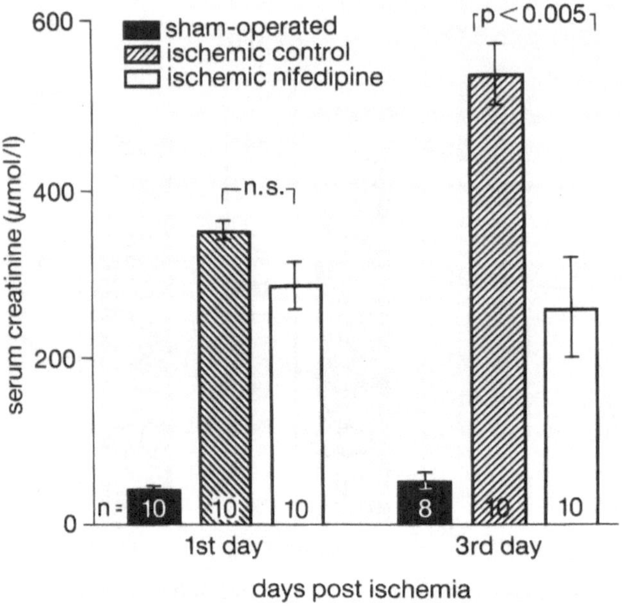

a

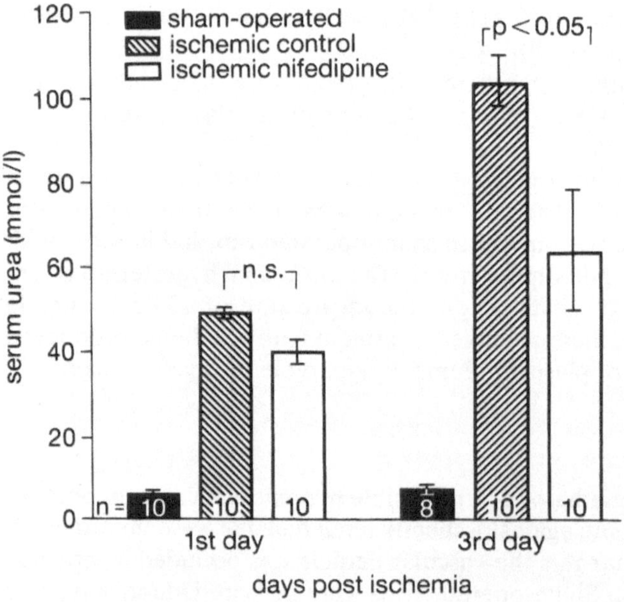

b

Fig. 3a,b. Improvement of recovery of post-ischemic function of Wistar rat kidneys. Animals were uninephrectomized 4 days prior to 1 h of normothermic ischemia by clipping and unclipping the renal pedicle. Serum creatinine (**a**) and serum urea (**b**) were measured on the 1st and 3rd days after ischemia. Nifedipine treatment was continued from uninephrectomy to the end of the experiment. A single dose of 10 mg/kg p.o. was added before clipping

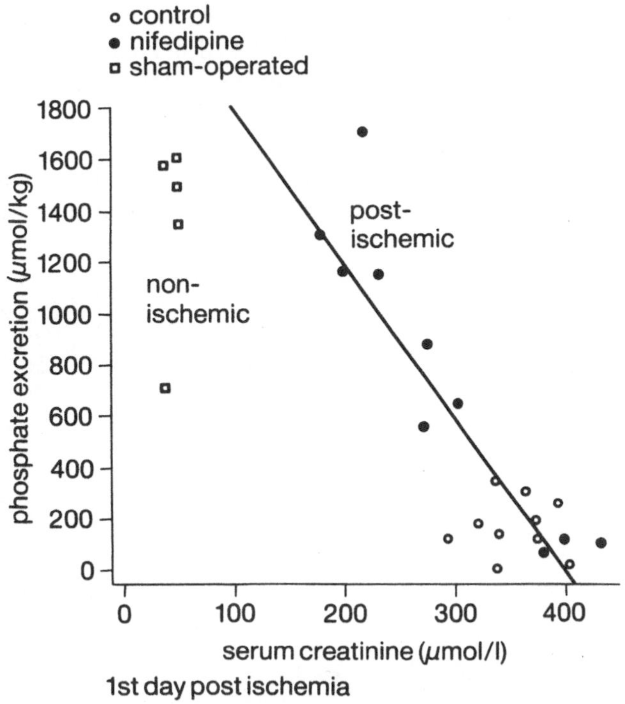

Fig. 4. Improvement of post-ischemic renal function by nifedipine. For details, see legend to Fig. 3

substantial further rise was observed (Fig. 3). Values observed in animals on nifedipine were also markedly increased, but showed no significant progression between days 1 and 3 after ischemia, so that maximum values of only one-half of those of untreated controls were observed (Fig. 3). It is evident from Fig. 4 that the excretory function of the post-ischemic kidney of rats treated with nifedipine was better than that of untreated rats in 7 of 10 animals even 1 day after ischemia.

These data show that a calcium antagonist may protect renal function during temporary ischemia even if the blood supply is completely cut off. Under similar conditions Hertle and Garthoff [21] found no protective effect of the noncalcium antagonistic vasodilator minoxidil.

Influence on Vascular Wall Structure

Dihydropyridine calcium antagonists are known to have good anti-hypertensive efficacy in the sodium-dependent Dahl hypertension [5, 6, 8, 22]. Salt-sensitive Dahl-"S" rats develop a malignant hypertension while on a diet containing 8% sodium chloride in their solid feed, which is characterized by a marked hypertensive arteriopathy with fibrinoid necrosis of the intima and intramural fibrin insudates.

The influence of nifedipine on arterial pathomorphology accompanying Dahl-S rat hypertension was studied by Luckhaus et al. [22]. Animals were followed individually by examining biopsies of mesenteric arterioles after 6–8 weeks of a high-salt diet and comparing them with appropriate material at necropsy after 6 further weeks on a high-salt diet which also contained 300 ppm of nifedipine. At the time of biopsy, animals were severely hypertensive (238 ± 12 mmHg) and after 6 weeks of nifedipine, in spite of continued sodium load, systolic blood

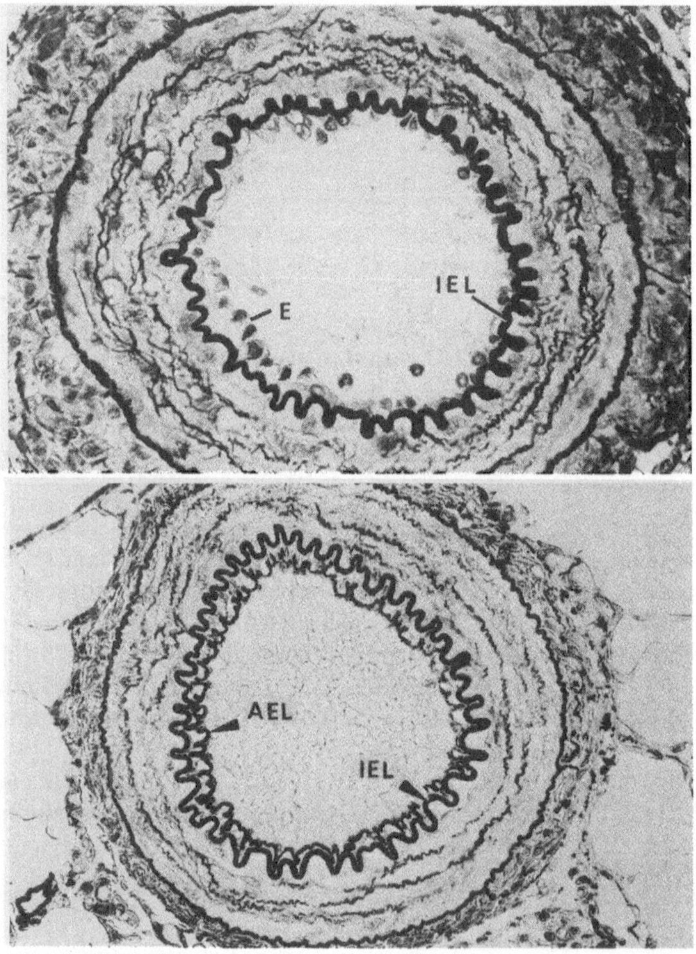

a

b

Fig. 5a,b. Histology of a small mesenteric artery from biopsy material of a hypertensive Dahl-S rat before (**a**) and after (**b**) 6 weeks of nifedipine treatment. Systolic blood pressure was 255 mmHg before and 135 mmHg after nifedipine. **a** Early hypertensive lesions can be seen including endothelial damage, focal plasma insudation causing distension of the subendothelial space (*E*, endothelial cell layer). A predominantly mononuclear periarteritis is seen. After nifedipine, in the arteries of the same animal (**b**), no intimal fibrin depositions are present, periarteritis is reduced, and an additional elastic lamina has formed (*AEL*) on the luminal side of the original internal elastic lamina (*IEL*). From [21], with permission pressure was 164 ± 7 mmHg. In the untreated control group it was 223 ± 9 mmHg in this latter instance.

At biopsy immediately before commencing nifedipine treatment, hypertensive arteriopathy was seen to consist of endothelial damage with distension of the subendothelial space due to fibrin insudation, occasional rupture of the internal elastic lamina, as well as myointimal proliferation, and a mononuclear periarteritis. After 6 weeks of nifedipine treatment, the same individuals exhibited a reduction of periarteritis and prominent signs of endothelial reorganisation with intimal fibrin resorption. An additional internal elastic lamina with microscopically intact endothelium had formed on the luminal side of the original internal elastic lamina (Fig. 5). Although the narrowing of the vascular lumen remained, signs of healing of previously existing arterial hypertensive damage were demonstrated.

That reversal of hypertensive tissue damage may not be entirely dependent on the blood pressure lowering effect of calcium antagonists was demonstrated by Kazda et al. [23]. In old SHRSP, which are not very sensitive to calcium antagonists, nitrendipine (for 9 weeks) and nimodipine (for 6 weeks) had practically no influence on high blood pressure, but no mortality and a marked reduction in the incidence of primarily cerebral and cardiac, but also kidney, lesions was noted (Table 1). The mortality in the two control groups was 60% and 25% (Table 1).

When the period of treatment with nimodipine was prolonged to 25 weeks, 55% of the SHRSP still survived, all of them without brain lesions and most without cardiac lesions, whereas mortality in the control group was 100% (Table 1).

From these results one may conclude that reduction of blood pressure is not an essential prerequisite for the tissue protective action of calcium antagonists, at least in the models of hypertension studied until now.

Table 1. Number of male hypertensive rats (SHRSP) surviving and with brain, kidney, or cardiac lesions. Shown are approximately 5-month-old animals which were either treated with calcium antagonists or not

Treatment	Initial *n*	Survived	Lesions		
			Brain	Kidneys	Heart
9 weeks on 8% NaCl					
control	16	5	12	16	11
nitrendipine	16	16	1	14	0
6 weeks on 8% NaCl					
control	19	14	7	19	18
nimodipine	19	19	0	13	10
25 weeks on 8% NaCl					
control	20	0	11/11[a]	11/11	10/11
nimodipine	20	11	0/11	8/11	2/11

[a] In the upper two panels, histology was done on all animals, both after spontaneous death or after killing at the end of the study. In the lower panel, only 11 of 20 animals killed by hypertension were available for histology. In the nimodipine group only the 11 surviving animals were examined histologically

Influence on Tissue Calcium

The role of tissue calcium in ischemic and hypertensive cell damage has been extensively studied by Fleckenstein and colleagues [2, 3]. In a more recent study Kazda et al. [24] investigated the influence of nimodipine, a calcium antagonist with a relatively weak peripheral vasodilator effect, on tissue and serum calcium. This was done in comparison with the effect of removing parathyroid hormone by parathyroidectomy (PTX).

At the onset of this experiment, male SHRSP were 7 months of age and were observed for 14 more weeks. They were fed a normal sodium chow, containing less than 1% sodium chloride. In controls, nimodipine-treated, and PTX rats, systolic blood pressure continued to rise, reaching values of 260–280 mmHg on the day of killing. All control rats developed neurological symptoms such as transitory convulsions, hemi- or paraplegia, and high irritability. In sharp contrast, neither nimodipine-treated nor PTX rats had any neurological symptoms until the end of the experiment.

The kidneys and brains of these animals, as well as those of Wistar Kyoto (WKY) and young SHRSP were minced and their total calcium content was measured by atomic absorption spectometry. Serum ionized calcium was measured in the first three groups.

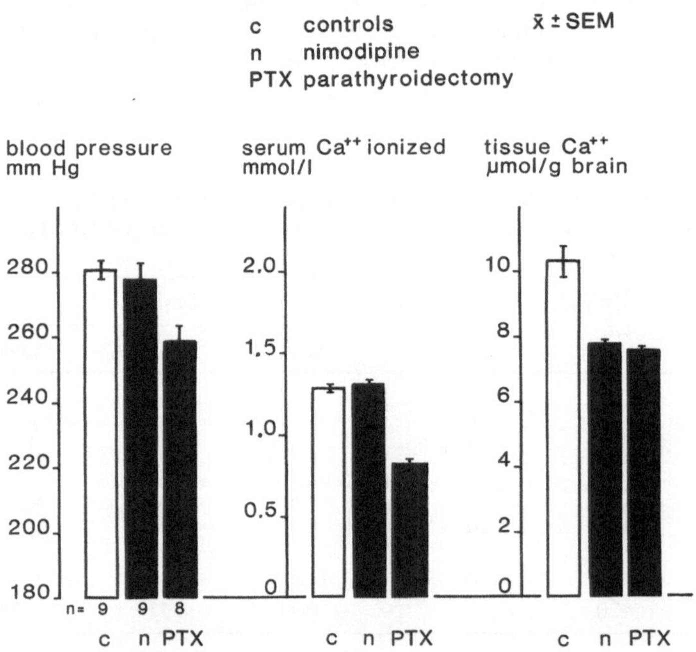

Fig. 6. Influence of nimodipine (14 weeks of treatment) or parathyroidectomy (*PTX*) on blood pressure, serum ionized calcium, and brain calcium of male, old (42 weeks) stroke-prone spontaneously hypertensive rats

The tissue calcium level in kidneys of young SHRSP, WKY, and PTX was between 11 and 12 μmol/g tissue. It was increased to approximately 20 μmol/g in controls, but largely normalized to a mean value around 14 μmol/kg in animals treated with nimodipine [23]. Brain calcium was lowest in WKY (approx. 6 μmol/g). In old (control) SHRSP it was increased to values exceeding 10 μmol/g. Here both PTX and nimodipine normalized tissue calcium to values between 7 and 8 μmol/g ([23], see also Fig. 6). Serum ionized calcium was decreased by PTX, but unaltered by nimodipine (Fig. 6).

These results suggest that the prevention of neurological symptoms is probably more related to the reduction of tissue calcium overload than to lowering high blood pressure. This contention may be extended to the life-saving effect of nimodipine in salt-overloaded young SHRSP, which is partially mimicked by PTX, too [24].

Conclusions

The results presented here demonstrate that dihydropyridine calcium antagonists can improve the function of the hypertensive kidney as well as the functional recovery of the post-ischemic kidney. The former may be due mostly to a direct inhibition of tubular sodium reabsorption and the latter probably to a prevention of ischemic renal cell death. On the other hand, in malignant hypertensive arteriopathy of Dahl-S rats, nifedipine induces repair processes leading to regeneration of the vascular endothelium and a fibroelastic stratification of the intima. Such effects and the reduction of hypertension-associated tissue calcium overload may be more important to the reduction in morbidity and mortality observed in hypertensive rats after long-term treatment with dihydropyridine calcium antagonists than the depressor effect of these agents.

References

1. Aoki A, Yoshida T, Kato S, Tazumi K, Sato I, Takikawa K, Hotta K (1976) Hypertensive action and increased plasma renin activity by Ca^{2+} antagonist (nifedipine) in hypertensive patients. Jpn Heart J 17: 479–484
2. Fleckenstein A (1968) Myokardstoffwechsel und Nekrose. In: Heilmeyer L, Holtmeier HS (eds) Herzinfarkt und Schock. Thiem, Stuttgart, pp 94–109
3. Fleckenstein A, Frey M, von Witzleben H (1983) Vascular calcium overload—a pathogenic factor in arteriosclerosis and its neutralisation by calcium antagonists. In: Kaltenbach M, Neufeld HN (eds) New therapy of ischemic heart disease and hypertension. 5th Int. Adalat Symposium. Excerpta Medica, Amsterdam, pp 36–52
4. Fleckenstein A, Frey M, Zorn S, Fleckenstein-Grün G (1985) Experimental basis of the long-term therapy of arterial hypertension with calcium antagonists. Am J Cardiol 56: 3H–14H
5. Garthoff B, Kazda S (1981) Calcium antagonist nifedipine normalizes high blood pressure and prevents mortality in salt loaded DS substrain of Dahl rats. Eur J Pharmacol 74: 111–113
6. Luckhaus G, Garthoff B, Kazda S (1982) Prevention of hypertensive vasculopathy by nifedipine in salt loaded Dahl rats. Arzneim Forsch/Drug Res 32: 1421–1425

7. Kazda S, Garthoff B, Luckhaus G (1983) Calcium antagonists prevent brain damage in stroke-prone spontaneously hypertensive rats independent of their effect on blood pressure. J Cerebral Blood Flow Metab 3 (Supp 1): S526–S527
8. Kazda S (1986) Effects of nitrendipine on vascular integrity. Am J Cardiol 58: 31D–31D
9. Garthoff B, Kazda S, Knorr A, Thomas G (1983) Factors involved in the antihypertensive action of calcium antagonists. Hypertension 5 (Suppl II): II34–II38
10. Koch-Weser J (1974) Vasodilating drugs in the treatment of hypertension. Arch Int Med 133: 1017–1019
11. Klütsch K, Schmidt P, Grosswendt J (1972) Der Einfluβ von BAY A 1040 auf die Nierenfunktion des Hypertonikers. Arzneim Forsch/Drug Res 22: 377–380
12. Loutzenhiser R, Epstein M (1985) Effect of calcium antagonists on renal hemodynamics. Am J Physiol 249: F619–F629
13. Steinhausen M, Flemming JT, Holz FG, Parekh N (1987) Nitrendipine and the pressure-dependent vasodilation of vessels in the hydronephrotic kidney. J Cardiovasc Pharmacol 9 (Suppl 1): S39–S47
14. Johns EJ (1984) The effect of acute administration of diltiazem and nifedipine on the function of denervated rat kidney. Br J Pharmacol 82 (Suppl): 328P
15. Johns EJ (1985) The influence of diltiazem and nifedipine on renal function in the rat. Br J Pharmacol 84: 707–713
16. Johns EJ, Manitius J (1987) The renal actions of nitrendipine and its influence on the neural regulation of caclium and sodium reabsorption in the rat. J Cardiovasc Pharmacol 9 (Suppl 1): S49–S56
17. Häberle DA, Kawata T, Davis JM (1987) The site of action of nitrendipine in the rat kidney. J Cardiovasc Pharmacol 9 (Suppl 1): S17–S23
18. Lederballe Pedersen O, Krusell LR, Christensen CK, Jaspersen LT, Thomsen K (1986) Effects of calcium antagonists in essential hypertension with special reference to renal function. In: Aoki K (ed) Essential hypertension: calcium mechanisms and treatment. Springer, Tokyo Berlin Heidelberg New York, pp 149–160
19. DiBona GF (1987) Effects of vasodilator antihypertensive agents on renal function. J Cardiovasc Pharmacol 9 (Suppl 1): S14–S16
20. Giebisch G, Guckian VA, Klein-Robbenhaar G, Klein-Robbenhaar MT (1987) Renal clearance and micropuncture studies of nisoldipine effects in spontaneously hypertensive rats. J Cardiovasc Pharmacol (Suppl 1): S24–S31
21. Hertle L, Garthoff B (1985) Calcium channel blocker nisoldipine limits ischemic damage in rat kidney. J Urology 134: 1251–1254
22. Luckhaus G, Nash G, Garthoff B, Kazda S, Feller W (1985) Healing of malignant hypertensive arteriopathy Dahl rats by nifedipine. Arzneim Forsch/Drug Res 35: 115–121
23. Kazda S, Grunt M, Hirth C, Preis W, Stasch J-P (1987) Calcium antagonism and protection of tissue from calcium damage. J Hypertension 5 (Suppl 4): S37–S42
24. Kazda S, Garthoff B, Hirth C, Preis W, Stasch J-P (1986) Parathyroidectomy mimics the protective effect of the calcium antagonist nimodipine in salt-loaded stroke-prone spontaneously hypertensive rats. J Hypertension 4 (Suppl 3): S483–S485

Hemodynamic and Hormonal Responses to Stimuli in Spontaneously Hypertensive Rats: Effects of Chronic Diltiazem Treatment*

JIN YAMAMOTO and HIROAKI MATSUBARA[1]

Summary. We have previously shown the enhanced reactivity of spontaneously hypertensive rats (SHR) to acute shaking stimuli. This study evaluates the effects of 6-weeks' diltiazem treatment on hemodynamic and hormonal responses to this stress in SHR as compared with Wistar-Kyoto rats (WKY). Hemodynamics and plasma catecholamine and atrial natriuretic peptide (ANP) levels were measured at rest and during stress. Diltiazem treatment lowered mean arterial pressure (MAP) and reduced left ventricular weight in SHR. With diltiazem, cardiac index (CI) and regional blood flow were not altered, while total peripheral resistance index (TPRI) and skeletal muscle and renal vascular resistances were decreased. Diltiazem also decreased MAP and TPRI and several vascular resistances during stress. When percentage changes from baseline were compared, these effects were not significant. Diltiazem had no hemodynamic effects in WKY. Diltiazem did not affect resting plasma catecholamine levels in either strain. Diltiazem tended to decrease the already increased resting level of plasma ANP in SHR. In both strains, these hormone levels increased many times with stress, but diltiazem did not influence their levels during stress. The results suggest that calcium mechanisms contribute more importantly to increased vascular tone at rest and during stress in SHR. Diltiazem treatment seems to alter stressed hemodynamics mainly by reducing the baseline values for MAP and TPRI and several regional vascular resistances, consequently reducing their peak values during stress, in SHR.

Key words: Shaker stress—Systemic and regional hemodynamics—Plasma catecholamines—Plasma atrial natriuretic peptide—Wistar-Kyoto rats

Introduction

Spontaneously hypertensive rats (Aoki SHR) and patients with essential hypertension show similar hemodynamic patterns [1–3]. It was recently demonstrated in this laboratory that elevated blood pressure was accompanied by increased total peripheral resistance index (TPRI) and no changes in cardiac index (CI) and heart rate (HR) in 8-, 24-, 48-, and 96-week-old SHR, compared with age-

*Supported in part by a Grant-in-aid for Scientific Research (C-62570408) from the Ministry of Education, Science and Culture, Japan.
[1]National Cardiovascular Center Research Institute, Laboratory for Hemodynamics, Fujishirodai 5, Suita, Osaka 565, Japan

matched Wistar-Kyoto rats (WKY) [4]. Regional blood flow was unaltered for the most part, and therefore regional vascular resistance was almost uniformly increased in the SHR. Generalized vasoconstriction thus prevails in SHR hypertension, as in human essential hypertension. Also, SHR and hypertensive patients show hyperreactivity to a variety of stimuli [3–7]. Hemodynamic and sympatho-adrenal reactivity to acute shaking stimuli, which evoked circulatory changes similar to a defense reaction, were enhanced in SHR, regardless of age [4].

It is reasonable that calcium channel blockers are becoming increasingly used in the therapy of hypertension, considering that transmembrane calcium influx plays a pivotal role in generalized vascular smooth muscle constriction in hypertensive disease [8, 9]. Since so-called calcium blockers consist of structurally different groups of compounds that inhibit calcium transport with different properties [10], each member of this category of drugs must be studied in detail. Long-term antihypertensive effects of diltiazem have been examined clinically [8, 9, 11, 12] and experimentally [13–15]; however, there is a paucity of data on the effects of long-term diltiazem treatment on stressed hemodynamics in the hyperresponsive SHR model. We investigated the possible effects of chronic diltiazem treatment on hemodynamics at rest and during stress in SHR and WKY.

Atrial natriuretic peptide (ANP) [16] is found to be released into the plasma with progression of hypertension and associated left ventricular (LV) hypertrophy [17–20], and with stressful stimuli and volume loading [21, 22]. We also investigated plasma ANP levels in untreated and diltiazem-treated rats at rest and during stress.

Methods

Male SHR and WKY (Charles-River) were kept in an animal facility under standardized conditions. When the rats were 10 weeks old, diltiazem administration was commenced in the drinking water bottles (500 mg/l), which were daily changed and protected from light. Rats given only tap water were used as controls. Food was provided ad libitum. After 6 weeks of treatment with or without diltiazem, i.e., at 16 weeks of age, the rats were subjected to experiments, as described previously [4, 7]. At 1700 to 1900 h, under anesthetization with ether, tip-tapered polyethylene catheters (PE-50) were placed in the femoral artery and vein, and in the left ventricle (LV) through the right carotid artery. All catheters were brought out at the nape. In addition, several normal male Wistar rats were prepared with arterial and venous catheters and were used as donors for blood transfusion. The next morning, after an 18-to-20-h period of recovery, the conscious rat was placed in a small, nonconfining cage, which was mounted and fixed on the tabletop of a shaker. Catheters were connected to Statham transducers and a polygraph, by which mean arterial pressure (MAP), LV pressure, and heart rate (HR) were recorded directly.

Hemodynamics

The radioactive microsphere method was used to measure cardiac output and regional blood flow, as described previously [4, 7, 23]. In short, 15 ± 3 μm radioactive microspheres labelled with ^{141}Ce or ^{46}Sc (New England Nuclear, Boston, MA, USA) were suspended in 0.9% saline with 0.01% Tween 80. Immediately after agitation, 0.05 ml of this suspension containing 40000–100000 microspheres were flushed into the left ventricle with 0.4 ml of fresh blood obtained from donor rats, over a 20-s period. Starting 10 s before this injection, reference blood was taken using a Harvard pump from the femoral artery catheter at a rate of 0.93 ml/min over a 50-s period. Blood loss was immediately restored by donor blood.

After a restabilization, the rats were shaken at a rate of 90 oscillations per min (horizontal displacement of about 4 cm) for 5 min [4, 7]. This shaker stress invoked responses that combined emotional arousal and consequent behavioral excitation. Since circulatory variables were most stable at between 2 and 3 min, hemodynamics were measured again during this period, using a different batch of microspheres, as already described.

At completion of the experiment, the rats were killed with an overdose of pentobarbital, and tissue and organs were removed. Skeletal muscle samples were taken from the right leg. Tissue and reference samples were counted in a computerized γ counter (LKB, Stockholm, Sweden). Hemodynamic variables were calculated using the reference and tissue sample radioactivity [4, 7, 23]. Measurements showing more than a 10% difference in blood flow between the right and left kidneys were eliminated.

Plasma Levels of Catecholamines and ANP

SHR and WKY were prepared with catheters and were then subjected to a protocol similar to that described above. Arterial blood samples were withdrawn using the cross circulation technique at rest and between 2 and 4 min of stress imposition [4, 7]. Blood (3 ml) was taken from the arterial line of the experimental rats into an ice-chilled tube containing aprotinin (5000 IU/ml) and EDTA-Na_2 (1 mg/ml), while blood was being transfused simultaneously from the femoral arterial catheter of the donor rat to the femoral venous catheter of the experimental rat by way of a Gilson pump.

Plasma catecholamine concentrations were determined by the HPLC-THI method [4, 7]. Plasma immunoreactive ANP levels were determined with radioimmunoassay. Plasma samples were applied to Sep-Pack C18 cartridges (Water Associations) that were activated with 10 ml of ethanol and preequilibrated with 4% acetic acid. The cartridges were then washed with 5 ml of 4% acetate. The absorbed substances were eluted with 1 ml of 90% methanol mixed with 4% acetic acid. This solution was dried. Using anti-alpha-ANP serum (Peninsula Laboratories, Belmont, CA, USA) and ^{125} I-labelled and nonlabelled alpha-ANP (Amersham International plc, Amersham, Buckinghamshire, England), subsequent assay procedures were performed, as described elsewhere [24].

Table 1. Body and heart weights in rats, with or without diltiazem treatment

Variables	Untreated		Diltiazem-treated	
	SHR (n = 11)	WKY (n = 11)	SHR (n = 11)	WKY (n = 11)
BW (g)	301 ± 4	315 ± 3	285 ± 4	311 ± 7
LV (mg)	872 ± 11[a]	780 ± 13	791 ± 13[b]	765 ± 17
LV/BW (mg/g)	2.90 ± 0.04[a]	2.48 ± 0.04	2.77 ± 0.04[c]	2.45 ± 0.04
RV (mg)	0.17 ± 0.01	0.17 ± 0.01	0.15 ± 0.01	0.17 ± 0.01
RV/BW (mg/g)	0.57 ± 0.02	0.55 ± 0.03	0.53 ± 0.02	0.53 ± 0.02

Values are means ± SE
SHR, hypertensive; WKY, normotensive; BW, body weight; LV, left ventricle; RV, right ventricle
[a] $P < 0.05$, compared with untreated WKY
[b] $P < 0.05$, compared with untreated SHR
[c] $P < 0.05$, compared with diltiazem-treated WKY

Statistics

Analysis of variance (ANOVA) and the Bonferroni method were used. Statistical significance was set at $P < 0.05$. All values are presented as means ± SE.

Results

Heart Weights

Body weight (BW) was not different among the groups (Table 1). The absolute LV weight was greater in untreated SHR than in untreated WKY. Importantly, this value was significantly ($P < 0.05$) less in diltiazem-treated SHR than in untreated SHR. The LV/BW ratio was significantly ($P < 0.05$) larger in untreated SHR than in untreated WKY. This ratio was not significantly decreased in diltiazem-treated SHR compared with untreated SHR, but there was a tendency toward a decrease ($0.05 < P < 0.10$). The right ventricular weight (RV) and the RV/BW ratio showed no difference among any groups.

Resting Baseline Systemic Hemodynamics

As shown in Fig. 1, at rest the untreated SHR compared with untreated WKY showed a significant elevation in MAP (148 ± 3 vs 114 ± 3 mmHg, $P < 0.01$) resulting from a significant increase in TPRI (5.35 ± 0.21 vs 3.99 ± 0.15 mmHg ml^{-1} min^{-1} 100 g^{-1}, $P < 0.01$), while CI remained unchanged. There was no significant difference in HR between untreated SHR and untreated WKY. MAP in diltiazem-treated SHR was significantly lower than that in untreated SHR (133 ± 4 vs 148 ± 3 mmHg, $P < 0.01$); this decrease in diltiazem-treated SHR was associated with a decrease in TPRI (4.48 ± 0.23 vs 5.35 ± 0.21 mmHg ml^{-1}

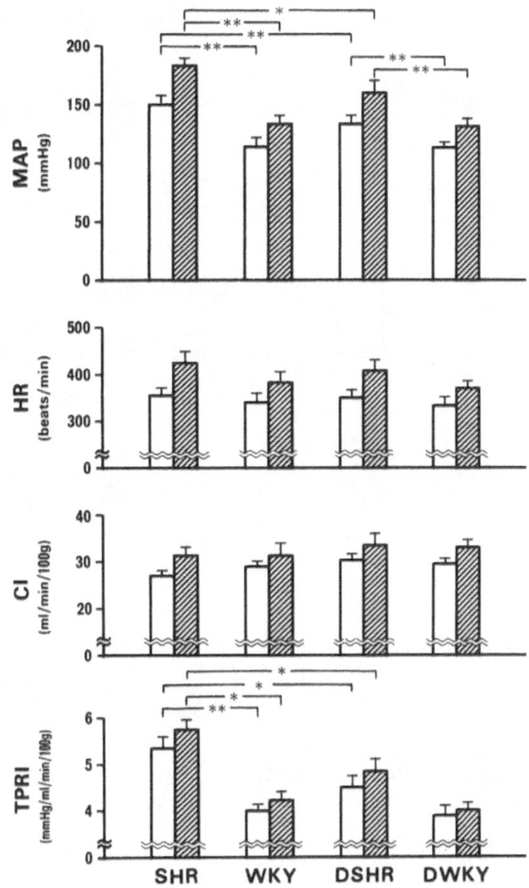

Fig. 1. Systemic hemodynamics at rest and during acute stress in hypertensive (*SHR*) and normotensive (*WKY*) rats. *MAP*, mean arterial pressure; *HR*, heart rate; *CI*, cardiac index; *TPRI*, total peripheral resistance index; *DSHR*, diltiazem-treated SHR; *DWKY*, diltiazem-treated WKY. *Open bars* represent data at rest, and *hatched bars* represent data during stress. Data are means ± SE. *, $P<0.05$; **, $P<0.01$

min^{-1} 100 g^{-1}, $P<0.05$) with no change in CI and HR. Diltiazem treatment had no effects on MAP and other hemodynamic variables in the WKY.

Systemic hemodynamic Response to Stress

In response to short-term shaker stress, all circulatory varaiables were, on average, increased (Fig. 1). During stress, the untreated SHR compared with the corresponding WKY showed a significantly greater MAP (183 ± 5 vs 133 ± 4 mmHg, $P<0.01$) and TPRI (5.78 ± 0.29 vs 4.21 ± 0.15 mmHg ml^{-1} min^{-1} 100 g^{-1}, $P<0.05$). Notably, the MAP during stress was significantly lower in diltiazem-treated SHR than in untreated SHR (160 ± 6 vs 183 ± 5 mmHg, $P<0.05$). The TPRI during stress was also significantly smaller in diltiazem-

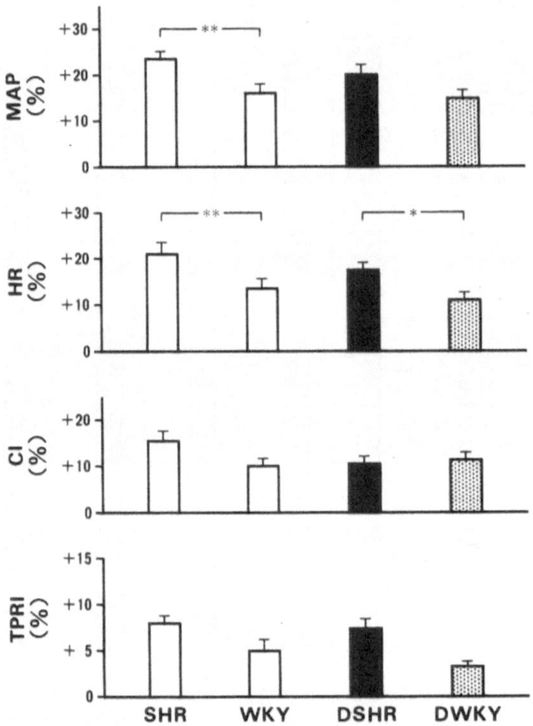

Fig. 2. Systemic hemodynamic response to stress expressed as percentage change from resting values. Abbreviations are as shown in Fig. 1. Data are means ± SE

treated SHR than in untreated SHR (4.83 ± 0.30 vs 5.78 ± 0.26 mmHg ml^{-1} min^{-1} 100 g^{-1}, $P<0.05$). However, CI and HR during stress did not significantly differ among the four groups of rats.

When the percentage change was used as a measure of response (Fig. 2), significantly enhanced responses were evident in MAP and HR in the untreated SHR compared with untreated WKY. However, the percentage changes in MAP and other variables were not significantly different between the diltiazem-treated and untreated groups from either strain, though there tended to be somewhat smaller percentage changes in the diltiazem-treated SHR.

Resting Regional hemodynamics

There were no significant differences in resting baseline blood flow to the splanchnic (gastrointestinal tract, spleen, and pancreas), renal, myocardial (LV), cerebral, skeletal muscle, and cutaneous vascular beds between untreated SHR and untreated WKY (Fig. 3). This was true for each of the splanchnic organs, RV and testes (data not shown). Diltiazem treatment did not significantly alter any of the regional blood flows in either strain. Resting vascular resistance through the splanchnic, renal, cerebrovascular, skeletal muscle, and cutaneous beds at rest was significantly higher in untreated SHR than in untreated WKY (Fig. 4). Diltiazem treatment significantly decreased vascular resistance through the skeletal muscle vasculature at rest in the treated SHR, compared with untreated SHR (11.4 ± 1.0 vs 15.3 ± 1.2 mmHg ml^{-1} min^{-1} 100 g^{-1}, $P<0.05$). Diltiazem treatment also significantly decreased resting renal vascular resistance in SHR (0.18 ± 0.01 vs 0.22 ± 0.01 mmHg ml^{-1} min^{-1} 100 g^{-1}, $P<0.05$). Vascular resistance through the splanchnic and cerebral circulations tended to be decreased in the diltiazem-treated SHR. In contrast, diltiazem treatment had no effects on resting regional hemodynamics in WKY.

Regional Hemodynamics Response to Stress

With acute shaker stress, cerebral, myocardial (LV), and skeletal muscle blood flows were increased, and their vascular resistance was decreased or at least unchanged in all groups of rats (Figs. 3 and 4). This was also so for RV myocardium (data not shown). With respect to absolute blood flow during stress, there were no differences between strains, except for coronary circulation. The LV myocardial blood fiow during stress was significantly greater in untreated SHR than in untreated WKY (813 ± 45 vs 613 ± 25 ml^{-1} min^{-1} 100 g^{-1}, $P<0.05$); this was true for the RV myocardium (552 ± 31 vs 439 ± 36 ml min^{-1} 100 g^{-1}, $P<0.05$). Diltiazem therapy did not significantly alter cerebral, myocardial, and skeletal muscle blood flow during stress in either strain (Fig. 3). Vascular resistance through these vascular beds during stress was not significantly different between untreated SHR and untreated WKY (Fig. 4). Diltiazem treatment did not significantly influence vascular resistance through cerebral and myocardial vasculatures during stress in both strains. However, skeletal muscle vascular resistance during stress was significantly decreased in the diltiazem-treated SHR, as compared with untreated SHR (10.6 ± 1.3 vs 13.2 ± 1.2 mmHg ml^{-1} min^{-1} 100 g^{-1}, $P<0.05$). When percentage changes were compared, the responses in blood flow and vascular resistance for the cerebral, myocardial, and skeletal muscle vasculatures were all significantly greater in untreated SHR than in untreated WKY (Figs. 5 and 6). Diltiazem treatment did not significantly affect these percentage change responses in both strains.

Acute shaker stress significantly decreased splanchnic, renal, and cutaneous blood flow, while it increased their vascular resistance markedly (Figs. 3 and 4). Similar changes occurred with regard to each of the splanchnic organs and testes (data not shown). Blood flow to the splanchnic, renal, and cutaneous vasculatures during stress did not significantly differ with respect to strain or to diltiazem treatment. Vascular resistance through these three vascular beds during stress was significantly greater in untreated SHR than in untreated WKY (Fig. 4). Notably, during stress, the diltiazem-treated SHR, compared with untreated SHR, showed significant decreases in splanchnic vascular resistance (1.15 ± 0.09 vs. 1.45 ± 0.09 mmHg ml^{-1} ml^{-1} 100 g^{-1}, $P<0.0.05$) and also in renal vascular resistance (0.25 ± 0.02 vs 0.31 ± 0.02 mmHg ml^{-1} min^{-1} 100 g^{-1}, $P<0.05$). No

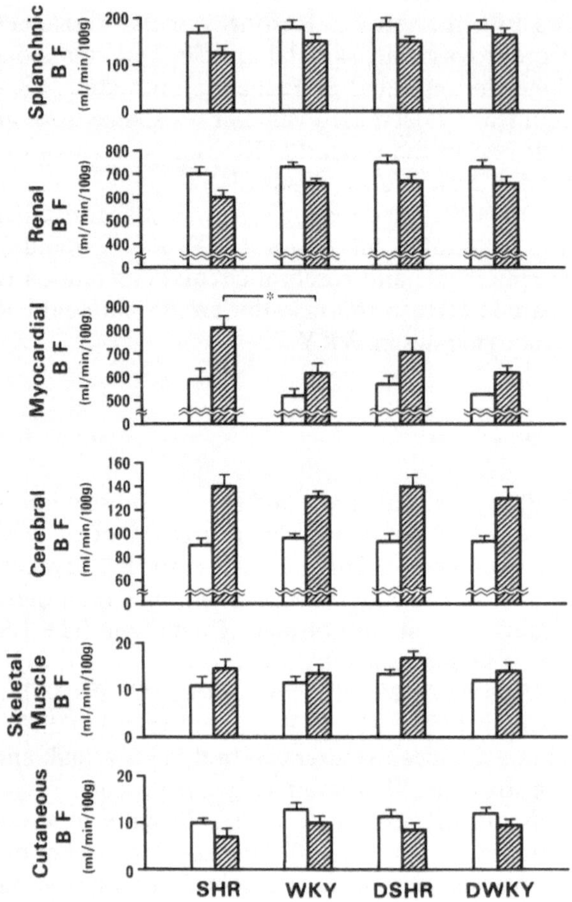

Fig. 3. Regional blood flows at rest and during acute stress in hypertensive (*SHR*) and normotensive (*WKY*) rats. *DSHR*, diltiazem-treated SHR; *DWKY*, diltiazem-treated WKY; *BF*, blood flow; *myocardial*, left ventricular. *Open bars* represent data at rest, and *hatched bars* represent data during stress. Data are means ± SE. *, $P<0.05$

difference was noted in cutaneous vascular resistance. When the percentage changes from baseline were compared, the untreated SHR, compared with untreated WKY, showed a significantly greater decrease in splanchnic blood flow, associated with a greater increase in splanchnic vascular resistance (Figs. 5 and 6). They also exhibited a greater increase in renal vascular resistance. Regarding diltiazem's effect, comparison of the percentage changes between diltiazem-treated and untreated rats reached no statistically significant differences in both strains, although there was a trend toward a decrease in percentage changes in most of the vascular beds tested in SHR (Figs. 5 and 6).

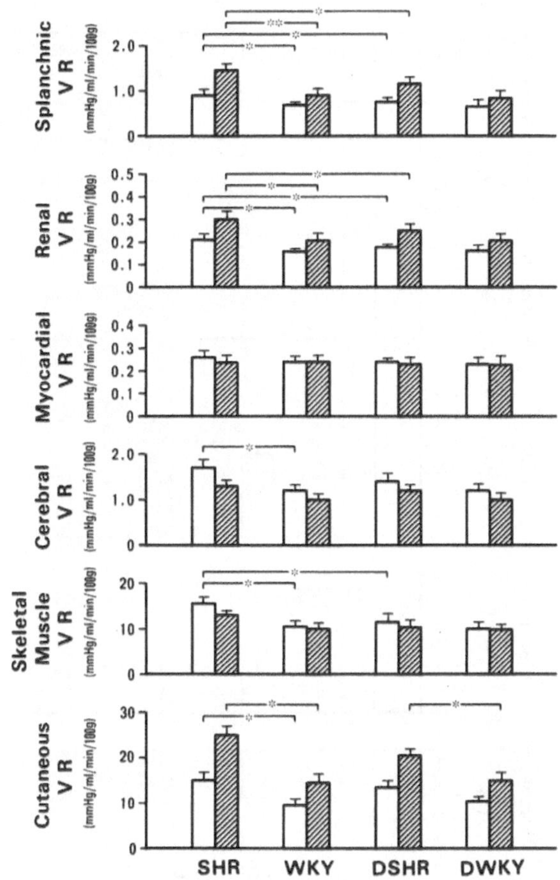

Fig. 4. Regional vascular resistance at rest and during acute stress in hypertensive (*SHR*) and normotensive (*WKY*) rats. *DSHR*, diltiazem-treated SHR; *DWKY*, diltiazem-treated WKY; *VR*, vascular resistance; *myocardial*, left ventricular. *Open bars* represent data at rest, and *hatched bars* represent data during stress. Data are means ± SE. *, **P** < 0.05; **, $P < 0.01$

Plasma Catecholamines and ANP

There were no significant differences regarding strain and diltiazem's effect with respect to the resting plasma levels of epinephrine and norepinephrine (Fig. 7). Plasma catecholamines showed a manyfold increase with acute shaker stress. The untreated SHR, compared with untreated WKY, displayed significantly ($P < 0.05$) higher plasma levels for both norepinephrine and epinephrine during stress. Effects of diltiazem treatment on these levels during stress reached no statistically significant level in either strain.

Resting plasma immunoreactive ANP levels in untreated SHR were significantly ($P < 0.05$) greater than those in untreated WKY (47.8 ± 6.0 pg/ml vs

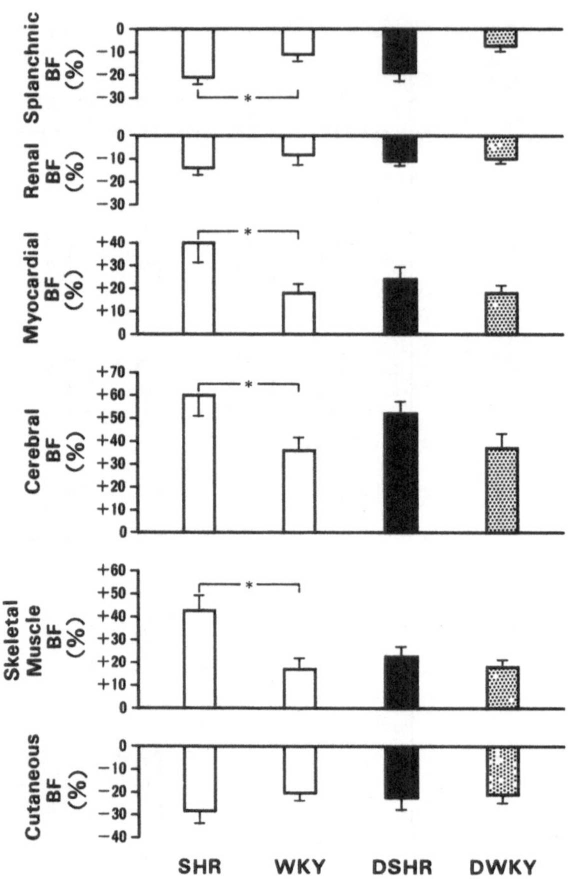

Fig. 5. Regional blood flow response to stress expressed as percentage change from resting values. *DSHR*, diltiazem-treated hypertensive (*SHR*); *DWKY*, diltiazem-treated normotensive (*WKY*); *BF*, blood flow; *myocardial*, left ventricular. Data are means ± SE. *, $P<0.05$

28.7 ± 4.3 pg/ml, $P<0.05$) (Fig. 7). Diltiazem treatment did not significantly decrease resting plasma ANP levels in either strain, but there was an appreciable trend toward a decrease in diltiazem-treated SHR compared with untreated SHR (36.2 ± 3.6 pg/ml vs 47.8 ± 6.0 pg/ml, $0.05<P<0.10$). Exposure to acute stress produced a two- to six fold increase in plasma ANP. Plasma ANP levels during stress were widely scattered. Nonetheless, these values of untreated SHR were significantly greater than those of untreated WKY (226 ± 89 pg/ml vs 69 ± 13 pg/ml, $P<0.05$). Comparison of ANP levels during stress between diltiazem-treated and untreated SHR achieved no statistically significant difference, although the mean value was decreased by 39.8% in diltiazem-treated SHR.

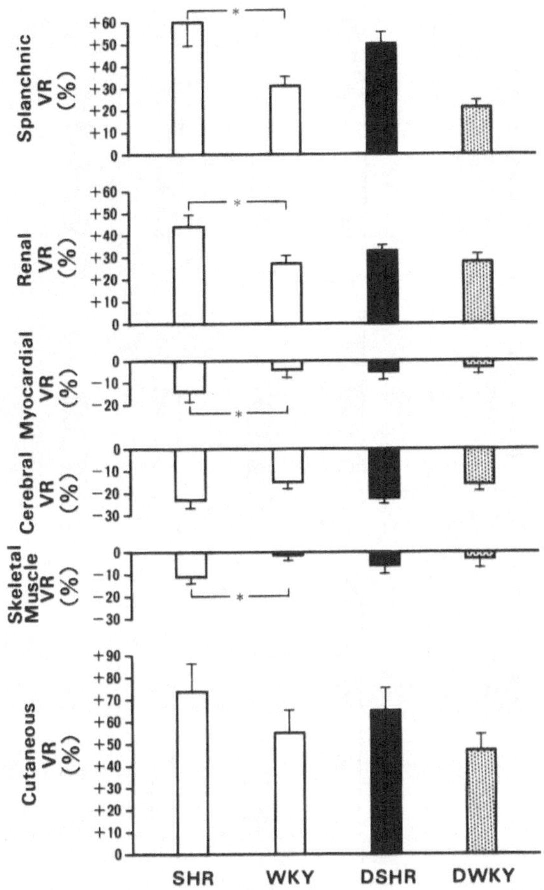

Fig. 6. Regional vascular resistance response to stress expressed as percentage change from resting values. *DSHR*, diltiazem-treated hypertensive (*SHR*); *DWKY*, diltiazem-treated normotensive (*WKY*); *VR*, vascular resistance; *myocardial*, left ventricular. Data are means ± SE. *, $P<0.05$

Discussion

The systemic and regional hemodynamic profile we obtained in conscious, untreated SHR and WKY in the resting state largely confirmes our own [4, 7] and others' [2] previous observations. Six weeks' treatment with diltiazem led to a mild-to-moderate hypotensive effect (15 mmHg or 10.1% decrease in MAP) in SHR alone. This was brought about mainly by an decrease in TPRI associated with no change in CI and HR. These changes in SHR are the same as those seen in hypertensive patients [12]. None of the variables were altered in the WKY with diltiazem treatment. The observed moderate blood pressure reduction is similar, but the associated systemic hemodynamic changes are somewhat dissimilar, to experimental data of Natsume et al. [14]. The slightly but significantly

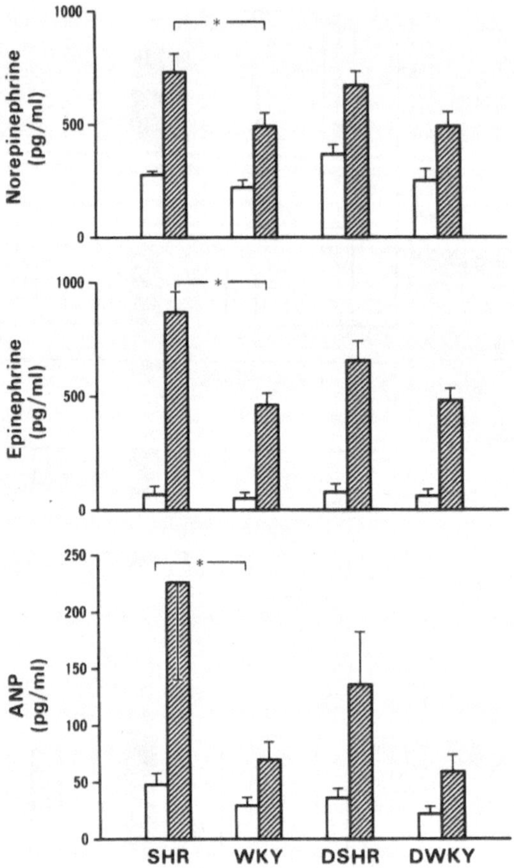

Fig. 7. Plasma levels of catecholamines and atrial natriuretic peptide (*ANP*). *DSHR*, diltiazem-treated hypertensive (*SHR*); *DWKY*, diltiazem-treated normotensive (*WKY*). Number of rats: SHR, $n=9$; DSHR, $n=10$; WKY, $n=9$; DWKY, $n=9$. *Open bars* represent resting data, and *hatched bars* represent data during stress. Data are means ± SE. *, $P<0.05$

reduced LV weight associated with the insignificantly reduced LV/BW ratio seen in the diltiazem-treated SHR may reflect a partial regression of LV hypertrophy resulting mainly from an antihypertensive effect of diltiazem. Tubau et al. observed a significant decrease in the LV weight as well as the heart/BW ratio after 6 months of diltiazem treatment [15], while Natsume et al. found no decrease in the LV/BW ratio following 3 weeks of diltiazem treatment [14].

In the present study, the influence of chronic diltiazem treatment on resting regional blood flow was not remarkable in both SHR and WKY. With diltiazem treatment, in parallel with a decrease in TPRI, resting regional vascular resistance was, on average, decreased in most vascular beds studied and was decreased significantly in the skeletal muscle and renal beds in the SHR. There were no such changes in the WKY. Earlier studies demonstrated that short-term administration of diltiazem into normal rats produced a dramatic increase in systemic regional blood flow to the coronary and several other vascular beds [25]. Such selected blood flow changes were not obtained here. The largely unaltered CI and regional blood flow together with the balanced decrease in overall and regional vascular resistance seen in our study may be explained by adaptation and adjustment of baroreflex and other homeostatic mechanisms to a new steady state occurring during long-term treatment [8]. The initially appreciable diuretic effects of diltiazem [13] may not persist by similar mechanisms. Our finding is at some variance with the data of Natsume et al. showing reduced renal blood flow and augmented coronary flow in SHR treated with diltiazem for 3 weeks [14]. The discrepancy seen in systemic and regional hemodynamics between their and our results may relate partly to differences in treatment periods and experimental designs.

In general, shaker stress increased all MAP, HR, CI, and TPRI, with a redistribution of blood flow from splanchnic, renal, testicular, and cutaneous vasculatures to cerebral, myocardial, and skeletal muscle vasculatures, a response seen typically in the denfense reaction [3]. The present result of more or less larger responses in MAP, HR, and several regional flow and vascular resistances in SHR compared with WKY confirms our prior work [4, 7].

Diltiazem treatment decreased MAP and TPRI during stress in SHR. Diltiazem treatment also decreased splanchnic and renal vascular resistances during stress in SHR. In contrast, no effects were noted in WKY. Therefore, it seems that calcium-related mechanisms play a more important role in increased vascular tone not only at rest but also during stress in SHR, compared with WKY. On the other hand, when percentage changes from baseline were compared, these effects of diltiazem were not significant in SHR, though there was a subtle tendency toward a reduction. Similar observations were made when hypertensive patients receiving short- or long-term therapy with calcium blockers were subjected to acute stress [26–28]. In these patients, the peak blood pressure during stress was lowered, but absolute or percentage increments of blood pressure from baseline usually were not much influenced.

The result of unaltered resting plasma catecholamine levels and enhanced plasma catecholamine levels during stress in untreated SHR in comparison with untreated WKY confirms previous data including our own [2–4, 6, 7]. The lack of increased catecholamines observed in diltiazem-treated SHR is consistent with the finding in patients receiving diltiazem for 4 weeks [12]. This may not be a unique feature of diltiazem. Dihydropyridine calcium antagonists also tend to produce a less pronounced increase in plasma catecholamine levels with continuation of therapy [8].

Our observation of increased circulating ANP levels in 16-week-old SHR compared with WKY supports several earlier reports [17–20]; however, a contradictory report showing unchanged ANP levels has also been presented [22]. In general, enhanced ANP levels seem to become evident as a function of age in SHR. The increased ANP level is viewed as a manifestation of the excessive cardiac overload and associated LV hypertrophy that develop with the progression of hypertension. In our study, diltiazem treatment tended to decrease resting plasma ANP levels. Recent studies demonstrated that long-term treatment with nisoldipine virtually normalized hypertension, LV hypertrophy, and plasma ANP levels, all in SHR [29]. In this regard, diltiazem in the dosage used here seems to exert a much milder effect. Plasma ANP concentrations increased immediately with short-term stress, as did plasma catecholamines. The seemingly greater, although not significant, values for plasma ANP during stress in the SHR in comparison with WKY were observed here. In a recent study, a similar quick release of ANP was found in both SHR and WKY following acute volume loading; however, the relationship between right atrial pressure and plasma ANP levels did not differ between the two strains [21]. We do not have data concerning left and right atrial pressure levels, major determinants of ANP secretion [24, 30], during shaker stress. It is possible that SHR has a higher atrial pressure during stress. The increased circulating ANP may have counteracted stress-associated vasoconstrictor and hence pressor responses in both strains. Obviously, this counteraction was not enough to prevent the responses. The extent to which ANP played a counteractive role is unclear. Chronic diltiazem treatment here did not significantly alter the stressed values of plasma ANP in either strain.

References

1. Okamoto K, Aoki K (1963) Development of a strain of spontaneously hypertensive rats. Jpn Circ J 27: 282–293
2. Trippodo NC, Frohlich ED (1981) Similarities of genetic (spontaneous) hypertension: man and rat. Circ Res 48: 309–319
3. Folkow B (1982) Physiological aspects of primary hypertension. Physiol Rev 62: 347–504
4. Yamamoto J, Nakai M, Natsume M (1987) Cardiovascular responses to acute stress in young-to-old spontaneously hypertensive rats. Hypertension 9: 362–370
5. Hallbäck M, Folkow B (1974) Cardiovascular responses to acute mental stress in spontaneously hypertensive rats. Acta Physiol Scand 90: 684–698
6. McCarty R (1983) Stress, behavior and experimental hypertension. Neurosci Biobehav Rev 7: 493–502
7. Yamamoto J, Akabane S, Yoshimi H, Nakai M, Ikeda M (1985) Effects of taurine on stress-evoked hemodynamic and plasma catecholamine changes in spontaneously hypertensive rats. Hypertension 7: 913–922
8. Bühler FR, Kiowski W (1987) Calcium antagonists in hypertension. J Hypertension 5 (Suppl 3): S3–S10
9. Aoki K (1986) Calcium membrane theory of essential hypertension. In: Aoki K (ed) Essential hypertension. Springer, Tokyo Berlin Heidelberg New York, pp 223–242
10. Ohya Y, Terada K, Satoh S, Fujiwara T, Nagao T, Komori K, Nozaki M, Kuriyama H (1986) Actions of calcium antagonist on smooth muscle cells of vascular tissues: current knowledge on actions of Ca antagonist. In: Aoki K (ed) Essential hypertension. Springer, Tokyo Berlin Heidelberg New York, pp 81–94
11. Yamakado T, Oonishi N, Kondo S, Noziri A, Nakano T, Takezawa H (1983) Effects of diltiazem on cardiovascular responses during exercise in systemic hypertension and comparison with propranolol. Am J Cardiol 52: 1023–1027
12. Amodeo C, Kobrin I, Ventura HO, Messerli FH, Frohlich ED (1986) Immediate and short-term hemodynamic effects of diltiazem in patients with hypertension. Circulation 73: 108–113

13. Narita H, Nagao T, Yabana H, Yamaguchi I (1983) Hypotensive and diuretic actions of diltiazem in spontaneously hypertensive and Wistar Kyoto rats. J Pharmacol Exp Ther 227: 472–477
14. Natsume T, Gallo AJ, Pegram BL, Frohlich ED (1985) Hemodynamic effects of prolonged treatment with diltiazem in conscious normotensive and spontaneously hypertensive rats. Clin Exp Hypertens A7(10): 1471–1479
15. Tubau JF, Wikman-Coffelt J, Massie BM, Szlachcic J, Parmley WW, Sievers R, Henderson S (1987) Diltiazem prevents hypertrophy progression, preserves systolic function, and normalises myocardial oxygen utilization in the spontaneously hypertensive rats. Cardiovasc Res 21: 606–614
16. De Bold AJ (1987) On the shoulders of giants: the discovery of atrial natriuretic factor. Can J Physiol Pharmacol 65: 2007–2012
17. Imada T, Takayanagi R, Inagami T (1985) Changes in the content of atrial natriuretic factor with the progression of hypertension in spontaneously hypertensive rats. Biochem Biophys Res Commun 133: 759–795
18. Morii N, Nakao K, Kihara M, Sugawara A, Sakamoto M, Yamori Y, Imura H (1986) Decreased content in left atrium and increased plasma concentration of atrial natriuretic polypeptide in spontaneously hypertensive rats (SHR) and SHR stroke-prone. Biochem Biophys Res Commun 135: 74–81
19. Gutkowska J, Horky K, Lachance C, Racz K, Garcia R, Thibault G, Kuchel O, Genest J, Cantin M (1986) Atrial natriuretic factor in spontaneously hypertensive rats. Hypertension 8 (Suppl I): I–137–I–140
20. Kato J, Kida O, Nakamura S, Sasaki A, Kodama K, Tanaka K (1987) Atrial natriuretic polypeptide (ANP) in the development of apontaneously hypertensive rats (SHR) and stroke-prone SHR (SHRSP). Biochem Biophys Res Commun 143: 316–322
21. Hoky K, Gutkowska J, Garcia R, Thibault G, Genest J, Cantin M (1985) Effect of different anesthetics on immunoreactive atrial natriuretic factor concentration in rat plasma. Biochem Biophys Res Commun 129: 651–657
22. Haass M, Zamir N, Zukowska-Groject Z (1987) Circulating atrial natriuretic peptides in spontaneously hypertensive rats: altered secretion in early hypertension. J Cardiovasc Pharmacol 10 (Suppl 5): S28–S33
23. Yamamoto J, Yamane Y, Umeda Y, Yoshioka T, Nakai M, Ikeda M (1984) Cardiovascular hemodynamics and vasopressin blockade in DOCA-salt hypertensive rats. Hypertension 6: 397–407
24. Matsubara H, Nishikawa M, Umeda Y, Taniguchi T, Iwasaka T, Kurimoto T, Yamane Y, Inada M (1987) The role of atrial pressure in secreting natriuretic polypeptides. Am Heart J 113: 1457–1463
25. Flaim SF, Annibali JA, Newman ED, Zelis R (1982) Effect of diltiazem on the cardiocirculatory response to exercise in the conscious rat. J Pharmacol Exp Ther 223: 624–630
26. Harris L, Dargie HJ, Glynch PG, Bulpitt CJ, Krikler DM (1982) Blood pressure and heart rate in patients with ischaemic heart disease receiving nifedipine and propranolol. Br Med J 284: 1148–1151
27. Midtbo K, Hals O (1986) Influence of verapamil and nifedipine on the pressor response to isometric exertion in hypertensive patients. Curr Ther Res 40: 326–332
28. Schmieder RE, Rueddel H, Neus H, Messerli FH, Von Eiff AW (1987) Disparate hemodynamic responses to mental challenge after antihypertensive therapy with beta-blockers and calcium entry blockers. Am J Med 82: 11–16
29. Stasch JP, Kazda S, Hirth C, Morich F (1987) Role of nisoldipine on blood pressure, cardiac hypertrophy, and atrial natriuretic peptides in spontaneously hypertensive rats. Hypertension 10: 303–307
30. Haass M, Zukowska-Groject Z, Kopin IJ, Zamir N (1987) Role of autonomic nervous system and vasoactive hormones in the release of atrial natriuretic peptides in conscious rats. J Cardiovasc Pharmacol 10: 424–432

Mechanism and Consequences of Cellular Calcium Elevation in Hypertension

Sophie Koutouzov, Maryvonne Baudouin-Legros, Sylvie Durant, Jean-Luc Paquet, Pierre Marche, and Philippe Meyer[1]

Summary. The physiological responses displayed by platelets and arterial vascular smooth muscle cells of spontaneously hypertensive rats (SHR) are markedly enhanced compared to those of normotensive controls (WKY). These cellular responses, aggregation and release of serotonin for platelets and growth for smooth muscle cells are mainly controlled by an elevation of intracellular Ca^{2+} which can occur by Ca^{2+} influx, Ca^{2+} mobilization or both. Since phosphoinositide metabolism is involved in calcium-mediated cellular responses, the metabolism of these lipids was investigated in platelets and cultured vascular smooth muscle cells. Our results show that the early thrombin-induced phosphoinositide metabolism, when monitored as changes in ^{32}P-phosphatidic acid, was significantly higher in SHR than in WKY. Furthermore, under stimulation with angiotensin II or NaF-$AlCl_3$, the release of ^{3}H-inositol phosphates was enhanced in the smooth muscle cells of the hypertensive strain. Both results clearly indicate that phospholipase C activity is increased in platelets and vascular smooth muscle cells of SHR. This hypersensitivity of phospholipase C suggests that genetic alterations in phosphoinositide metabolism may play a role in the increased peripheral arteriolar tone and vascular resistance which characterize hypertension and may be important in the etiology of essential hypertension.

Key words: Phospholipase C—Hypertension—Platelet—Smooth muscle

Introduction

Elevation of free intracellular Ca^{2+} is the triggering factor of hypertension, as it results in increased vascular tone and contractility. Various observations, summarized in the present chapter, favor the concept that in primary hypertension the increase in ionized cytosolic calcium arises largely from translocation from intracellular storage sites. In addition, the molecular mechanisms responsible for this phenomenon, which involve the membrane phosphoinositide metabolism, may also contribute to vascular hyperplasia, i.e., to the chronic phase of hypertension. This chapter summarizes our contribution in this respect both in animals and in humans.

[1] Department of Pharmacology, U7 INSERM/UA 318 CNRS, Hôpital Necker, 161 rue de Sèvres, 75015 Paris, France

Spontaneously Hypertensive Rats

Experiments were performed on platelets and on cultured smooth muscle cells (SMC) in spontaneously hypertensive rats (Aoki SHR).

Platelets

Thrombin-Induced Physiological Responses

1. Aggregation. SHR platelets were compared to Wistar-Kyoto rat (WKY) control platelets. When platelet suspensions from both strains were allowed to aggregate, the velocity of aggregation was higher in SHR than in WKY (91 ± 3 vs 61 ± 6 mm/min; $n = 8$, $P < 0.01$, with thrombin 0.3 U/ml).

2. Serotonin release. The time course of [^{3}H]serotonin release by platelets stimulated with thrombin (0.3 U/ml) showed an increase within the first 30–60 s and a plateau thereafter. At all times studied, from 10 up to 2 min, SHR platelets exhibited release values about 2.5 times higher than the respective values produced by WKY platelets. When studied as a function of the dose of thrombin, the [^{3}H]serotonin release (after 20 s of stimulation) was barely detectable for thrombin concentrations below 0.1 U/ml, but it increased in a dose-dependent manner between 0.2 and 0.5 U/ml and reached a maximum of 70%–80% at thrombin 1 U/ml. When stimulated by thrombin concentrations of 0.2–0.5 U/ml, SHR platelets released significantly more [^{3}H]serotonin than did WKY platelets; the difference was no longer significant at thrombin 1 U/ml.

^{32}P-Labeling of Lipids in Unstimulated Platelets: Kinetics and Distribution

Under the experimental conditions used, five major lipids incorporated the ^{32}P-label: PIP_2, PIP, PI, PA, and an as yet unidentified inositol-containing glycerophospholipid designated X.

Thrombin-Induced ^{32}P-Labeled Lipid Changes

The comparison between SHR and WKY of ^{32}P-labeled phosphoinositide changes brought about by exposure to thrombin (0.3 U/ml) for various times is made as follows. A significant loss (10%–15%) of the radioactivity associated with PIP_2 was constantly observed within 20–30 s after stimulation and was followed by an increased incorporation of the label at longer times. There was no significant difference between SHR and WKY in the extent of the [^{32}P]PIP_2 breakdown. At 20 s [^{32}P]PIP_2 represented $88\% \pm 3\%$ and $88\% \pm 4\%$ ($n = 9$) of values in unstimulated samples for SHR and WKY, respectively. By contrast, after the first 30 s of stimulation, the resynthesis of [^{32}P]PIP_2 was significantly accelerated for SHR. At 60 s, [^{32}P]PIP_2 represented $115\% \pm 5\%$ and $98\% \pm 3\%$ ($n = 4$, $P < 0.05$) of control values (obtained in the absence of thrombin) for SHR and WKY, respectively. Within the first 60 s of thrombin stimulation, [^{32}P]PIP increased regularly, and from 20 to 60 s this increase was significantly

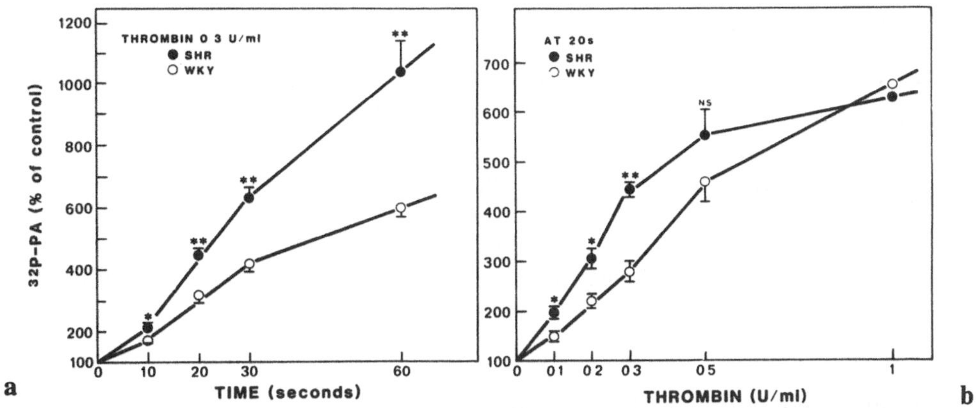

Fig. 1a,b. Kinetic (**a**) and dose-response (**b**) studies of thrombin-induced phosphatidic acid (*PA*) formation in platelets of SHR and WKY. ^{32}P-labeled washed platelet suspensions from SHR and WKY were stimulated with thrombin (0.3 U/ml) or buffer (control) for the designated times (**a**) or with various doses of thrombin for 20 s (**b**). Results were expressed as the percentage of [^{32}P] PA in stimulated samples compared with that in controls. *Single* ($P<0.01$) and *double* ($P<0.001$) *asterisks* indicate a significant difference compared with values for WKY

higher in SHR than in WKY. At 30 and 60 s, [^{32}P]PIP values represented, respectively, 132% ± 2% and 143% ± 3% of control values for SHR vs 115% ± 2% and 128% ± 3% of control values of WKY ($n=3$, $P<0.01$). During the first minute of activation, neither [^{32}P]PI nor [^{32}P]X varies significantly from the respective basal values (without thrombin). In contrast with the slight variations in [^{32}P]PIP$_2$ (expressed in terms of both ^{32}P-labeled lipid changes and comparison of SHR vs WKY), there was considerable variation in the early thrombin-induced increase of [^{32}P]PA between the two strains.

The kinetics of [^{32}P]PA formation obtained with SHR and WKY when platelets were stimulated with thrombin (0.3 U/ml) are represented in Fig. 1 which shows that [^{32}P]PA formation increased regularly within 60 s. The data clearly indicate that at all the times studied (from 10 to 60 s), the increase in [^{32}P]PA was significantly higher in SHR platelets than in WKY platelets. Thus, at 20 s, [^{32}P]PA values for SHR platelets were about 1.5 times higher than those for WKY platelets. [^{32}P]PA formation was further studied as a function of thrombin concentration after 20 s of stimulation. When expressed as a percentage of control value (i.e., the [^{32}P]PA value obtained in unstimulated platelets), [^{32}P]PA increased in a dose-dependent manner up to 0.5 U/ml and reached a plateau thereafter to attain a maximum of about 600% at 1 U/ml. Comparison of the dose-response curves obtained with platelets of SHR and WKY reveals that at concentrations up to 0.5 U/ml the thrombin-induced [^{32}P]PA increase was significantly higher (about 1.5 times) in SHR than in WKY. At 0.5 U/ml, although [^{32}P]PA values were higher in SHR than in WKY, the difference was not statistically significant. The [^{32}P]PA increases obtained after platelet stimulation with thrombin (1 U/ml) were similar in SHR and WKY.

Smooth Muscle Cells (SMC)

SHR aortic SMC were cultured on RPMI 640 in the presence of fetal calf serum. Their growth was estimated by microscopical counting and [^{3}H]thymidine uptake measurements. Phospholipase C activity was determined by inositide phosphate measurements. Prelabeling with [^{3}H]myoinositol was performed over 72 h. Separation of the compounds was done with the use of AGX-8 resin.

Growth of SHR aortic SMC is accelerated 0.5- to two fold as compared with WKY controls. The responsiveness to angiotensin II and NaF-$AlCl_3$ after 30–60 s was increased two- to three-fold in SHR cells as compared with WKY cells.

Human Essential Hypertension

Subjects and Methods

Seventeen untreated patients from 30 to 67 years old (mean 47 ± 3 years) 10 women and 7 men with moderate essential hypertension, were studied. Their systolic and diastolic supine blood pressures were 165 ± 5 and 99 ± 3 mmHg, respectively. These patients had normal renal function and no cardiac insufficiency. They were compared with a group of 17 sex- and age-matched normotensive healthy volunteers (11 women, 6 men) from 26 to 55 years old (mean 41 ± 3 years), whose systolic and diastolic blood pressures were 119 ± 4 and 81 ± 3 mmHg, respectively. None of the patients or controls were on medication. All had a normal diet providing a mean NaCl intake of 100 nmol/day. The subjects did not fast before blood sampling.

After blood pressure was measured, platelet-rich plasma (PRP) was obtained by 10 min of centrifugation (100 *g*, 20°C). For ^{32}P-labeling of platelets, PRP ($3–6 \times 10^8$ platelets per ml) was incubated for 90 min at 37°C with carrier-free [^{32}P]orthophosphate (10 μCi per 10^8 cells); aspirin (0.5 m*M* final) was included in the medium during the last 5 min of labeling. Then PRP was layered onto a metrizamide gradient and centrifuged for 15 min at 1000 *g*. The metrizamide gradient platelets thus obtained were washed by repeating the procedure on a similar metrizamide graident. The final MGP suspension was adjusted to 5×10^8 cells/ml in buffer containing: NaCl 140 m*M*, KCl 3 m*M*, KCl 3 m*M*, $NaHCO_3$ 12 m*M*, glucose 10 m*M*, $MgCl_2$ 0.5 m*M*, phosphocreatine 2 m*M*, creatine phosphokinase 20 U/ml, pH 7.4, 300 mOsm.

Samples (0.4 ml) of a suspension of ^{32}P-labeled washed platelets was transferred to the aggregometer cuvette, stirred for 2 min at 37°C after addition of $CaCl_2$ (1.3 m*M* final) and then exposed to buffer (for unstimulated control platelets) or thrombin for 20 s. Labelled phospholipids were separated by thin-layer chromatography and analysed for radioactivity or for their phosphorus content, as already detailed.

Results

Unstimulated Platelets

No significant difference could be found between hypertensive and normotensive individuals concerning the amount and distribution of the various major phospholipids in washed platelets.

Stimulated Platelets

In order to study the biochemical event that results from the interaction between the agoinst and its receptor, [^{32}P]lipid variations were measured 20 s after thrombin addition. Under these conditions, the low extent of thrombin-induced changes in ^{32}P-labeled PI-P_2, PI-P, and PI was such that the values obtained in hypertensives did not differ significantly from those of normotensives. In contrast, the determination of thrombin-induced [^{32}P]PA changes, expressed as a function of the dose of thrombin, indicated that the rise in [^{32}P]PA was higher in platelets isolated from hypertensive patients than in normotensives. The difference between the two groups of subjects—about 30% at the lowest doses of thrombin (25 and 50 mU/ml)—was no more significant at doses of 50 and 75 mU/ml. There is an overlap between the two groups' studies. However, at the lowest dose of thrombin (25 mU/ml), 10 patients out of 17 (~60%), compared with 4 out of 17 (~25%) in the normotensive group, displayed a [^{32}P]PA increase greater than 100%. No significant correlation could be found between blood pressure (either systolic or diastolic) and the thrombin (25 mU/ml)-induced [^{32}P]PA enhancement.

The reproducibility of the thrombin-induced [^{32}P]PA formation was checked by measuring this parameter several times on the same subject, at time intervals varying from 1 to 3 months. The results obtained from five subjects (three hypertensive and two normotensive) indicated that thrombin (25 mU/ml)-induced [^{32}P]PA formation was reproducible at $\pm 9\%$.

Five patients had increased level(s) of either cholesterol or triglycerides, or both. Thrombin-induced [^{32}P]PA formation did not appear to be influenced by these lipid changes.

Discussion

The turnover of phosphoinositides appears to be impaired and the activity of phosphoinositide-specific phospholipase C, to be enhanced in various tissues of SHR compared with those of normotensive control WKY. These membrane changes have been reported by other laboratories and ourselves in various tissues; red blood cells, renal cotext, cultured vascular SMC, arterial wall, and circulating platelets [1–6].

The products of phospholipase C reaction may participate in the pathogenesis of hypertension. Inositol trisphosphate (IP_3) on the one hand is responsible for intracellular Ca^{2+} translocation from storage sites, and diacylglycerol (DG) on the other induces cell activation and promotes cell growth, differentiation, and division. It is thus possible that phospholipase C products are responsible for the arterial cell contraction and arterial cell hyperplasia which constitute the major changes leading to the increase in blood pressure. The interest of these findings is further increased by the fact that several steps of phosphoinositide metabolism are expressed by oncogenes which also govern mitogenic reactions, providing an interesting similarity between hypertension and cancer [7].

However, phospholipase C hyperreactivity occurs in approximately half of the investigated patients and the above-described membrane alterations also affect only a limited number of patients, suggesting the presence of other pathogenic mechanisms and confirming the heterogeneity of human essential hypertension. An enhanced platelet cAMP response to PGE_1 in essential hypertension has been reported. One may thus conceive that the hypersensitivity of the two transducing systems, which through adenylate cyclase and phospholipase C control the cell reactivity, is related to a dysfunction of a regulatory G-protein, which deserve further investigation.

References

1. Remmal A, Koutouzov S, Girard A, Meyer P, Marche P (1988) Defective phosphoinositide metabolism in primary hypertension. Experientia 44: 133–137
2. Koutouzov S, Remmal A, Marche P, Meyer P (1987) Hypersensitivity of phospholipase C in platelets of spontaneously hypertensive rats. Hypertension 10: 497–504
3. Marche P, Koutouzov S, Girard A, Barbier P, Meyer P (to be published) Hyperresponsiveness of platelet phospholipase C in essential hypertension. J Vasc Med Biol
4. Uehara Y, Ishii M, Ishimitsu T, Sugimoto T (1988) Enhanced phospholipase C activity in the vascular wall of spontaneously hypertensive rats. Hypertension 11: 28–33
5. Kato H, Takenawa T (1987) Phospholipase C activation and diacylglycerol kinase inactivation lead to an increase in diacylglycerol content in spontaneously hypertensive rat. Biochem Biophys Res Commun 146: 1419–1424
6. Paquet JL, Baudouin-Legros M, Marche P, Meyer P (to be published) Enhanced proliferating activity of cultured smooth muscle cells from SHR. Am J Hypertens
7. Meyer P (1987) Increased intracellular calcium: from hypertension to cancer. J Hypertens (Suppl IV) 5: S3–S4

IV Calcium, Magnesium, and Calcium Antagonists in Human Hypertension

Pressor Effects of Calcium Infusion in the Absence and Presence of a Calcium Antagonist in Subjects with Normotension and Hypertension

Takahiro Suzuki, Kyuzo Aoki, and Koichi Miyagawa[1]

Summary. The effects of intravenous infusion of calcium at 7.5 mg/kg per hour for 1 h were investigated in normotensives and hypertensives, ages 43–83 years, in the absence and presence of the calcium antagonist, verapamil. The infused calcium elevated blood pressure from 125 ± 14/62 ± 6 to 143 ±13/69 ± 7 mmHg (mean ± SD) and from 173 ± 15/81 ± 6 to 202 ± 23/92 ± 9 mmHg with the increase in serum concentration of calcium from 8.9 ± 0.4 to 12.9 ± 0.6 mg/dl and from 8.9 ± 0.5 to 12.6 ± 0.6 mg/dl in normotensives ($n = 20$) and hypertensives ($n = 12$), respectively. In the hemodynamic studies, the calcium infusion significantly raised blood pressure and total peripheral vascular resistance and decreased heart rate with no change in cardiac output, in normotensives ($n = 8$) and hypertensives ($n = 8$). The absolute elevation in both blood pressure and vascular resistance was significantly greater in hypertensives than in normotensives, but the percentage elevation was not significantly different in the two groups. The infusion of verapamil at a rate of 0.15 mg/kg per hour inhibited the elevation in both blood pressure and vascular resistance during the calcium infusion in normotensives ($n = 6$) and hypertensives ($n = 6$).

The rise in blood pressure during calcium infusion was caused by elevated vascular resistance, which was inhibited by the calcium antagonist. These results suggest that infused calcium may affect arterial smooth muscle, inducing contraction of vascular smooth muscle, leading to the elevation of vascular resistance, and producing a rise in blood pressure. In conclusion, the calcium infusion raised blood pressure due to elevated vascular resistance, which was inhibited by a calcium antagonist in normotensives and hypertensives.

Key words: Calcium infusion—Serum concentration of calcium—Blood pressure—Calcium antagonist

Introduction

Recent reports of hypertension mechanisms have concentrated on the causal relationship between calcium and hypertension [1–20]. In an epidemiological survey on hypertension, a positive correlation was found between serum calcium concentration and blood pressure [8]. In contrast, another study reported a lower concentration of serum ionized calcium in hypertensives than in normotensives [9]. Moreover, it was reported that oral calcium supplementation represents a

[1] Second Department of Internal Medicine, Nagoya City University Medical School, Mizuho-ku, Nagoya 467, Japan

nonpharmacologic intervention that lowers blood pressure in patients with mild to moderate essential hypertension [11]. However, no clinically impressive effect of oral calcium supplementation has been reported in mild essential hypertensives [13–16], and calcium supplementation does raise the blood pressure in some hypertensive patients [12]. Thus, the studies of the effect of oral calcium supplementation on blood pressure have yielded conflicting results in both normotensive and hypertensive subjects. Therefore, more detailed investigations are necessary on the role of calcium in blood pressure regulation, and to determine whether calcium has a hypotensive action or a hypertensive action in humans.

It has been hypothesized that intravenously infused calcium enters the arterial wall through the intima of blood vessels and increases the calcium concentration in the cytosol of arterial smooth muscle, possibly inducing arterial vasoconstriction causing hypertension. In addition, calcium antagonists inhibit calcium influx through the cell membrane and calcium release from the sarcoplasmic reticulum and cell membrane [1, 2], which may cause a decrease in concentration of calcium in the cytosol and inhibit the rise in blood pressure during calcium infusion. To demonstrate the effect of hypercalcemia on vasoconstriction in normotensive and hypertensive subjects, we investigated the hemodynamic responses to intravenously infused calcium in the absence and presence of a calcium antagonist.

Methods

Subjects

Thirty-two subjects including 20 normotensives and 12 hypertensives, aged 43–83 years, were studied (Table 1). No subjects had evidence of myocardial infarction, heart failure, atrial fibrillation, renal disease or endocrine disorder. All subjects either had no previous treatment or were withdrawn from antihyperten-

Table 1. Clinical findings of subjects with normotension and hypertension [20]

	Normotension (n = 20)	Hypertension (n = 12)
Male/Female	10/10	7/5
Age (yr)	66 ± 11	63 ± 11
Height (cm)	149 ± 7	150 ± 5
Weight (kg)	47 ± 8	51 ± 8
Blood pressure (mmHg)	125 ± 14/62 ± 6	173 ± 15*/81 ± 6*
Heart rate (beats/min)	67 ± 10	67 ± 7
Serum creatinine (mg/dl)	1.2 ± 0.2	1.2 ± 0.1
Serum total protein (g/dl)	7.2 ± 0.6	7.3 ± 0.7
Serum albumin (g/dl)	4.2 ± 0.5	4.2 ± 0.4

Values are mean ± SD
* $P < 0.01$, compared with values in subjects with normotension

sive drugs at least 4 weeks prior to the study. They were hospitalized and placed on a diet containing 140 mmol sodium (8g NaCl) and 15 mmol (600 mg) calcium daily. Their serum calcium concentration was in the normal range between 8.4 and 10.4 mg/dl. Hypertension was defined by multiple measurements (by mercury sphygmomanometer) of systolic blood pressure over 150 mmHg or diastolic blood pressure over 90 mmHg after 5 min in the supine position in the hospital.

Study Protocol

The subjects rested in the supine position in a quiet room. A polyethylene catheter was inserted into antecubital vein for the intravenous infusion of test substances, and a 5% dextrose solution was continuously infused at the rate of 50 ml/h for 1 h of equilibration. A second catheter was introduced into right brachial artery, for direct monitoring of intra-arterial pressure and obtaining blood samples, and a heparin solution (two units of aqueous heparin per 1 ml of 0.9% normal saline) was constantly infused at the rate of 3 ml/h.

One normotensive subject, a 49-year-old male had infusions of 8.5% calcium gluconate solution at rates of 3.75, 7.5, and 15.0 mg calcium/kg per hour for 1 h. The infusions elevated blood pressure dose-dependently (Fig. 1). Thus, in the following studies, a calcium infusion rate of 7.5 mg/kg/h was used. The calcium

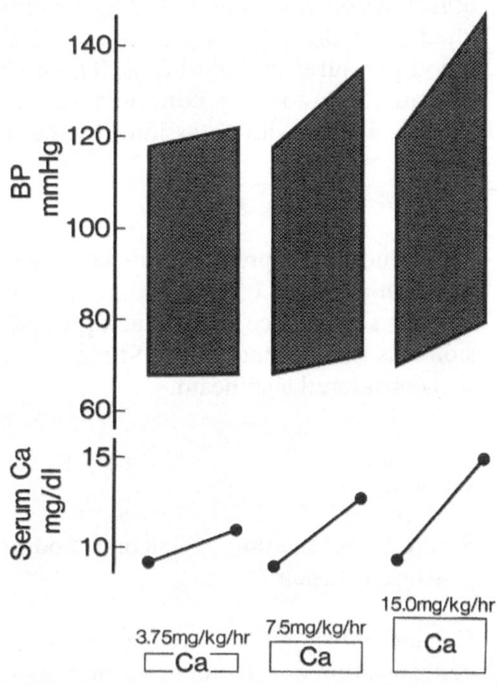

Fig. 1. The dose-dependent pressor effects of calcium (*Ca*) infusion. Blood pressure (*BP*) and serum calcium concentration (*Serum Ca*) before and after infusion of calcium at 3.75, 7.5, and 15.0 mg/kg per hour, each for 1 h. The rise in BP is accompanied by a parallel increase in serum calcium concentration

infusion was continued for 1 h, and blood pressure and heart rate were measured and blood samples were collected before and at the end of calcium infusion in 20 normotensives and 12 hypertensives. For the hemodynamic studies, cardiac output as well as blood pressure and heart rate were determined in 8 normotensives and 8 hypertensives. The hemodynamic determinations were performed before and at the end of the infusion of calcium, or calcium with the calcium antagonist verapamil (verapamil hydrochloride 0.2 mg/ml in 5% dextrose solution) at the rate of 0.15 mg/kg per hour in 6 normotensives and 6 hypertensives. A constant infusion pump was used for the infusions of calcium and verapamil (Terumo syringe pump, STC-521).

Measurement

Blood pressure was measured by direct and continuous recording of intra–arterial pressure, using a pressure transducer (Statham P23ID) and polygraph system (Nihon Kohden RM-6000); measurements from 10 consecutive pulsations were averaged. Heart rate was measured by a continuously monitored electrocardiogram. Cardiac output was measured in duplicate by the dye-dilusion technique, with a bolus injection of 10 mg of indocyanine green followed by 10 ml of 5% dextrose solution through the venous cathether, withdrawing blood through a cuvette densitometer from the brachial artery, using a cardiac computer (Erma optical works Dye Mac EW-90). Total peripheral vascular resistance was calculated from the following formula: total peripheral vascular resistance = mean blood pressure/cardiac output. The collected blood samples were used for the determination of the concentration of serum calcium (by OCPC method, Wako), sodium and potassium (by the flame photometric method).

Statistics

The values are expressed as mean ± SD. The statistical significance of differences between averages of both groups was determined by Student's un-paired *t*-test, and the significance within each group between values of before and after infusion was determined by the Student's paired *t*-test. A *P* value of less than 0.05 was considered significant.

Results

Serum Concentration of Calcium, Sodium and Potassium before and after Calcium Infusion

Calcium infusion of 7.5 mg/kg per hour for 1 h significantly increased the serum concentration of calcium in normotensives and hypertensives ($P < 0.01$), but did not alter the serum concentration of sodium or potassium in the two groups (Table 2).

Table 2. Serum concentration of calcium, sodium and potassium before and after calcium infusion [20]

Calcium infusion	Normotension (n = 20)		Hypertension (n = 12)	
	Before	After	Before	After
Serum calcium (mg/dl)	8.9 ± 0.4	12.9 ± 0.6*	8.9 ± 0.5	12.6 ± 0.6*
Serum sodium (mEq/l)	143 ± 5	144 ± 3	145 ± 3	146 ± 3
Serum potassium (mEq/l)	4.1 ± 0.4	4.1 ± 0.4	4.0 ± 0.4	4.1 ± 0.4

Values are mean ± SD
* $P < 0.01$, compared with values before calcium infusion

Dose-dependent Pressor Effects of Calcium Infusion

Calcium infusion at rates of 3.75, 7.5, and 15.0 mg/kg per hour for 1 h raised blood pressure from 118/68 to 122/68, from 118/68 to 136/72 from 120/70 148/80 mmHg with an increased serum concentration of calcium from 9.2 to 11.0, from 9.0 to 12.8 and from 9.4 to 15.0 mg/dl, respectively (Fig. 1).

Effects of Calcium Infusion of Blood Pressure and Heart Rate

Calcium infusion at the rate of 7.5 mg/kg per hour raised blood pressure from 125 ± 14/62 ±6 to 143 ± 13/69 ± 7 mmHg ($P<0.01$) in normotensives ($n=20$) and from 173 ± 15/81 ± 6 to 202 ± 23/92 ± 9 mmHg ($P<0.01$) in hypertensives ($n=12$). The elevation in blood pressure from baseline value was significantly higher in hypertensives (29 ± 10/11 ± 4 mmHg) than in normotensives (18 ± 5/7 ± 3 mmHg, $P<0.01/P<0.01$). The percentage rise in blood pressure was not significantly different between normotensives (15 ± 5/12 ± 5%) and hypertensives (17 ± 5/14 ± 4%). The infusion decreased heart rate from 67 ± 10 to 65 ± 10 ($P<0.01$) and from 67 ± 7 to 64 ± 7 beats/min ($P<0.01$) in normotensives and hypertensives, respectively. The absolute decrease and percentage decrease in heart rate did not differ significantly in the two groups.

Hemodynamic Effects of Calcium Infusion:

Calcium infusion significantly raised blood pressure ($P<0.01$) and total peripheral vascular resistance ($P<0.01$) and significantly decreased heart rate ($P<0.01$) with no change in cardiac output in normotensives and hypertensives (Table 3, Fig. 2). The elevations in systolic, diastolic and mean blood pressure were significantly greater in hypertensives than in normotensives ($P<0.01$). The changes in heart rate and cardiac output did not differ in the two groups. The elevation of vascular resistance was significantly higher in hypertensives than in normotensives ($P<0.05$) (Table 3, Fig. 3). The percentage changes in blood pressure, heart rate, cardiac output and vascular resistance showed to significant differences in normotensives and hypertensives (Table 3, Fig. 4).

Table 3. Hemodynamic effects of calcium infusion in subjects with normotension and hypertension [20]

Calcium infusion	Normotension (n = 8)			Hypertension (n = 8)		
	Before	After	Δ (%)	Before	After	Δ (%)
SBP (mmHg)	121 ± 13	140 ± 13*	20 ± 4 (16 ± 4)	174 ± 9	206 ± 13*	32 ± 8†† (18 ± 4)
DBP (mmHg)	62 ± 4	69 ± 5*	7 ± 3 (12 ± 5)	82 ± 6	94 ± 9*	13 ± 4†† (15 ± 4)
MBP (mmHg)	84 ± 6	95 ± 8*	12 ± 4 (14 ± 4)	114 ± 7	133 ± 11*	19 ± 4†† (17 ± 3)
HR (beats/min)	64 ± 6	61 ± 7*	−2 ± 1 (−4 ± 2)	66 ± 7	63 ± 7*	−2 ± 2 (−4 ± 3)
CO (l/min)	4.34 ± 0.73	4.27 ± 0.78	−0.07 ± 0.09 (−2 ± 2)	4.77 ± 0.56	4.65 ± 0.62	−0.12 ± 0.12 (−3 ± 3)
TPR (mmHg/l/min)	19.8 ± 3.3	22.9 ± 4.0*	3.1 ± 1.1 (16 ± 4)	24.3 ± 3.3	29.2 ± 4.6*	4.9 ± 1.6† (20 ± 5)

Values are mean ± SD
* $P < 0.01$, compared with values before calcium infusion
† $P < 0.05$, †† $P < 0.01$, compared with values in normotension
Δ = change, % = percentage of change; SBP, systolic blood pressure; DBP, diastolic blood pressure; MBP, mean blood pressure; HR, heart rate; CO, cardiac output; TPR, total peripheral vascular resistance

Fig. 2. Effects of calcium (*Ca*) infusion on blood pressure (*BP*), heart rate (*HR*), cardiac output (*CO*) and total peripheral vascular resistance (*TPR*) in normotensives ($n = 8$) and hypertensives ($n = 8$). Calcium infusion elevated BP and TPR and decreased HR with no change in CO in either groups. Values are mean ± SD. $^{**}P < 0.01$, compared to values before calcium infusion [20]

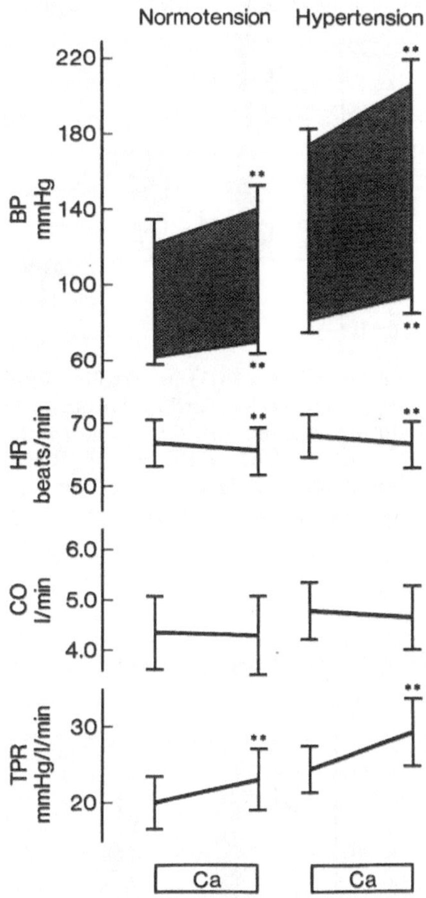

Hemodynamic Effects of Calcium Infusion in combination with Calcium Antagonist

The infusion of calcium raised blood pressure ($P < 0.01$) and total peripheral vascular resistance ($P < 0.01$), and reduced heart rate ($P < 0.01$) with no change in cardiac output in normotensives and hypertensives. The calcium antagonist verapamil significantly inhibited the rise in both blood pressure ($P < 0.01$) and vascular resistance ($P < 0.01$) during calcium infusion in both groups, but there was no significant decrease in cardiac output either in the absence or presence of verapamil (Fig. 5). The absolute rise and percentage rise of both blood pressure and vascular resistance induced by calcium infusion were significantly greater ($P < 0.01$) than those induced by calcium in combination with verapamil in normotensives and hypertensives, but the changes in heart rate and cardiac output were not greater. The absolute and percentage changes in blood pressure, heart rate, cardiac output, and vascular resistance induced by calcium with verapamil were not significantly different between normotensives and hypertensives.

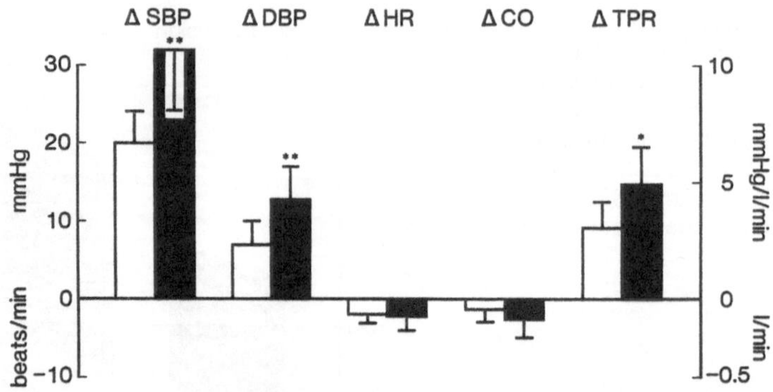

Fig. 3. Changes (*Δ*) in systolic blood pressure (*SBP*), diastolic blood pressure (*DBP*), heart rate (*HR*), cardiac output (*CO*), and total peripheral vascular resistance (*TPR*) during calcium infusion in normotensives (□: $n=8$) and hypertensives (■: $n=8$). Changes in SBP, DBP, and TPR were greater in hypertensives than in normotensives, but changes of HR and CO did not differ in the two groups. Values are mean ± SD. $^{*}P<0.05$, $^{**}P<0.01$, compared to values in normotensives

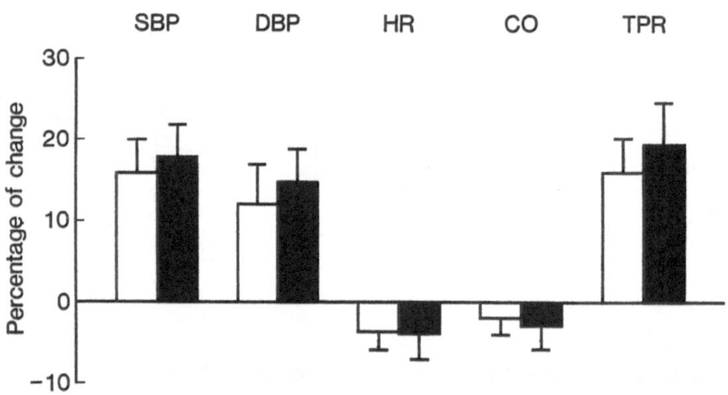

Fig. 4. Percentage change in systolic blood pressure (*SBP*), diastolic blood pressure (*DBP*), heart rate (*HR*), cardiac output (*CO*), and total peripheral vascular resistance (*TPR*) during calcium infusion in normotensives (□: $n = 8$) and hypertensives (■: $n = 8$). Percentage change in SBP, DBP, HR, CO, and TPR did not differ in the two groups. Values are mean ± SD

Serum calcium concentration was increased from 8.9 ± 0.4 to 13.0 ± 0.6 and from 8.9 ± 0.5 to 12.9 ± 0.3 mg/dl in normotensives and from 9.0 ± 0.4 to 12.9 ± 0.6 and from 8.8 ± 0.2 to 12.8 ± 0.6 mg/dl in hypertensives, by infusion of calcium and by infusion of calcium with verapamil, respectively. The increased in serum calcium concentration was similar in both infusions.

Table 4. Hemodynamic effects of calcium infusion in combination with calcium antagonist [20]

	Normotension (n = 6)			Hypertension (n = 6)		
	Before	After	Δ (%)	Before	After	Δ (%)
Calcium						
SBP (mmHg)	123 ± 11	143 ± 11**	19 ± 4 (16 ± 4)	177 ± 5	212 ± 6**	36 ± 2 (20 ± 1)
DBP (mmHg)	63 ± 5	71 ± 7**	8 ± 3 (12 ± 5)	83 ± 6	97 ± 9**	14 ± 4 (16 ± 4)
MBP (mmHg)	85 ± 6	97 ± 8**	12 ± 4 (13 ± 5)	117 ± 5	138 ± 8**	21 ± 3 (18 ± 1)
HR (beats/min)	64 ± 4	61 ± 3**	−3 ± 1 (−4 ± 1)	66 ± 7	63 ± 8**	−3 ± 1 (−5 ± 2)
CO (l/min)	4.25 ± 0.59	4.16 ± 0.60	−0.10 ± 0.07 (−2 ± 2)	4.64 ± 0.58	4.52 ± 0.66	−0.12 ± 0.13 (−3 ± 3)
TPR (mmHg/l/min)	20.4 ± 3.1	23.8 ± 3.6**	3.4 ± 1.0 (17 ± 4)	25.5 ± 2.6	31.0 ± 3.6**	5.5 ± 1.2 (22 ± 3)
Calcium + verapamil						
SBP (mmHg)	127 ± 9	132 ± 9*	5 ± 3†† (4 ± 2)††	175 ± 10	187 ± 15*	11 ± 7†† (6 ± 4)††
DBP (mmHg)	65 ± 6	67 ± 6*	2 ± 1†† (3 ± 2)††	82 ± 9	84 ± 11	2 ± 4†† (2 ± 6)††
MBP (mmHg)	87 ± 6	90 ± 6*	3 ± 1†† (3 ± 1)††	115 ± 8	120 ± 12	5 ± 5†† (4 ± 4)††
HR (beats/min)	65 ± 4	61 ± 4**	−4 ± 2 (−6 ± 3)	65 ± 7	62 ± 6**	−3 ± 1 (−5 ± 3)
CO (l/min)	4.26 ± 0.57	4.26 ± 0.48	−0.01 ± 0.17 (0 ± 4)	4.58 ± 0.66	4.57 ± 0.67	−0.01 ± 0.10 (0 ± 2)
TPR (mmHg/l/min)	20.9 ± 3.7	21.5 ± 3.3	0.6 ± 1.1†† (3 ± 5)††	25.4 ± 2.1	26.6 ± 2.3	1.2 ± 1.1†† (5 ± 4)††

Values are mean ± SD

* $P < 0.05$, ** $P < 0.01$, compared with values before calcium infusion

†† $P < 0.01$, compared with values in calcium infusion alone

Δ = change, % = percentage of change; SBP, systolic blood pressure; DBP, diastolic blood pressure; MBP, mean blood pressure; HR, heart rate; CO, cardiac output; TPR, total peripheral vascular resistance

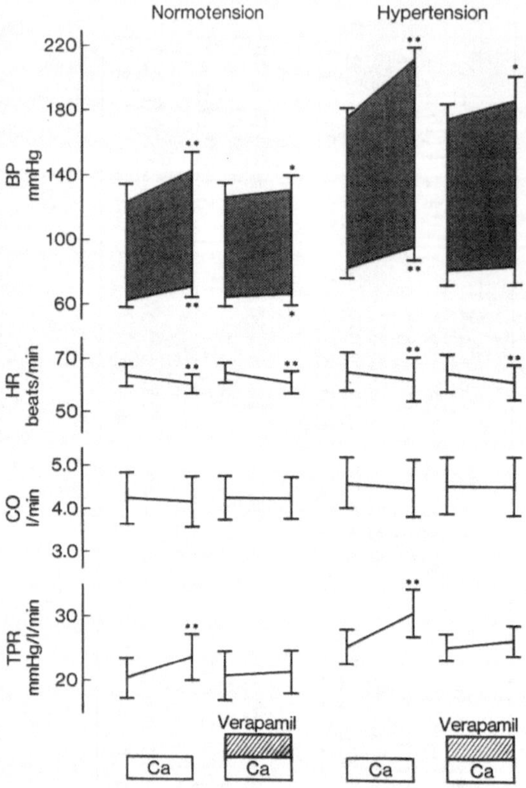

Fig. 5. Effects of calcium (*Ca*) infusion on hemodynamics in the absence and presence of calcium antagonist. Calcium antagonist, verapamil, inhibited the rise in blood pressure (*BP*) and total peripheral vascular resistance (*TPR*) during calcium infusion in normotensives ($n = 6$) and hypertensives ($n = 6$), but there was no significant decrease in cardiac output (*CO*) either in the absence or presence of verapamil. Values are mean ± SD. $^*P < 0.05$, $^{**}P < 0.01$, compared to the values before calcium infusion in the absence or presence of verapamil

Discussion

Our results have confirmed the early findings [17–19] that an intravenous infusion of calcium raises blood pressure in normotensives and hypertensives, and that the rise in blood pressure is accompanied by a parallel increase in the serum concentration of calcium. In addition, in this study we clearly demonstrated that the rise in blood pressure was caused by the elevation of total peripheral vascular resistance, which was inhibited by the calcium antagonist verapamil; and the absolute elevations in blood pressure and vascular resistance were greater in hypertensives than in normotensives, but the percentage rises were not greater in blood pressure or vascular resistance.

In this study, calcium infusion at the rate of 7.5 mg/kg per hour for 1 h raised blood pressure by 18/7 and 29/11 mmHg with an increased serum concentration of calcium of 4.0 and 3.7 mg/dl in normotensives and hypertensives, respectively. Marone et al. [17] reported a rise in blood pressure of 40/15 mmHg during calcium infusion at the rate of 5 mg/kg per hour for 3 h with an increase in serum calcium concentration of 4.3 mg/dl. Bianchetti et al. [18] reported that calcium infusion at the rate of 2 mg/kg per hour for 2 h caused an elevation in blood pressure of 2/1 and 11/5 mmHg with an increase in serum calcium concentration of 1.0 mg/dl in normotensives and hypertensives, respectively. Ellison et al. [19] reported an elevation of blood pressure of 10/1 and 10/5 mmHg by calcium infusion at the rate of 3.75 mg/kg per hour for 3 h with an increased concentration of serum ionized calcium of 0.5 and 0.7 mEq/l in normotensives and hypertensives, respectively. These findings indicate that an increase in serum concentration of calcium causes a rise in blood pressure in both normotensives and hypertensives.

Our hemodynamic studies demonstrated that calcium infusion raised blood pressure and vascular resistance, but it did not change cardiac output. Moreover, we observed that the infusion of the calcium antagonist verapamil at the rate of 0.15 mg/kg per hour almost completely inhibited the pressor effect of infused calcium, which was associated with a depressed elevation of vascular resistance. A possible mechanism for the rise in blood pressure by acute hypercalcemia during intravenous infusion of calcium may be explained as the following:

1. An increase in the serum calcium concentration causes the calcium ion to enter the arterial wall through the intima of arterial vessels and then enters the cytosol of arterial smooth muscle through calcium pathways in the cell membrane.
2. This calcium then brings about an increase in intracellular calcium, which induces contraction of arterial smooth muscle.
3. The arterial contraction brings about a reduction of arterial lumen through vasoconstriction, which causes elevation of total peripheral vascular resistance.
4. The elevation of vascular resistance raises blood pressure.

Marone et al. [17] and Bianchetti et al. [18] studied the possible hormonal contribution on hemodynamic responses to acute hypercalcemic hypertension. Calcium infusion slightly, but significantly, increased plasma catecholamine levels. They suggested that enhanced catecholamine release might participate in the pressor response to infused calcium. McCarron [7] and Ellison et al. [19] studied the relationship between parathyroid hormone homeostasis and blood pressure, and suggested that acute suppression of parathyroid hormone release during calcium infusion might diminish the vasodilative effects of this hormone. These reports suggest that hormonal homeostasis may also play a mild contributory role in acute hypercalcemic hypertension.

Our results showed that the absolute elevations in both blood pressure and vascular resistance induced by calcium infusion were significantly greater in hypertensives than in normotensives, but the percentage rises in blood pressure and vascular resistance were not significally different in both groups. Ellison et al. [19] noted that calcium infusion caused a greater absolute rise in blood pressure in subjects with essential hypertension than in normal subjects, and concluded that vascular reactivity to infused pressors is enhanced in hypertensive subjects. Bianchetti et al. [18] reported that blood pressure rose in response to a lower dose of calcium infusion (2 mg/kg per hour) in hypertensives than in normotensives, however, the slopes of the correlations between blood pressure and serum calcium concentration during calcium infusion did not differ significantly between normal and hypertensive subjects. These findings support the hyperresponses of blood pressure and vascular resistance to infused calcium in hypertensives. A possible explanation for the hyperesponses is that the vasoconstrictive responses to calcium are enhaned in hypertensives, due to either increased permiability of calcium in the membrane of arterial smooth muscle or a structural alteration of blood vessels, such as increased wall thickness or narrowed lumen in hypertensives [21, 22]. However, the hypersensitivity of the vasoconstrictive response to calcium is not entirely accepted in subjects with essential hypertension [23].

There are few hemodynamic studies concerning the infusion of calcium with a calcium antagonist. The present results indicated that the infusion of verapamil at the rate of 0.15 mg/kg per hour inhibited the pressor effect of the infused calcium by 10% in normotensives and 14% in hypertensives, which was associated with a depressed elevation of vascular resistance. Robinson and Phillips [24] reported that the increased calcium concentration of 0.5 mmol/l in forearm venous blood, induced by calcium infusion into the brachial artery, caused a 35% reduction of the vasodilator response to verapamil in normotensive subjects. These findings suggest an important role for calcium in the regulation of cell membrane function in the vascular smooth muscle from resistance vessels, and support the view that membrane abnormalities of arterial smooth muscle may lead to hypertension in patients with essential hypertension.

In conclusion, we have shown that the intravenous infusion of calcium raised blood pressure due to an elevated total peripheral vascular resistance in both normotensives and hypertensives. A calcium antagonist inhibited the rise in blood pressure during calcium infusion. The infused calcium may affect vascular smooth muscle, which induces contraction of arterial smooth muscle leading to an elevation in vascular resistance and a rise in blood pressure.

References

1. Aoki K (1986) Calcium membrane theory of essential hypertension. In: Aoki K (ed) Essential hypertension: calcium mechanisms and treatment. Springer, Tokyo Berlin Heidelberg New York London Paris, pp 223–242
2. Aoki K, Kawaguchi Y, Sato K, Kondo S, Yamamoto M (1982) Clinical and pharmacological properties of calcium antagonists in essential hypertension in humans and spontaneously hypertensive rats. J Cardiovasc Pharmacol 4 (Suppl III): III–298–302
3. Bohr DF (1963) Vascular smooth muscle: dual effect of calcium. Science 139: 597–599
4. Robinson BF (1984) Altered calcium handling as a cause of primary hypertension. J Hypertens 2: 453–460
5. Lau K, Eby B (1985) The role of calcium in genetic hypertension. Hypertension 7: 657–667
6. Postnov YV, Orlov SN (1985) Ion transport across plasma membrane in primary hypertension. Physiol Rev 65: 904–945

7. McCarron DA (1982) Low serum concentrations of ionized calcium in patients with hypertension. N Engl J Med 307: 226–228
8. Kesteloot H, Geboers J, Hoof RV (1983) Epidemiological study of the relationship between calcium and blood pressure. Hypertension 5(Suppl II): II52–56
9. Resnick LM, Laragh JH, Sealey JE, Alderman MH (1983) Divalent cations in essential hypertension: relations between serum ionized calcium, magnesium, and plasma renin activity. N Engl J Med 309: 888–891
10. McCarron DA, Morris CD, Henry HJ, Stanton JL (1984) Blood pressure and nutrient intake in the United States. Science 224: 1392–1398
11. McCarron DA, Morris CD (1985) Blood pressure response to oral calcium in persons with mild to moderate hypertension. Ann Intern Med 103: 825–831
12. Meese RB, Gonzales DG, Casparian JM, Ram CVS, Pak CM, Kaplan NM (1987) The inconsistent effects of calcium supplements upon blood pressure in primary hypertension. Am J Med Sci 294: 219–224
13. Cappuccio FP, Markandu ND, Singer DRJ, Smith SJ, Shore AC, MacGregor GA (1987) Does oral calcium supplementation lower high blood pressure? A double blind study. J Hypertens 5: 67–71
14. Lyle RM, Melby CL, Hyner GL, Edmondson JW, Miller JZ, Weinberger MH (1987) Blood pressure and metabolic effects of calcium supplementation in normotensive white and black men. J Am Med Assoc 257: 1772–1776
15. Siani A, Strazzullo P, Gulielmi S, Pacioni D, Giacco A, Iacono R, Mancini M (1988) Controlled trial of low calcium versus high calcium intake in mild hypertension. J Hypertens 6: 253–256
16. Zoccali C, Mallamaci F, Delfino D, Ciccarelli M, Parlongo S, Iellamo D, Moscato D, Maggiore Q (1988) Double-blind randomized, crossover trial of calcium supplementation in essential hypertension. J Hypertens 6: 451–455
17. Marone C, Beretta-Piccoli C, Weidmann P (1981) Acute hypercalcemic hypertension in man: Role of hemodynamics, catecholamines, and renin. Kidney Int 20: 92–96
18. Bianchetti MG, Beretta-Piccoli C, Weidmann P, Link L, Boehringer K, Ferrier C, Morton JJ (1983) Calcium and blood pressure regulation in normal and hypertensive subjects. Hypertension 5 (Suppl II): II57–65
19. Ellison DH, Shneidman R, Morris C, McCarron DA (1986) Effects of calcium infusion on blood pressure in hypertensive and normotensive humans. Hypertension 8: 497–505
20. Suzuki T, Aoki K (1988) Hypertensive effects of calcium infusion in subjects with normotension and hypertension. J Hypertens 6: 1003–1008
21. Webb RC, Bohr DF (1981) Recent advances in the pathogenesis of hypertension: consideration of structural, functional and metabolic vascular abnormalities resulting in elevated vascular resistance. Am Heart J 102: 251–264
22. Folkow B (1987) Structure and function of the arteries in hypertension. Am Heart J 114: 938–948
23. Aalkjaer C, Heagerty AM, Petersen KK, Swales JD, Mulvany MJ (1987) Evidence for a negative correlation between verapamil sensitivity and calcium sensitivity in isolated human resistance vessels. J Hypertens 5 (Suppl V): V–157–159
24. Robinson BF, Phillips RJW (1984) Effect of small increments in plasma calcium concentration on the responsiveness of forearm resistance vessels to verapamil in normal subjects. Clin Sci 67: 613–618

Effects of Intravenous Infusion of Magnesium on Hemodynamics in Normotensives and Hypertensives

KOICHI MIYAGAWA, TAKAHIRO SUZUKI, and KYUZO AOKI[1]

Summary. The effects of continuous intravenous infusion of magnesium on hemodynamics were studied in 5 normotensives and 7 hypertensives. Intravenous infusion of magnesium sulfate at a rate of 0.1 mg/kg per minute for 60 min significantly increased the serum concentration of magnesium from 2.4 ± 0.1 to 4.7 ± 0.5 mg/dl in normotensives and from 2.3 ± 0.2 to 4.5 ± 0.4 mg/dl in hypertensives ($M \pm SD$, $P < 0.01$). There was, however, no change in blood pressure, heart rate, cardiac output or vascular resistance. The results obtained suggest that an increased concentration of serum magnesium may not change blood pressure, cardiac output or vascular resistance in humans with normotension and essential hypertension:

Key words: Blood pressure—Cardiac output—Hypertension—Magnesium—Normotension—Vascular resistance

Introduction

A causal relationship between a decreased concentration of serum magnesium ion (magnesium) and hypertension has been reported [1–3]. This hypothesis has been supported by the following reports [3–8].

A significant decrease in blood pressure of 12/8 mmHg was achieved by administration of a magnesium supplement in hypertensive patients during long-term thiazide treatment [3]. Intravenous administration of magnesium sulfate was used in treating hypertensive emergency patients [4, 5], whereby the magnesium induced vasodilation and decreased blood pressure [5]. Intra-arterial infusion of magnesium sulfate resulted in an increased forearm blood flow and a decreased forearm vascular resistance due to the vasodilating effect of magnesium, which was more pronounced in patients with essential hypertension [6]. Administration of magnesium prevented an increase in blood pressure during calcium infusion [7]. In addition, low serum magnesium with high serum calcium was reported in humans with certain types of hypertension [8]. Increasing dietary magnesium was shown to decrease blood pressure [3]. A high incidence of

[1] Second Department of Internal Medicine, Nagoya City University Medical School, Mizuho-ku, Nagoya 467, Japan

hypertension was observed in geographic areas with soft drinking water or magnesium-poor soil [1], and there were a number of hypertensive patients associated with hypomagnesemia [9].

Acute hypermagnesemia due to magnesium infusion in the prevention of premature delivery might be the result of the displacement of calcium from its binding sites in uterine smooth muscle, which thereby reduces its contractility [10, 11].

In animal experiments, acute intravenous infusion of magnesium lowered the blood pressure in rats with mineralocorticoid-salt (DOCA-salt) hypertension but not in rats with normotension or renovascular hypertension [12, 13]. Intravenous administration of magnesium lowered heart rate, cardiac output, and aortic arterial pressure, but did not change vascular resistance in dogs [14]. A magnesium-deficient diet elevated blood pressure in rats compared to rats on a standard diet [1], suggesting that magnesium depletion contributes to the genesis of specific types of hypertension under certain circumstances [4]. Several possible explanations for the hypotensive mechanism of magnesium have been suggested. Magnesium may bind competitively to the same sites as calcium in vascular smooth muscle. Magnesium may compete with calcium for a binding site, or it may inhibit the influx of calcium, or displace calcium from intracellular binding sites of calcium [10, 11]. Magnesium may inhibit contraction of arterial smooth muscle, leading to vasodilation and reduction of blood pressure.

Interestingly, it has been reported that magnesium administration does not appear to decrease blood pressure in patients with essential hypertension [15, 16], or in rats with normotension or renovascular hypertension [13]. In addition, despite a significant increase in the plasma concentration of magnesium and a significant increase in the urinary excretion of magnesium while taking magnesium aspartate, blood pressure was not reduced compared with treatment placebo values or the values before treatment [16]. Thus, it is not clear from these reports if magnesium has an antihypertensive effect in humans with essential hypertension. Also, whether or not magnesium had a direct effect on blood vessels or whether the fall in blood pressure by magnesium administration was due to a reduction in intracerebral pressure were also unclear [16]. Unfortunately, there are few in vivo studies examining the effects of hypermagnesemia on hemodynamics.

We carried out a study to investigate whether or not blood pressure is lowered by intravenous infusion of magnesium, and to determine the effects of this infusion on heart rate, cardiac output, and vascular resistance in subjects with normotension and patients with essential hypertension.

Methods

Subjects

Twelve subjects including 5 volunteers with normotension, aged 40–70 years, and 7 patients with essential hypertension, aged 40–80 years, participated in this study. They were hospitalized and placed on a diet containing 8 g/day NaCl,

Table 1. Clinical findings of normotensives and hypertensives

	Normotensives	Hypertensives
No. of cases	5	7
Male : Female	3 : 2	3 : 4
Age (years)	72 ± 6	75 ± 2
Body weight (kg)	55 ± 6	44 ± 6
Height (cm)	153 ± 8	150 ± 6
Body surface area (m^2)	1.51 ± 0.10	1.35 ± 0.11
BUN (mg/dl)	18.0 ± 3.2	15.6 ± 3.8
Cre (mg/dl)	1.5 ± 0.2	1.2 ± 0.1

Values represent mean ± SD
BUN, serum blood urea nitrogen; Cre, serum creatinine

600 mg/day calcium, and 350 mg/day magnesium. The serum magnesium concentration was within the normal range, 2.4 mg/dl in normotensives and 2.3 mg/dl in hypertensives; the calcium concentration was 9.6 and 9.4 mg/dl in normotensives and hypertensives, respectively.

Study Protocol

All subjects discontinued medications including antihypertensive drugs for 4 weeks during the observation period. Following the period, blood pressures were multiply measured by the auscultation method using a mercury sphygmomanometer, after 5 min in the supine position in the hospital. Hypertension was defined by the values of multiple measurements as systolic blood pressure of greater than 150 mmHg or diastolic pressure over 90 mmHg. No subjects had evidence of secondary hypertension, heart failure, myocardial infarction, or renal impairment (Table 1).

The subjects were instructed to remain in the supine position in a quiet room. Then, a polyethylene venous catheter for infusion of dextrose and magnesium solution and another arterial catheter for recording of blood pressure and collecting arterial blood samples were percutaneously inserted into the antecubital vein and brachial artery, respectively. After the catheterization, intravenous infusion of 5% dextrose was started at the rate of 50 ml/hr through the venous catheter.

Following the preparation, the subjects rested in the supine position for 1 h as an equilibration period. Then, magnesium sulfate as ionized magnesium was infused at the rate of 0.1 mg/kg per minute for 60 min. Intra-arterial blood pressure, heart rate, and cardiac output were measured and blood samples were collected before and at the end of magnesium infusion.

Measurements

Intra-arterial blood pressure was continuously recorded through the arterial catheter attached to a pressure transducer (Statham, P23ID). The heart rate was measured by a electrocardiographic monitor with a polygraph system (Nihon

Kohden, RM-6000). Cardiac output was determined by the dye-dilution method, by bolus injection of 10 mg indocyanine green followed by 10 ml of 5% dextrose solution through the venous catheter; blood was withdrawn through a cuvette densitometer from the brachial artery, using a cardiac computer (Erma optical works, Dye Mac EW-90) [17]. Stroke volume was calculated by dividing the obtained cardiac output by the heart rate. Total peripheral vascular resistance was calculated by dividing the mean arterial blood pressure by the cardiac output. Serum concentration of sodium and potassium were assayed by the flame photometric method (Hitachi 775), chloride by the argentum electrode method (Jookoo, C-200AP), and calcium and magnesium by spectrophotometry (Hitachi U-1080).

Data Analysis

Paired Student's *t*-test was used for statistical analysis to compare the values obtained before and after the infusion of magnesium. A P value of less than 0.05 was considered significant.

Results

Effects of Magnesium Infusion on Serum Concentration of Electrolytes

Magnesium infusion of 0.1 mg/kg per minute for 60 min significantly increased the concentration of serum magnesium from 2.4 ± 0.1 to 4.7 ± 0.5 mg/dl in normotensives and from 2.3 ± 0.2 to 4.5 ± 0.4 mg/dl in hypertensives ($M \pm SD$ $P < 0.01$), but the increase in serum magnesium did not effect the serum concentrations of calcium, sodium, potassium, chloride, or phosphate in either group (Table 2).

Hemodynamic Effects of Magnesium Infusion

Magnesium infusion did not alter the hemodynamic indexes of systolic and diastolic blood pressure, heart rate, cardiac output, stroke volume, or vascular

Table 2. Concentration of serum electrolytes before and after magnesium infusion

	Normotensives		Hypertensives	
	Before	After	Before	After
Mg (mg/dl)	2.4 ± 0.1	4.7 ± 0.5*	2.3 ± 0.2	4.5 ± 0.4*
Ca (mg/dl)	9.6 ± 0.3	9.6 ± 0.2	9.4 ± 0.4	9.5 ± 0.5
Na (mg/dl)	146 ± 2.3	143 ± 1.4	146 ± 2.3	145 ± 2.9
K (mg/dl)	4.3 ± 0.3	4.1 ± 0.3	4.4 ± 0.1	4.3 ± 0.3
Cl (mg/dl)	108 ± 1.3	108 ± 2.3	107 ± 4.2	108 ± 3.8
P (mg/dl)	3.2 ± 0.4	3.3 ± 0.3	3.3 ± 0.5	3.4 ± 0.4

Values represent mean ± SD
* $P < 0.01$

resistance in normotensives. In hypertensives there was small but not significant decrease in systolic and diastolic blood pressure from 175 ± 21 to 170 ± 26 mmHg and 78 ± 10 to 75 ± 12 mmHg (mean ± SD), respectively (Table 3, Fig. 1).

Magnesium given as an intravenous infusion was well tolerated and did not cause any adverse signs or symptoms of hypermagnesemia, such as nausea, bradycardia, or respiratory hypofunction.

Table 3. Hemodynamics before and after infusion of magnesium in normotensives and hypertensives

	Normotensives		Hypertensives	
	Before	After	Before	After
SBP (mmHg)	134 ± 8	134 ± 8	175 ± 21	170 ± 26
DBP (mmHg)	66 ± 9	66 ± 9	78 ± 10	75 ± 12
MBP (mmHg)	89 ± 7	89 ± 7	110 ± 13	107 ± 16
HR (bpm)	61 ± 4	62 ± 4	76 ± 14	75 ± 14
CO (l/min)	4.1 ± 0.6	4.1 ± 0.7	4.0 ± 1.4	4.2 ± 1.2
SV (ml)	67 ± 12	66 ± 11	53 ± 15	57 ± 14
TPR (mmHg/l/min)	22.5 ± 4.4	22.7 ± 4.8	30.5 ± 11.7	27.1 ± 9.5

Values represent mean ± SD
SBP, systolic blood pressure; DBP, diastolic blood pressure; MBP, mean blood pressure; CO, cardiac output; SV, stroke volume; TPR, total peripheral resistance

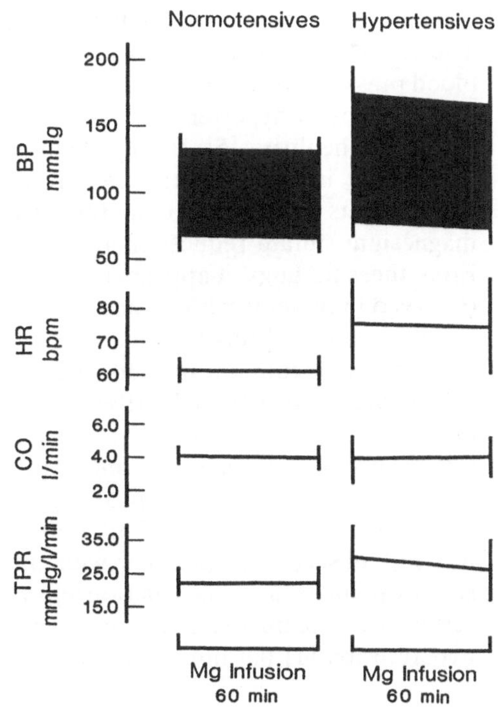

Fig. 1. Blood pressure (*BP*), heart rate (*HR*), cardiac output (*CO*), and total peripheral resistance (*TPR*) during magnesium sulfate infusion at a rate of 0.1 mg/kg per minute in subjects with normotension ($n = 7$, left) and patients with essential hypertension ($n = 5$, right). Values = mean ± SD

Discussion

Our findings show that an increased concentration of serum magnesium by intravenous infusion of magnesium sulfate does not change blood pressure, heart rate, cardiac output, or vascular resistance in untreated patients with mild to moderate essential hypertension. These results indicate that an increase in plasma magnesium concentration by an increase in the intake of magnesium, does not cause any change in blood pressure [16]. These results are in disagreement with a study by Dyckner and Wester [3], in which a significant fall in blood pressure was observed during magnesium aspartate supplementation given to patients who were already treated with thiazide diuretics and who tended to have plasma magnesium concentrations below the normal range. Ji et al. [6] demonstrated that magnesium sulfate infused into the arterial circulation of the forearm was a potent vasodilator. Therefore, no definite conclusion can be drawn about the effect of magnesium supplementation in the lowering of blood pressure [16].

We found no difference in the serum magnesium concentration of subjects with normotension and patients with essential hypertension. In particular, there was no significant correlation between changes in blood pressure and increased serum concentration of magnesium or the initial serum concentration of magnesium. The results support the claim of an inverse relationship of serum magnesium concentration and blood pressure [9]. Kesteloot et al. [18] showed a negative relationship between 24-h urinary magnesium excretion and blood pressure in Belgium, but a positive relationship in the People's Republic of China. The relationship between 24-h urinary excretion of magnesium and blood pressure is complex, and varying results have been obtained in different populations [18]. Therefore, no general conclusions can be drawn on the role of magnesium in blood pressure regulation. Intravenous magnesium sulfate lowers blood pressure in patients with hypertensive crisis, particularly in eclampsia [4] and in patients with acute nephritis [5]. Both eclampsia and acute nephritis are associated with an increase in intracerebral pressure and brain edema, but are not associated with patients with mild to moderate essential hypertension. It is well known that magnesium sulfate reduces increased intracerebral pressure and brain edema. From these findings, it appears that there is no consistent antihypertensive effect observed in patients with essential hypertension [5, 16, 19].

Altura et al. [1] reported that rats maintained on a diet moderate or severely deficient in magnesium showed a significant rise in arterial blood pressure compared with control animals. It was observed that the greater the degree of dietary magnesium deficiency, the greater the reduction in microvascular lumen size. The serum magnesium concentration in the moderately and severely magnesium deficient rats was approximately 70 and 30 percent, respectively, of the normotensive controls [1]. This reduction in serum magnesium concentration was not observed in patients with untreated essential hypertension. Several studies point to a causal relationship between the decreased concentrations of serum magnesium ion and some types of hypertension, magnesium-deficient hypertension [1] but not essential hypertension [13, 16]. Experimental studies-showed that acute hypermagnesemia inhibited the spontaneous tone of arteries and veins both in vitro and in intact animals, and decreased vascular resistance to blood flow [1]. Those findings suggested that extracellular magnesium plays a critical role in the regulation of vascular tone through the contractile reactivity of vascular smooth muscle [1]. However, we have not observed a significant decrease in the serum concentration of magnesium in untreated patients with mild to moderate essential hypertension. Thus, our data was not in support of a causal relationship between decreased serum concentrations of magnesium and blood pressure in essential hypertension.

In conclusion, despite a significant increase in serum magnesium concentration there was no reduction in blood pressure, cardiac output, or vascular resistance, suggesting that serum magnesium concentration does not play a role in the regulation of blood pressure in subjects with normotension or in patients with essential hypertension during intravenous infusion of magnesium.

References

1. Altura M, Altura T, Gebrewold A (1984) Magnesium deficiency and hypertension: correlation between magnesium-deficient diets and microcirculatory changes in situ. Science 223: 1315–1317
2. Iseri LT, French JH (1984) Magnesium: nature's physiologic calcium blocker. Am Heart J 108: 188–193
3. Dyckner T, Wester PO (1983) Effect of magnesium on blood pressure. Br Med J 286: 1847–1849
4. Rogers SF, Flowers CE Jr, Alexander JA (1969) Aggressive toxemia management. Obst Gynecol 33: 724–728
5. Winkler AW, Smith PK, Hoff EH (1942) Intravenous magnesium sulfate in the treatment of nephrotic convulsion in adults. J Clin Invest 21: 207–216
6. Ji BH, Erne P, Kiowski W, Bühler FR, Bolli P (1983) Magnesium-induced vasodilation is comparable to that induced by calcium entry blockade. J Hypertens 1 (Suppl II): 368–371
7. Zawada ET Jr, TerWee JA, McClung DE (1987) Magnesium prevents acute hypercalcemic hypertension. Nephron 47: 109–114
8. Resnick LM, Laragh JH, Sealey JE, Alderman MH (1983) Divalent cations in essential hypertension; relations between serum ionized calcium, magnesium, and plasma renin activity. N Engl J Med 309: 888–891
9. Petersen B, Schrell M, Christiansen C, Transbol I (1977) Serum and erythrocyte magnesium in normal elderly Danish people. Acta Med Scand 201: 31–34
10. Levine BS, Coburn JW (1983) Magnesium, the mimic/antagonist of calcium. N Engl J Med 310: 1253–1255
11. Cholst IN, Steinberg SF, Tropper PJ, Fox HE, Segre GV, Bilezikan JP (1984) The influence of hypermagnesemia on serum calcium and parathyroid hormone levels in human subjects. N Engl J Med 310: 1221–1225
12. DiPette DJ, Simpson K, Guntupalli J (1987) Systemic and regional hemodynamic effect of acute magnesium administration in the normotensive and hypertensive state. Magnesium 6: 136–149
13. Dipette DJ, Simpson K, Rogers A, Holland OB (1988) Haemodynamic response to magnesium administration in mineralocorticoid-salt and two-kidney, one clip renovascular hypertension. J Hypertens 6: 413–417
14. Friedman HS, Nguyen TN, Mokraoui AM, Barbour RL, Murakawa T, Altura BM (1987) Effects of magnesium chloride on cardiovascular hemodynamics in the neurally intact dog. J Pharmacol Exp 243: 126–130

15. Mordes JP, Wocker WEC (1978) Excess magnesium. Pharmacol Rev 29: 273–300
16. Cappuccio FP, Markandu ND, Beynon GW, Shore AC, Sampson B, MacGregor GA (1985) Lack of effect or oral magnesium on high blood pressure: a double blind study. Br Med J 291: 235–238
17. Suzuki T, Aoki K (1988) Hypertensive effects of calcium infusion in subjects with normotension and hypertension. J Hypertens 6: 1003–1008
18. Kesteloot H, Geboers J, Huang DX (1986) Epidemiological studies on the role of sodium, potassium, calcium, and magnesium in hypertension. In: Aoki K (ed) Essential hypertension. Springer, Tokyo Berlin Heidelberg New York London Paris, pp 179–190
19. Randall RE Jr, Cohen MD, Spray CC Jr, Rossmeisl EC (1964) Hypermagnesemia in renal failure. Ann Intern Med 61: 73–88

Marked Blood Pressure Responses to Norepinephrine, Epinephrine, and Angiotensin II in Borderline Hypertension with a Parental History of Hypertension

KYZUO AOKI and KOICHI SATO[1]

Summary. The hemodynamic responses to norepinephrine, epinephrine, and angiotensin II were investigated in borderline essential hypertensives with a positive parental history of hypertension and normotensives. Intravenous infusion of norepinephrine at the rate of 0.2 μg/kg per minute was administered for 5 min to 12 normotensives and 14 hypertensives, and angiotensin II, at the rate of 0.02 μg/kg per minute, was given for 5 min to 14 subjects of both groups. Both norepinephrine and angiotensin II raised the blood pressure and total vascular resistance with accompanying reductions in stroke volume and cardiac output, which indicated that the rise in blood pressure was caused by the elevation of vascular resistance. Epinephrine infusion (0.2 μg/kg per minute for 5 min) in 10 subjects in both groups raised blood pressure with an increased stroke volume and cardiac output associated with a reduction in vascular resistance, which indicated the rise in blood pressure was caused by the increased cardiac output. The absolute rise in blood pressure during infusion of these three agonists was greater in the hypertensives than in the normotensives. The percentage rise in both stroke volume and cardiac output and the percentage reduction of vascular resistance during epinephrine infusion were smaller in the hypertensives than in the normotensives, but not during norepinephrine or angiotensin II infusion. In conclusion, the marked blood pressure responses to the agonists were observed in borderline essential hypertensives, which suggests that the hyperresponses may possibly carry an abnormality in the contractile responses of vascular smooth muscle. This abnormality may be transferred by a hypertension gene from parents to offspring.

Key words: Hemodynamic—Essential hypertension—Hyperresponse—Hypersensitivity —β-Adrenoceptor

Introduction

The first demonstration of hyperresponses to infused norepinephrine in adolescents with a positive parental history of hypertension was carried out in Doyle's laboratory [1]. Studies have reported that young adult normotensive sons of hypertensive parents have a significantly greater vascular response to norepinephrine than young men of a similar age whose parents have normal blood pressure [1]. Doyle et al. [1] suggested that heightened vascular reactivity may depend on a specific defect which is inherited as a dominant characteristic. This

[1] Second Department of Internal Medicine, Nagoya City University Medical School, Mizuho-ku, Nagoya 467, Japan

defect is involved in the pathogenesis of essential hypertension [1].

Recently, Bianchetti et al. [2] have demonstrated that an exaggerated cardiovascular pressor response to infused norepinephrine occurs in normotensive members of hypertensive families compared with that in members of normotensive families. In addition, the following disturbances have been reported in offspring of hypertensive parents: (1) an enhanced sensitivity to forearm vascular resistance to infused norepinephrine [2], (2) an increased responsiveness of blood pressure to mental stress [3], and (3) an increased contractile response of arterial vessels to calcium agonist [4]. Also, it has been reported that there are exaggerated blood pressure responses to exercise in normotensive adolescents [5, 6] and borderline hypertensives [7, 8] with a positive parental history of hypertension. These findings suggest that the disease of essential hypertesnion may be predisposed by certain familial abnormalities in blood pressure control mechanisms in cardiovascular muscle cells.

On the other hand, several studies have reported that essential hypertensive adolescents with normotension did not have any exaggerated blood pressure responses to stimuli [9, 10]. The present study was designed to determine whether or not heightened responses of blood pressure to norepinephrine, epinephrine, and angiotensin II occur in borderline hypertensive people with a positive parental history of hypertension. Additionally the hyperresponse was reviewed in subjects with a family history of hypertension.

Methods

Subjects

Seventy-four middle-aged male subjects, aged 30–45 years, including 38 borderline essential hypertensives and 36 sex- and age-matched normotensive volunteers participated in this study. The subjects completed questionnaries on their medical history including the health status of both biological parents. A positive parental history of hypertension (PH+) was defined as one or both biological parents having a diagnosis of or medication for essential hypertension. The borderline hypertensives in this study only included the subjects which were Ph+. Secondary forms of hypertension were excluded by the typical tests, including serum concentration of creatinine, sodium, and potassium as well as conventional urine examination, and the determination of aldosterone and catecholamine concentration in a few subjects. The normotensive volunteers had neither a family history of hypertension of either parent (Ph−) nor any known previous episodes of hypertension. They had a normal blood pressure recorded once in a year preceding the usual tests including measurement of blood pressure.

Study Protocol

All were informed of the investigative nature of the study. On the morning of the study subjects were seated on a chair in a temperature-controlled laboratory for 30 min, and they were instructed to remain in the supine position. Then, a polyethylene venous catheter for infusion was percutaneously inserted into antecubital vein. An infusion of 5% dextrose at the rate of 50 ml/h was continued during a 60-minute equilibration period. Following the equilibration period, the resting baseline of blood pressure, heart rate, and cardiac output was measured. Then, the following agonists were infused for testing responses: (1) Norepinephnine was infused at 0.2 μg/kg per minute for 5 minutes to 12 normotensives and 14 hypertensives, (2) epinephrine was infused at 0.2 μg/kg per minute for 5 min to 10 normotensives and 10 hypertensives, and (3) angiotensin II was infused at 0.02 μg/kg per minute for 5 min to 14 subjects in both groups. At the end of agonist infusion, blood pressure, heart rate, and cardiac output were measured.

Measurements

Blood pressure (BP) of the upper right arm was measured several times by the auscultation method, using a mercury sphygmomanometer. Heart rate (HR) was continuously monitored using an electrocardiogram. Cardiac output was determined in duplicate by the earpiece dye-dilution method, with a bolus injection of 5 mg indocyanine green followed by 10 ml of 5% dextrose solution through the venous cathether, and calculated on a cardiac computer (Erma optical works, Dye-Mac Ed-90).

Mean blood pressure was calculated as diastolic BP plus 1/3 pulse pressure. Total peripheral vascular resistance was calculated as mean BP divided by cardiac output, which was expressed in an arbitrary unit. Stroke volume was calculated by dividing cardiac output by HR. Stroke volume, cardiac output, and total vascular resistance were corrected for body surface area, and presented as stroke index (ml/m^2), cardiac index (l/m^2), and total vascular resistance index (Um^2), respectively.

Statistical analysis

Paired student's *t*-test was used for statistical analysis to compare the values obtained before and after the infusion of agonist. A P value less than 0.05 was considered significant.

Results

Baseline Hemodynamics

There were no significant differences in age, body weight, or height across the six sub-groups (Tables 1–3). Systolic BP, diastolic BP, and mean BP was higher at rest in the borderline hypertensive groups than in the normotensive groups ($P < 0.01$) (Tables 1–3). Heart rate was lower at rest in the normotensive groups than in the borderline hypertensive groups, however the differences between the two was small and not significant. Stroke volume in the normotensives did not differ from the borderline hypertensives. Cardiac output was slightly greater at rest in the hypertensives than in the normotensives, however the differences was not significant. Total peripheral vascular resistance was significantly higher in the hypertensives than in the normotensives ($P < 0.05$).

Table 1. Clinical findings of normotensives and borderline hypertensives in the study of infusion of norepinephrine

	Infusion of norepinephrine	
	Normotensives (n = 12)	Hypertensives (n = 14)
Age (years)	37 ± 6	38 ± 6
Weight (kg)	63 ± 3	61 ± 5
Height (cm)	168 ± 4	166 ± 4
BSA (m^2)	1.7 ± 0.1	1.7 ± 0.1
SBP (mmHg)	104 ± 9	138 ± 9*
DBP (mmHg)	63 ± 5	86 ± 7*
HR (beats/min)	60 ± 7	63 ± 8
S creat (mg/dl)	1.0 ± 0.1	1.0 ± 0.1

BSA, body surface area; S creat, serum creatinine
* $P < 0.01$

Table 2. Clinical findings of normotensives and borderline hypertensives in the study of infusion of epinephrine

	Infusion of epinephrine	
	Normotensives (n = 10)	Hypertensives (n = 10)
Age (years)	36 ± 5	36 ± 4
Weight (kg)	62 ± 4	63 ± 7
Height (cm)	173 ± 5	168 ± 6
BSA (m^2)	1.7 ± 0.1	1.7 ± 0.1
SBP (mmHg)	110 ± 7	134 ± 10*
DBP (mmHg)	62 ± 5	85 ± 8*
HR (beats/min)	61 ± 6	64 ± 10
S creat (mg/dl)	1.0 ± 0.2	1.0 ± 0.1

BSA, body surface area; S creat, serum creatinine
* $P < 0.01$

Table 3. Clinical findings of normotensives and borderline hypertensives in the study of infusion of angiotensin II

	Infusion of angiotensin II	
	Normotensives (n = 14)	Hypertensives (n = 14)
Age (years)	31 ± 7	35 ± 6
Weight (kg)	63 ± 5	67 ± 10
Height (cm)	171 ± 5	168 ± 7
BSA (m^2)	1.7 ± 0.1	1.8 ± 0.2
SBP (mmHg)	111 ± 6	133 ± 7*
DBP (mmHg)	61 ± 5	86 ± 6*
HR (beats/min)	60 ± 6	67 ± 9
S creat (mg/dl)	1.0 ± 0.1	1.0 ± 0.1

BSA, body surface area; S creat, serum creatinine
* $P < 0.01$

Responses to Norepinephrine

Infusion of norepinephrine increased BP from an average of 104 ± 9/63 ± 5 to 122 ± 10/80 ± 8 mmHg in the normotensives, and 138 ± 9/86 ± 7 to 171 ± 11/114 ± 7 mmHg in the borderline hypertensives ($P < 0.001$) (Figs. 1, 2). The infusion decreased HR from 60 ± 7 to 55 ± 7 beats/min ($P < 0.01$) in the normotensives and from 63 ± 8 to 58 ± 7 beats/min ($P < 0.001$) in the hypertensives. The infusion decreased cardiac output from 2.8 ± 0.3 to 2.4 ± 0.4 l/m² in the normotensives ($P < 0.001$) and from 3.0 ± 0.2 to 2.8 ± 0.3 l/m² in the hypertensives ($P < 0.01$), and it increased total vascular resistance from 28 ± 3 to 39 ± 6 and 35 ± 4 to 49 ± 5 Um² in the normotensives and hypertensives ($P < 0.001$), respectively (Figs. 1, 2).

Responses to Epinephrine

The infusion of epinephrine increased BP (110 ± 7/62 ± 5 to 127 ± 9/67 ± 5 mmHg in normotensives; 134 ± 10/85 ± 8 to 159 ± 12/98 ± 12 mmHg in hypertensives) and cardiac output (3.1 ± 0.5 to 5.0 ± 0.6 l/m² in normotensives; 3.3 ± 0.8 to 4.3 ± 1.0 l/m² in hypertensives) with an reduction of vascular resistance in the normotensives (26 ± 5 to 18 ± 2 Um²) and in borderline hypertensives (32 ± 5 to 29 ± 5 Um²). However the infused epinephrine increased stroke volume in the normotensives (52 ± 10 to 64 ± 11 ml/m²), but not significantly in the hypertensives (52 ± 7 to 54 ± 9 ml/m²) (Figs. 1, 2).

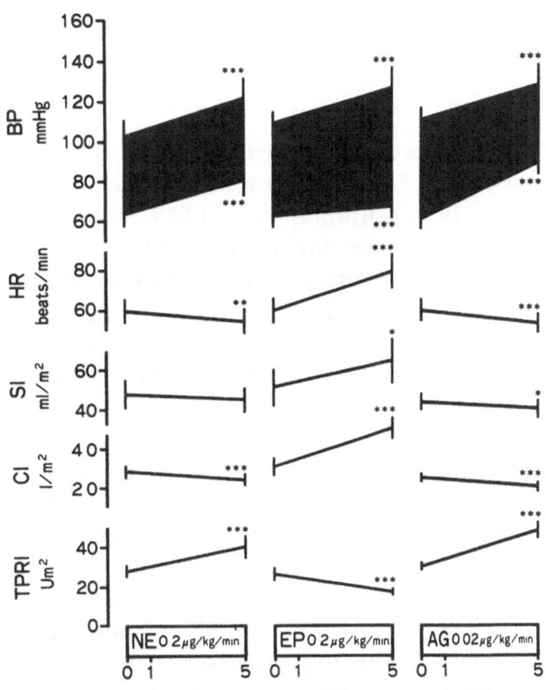

Fig. 1. Hemodynamic responses to norepinephrine (*NE*), epinephrine (*EP*), and angiotensin II (*AG*) in normotensives. *BP*, blood pressure; *SI*, stroke index; *CI*, cardiac index; *TPRI*, total vascular resistance. *$P < 0.05$, **$P < 0.01$, ***$P < 0.001$

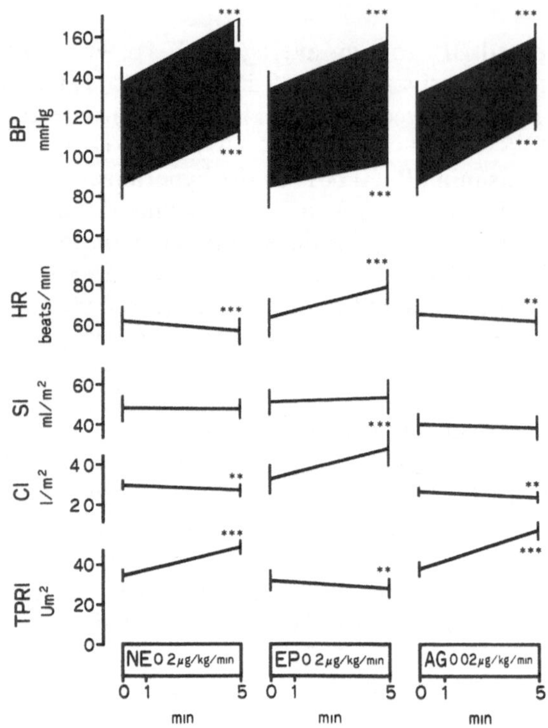

Fig. 2. Hemodynamic responses to norephinephrine (*NE*), epinephrine (*EP*), and angiotensin II (*AG*) in borderline hypertensives. See abbreviations in Fig. 1

Responses to Angiotensin II

The infusion of angiotensin II elevated BP (116 ± 6/61 ± 5 to 129 ± 11/89 ± 5 mmHg in normotensives; 133 ± 7/86 ± 6 to 160 ± 8/119 ± 5 mmHg in hypertensives) and vascular resistance (31 ± 3 to 49 ± 5 Um² and 38 ± 5 to 57 ± 7 Um²), however the infusion decreased HR (60 ± 6 to 53 ± 5 beats/min; 67 ± 9 to 63 ± 7 beats/min) and cardiac output (2.6 ± 0.2 to 2.1 ± 0.2 and 2.7 ± 0.3 to 2.4 ± 0.3 l/m²) in both groups. The infusion decreased stroke volume in the normotensives (43 ± 5 to 40 ± 6 ml/m²) but not significantly in the hypertensives (41 ± 6 to 39 ± 7 ml/m²) (Figs. 1, 2).

Comparison of Absolute Changes between Normotensives and Hypertensives

The increase in BP of 33 ± 7/28 ± 5 mmHg in the hypertensives during norepinephrine infusion was greater than that of 19 ± 10/18 ± 6 mmHg in the normotensives, however the changes in HR, cardiac output, and vascular resistance were not greater (Fig. 3, *upper column*). During epinephrine infusion there was an increase in BP of 17 ± 8/4 ± 7 mmHg in the normotensives and 25 ± 10/12 ± 6 mmHg in the hypertensives. The increase in diastolic BP was greater in the hypertensives than in the normotensives ($P < 0.01$), but the increase in systolic BP was not different between both groups. The hypertensives had smaller increases in stroke volume and cardiac output with a small reduction of vascular resistance ($P < 0.05$) (Fig. 3, *middle column*).

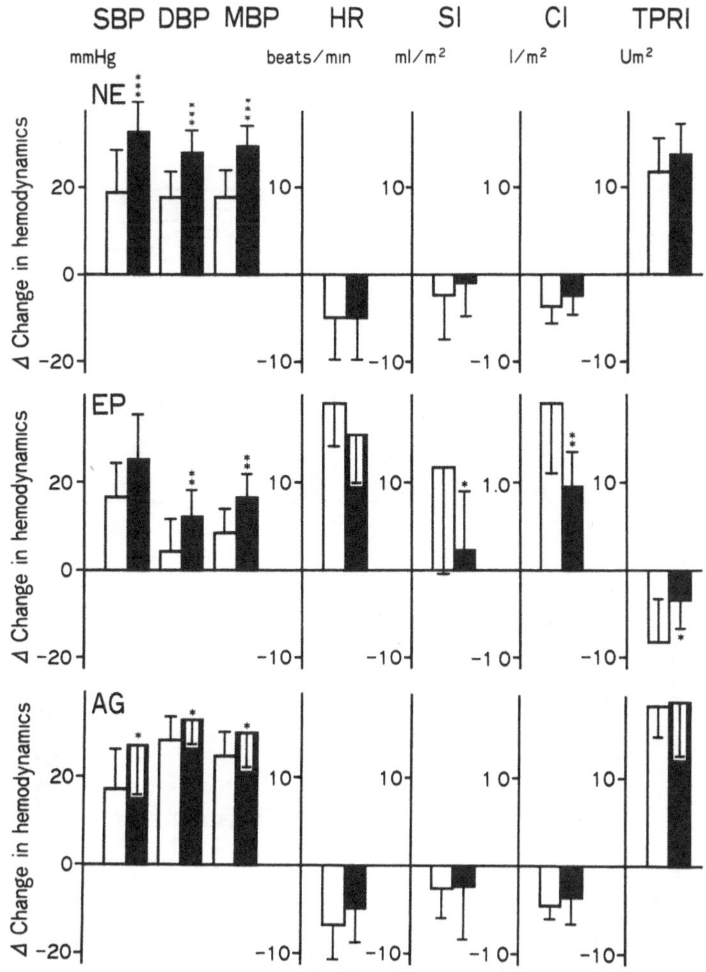

Fig. 3. Comparison of absolute hemodynamic changes to norepinephrine (*NE*), epinephrine (*EP*), and angiotensin II (*AG*) between normotensives (□) and hypertensives (■). *SBP*, systolic blood pressure; *DBP*, diastolic blood pressure; *MBP*, mean blood pressure; *SI*, stroke volume index; *CI*, cardiac output index; *TPRI*, total peripheral vascular resistance index. $*P<0.05$, $**P<0.01$, $***P<0.001$

During angiotensin II infusion, the increase in systolic BP was greater in the hypertensives than in the normotensives (27 ± 12/33 ± 6 vs 17 ± 9/28 ± 5 mmHg). The changes in HR, stroke volume, cardiac output, and vascular resistance did not differ between the two groups (Fig. 3, *lower column*).

Percent Change in Normotensives and Hypertensives

The comparisons of percentage change in BP, HR, stroke volume, cardiac output, and vascular resistance are shown in Fig. 4. The percentage increases in BP (19 ± 10/29 ± 10% vs 24 ± 6/33 ± 7%, in the normotensive vs the hypertensives)

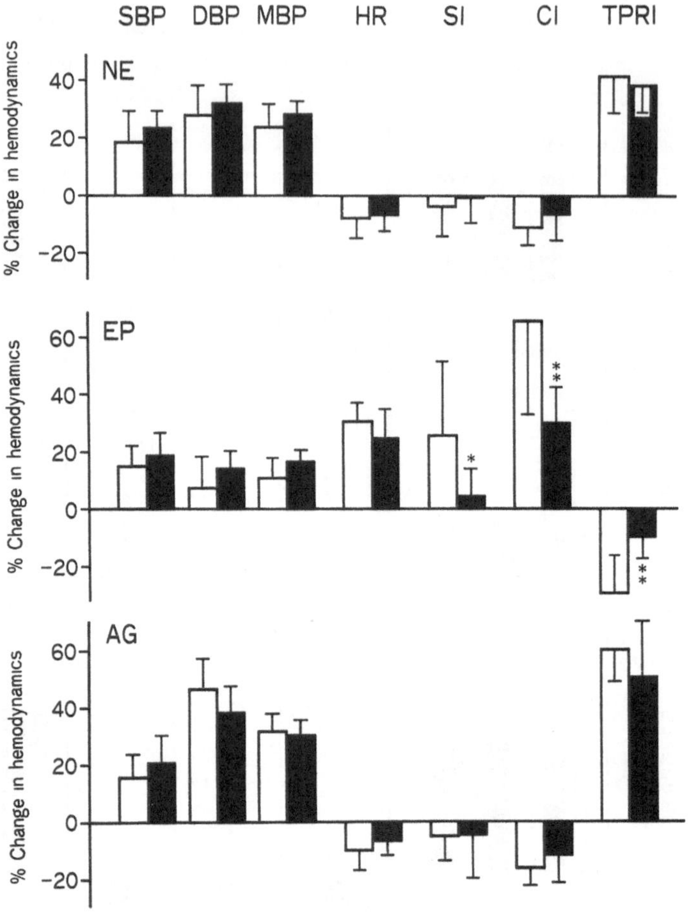

Fig. 4. Comparison of percentage hemodynamic changes to norepinephrine (*NE*), epinephrine (*EP*), and angiotensin II (*AG*) between normotensives (□) and hypertensives (■). *SBP*, systolic blood pressure; *DBP*, diastolic blood pressure, *MBP*, mean blood pressure; *SI*, stroke volume index; *CI*, cardiac output index; *TPRI*, total peripheral vascular resistance index. $*P<0.05$, $**P<0.01$, $***P<0.001$

and vascular resistance (43 ± 14% vs 40 ± 11%) during norepinephrine infusion did not differ between the normotensives and the hypertensives. The percentage decreases in HR (−8 ± 8 vs −8 ± 7), stroke volume (−5 ± 10% vs 0 ± 9%), and cardiac output (−11 ± 7% vs − ± 9%) did not differed between the two groups (Fig. 4, *upper column*).

During epinephrine infusion the percentage increases in BP (15 ± 7/8 ± 12% vs 19 ± 9/14 ± 7%) and HR (31 ± 6 vs 25 ± 11%) did not differ between the normotensives and the hypertensives. However, the percentage increases in stroke volume (26 ± 25% vs 5 ± 10%, $P<0.05$) and cardiac output (65 ± 34% vs 30 ± 13%, $P<0.01$) were significantly smaller in the hypertensives than in the normotensives. The percentage reduction of vascular resistance was significantly smaller in the hypertensives than in the normotensives ($-30 \pm 15\%$ vs $-10 \pm 9\%$, $P < 0.01$) (Fig. 4, *middle column*).

During angiotensin II infusion, the percentage changes in BP, HR, stroke volume, cardiac output, and vascular resistance did not differ significantly between the two groups (Fig. 4, *lower column*).

Discussion

These results demonstrated that intravenous infusions of norepinephrine and angiotensin II (Ang II) raised both BP and vascular resistance with accompanying reductions in stroke volume and cardiac output, and that epinephrine infusion raised BP with increased stroke volume and cardiac output associated with a reduction of vascular resistance in both normotensives and borderline essential hypertensives. The results indicate that the rise in BP during infusion of norepinephrine and Ang II was caused by an elevation in vascular resistance, in contrast the rise during epinephrine infusion was caused by an increase in cardiac output.

In addition, we observed that the absolute rises in BP were greater in the hypertensives than in the normotensives during infusions of norepinephrine and Ang II, but the absolute changes in stroke volume, cardiac output, and vascular resistance were not greater. The percentage changes in all hemodynamic indexes did not differ in the two groups during infusions of norepinephrine and Ang II. However, during epinephrine infusion, the absolute increase in BP was greater in the hypertensives, but both the increases in stroke volume and cardiac output and the absolute reduction of vascular resistance were smaller in the hypertensives. The percentage increases in stroke volume and cardiac output and the reduction in vascular resistance were smaller in the hypertensives.

Responses to Norepinephrine and Angiotensin II

Hyperresponses of BP to norepinephrine and Ang II with no hypersensitivities to the agonist were observed in the borderline essential hypertensives in this study. Egan et al. [11] have reported that sensitivities of forearm blood flow and forearm vascular resistance to infusions of norepinephrine and Ang II, defined as the percentage increase in the lowest dose of each agonist, are similar in the mild essential hypertensives and normotensives, suggesting a structural vascular abnormality in the hypertensives. Blood flow is maintained normally in the hypertensives, however albeit workload of the heart and vessels is higher, consequently the arteriolar cross-sectional area is reduced by 4% in the mild or borderline essential hypertensives. The reduction in arterial lumen might cause the elevated resistance in the mild as well as borderline essential hypertensives [11]. The marked BP responses without hypersensitivities might be induced by the reduction of arterial lumen with an increased wall/lumen ratio, which might be due to an overcontraction of vascular smooth muscle [11].

An exaggerated blood pressure response to infused norepinephrine occurred in normotensive members of hypertensive families, with no accompanying hypertensive structural change. The marked BP response was characterized by a decrease in the norepinephrine pressor dose, a shift in the relation between norepinephrine induced a rise in BP and plasma norepinephrine concentrations, and an enhanced BP response to plasma norepinephrine concentration over 20 mol/l [2]. The relationship of norepinephrine infusion rates and plasma concentrations of norepinephrine was similar in both groups [2]. Therefore, the hyperresponses could not be explained by altered metabolism of circulating norepinephrine or by structural changes in the blood vessel walls. The hyperresponse in hypertensive families may be due to a hypertension gene which may carry an abnormality in the regulation of contractile response of vascular smooth muscle cells and a mechanism which could attract enough calcium into cells to cause an increase in vascular tone.

Responses to Epinephrine

Epinephrine infusion decreased diastolic BP and vascular resistance in borderline essential hypertensives and normotensives, and increased stroke volume and cardiac output in both groups in this study.

Unlike norepinephrine, epinephrine reduced vascular resistance, which may result from a relaxant response of arterial smooth muscle to epinephrine through stimulation of β-adrenoceptors. This differing effect of norepinephrine and epinephrine on hemodynamics could be explained by the fact that epinephrine is a more potent β-adrenoceptor stimulant than norepinephrine. The reduction of vascular resistance by epinephrine was significantly smaller in the hypertensives (-3 Um^2) than normotensives (-8 Um^2). Therefore, the observed results suggest that the relaxant effect of epinephrine on β-adrenoceptors in arterial smooth muscle may be diminished in the hypertensives. Fujimoto et al. [12] have demonstrated a reduced relaxant response of the muscle to β-agonists in spontaneously hypertensive rats (SHR). Thus, it may be that the response of vasodilation to epinephrine is reduced in humans with essential hypertension as well as in rats with spontaneous hypertension (SHR).

Lande et al. [13] have reported that a small and equal dose of epinephrine elicits a more pronounced forearm vasodilation through β-adrenoceptor stimulation in mild essential hypertensives than in normotensives. The results may be explained by an up-regulation of β-adrenoceptors in vascular smooth muscle in the hypertensives, who are hyper-responders to adrenergic stimulation. Thus, it may be conclusive that β-adrenoceptor activity is decreased in vascular smooth muscle of human essential hypertensives. We observed that the percentage changes in stroke volume and cardiac output in response to epinephrine were significantly smaller in the hypertensives than in the normotensives. The smaller percentage changes may be explained by the fact that the hypertensives have a diminished β-adrenoceptor-mediated inotropic effect of epinephrine on the heart.

Family History of Hypertension

The possible etiological role of factors such as altered vascular reactivity has been evaluated by comparing normotensive individuals with and without a biological parental history of hypertension [1–4, 8, 14, 15]. In normotensives, dynamic exercise elevates systolic BP, whereas the diastolic BP remains the same or falls slightly during dynamic exercise [2, 8, 15–19]. However, dynamic exercise elevates systolic BP and diastolic BP in hypertensives. The rise tends to be greater in hypertensives than in normotensives [2, 8, 15–19]. An exaggerated BP response to infused norepinephrine occurs in normotensive members of hypertensive families [2]. The sensitivity of forearm blood flow to norepinephrine also increases in normotensive sons of hypertensive parents [1]. These results indicate that the exaggerated BP responses to stimuli may be present as a common familial disturbance in normotensive adolescents; these may be the pathophysiological changes that precede sustained BP elevation in essential (gene) hypertensives [8, 14].

Probability of Future Hypertension

In a study of Dlin et al. [6], one group showed a marked BP response to exercise and the other group showed a normal response in normotensive individuals. After a 6-year follow-up period a number of the individuals in the high response group (12%) developed hypertension, whereas none from the normal response group developed hypertension [6]. If there is indeed a process in which hypertension becomes apparent only under stressful conditions, this would suggest that cardiovascular alterations are underway before the appearance of an elevated resting BP in essential hypertensives.

The reviewed data to assess the probability of future hypertension in subjects with a "hypertensive" response to exercise, indicate that the sensitivity of a hypertensive response to exercise for future hypertension varied between 16% and 60%, and the specificity between 53% and 95% [15, 16]. The variability in the reported blood pressure response is probably due to the heterogeneity of the population tested, including subjects with gene hypertension, environment hypertension (for example, obesity hypertension, salt hypertension), and disease hypertension (for example, renal hypertension, arteriosclerotic hypertension), as well as varying ages, sex, weight, and physical fitness [15, 16]. Moreover, the subjects may have a major hypertension gene. For the evaluation of hyperresponse and hypersensitivity of cardiovascular response in the heterogeneity of a syndrome of high blood pressure, subjects with hypertension should first be classified into gene hypertension those having the major hypertension gene, environment hypertension, and disease hypertension (see chapter 2 in this volume), and age, sex, weight matched control.

In conclusion, it is a fact that blood pressure tends to rise with age, but not in everyone. The higher a person's blood pressure, the higher the blood pressure of first degree relatives. Similarly, the lower the blood pressure, the lower the blood pressure of first degree relatives [16]. The findings suggest that marked blood pressure responses are present in hypertension gene hypertensives at the normotensive and early stages of hypertension, which is transferred from parents to offspring via the incomplete dominant form.

References

1. Doyle AE, Fraser JRE (1961) Essential hypertension and inheritance of vascular reactivity. Lancet ii: 509–511
2. Bianchetti MG, Weidmann P, Beretta-Piccoli C, Rupp U, Boehringer K, Link L, Ferrier C (1984) Disturbed noradrenergic blood pressure control in normotensive members of hypertensive families. Br Heart J 51: 306–311
3. Falkner B, Onesti G, Angelakos ET, Fernandes M, Langman C (1979) Cardiovascular response to mental stress in normal adolescents with hypertensive parents. Hemodynamics and mental stress in adolescents. Hypertension 1: 23–30
4. Aoki K, Asano M (1986) Effects of Bay K 8644 and nifedipine on femoral arteries of spontaneously hypertensive rats. Br J Pharmacol 88: 221–230
5. Molineux D, Steptoe A (1988) Exaggerated blood pressure responses to submaximal exercise in normotensive adolescents with a family history of hypertension. J Hypertens 6: 361–365
6. Dlin RA, Hanne N, Silverberg DS, Bar-Or O (1983) Follow-up of normotensive men with exaggerated blood pressure response to exercise. Am Heart J 106: 316–320
7. Aoki K, Sato K, Kondo S, Pyon C, Yamamoto M (1983) Increased response of blood pressure to rest and handgrip in subjects with essential hypertension. Jpn Circ J 47: 801–809
8. Aoki K, Kato S, Mochizuki A, Kawaguchi Y, Yamamoto M (1982) Abnormal response of blood pressure to Master's two-step exercise in patients with essential hypertension. Jpn Circ J 46: 261–266
9. Widimsky J, Jandova R, Ressel J (1981) Hemodynamic studies in juvenile hypertensives at rest and during supine exercise. Eur Heart J 2: 307–315
10. Schieken RM, Clarke WR, Lauer RM (1983) The cardiovascular response to exercise in children across the blood pressure distribution: the Muscatine Study. Hypertension 5: 71–78
11. Egan B, Schork N, Panis R, Hinderliter A (1988) Vascular structure enhances regional resistance responses in mild essential hypertension. J Hypertens 6: 41–48
12. Fujimoto S, Dohi Y, Aoki K, Matsuda T(1988) Beta-1 and beta-2 adrenoceptor-mediated relaxation responses in peripheral arteries from spontaneously hypertensive rat at prehypertensive and early hypertensive stages. J Hypertens 6: 543–550
13. Lande K, Kjeldsen ES, Os I, Westheim A, Hjermann I, Eide I, Gjesdal K (1988) Increased platelet and vascular smooth muscle reactivity to low-dose adrenaline infusion in mild essential hypertension. J Hypertens 6: 219–225
14. Aoki K, Sato K, Kondo S, Pyon C-B, Yamamoto M (1983) Increased response of blood pressure to rest and handgrip in subjects with essential hypertension. Jpn Circ J 47: 802–809
15. Benbassat J, Froom P (1986) Blood pressure response to exercise as a predictor of hypertension. Arch Intern Med 146: 2053–2055
16. Watt G (1986) Design and interpretation of studies comparing individuals with and without a family history of high blood pressure. J Hypertens 4: 1–7
17. Leibel B, Kobrin I, Ben-Ishay D (1982) Exercise testing in assessment of hypertension. Br Med J 285: 1535–1536
18. Brorson L, Wasir H, Sannerstedt R (1978) Haemodynamic effects of static and dynamic exercise in males with arterial hypertension of varying severity. Cariovasc Res 12: 269–275
19. Wolthuis RA, Froelicher VF, Fischer J, Triebwasser JH (1977) The response of healthy men to treadmill exercise. Circulation 55: 153–157

Responsiveness to Calcium Antagonists in Essential Hypertension: Pharmacokinetic and Neurohumoral Aspects

JOHN L. REID, RICHARD DONNELLY, PETER A. MEREDITH, and HENRY L. ELLIOTT[1]

Summary. There are considerable interindividual differences in the blood pressure response to treatment with a calcium antagonist, and this reflects both pharmacokinetic and pharmacodynamic variability. We have studied the acute and chronic responses to monotherapy with nifedipine ($n = 14$) and verapamil ($n = 14$) and measured hemodynamic and neuroendocrine indices. An integrated kinetic-dynamic modelling approach was used to give a mathematical description of the antihypertensive response in terms of the fall in blood pressure per unit drug concentration, and this index of responsiveness derived from acute dosing accurately predicted the responsiveness after 4–6 weeks. While the responsiveness to treatment was not influenced significantly by age, plasma renin activity, or plasma norepinephrine, the height of the pretreatment blood pressure was an important determinant of the response to both verapamil and nifedipine.

This integrated approach permits a rigorous appraisal of factors which influence the variability in the antihypertensive drug response. In particular, pharmacokinetic variability can be accounted for as well as effects due to time and dose-related changes in plasma drug concentration. We believe that a clearer understanding of factors determining the response to calcium antagonists will help in the rational choice of first line therapy in essential hypertension.

Key words: Calcium antagonists—Concentration effect analysis—Hypertension—Pharmacokinetics—Pharmacodynamics

Introduction

In recent years several new groups of antihypertensive drugs have been proposed as alternatives to a diuretic or beta blocker as first line treatment in essential hypertension [1]. There is particular support for the angiotensin-converting enzyme inhibitors and calcium antagonists on the grounds of their established efficacy and the increasing evidence of overall safety and good patient tolerance. Numerous clinical studies have documented the usefulness and patient acceptability of these agents both as monotherapy and in rational combinations with other antihypertensive drugs.

The wider selection of first line therapy has led to the proposal of individualising the choice of treatment rather than following a pragmatic, standardised step-

[1]Department of Materia Medica and Therapeutics, University of Glasgow, Glasgow, UK

care regimen [1]. The aim of this approach is to take into account a range of simple demographic factors (age, race, sex) together with other risk factors (lipids and smoking) and associated diseases (angina, heart failure, COAD, diabetes, etc) to develop a hierarchy of choices for the first line treatment in an individual patient. This is an important and laudable goal, but for many groups of drugs, including calcium antagonists, the factors which influence the antihypertensive response have not been clearly identified. Various claims that the response to a calcium antagonist is enhanced in the elderly [2] and reduced in blacks [3] or patients receiving diuretics [4] have been disputed [5, 6]. We have recently examined in an intensive and systematic manner factors determining the antihypertensive response to the calcium antagonists nifedipine [7] and verapamil [8].

Concentration-Effect Analysis and the Response to Antihypertensive Drugs

In general the factors influencing the drug response are pharmacokinetic or pharmacodynamic. Kinetic factors include influences on drug disposition and elimination over time, in addition to genetic and environmental effects. Dynamic variability will result from genetic factors, disease state, and the level of activity of the neural and endocrine mechanisms controlling blood pressure.

The response to an antihypertensive drug is not easy to define or to measure. Response implies a fall in blood pressure from elevated levels to normal or near normal. However, the definition of response must include some consideration of dose and time of the measurement after giving the dose. Also, acute effects may differ from long-term responses. Clinically, a response to a drug implies not only efficacy but also acceptability and absence of adverse effects.

We are endeavouring to describe the profile of the antihypertensive response in mathematical terms by integrating data on plasma drug concentrations (i.e., kinetics) and blood pressure reduction (i.e., dynamics) over a 12- or 24-h dosage interval [7, 8]. Such a mathematical approach allows us to derive the "responsiveness" of individual hypertensive patients in terms of fall in blood pressure (in mmHg) per unit change in drug concentration. Responsiveness thus defined differs from "response" in being a mathematical function which also takes account of time and dose-related changes in plasma drug concentration for an individual. Responsiveness does not, however, take account of other clinical aspects such as acceptability, side effects, or quality of life.

In the case of several classes of antihypertensive drugs, if an attempt is made to relate plasma levels after acute dosing to the effect on blood pressure either for individuals or groups of subjects, then no simple direct linear relationship emerges. However, in the case of the calcium antagonist nifedipine there is an apparent delay in the acute response, and a plot of the first dose concentration-effect relationship reveals an anticlockwise hysteresis loop [7]. A similar relationship has been observed after single doses of the alpha$_1$-adrenoceptor antagonist prazosin [9], and it has been proposed that the hysteresis effect results from a delay in the drug in the plasma reaching equilibrium at the effector site.

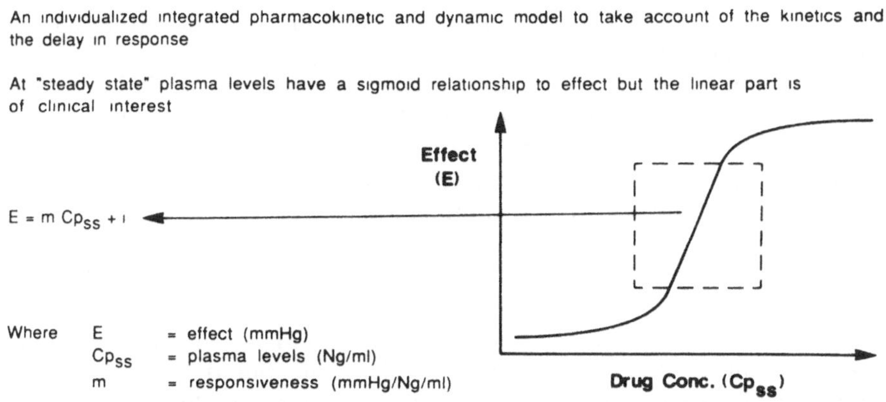

An individualized integrated pharmacokinetic and dynamic model to take account of the kinetics and the delay in response
At "steady state" plasma levels have a sigmoid relationship to effect but the linear part is of clinical interest
Effect (E)
E = m Cpss + i
Where E = effect (mmHg)
Cpss = plasma levels (Ng/ml)
m = responsiveness (mmHg/Ng/ml)
Drug Conc. (Cpss)

Fig. 1. Concentration effect analysis applied to individual responses to antihypertensive drugs

In pharmacokinetic terms, if the site of drug action is considered to be in a different "compartment" from the plasma, then the hysteresis reflects a delay in the drug concentrations reaching an equilibrium with the "effect" compartment. After long-term (chronic) dosing when steady state plasma levels have been achieved (C_{pss}), it can be assumed that the concentration in the effect compartment (C_e) is in equilibrium with that in plasma (C_{pss}).

The relationship between drug concentration and effect may be complex but will often fit a log-linear or exponential (E_{max}) model (Fig. 1). We have used this approach to analyze mathematically the response to antihypertensive drugs in individual patients. Firstly, the pharmacokinetics is analyzed using an appropriate compartmental model to derive conventional parameters, and then for the pharmacodynamic modelling an additional "effect" compartment is incorporated in the model [10, 11]. The measured effect, in this case blood pressure reduction, is then related to the drug concentration in the "effect" compartment by means of either the linear or E_{max} model. Statistical tests can be applied to determine whether the concentration-effect data are more appropriately fitted to a linear or an E_{max} relationship. In view of the established diurnal variation in blood pressure throughout the day, we have chosen to correct the blood pressure profile after drug dosing by subtracting the effect of placebo (recorded over 8 h on a separate study day). The placebo-corrected fall in blood pressure per unit change in drug concentration in the effect compartment is denoted by the slope of the linear function:

$$E = mC_e + i$$

where E is effect, C_e is drug concentration in the effect compartment and m is the slope, representing the "responsiveness" and having a unit of mmHg/ng/ml. In the case of the Langmuir E_{max} model:

$$E = \frac{E_{max}\, C_e}{C_e\,(50) + C_e}$$

E_{max} is the maximum possible effect and C_e (50) is the concentration required to produce 50% of E_{max}.

Using the linear model, we have made independent calculations of the slope of the concentration effect relationship (m) in the same individuals after the first dose and after 1 week and 6 weeks of regular therapy. There is a significant correlation between the responsiveness (m) calculated after acute and chronic dosing for both nifedipine and verapamil.

The derived measurements of responsiveness can be used to identify kinetic and dynamic factors which are associated with the interindividual variations in antihypertensive effect. We have identified wide intersubject variability in the responsiveness to both nifedipine and verapamil in patients with essential hypertension, but the differences cannot readily be attributed to simple demographic features like age or sex. Similarly, neurohumoral and endocrine markers such as plasma norepinephrine and plasma renin activity are not statistically (or clinically) important determinants of the responsiveness to calcium antagonists or $alpha_1$-blockers in individual patients. The single most important factor predicting responsiveness is the level of pretreatment blood pressure (i.e., untreated or placebo treated). It is increasingly acknowledged that pretreatment blood pressure is a major factor in determining the absolute fall in blood pressure as well as the responsiveness to a wide range of antihypertensive agents. We have found that the magnitude of the responsiveness to placebo is also correlated with levels of pretreatment blood pressure [12]. It should be noted that these relationships are independent of the statistical limitations of correlating two dependent variables, as when relating the actual change in blood pressure (BP) to starting or pretreatment pressure (BP) [13], and additionally they take account of drug concentrations and the effects of placebo.

We have developed the approach to concentration-effect analysis over several years. Our studies initially explored single doses of alpha-receptor antagonists in normotensive subjects [14], but recently we have undertaken a comprehensive study in hypertensive patients of acute and chronic dosing with alpha blockers, calcium antagonists, and an ACE inhibitor [15]. Studies have followed a common protocol and included groups of 10–14 patients with mild to moderate essential hypertension who had been off all therapy for at least 4 weeks. Patients initially received treatment with placebo for 2 weeks and attended the clinical research unit for a 10-h study period during which measurements of blood pressure, heart rate, and relevant neurohumoral indices were made before and after dosing. Similar studies were performed after the first dose of active treatment and after 4–6 weeks of long-term therapy. On the occasions when active treatment was administered, plasma drug levels were measured at frequent intervals to permit characterization in individual patients of the pharmacokinetic parameters and the integrated pharmacokinetic-pharmacodynamic relationships as described above. The responsiveness in individual patients after the first dose was compared with the responsiveness after 6 weeks' treatment, and additionally relationships between responsiveness and factors such as age, sex, plasma renin activity, and plasma catecholamines were examined.

Factors Determining the Responses to Calcium Antagonists

The responses in two groups of 14 patients with mild to moderate hypertension were studied. Blood pressure levels ranged from 160/90–210/115 and their ages from 33 to 79 years. Nifedipine was given as a 20-mg, delayed release, tablet formulation (Adalat Retard, Bayer UK Ltd). The first dose was 20 mg orally, and it was given for 6 weeks as 20 mg twice daily. Verapamil was given as a 120-mg tablet (Isoptin Knoll Ltd) and thereafter as 120 mg twice daily. At frequent intervals for 8 h after dosing, blood pressure and heart rate were measured using a semiautomatic sphygmomanometer (Datascope Accutorr) and venous blood samples collected for plasma levels of nifedipine and verapamil (by specific HPLC assays). The pharmacokinetic analysis and concentration effect analysis have been previously described in detail [7, 8] and are outlined earlier in this review. All results are expressed as mean ± SD.

In both studies the respective drugs were well tolerated, and no patient experienced significant adverse reactions. All the patients completed the study. Both nifedipine and verapamil lowered blood pressure acutely and chronically. There were wide interindividual variations in the magnitude and duration of the fall in blood pressure: the full results have been described in detail elsewhere [7, 8]. However, the responses of two individual patients, patients 4 and 9, have been selected as representative of the group receiving nifedipine and are presented in detail. These two subjects illustrate the variability in response between individuals and between study days, not only after active drug but also after placebo (Fig. 2). They highlight some of the problems described above in deciding what is the "response" to an antihypertensive drug. Baseline variation, placebo changes, and the marked difference between peak and trough responses and in the time of maximum response are illustrated (Fig. 2). There is no simple linear relationship between the fall in blood pressure and the plasma nifedipine concentration, as illustrated for patient 4 (Fig. 3); the response lagged behind the plasma concentration profile after acute dosing. However, when integrated kinetic-dynamic modelling of the concentration-effect relationships was undertaken for individual patients, a clear linear relationship could be established between the fall in blood pressure and the calculated drug concentration in the "effect" compartment (Fig. 3B). The slope of this relationship, m, represents the responsiveness and is defined in units of mmHg fall in blood pressure per unit drug concentration. The slope (m) differed between patients, reflecting interindividual differences in responsiveness (Fig. 3B). When the responsiveness derived from acute dosing was compared with that after 6 weeks, there was a statistically significant close correspondence. This is shown in Fig. 4 for patient 9. The mean (±SD) responsiveness for the group of 14 subjects after acute nifedipine was -0.48 ± 0.20 mmHg $ng^{-1}ml^{-1}$ and -0.49 ± 0.17 mmHg $ng^{-1}ml^{-1}$ after 6 weeks. A similar relationship was observed in the verapamil study, and there was a significant correlation between the responsiveness to the first dose (-0.13 ± 0.06 mmHg $ng^{-1}ml^{-1}$) and the resonsiveness after chronic (4 weeks') treatment (-0.12 ± 0.06 mmHg $ng^{-1}ml^{-1}$).

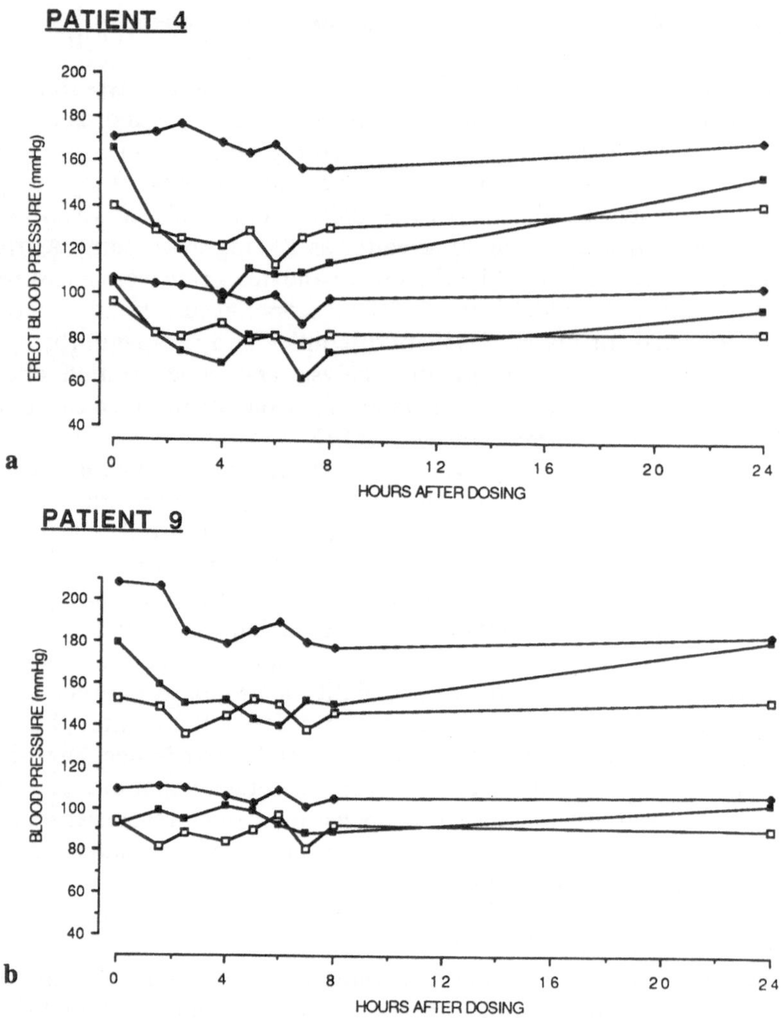

Fig. 2a,b. Individual blood pressure changes (erect systolic and diastolic) after placebo and nifedipine (20 mg) (retard formulation) in two essential hypertensives (**a** patient 4, **b** patient 9)

The influence of a range of factors on responsiveness was investigated. The only measurement which was signficantly (and positively) related to responsiveness was the level of starting blood pressure. This relationship was similar for both verapamil (not shown) and nifedipine (Fig. 5). Amongst those factors which did not appear to be related to the responsiveness to either drug were age and pretreatment plasma renin activity (Fig. 5).

Although these studies were performed on relatively small groups of patients, we did examine a wide age range, and there was no evidence of a clinically useful

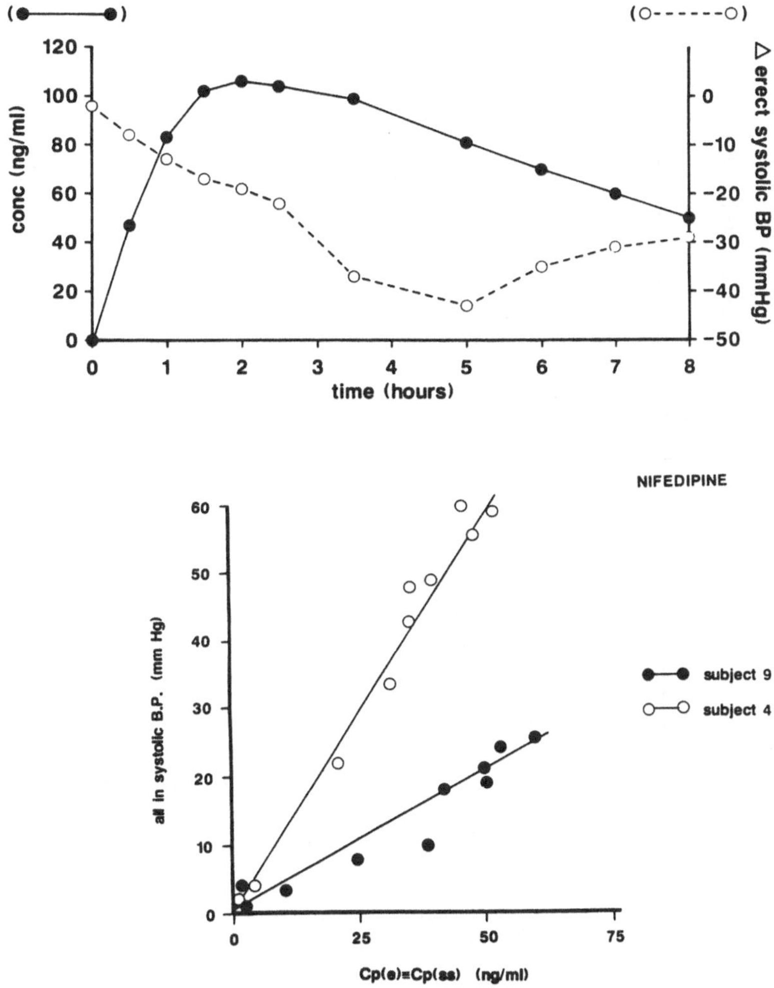

a

b

Fig. 3. a Concentration effect relationship in an individual patient when plasma levels and systolic blood pressure after nifedipine retard (20 mg) are compared. **b** Relationship between systolic blood pressure fall and calculated nifedipine concentration in the "effect" compartment in two individual patients

relationship between age and responsiveness. These studies have shown that the antihypertensive effect of a calcium antagonist is related to the drug concentration *within an individual* and that the response to treatment can be described mathematically by a linear function which incorporates kinetic as well as dynamic information in a single quantitative term. It appears from the present studies that the most important predictors of the response to a calcium antagonist during long-term therapy are the height of the pretreatment blood pressure [16, 17] and the magnitude of the acute (first dose) response.

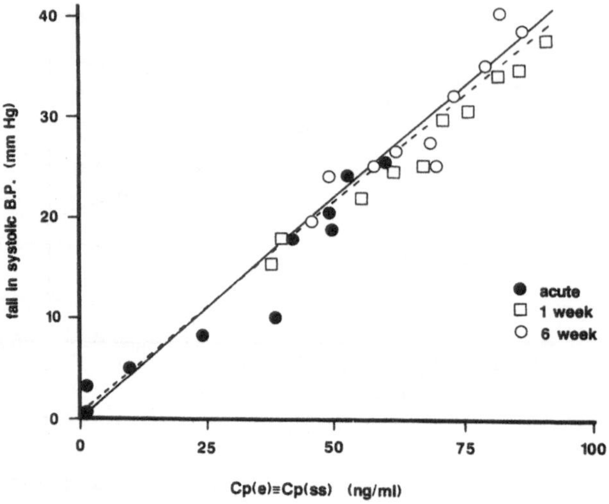

Fig. 4. Relationship between systolic blood pressure fall and nifedipine concentration in the "effect" compartment after the first dose of nifedipine (*solid circles*) and after chronic treatment for 6 weeks (*open circles*) in an individual patient

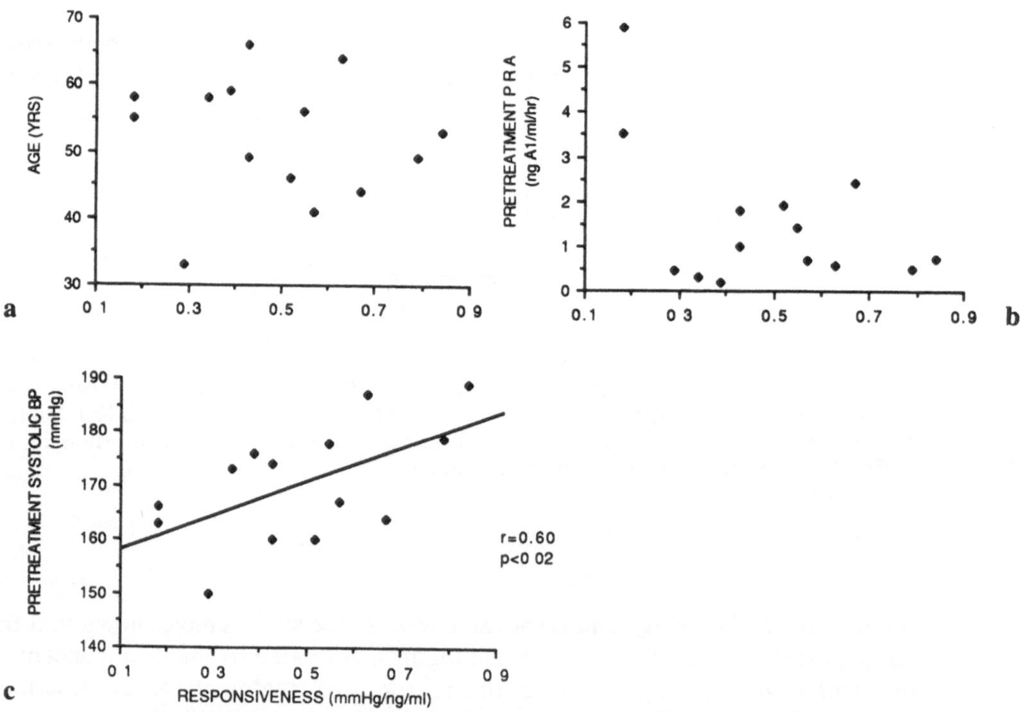

Fig. 5a–c. Relationship between "responsiveness" of individual hypertensive patients to nifedipine (mmHg $ng^{-1}ml^{-1}$) and age (**a**) pretreatment plasma renin activity (**b**), and pretreatment systolic blood pressure (**c**). Each point refers to an individual patient. Responsiveness is the slope of the linear relationship derived from the integrated concentration effect analysis

References

1. Reid JL (1988) Hypertension 1988 present challenges and future strategies. J Hypertension 6: 3–8
2. Buhler FR, Kiowski W (1987) Age and antihypertensive response to calcium antagonists. J Hypertension 5 (Suppl 4): S11–S114
3. Buhler FR, Hulthen L, Kiowski W, Boli P (1982) Greater antihypertensive efficacy of the calcium channel inhibitor in older and low renin patients. Clin Sci 63: 439S–442S
4. Cappuccio FP, Markandu ND, Tucker F, Sagnella GA, Macgregor GA (1986) Does a diuretic cause of further fall in blood pressure in hypertensive patients already on nifedipine? J Clin Hypertension 4: 346–353
5. Sever PS, Poulter NR (1987) Calcium antagonists and diuretics as combined therapy. J Hypertension 5 (Suppl 4): S123–S126
6. Ram CVS (1987) Calcium antagonists as antihypertensives are effective in all age groups. J Hypertension 5 (Suppl 4): S115–S118
7. Donnelly R, Elliott HL, Meredith PA, Reid JL (1987) Acute and chronic nifedipine in essential hypertension—concentration effect relationships. Clin Sci 73: 54
8. Meredith PA, Elliott HL, Ahmed JH, Reid JL (1987) Age and the antihypertensive efficacy of verapamil—an integrated pharmacokinetic-pharmacodynamic approach. J Hypertension 5 (Suppl 5): S219–S221
9. Elliott HL, Vincent J, Meredith PA, Reid JL (1988) Relationship between plasma prazosin concentrations and alpha antagonism in man: comparison of conventional and rate controlled (Oros) formulations. Clin Pharmacol Ther 43: 582–587
10. Kelman AW, Whiting B (1980) Modelling of drug response in individual subjects. J Pharmacokinetics Biopharmacol 8: 115–130
11. Holford NHG, Sheiner LB (1981) Understanding the dose effect relationship: clinical application of pharmacokinetic-pharmacodynamic modelling. Clin Pharmacokin 6: 429–453
12. Sumner DJ, Meredith PA, Howie CA, Elliott HL, Reid JL (1988) Initial blood pressure as a predictor of the response to antihypertensive therapy. Br J Clin Pharmacol 26(6): 715–720
13. Gill JS, Beavers DG, Zezulka AV, Davies P (1985) Relationship between initial blood pressure and its fall with treatment. Lancet 1: 567–569
14. Meredith PA, Elliott HL, Kelman AW, Reid JL (1985) A comparison of quinazoline alpha adrenoceptor antagonists by application of pharmacokinetic-pharmacodynamic modelling. J Cardiovasc Pharmacol 7: 532–537
15. Donnelly R, Elliott HL, Meredith PA, Reid JL (1987) Antihypertensive response and individual concentration effect relationships. J Hypertension 5: 766–767
16. Reid JL, Elliott HL (1989) Calcium antagonists in essential hypertension: clinical pharmacological aspects. In: Aoki K, Frohlich ED (eds) Calcium in essential hypertension. Academic Press, Tokyo San Diego New York London, pp 575–655
17. Aoki K, Sato K (1989) Hyperdynamic state in normotensives and hypertensives with calcium antagonists at rest. In: Aoki K, Frohlich ED (eds) Calcium in essential hypertension. Academic Press, Tokyo San Diego New York London, pp 537–574

Hemodynamic Effects of Calcium Antagonists at Rest and During Exercise in Essential Hypertension: A Comparison Between Verapamil, Diltiazem, Tiapamil, Nifedipine and Nisoldipine

PER LUND-JOHANSEN and PER OMVIK[1]

Summary. The cardinal hemodynamic disorder in established essential hypertension is an increased total peripheral resistance. In the starting phase of essential hypertension cardiac output during rest may be increased, but this is not the case during exercise when cardiac output tends to be subnormal.

Since calcium antagonists induce a reduction in blood pressure through vasodilatation, they are thought to be of special interest in the treatment of hypertension. However, since they also have a negative inotropic effect, knowledge about their effect on heart pump function during exercise appears important.

Central hemodynamics were studied invasively at rest and during exercise in 76 patients with mild to moderate essential hypertension and restudied after one year of treatment with verapamil ($n = 9$), tiapamil ($n = 19$), diltiazem ($n = 16$), nifedipine ($n = 15$), or nisoldipine ($n = 17$). After long-term treatment, blood pressure was reduced by about 10%–17% on the different regimes. The fall in blood pressure was associated with a reduction in total peripheral resistance in all drug regimes. Exercise heart rate was reduced approximately 10% with verapamil, diltiazem and tiapamil, but the reduction in heart rate was compensated by an increase in stroke volume, and thus the cardiac output was unchanged. No heart rate or stroke volume changes were seen during long-term treatment with nifedipine or nisoldipine.

Acute studies (tiapamil and nisoldipine) demonstrated that the initial drop in blood pressure—due to a marked and significant fall in total peripheral resistance—was associated with reflex tachycardia and increase in cardiac output. During long-term treatment the reflex tachycardia disappeared.

Side-effects were few with all regimes. In conclusion, all five calcium antagonists studied in these series reduced blood pressure at rest and during exercise through reduction in total peripheral resistance, without reduction in cardiac output. This hemodynamic profile is similar to that found with alpha-blockers and ACE inhibitors, but is in clear contrast with that observed with conventional beta-blockers.

Key words: Hypertension—Hemodynamics—Calcium antagonists—Verapamil—Diltiazem—Nifedipine—Nisoldipine—Tiapamil

Introduction

The hemodynamic mechanisms behind the increased blood pressure in essential hypertension differ widely and depend at least on the age of the patient and the

[1] Section of Cardiology, Medical Department, University of Bergen, School of Medicine, Haukeland Hospital, N–5021 Bergen, Norway

stage of the hypertensive disorder. The cardinal hemodynamic disturbance is the increased vascular resistance seen in most vascular beds, but the extent differs in various areas [1, 2]. The calcium ion seems to be an important factor in this process. In relatively young subjects (20–40 y) with mild hypertension, the typical hemodynamic pattern is an increased cardiac index during rest with almost normal total peripheral resistance. During exercise, however, total peripheral resistance is clearly increased and the heart pump function is disturbed with a subnormal stroke index—probably due to decreased compliance of the left ventricle [3]. Follow-up studies from our laboratory have shown that the hemodynamic pattern changes over time, with a fall in cardiac output and a progressive increase in total peripheral resistance [3].

Since calcium antagonists reduce blood pressure through vasodilatation, they are thought to be of particular interest in the treatment of hypertension. However, since they also have a negative inotropic effect, knowledge about the heart pump function during exercise seems important [4–6].

The purpose of this paper is to review briefly our hemodynamic results with five different calcium antagonists in patients with mild to moderate essential hypertension: verapamil, diltiazem, tiapamil, nifedipine, and nisoldipine. Some of these data have been published in detail previously [7–9]. The results will be compared briefly with our results obtained by beta-blockers, alpha-blockers and angiotensin converting enzyme inhibitors.

Materials and Methods

All patients had essential hypertension of mild or moderate degree and had been referred to our out-patient clinic and blood pressure had been measured repeatedly over at least 3 months, with a diastolic blood pressure of greater than 100 mmHg. Patients who had been on previous treatment were taken off drugs for at least 2 months before they were included in the study. All gave an informed consent.

The hemodynamic studies were performed on an out-patient basis between 9 a.m. and 1 p.m. Before the study the patients had a light breakfast without coffee or tea. Intra-arterial pressures [systolic (SAP), diastolic (DAP), and mean

Table 1. Patient profile before (1) and during (2) calcium antagonist therapy

Calcium antagonist	n	Age (years)		Casual BP (mmHg)		BSA (m^2)	Daily dose (mg)	
		Range	(mean)	1	2		Range	(mean)
Verapamil	9	35–55	(45)	166/106	143/92	2.01	120–240	(220)
Diltiazem	16	38–64	(52)	174/106	144/86	1.99	180–360	(278)
Tiapamil	19	19–64	(45)	166/106	148/92	1.97	600–1200	(980)
Nifedipine	15	20–64	(44)	160/104	140/94	1.99	40–80	(52)
Nisoldipine	17	29–61	(43)	167/108	141/88	2.01	10–40	(25)

BSA, body surface area

(MAP)] were measured continuously through an indwelling catheter in the brachial artery and heart rate (HR) was recorded by electrocardiogram. Cardiac output was measured by dye dilution (Cardiogreen) and oxygen consumption by the Douglas bag technique. Beckman gas analyzers were used for O_2 and CO_2 measurements. In some studies, body fluid volumes were measured by the isotope dilution technique [^{131}J human albumin for plasma volume (PV) and $^{35}SO_4$ for extracellular fluid volume (ECF)]. Cardiac index (CI), stroke index (SI) and total peripheral resistance index (TPRI) were calculated by conventional formulas. Hemodynamic recordings were made at supine rest, and sitting, and during steady state bicycle exercise at 50 w, 100 w, and 150 w (verapamil, diltiazem, nifedipine).

In the nisoldipine and tiapamil studies the acute effect of the first dose was studied and only one work load of 100 w was used. After the first hemodynamic study, the patients were started on monotherapy with one of the five calcium antagonists. After 9–12 months the hemodynamic study was repeated.

Drug Treatment and Side-Effects

The number of patients in each drug study, the age distribution, casual blood pressures, and the daily dose are given in Table 1. Generally the starting dose was approximately half of the later daily maintenance dose. The aim of treatment was a casual blood pressure of less than or equal to 140/90 mmHg with no side effects.

In general, the calcium antagonists were well tolerated. On verapamil and diltiazem no complaints were recorded. In particular, no patients had problems with constipation. For the other drugs the results were as follows: tiapamil: palpitations 1, flushing 1; nifedipine: palpitations 1, flushing 3; nisoldipine: palpitations 2, flushing 4, edema 4.

The palpitation and flushing complaints usually disappeared within the first 3 weeks. However, edema necessitated withdrawal of two patients in the nisoldipine group. Plasma volume and extracellular fluid volume (measured in the tiapamil, diltiazem and nisoldipine series) showed no significant changes.

Hemodynamic Results

The most important results are seen in Tables 2 and 3 and in Figs. 1–4.

Verapamil

At the first hemodynamic study, the following mean values were found at rest sitting: SAP/DAP 167/102 mmHg, CI 2.70 $1\ min^{-1}\ m^{-2}$ and TPRI 3848 dyn s $cm^{-5}\ m^2$. Blood pressures (SAP, DAP and MAP) were reduced by about 10% at rest, and slightly less during exercise. The blood pressure reduction was associated with a statistically significant reduction in TPRI at rest sitting (14%, $P < 0.001$) (Fig. 1). During exercise the reduction in resistance was modest (5%–7%). The heart rate was virtually unchanged during rest, but was decreased by about 8%–11% during exercise. This reduction in heart rate was completely compensated by an increase in stroke volume; CI was practically

Table 2. Hemodynamics at supine rest, sitting and during 100 w exercise before treatment (Mean values ± SD)

Calcium antagonist	Supine rest			Sitting			100 w		
	HR beats/min^{-1}	CI l min^{-1} m^{-2}	MAP mmHg	HR beats/min^{-1}	CI l min^{-1} m^{-2}	MAP mmHg	HR beats/min^{-1}	CI l min^{-1} m^{-2}	MAP mmHg
Verapamil	67.3 ± 11.8	3.30 ± 0.86	116.4 ± 9.0	72.7 ± 13.3	2.70 ± 0.45	127.4 ± 14.2	129.9 ± 24.0	7.20 ± 0.93	140.3 ± 22.6
Diltiazem	66.7 ± 11.1	2.97 ± 0.46	124.8 ± 9.5	68.9 ± 8.4	2.43 ± 0.42	137.0 ± 11.1	130.1 ± 16.4	6.76 ± 0.74	152.2 ± 18.5
Tiapamil	62.2 ± 11.6	2.92 ± 0.28	118.6 ± 10.1	68.2 ± 13.0	2.46 ± 0.27	129.1 ± 9.2	124.9 ± 16.2	6.84 ± 0.86	149.2 ± 14.0
Nifedipine	68.1 ± 12.5	3.28 ± 0.82	117.7 ± 9.0	72.5 ± 15.6	2.80 ± 0.82	126.4 ± 11.5	128.5 ± 23.6	7.31 ± 1.45	141.6 ± 14.1
Nisoldipine	68.2 ± 8.8	3.11 ± 0.58	121.1 ± 11.6	73.4 ± 9.8	2.60 ± 0.42	131.1 ± 11.7	130.8 ± 15.6	7.00 ± 0.80	154.4 ± 18.5

CI, cardiac index; MAP, mean arterial pressure

Table 3. Hemodynamic effects of the calcium antagonists

Calcium antagonist	Δ HR (%)			Δ MAP (%)			Δ CI (%)		
	Rest		Work 100 w	Rest		Work 100 w	Rest		Work 100 w
	Supine	Sitting		Supine	Sitting		Supine	Sitting	
Verapamil	−3	−8**	− 8**	−10**	−10**	− 7**	+0.5	+3	+1
Diltiazem	−9**	−4	−10***	−13***	−16***	−15***	−4	+5	−2
Tiapamil	−4	−7	− 8**	−10***	−11***	− 8***	−3	−4	−4
Nifedipine	+1	−2	+ 1	−17***	−16***	−11***	−4	−5	−5
Nisoldipine	+3	+2	− 3	−14***	−15***	−12***	+6	+4	−4

MAP, mean arterial pressure; CI, cardiac index

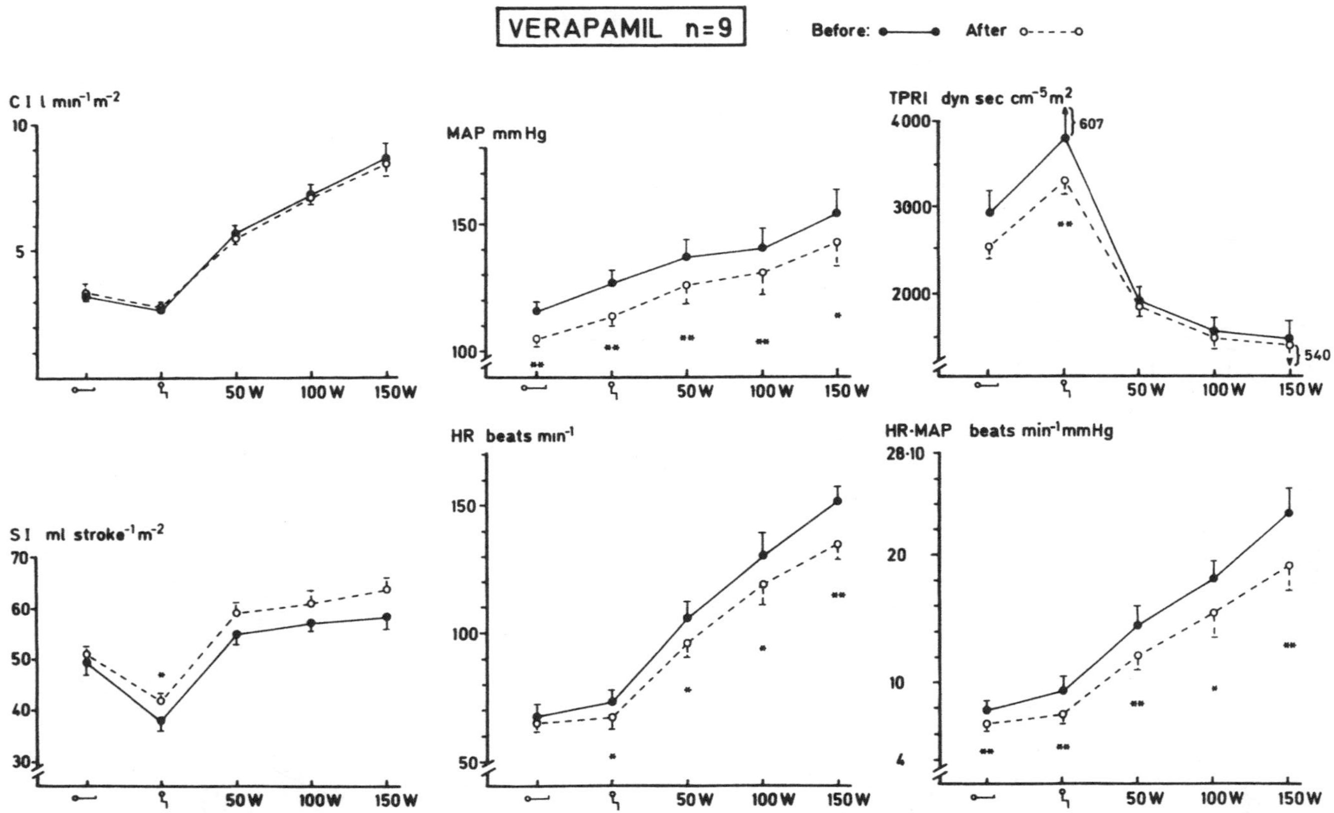

Fig. 1. Hemodynamic effects of verapamil at rest and during exercise before and after one year treatment. Mean values and SEM. *MAP* mean arterial pressure, *HR* heart rate, *CI* cardiac index, *SI* stroke index, *TPRI* total peripheral resistance index. * $P<0.05$, ** $P<0.01$, *** $P<0.001$. Note that the decrease in HR is compensated by an increase in SI. From Lund-Johansen [9]

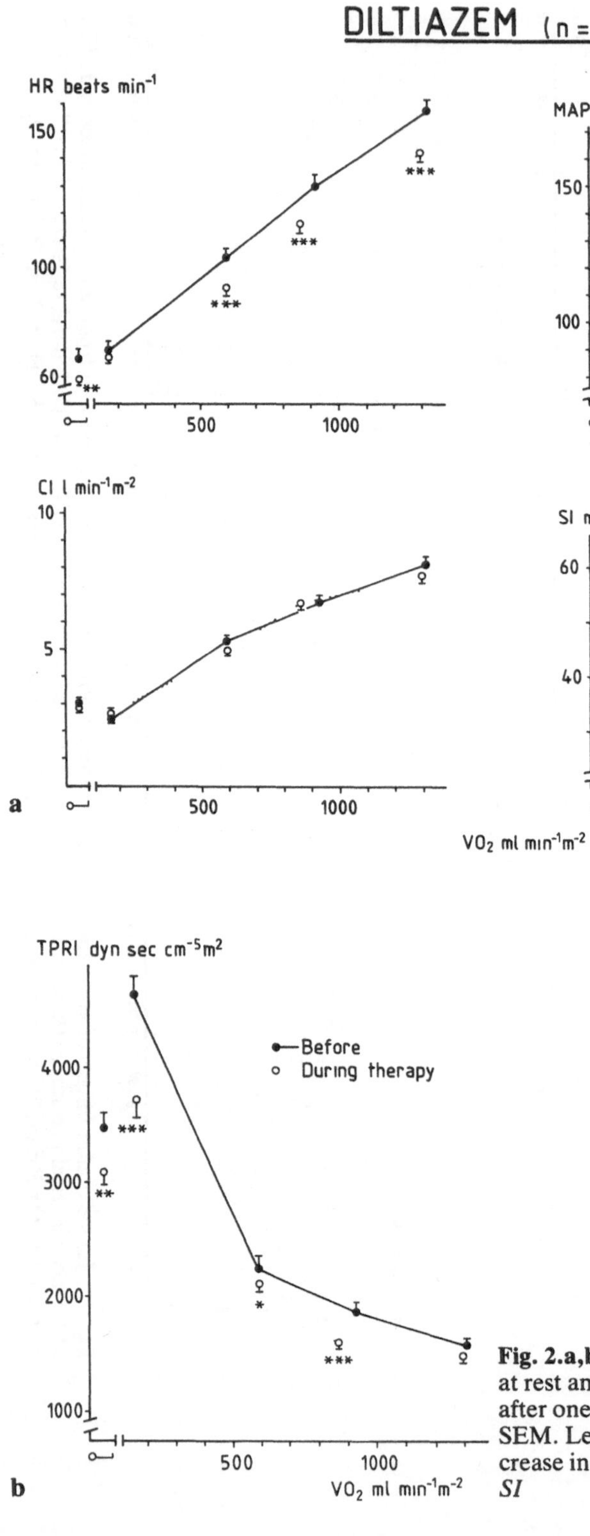

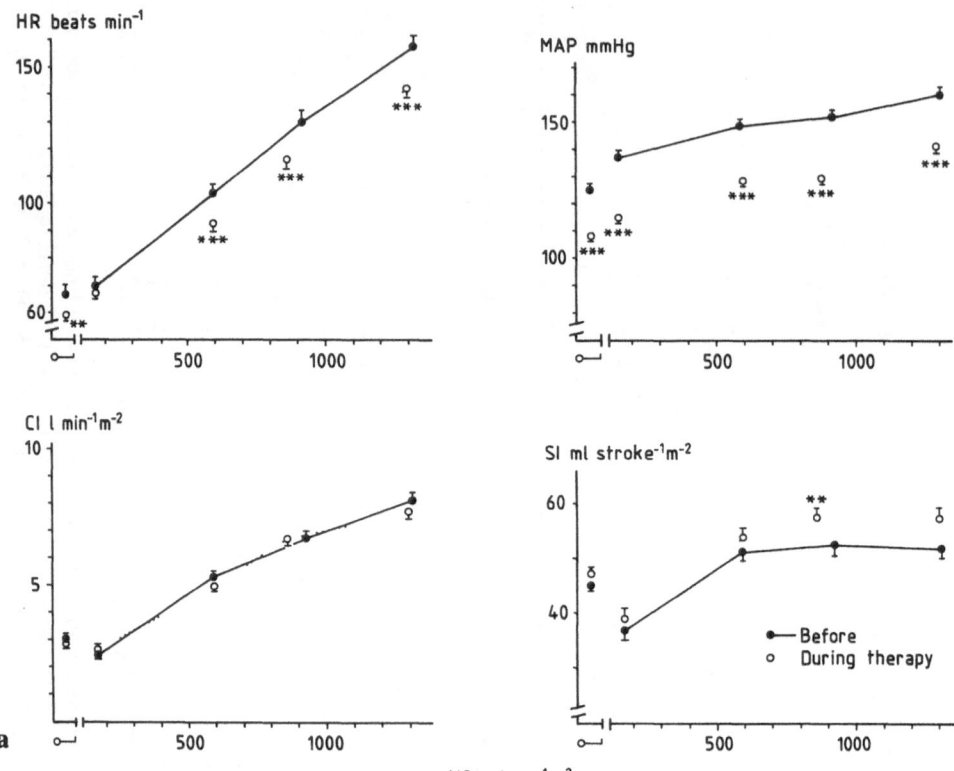

Fig. 2.a,b. Hemodynamic effects of diltiazem at rest and during exercise before and after one year of treatment. Mean values and SEM. Legend as in Fig 1. Note that the decrease in *HR* is compensated by an increase in *SI*

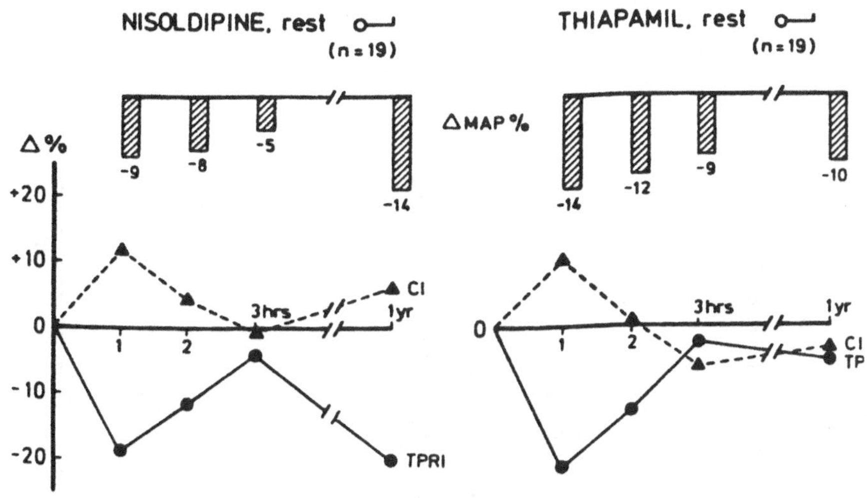

Fig. 3. Immediate and chronic hemodynamic effects of nisoldipine and tiapamil during supine rest. Legend as in Fig. 1. Note that with both drugs most of the acute effects disappear after three hours; and that during chronic treatment, nisoldipine was more effective in reducing *TPRI* than tiapamil

unchanged. It should be emphasized that the doses used in this study were relatively small. Studies by others using a daily dose of 320 mg or more have generally reported a greater reduction in blood pressure than the present study [10].

Diltiazem

The 16 patients participating in the diltiazem study had somewhat higher blood pressure and TPRI than patients participating in the other series. At the first hemodynamic study the intraarterially recorded BP in the sitting position was 183/108 mmHg and CI 2.43 $1\ min^{-1}\ m^{-2}$ and TPRI 4613 dyn s $cm^{-5}\ m^2$.

The hemodynamic profile of diltiazem (Figs. 2a, b) was relatively similar to the profile of verapamil, but the reduction in blood pressure was greater (13%–16% at rest and 12%–14% during exercise). Also diltiazem induced a fall in exercise HR. During 50, 100, and 150 w, the reduction in HR was consistently 10% ($P < 0.001$) for all situations. No patients developed bradycardia, however. The stroke index increased; during 100 w exercise the increase was 9% ($P < 0.01$). Due to the increase in SI, CI was virtually unchanged. The reduction in BP was due completely to the reduction in TPR, which was most pronounced at rest sitting (19%, $P < 0.001$).

Tiapamil

The 19 patients participating in the tiapamil trial had the following mean values at rest sitting before treatment was started: SAP/DAP 170/105 mmHg, CI 2.46 $1\ min^{-1}\ m^{-2}$, and TPRI 4269 dyn s $cm^{-5}\ m^2$.

Tiapamil (600 mg orally) induced a 21% reduction in TPRI after one hour, associated with a 7% increase in HR and an 11% ($P < 0.01$) increase in CI (Fig.

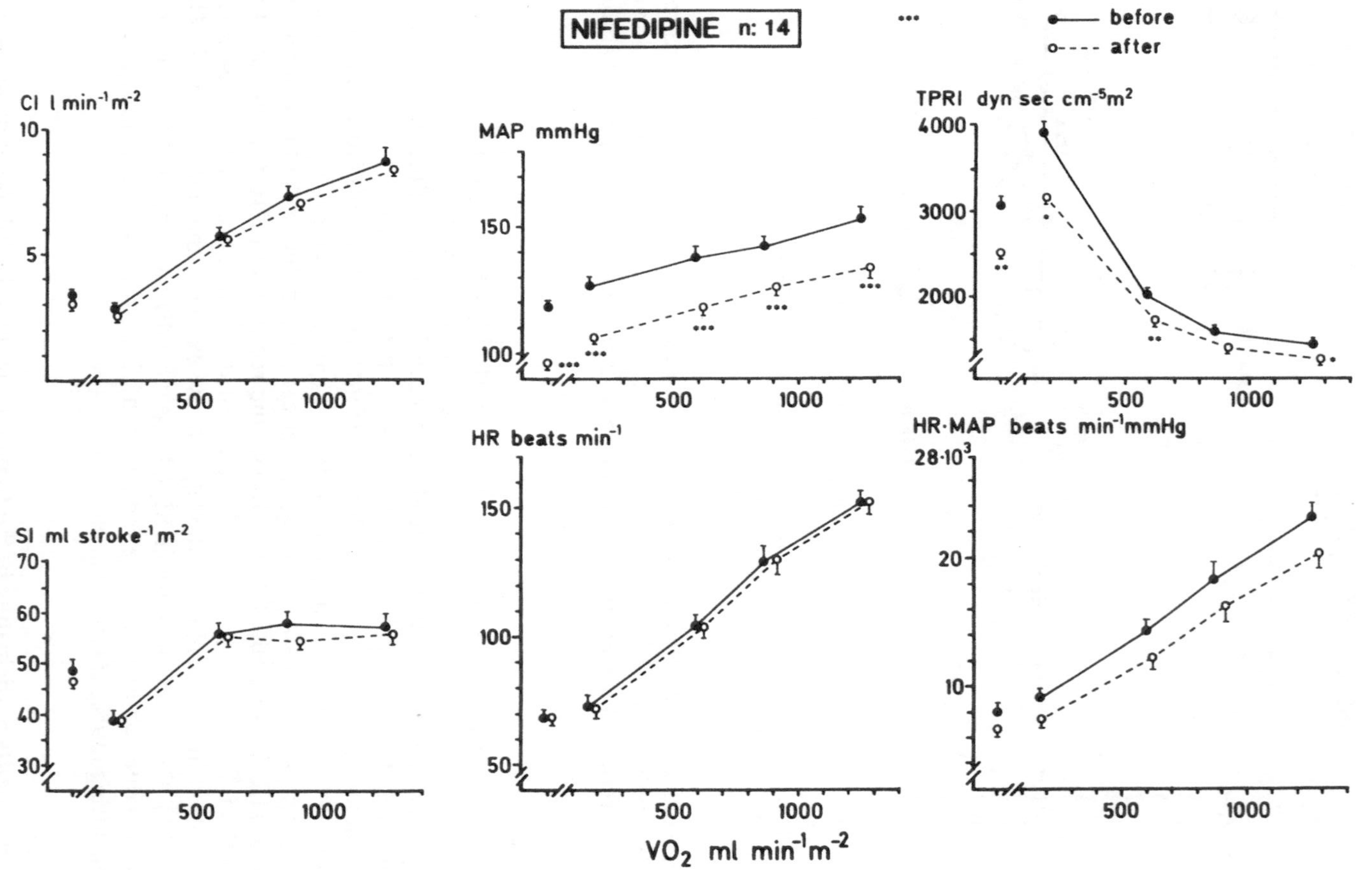

Fig. 4. Hemodynamic effects of nifedipine at rest and during exercise before and after one year of treatment. Legend as in Fig. 1. Note that *HR* is completely unchanged at rest as well as during exercise. From Lund-Johansen [8]

3). SAP, DAP and MAP all were reduced by about 14%. The effect levelled off during the subsequent three hours. During chronic treatment, casual BP fell in all 18 patients who were studied, from 166/106 mmHg after the placebo period to 148/92 mmHg shortly before the hemodynamic restudy.

Intraarterially recorded MAP was reduced by approximately 8%–12% in all situations ($P < 0.001$). The fall in blood pressure was associated with the reduction in TPRI, but this was modest (only 5%–6%, NS). In general, CI was practically unchanged. There were no significant reductions in HR during rest, but post-treatment exercise HR was reduced 8% ($P < 0.01$). The long-term reductions in TPRI on tiapamil was less (only 5%–6%, NS) than that achieved with diltiazem, nifedipine, and nisoldipine (see below).

The Dihydropyridine Derivatives (Nifedipine and Nisoldipine)

Nifedipine

Before treatment, the values for SAP/DAP, CI, and TPRI were almost identical to those in the verapamil series. All 15 patients studied achieved some reduction in BP. The fall in BP was approximately 17% during rest and 11%–15% during exercise. The changes in SAP, DAP, and MAP were quite similar. There were no changes in HR or stroke volume. The fall in BP was entirely due to the fall in TPRI of about 17%. The hemodynamic long-term profile of nifedipine is shown in Fig. 4.

Nisoldipine

Nisoldipine was studied acutely as well as chronically. The immediate and chronic effects are given in Fig. 3. One hour after the 10 mg oral administration there was a fall in TPRI of 19%. The fall in BP was partly counteracted by reflex tachycardia (HR increase of 9%) and increase in CI of 12%. Still, the immediate fall in BP was 9%. With continued treatment, the reflex tachycardia disappeared and one year after treatment was started, HR and CI were practically unchanged compared to pretreatment levels. The BP fall was 14% due to the reduction in TPRI of about 20%–18% at rest supine and sitting. During exercise the relative fall in BP was 12%, mainly due to the fall in TPRI.

Discussion

From the hemodynamic point of view, the adequate treatment of hypertension should lead to normalization of central hemodynamics—at rest as well as during exercise. The antihypertensive agents which have dominated treatment of hypertension during the last two decades—the diuretics and the beta-blockers—do not or do only partially restore normal hemodynamics. Thiazide diuretics induce chronic depression in plasma volume, and although they reduce total peripheral resistance, they are not ideal due to their many biochemical side effects [11].

Beta-blockers induce a chronic depression in cardiac output [12]. During rest, this reduction is modest, particularly for beta-blockers with intrinsic sympathicomimetic activity, but during exercise all beta-blockers induce a chronic depression in heart pump function [13, 14]. TPRI usually does not fall below pretreatment levels.

The results from our calcium antagonists series clearly show that all calcium antagonists reduced blood pressure through reduction in TPRI. In none of the series did we see a significant reduction in CI. It should be noted that the dihydropyridine derivatives on one hand, and verapamil, diltiazem, and tiapamil on the other differ with respect to the effect on HR. Verapamil, diltiazem, and tiapamil induced a 10% reduction in HR during exercise. This reduction in HR is less than what is seen on beta-blockers without intrinsic sympathicomimetic activity when HR is reduced by 25%–30% or 35–45 beats min^{-1} [12]. However, the reduction in HR seen with the calcium antagonists is completely compensated by the increase in stroke volume and no reduction in heart pump function measured as CI.

With the dihydropyridine derivatives, no changes in HR during rest or exercise were seen during chronic treatment. The acute reflex tachycardia seen with the dihydropyridine derivatives was originally thought to be a serious problem in the treatment of hypertension, but the results from our long-term study together with results from other laboratories have shown that the reflex tachycardia disappears over time. The hemodynamic results obtained in our verapamil series agree well with those seen in studies from other laboratories [10]. Studies of the regional circulation have demonstrated a fall also in forearm resistance [4]. Cody et al. [15] have also found maintenance of cardiac output during exercise on verapamil therapy.

Several studies have been performed on diltiazem [16–20]; Aoki et al. [16], Yamakado et al. [17] and Klein et al. [18] have studied diltiazem acutely and chronically. Also, in these studies it was shown that diltiazem reduces BP through a reduction in TPRI. Furthermore, studies of regional hemodynamics by Safar et al. [19] have shown a dilatation of the large and also of the small arteries. In a placebo controlled study, Pool et al. [20] found that diltiazem at 360 mg daily, reduced BP and HR and also increased the maximum duration of exercise.

There have been relatively few studies on tiapamil and we are not aware of any invasive exercise studies on this compound. In summary, these independent studies agree well that calcium antagonists like verapamil, diltiazem, and tiapamil reduce BP at rest and during exercise without a decrease in heart pump function or in exercise performance.

The mechanism behind the acute reduction in blood pressure with the dihydropyridine derivatives has been studied extensively. It is generally agreed that shortly after the first dose of nifedipine there is a marked fall in TPRI and also in resistance in different vascular beds [21]. With the newer drug, nisoldipine, we could demonstrate the same type of changes with the maximal effect being most dominant already after one hour. Good blood pressure control during exercise has been reported also with other dihydropyridine derivatives, like nitrendipine (20 mg daily) [22], and felodipine [23].

In comparison with beta-blockers, the calcium antagonists are clearly beneficial during endurance work. No effect on exercise time has been demonstrated for several calcium antagonists [14, 24]—in contrast to the beta-blockers when physical endurance is reduced [13, 14].

Additionally, other classes of antihypertensive agents are able to reduce total peripheral resistance without a reduction in heart pump function. Thus, in our series on the alpha-receptor blockers prazosin [25] and doxazosin [26], blood pressure was reduced entirely through the reduction in total peripheral resistance. In the alpha-blocker series a clear tendency for stroke volume to increase during exercise was observed and post-treatment exercise cardiac output tended to be above pretreatment levels.

Also angiotensin converting enzyme inhibitors, like enalapril and captopril, reduced blood pressure through vasodilatation without a reduction in heart pump function. In our series on enalapril (alone or in combination with hydrochlorothiazide) [27], the blood pressure reduction was induced entirely through the fall in total peripheral resistance. Post-treatment stroke volume did not increase, however.

In conclusion, all the calcium antagonists studied in the present series were efficient in reducing blood pressure. With the doses used, the reduction in blood pressure at rest and particularly during exercise was greater with diltiazem, nifedipine, and nisoldipine than with verapamil and tiapamil. The fall in pressure was due to the reduction in total peripheral resistance. There was no significant reduction in heart pump function either at rest or during exercise. In general, treatment with calcium antagonists is well tolerated, but ankle edema might be a significant side effect with the dihydropyridine derivatives. Peripheral edema is not associated with general fluid retention, but is due to the local effect on the vascular wall.

Most calcium antagonists reduce blood pressure rather abruptly and tend to induce acute reflex tachycardia [28–31]. Therefore treatment should be started with a low dose—and preferably with a long acting preparation. Such preparations are or will be available for nifedipine, diltiazem, and verapamil.

In the debate about the drug of first choice in the treatment of hypertension, the effects of the different classes of antihypertensive agents must be taken into consideration. Several animal models have shown that the calcium antagonists have an antisclerotic effect [32]. Whether the animal models have relevance to human arteriosclerosis is debatable. If an antisclerotic effect is demonstrated in man, this might—in addition to the favorable hemodynamic effect—be a strong case for the selection of a calcium antagonist as a drug of first choice in the treatment of hypertension. However, much more data are needed before this question can be answered.

References

1. Conway J (1984) Haemodynamic aspects of essential hypertension in humans. Physiol Rev 64: 617–660
2. Folkow B (1982) Physiological aspects of primary hypertension. Physiol Rev 62: 347–504

3. Lund-Johansen, P (1983) The haemodynamics of essential hypertension. In: Robertson JIS (ed) Handbook of hypertension, vol 1: clinical aspects of essential hypertension. Elsevier, Amsterdam, pp 151–173
4. Halperin AK, Cubeddu LX (1986) The role of calcium channel blockers in treatment of hypertension. Am Heart J 111: 363–382
5. McAllister RG Jr, Schloemer GL, Hamann SR (1986) Kinetics and dynamics of calcium entry antagonists in systemic hypertension. Am J Cardiol 57: D16–D21
6. Eichelbaum M, Echizen H (1984) Clinical pharmacology of calcium antagonists 4: A critical review. J Cardiovasc Pharmacol 6: S963–S967
7. Lund-Johansen P, Omvik P, Haugland H (1986) Acute and chronic haemodynamic effects of nisoldipine in essential hypertension at rest and during exercise. Acta Med Scand (Suppl) 714: 183–186
8. Lund-Johansen P, Omvik P (1983) Haemodynamic effects of nifedipine in essential hypertension at rest and during exercise. J Hypertens 1: 159–163
9. Lund-Johansen P (1984) Haemodynamic long-term effects of verapamil in essential hypertension at rest and during exercise. Acta Med. Scand (Suppl) 681: 109–115
10. Laragh JH (1986) Calcium antagonists in hypertension—focus on verapamil. A symposium. Am J Cardiol 57: D1–D107
11. Lund-Johansen P (1970) Haemodynamic changes in long-term diuretic therapy if essential hypertension. Acta Med Scand 187: 509–518
12. Lund-Johansen P (1984) Central haemodynamic effects of beta-blockers in hypertension. A comparison between atenolol, metoprolol, timolol, penbutolol, alprenolol, pindolol and bunitrolol. Eur Heart J (Suppl D) 4: D-1–12
13. Kaiser P (1984) Physical performance and muscle metabolism during beta-adrenergic blockade in man. Acta Physiol Scand (Suppl) 536: 1–44
14. Lundborg P, Åström H, Bengtsson C, Fellenius E, von Schenck H, Svensson L, Smith U (1981) Effect of beta-adrenoceptor blockade on exercise performance and metabolism. Clin Sci 61: 299–305
15. Cody RJ, Kubo SH, Covit AB, Müller FB, Lopez-Ovejero J, Laragh JH (1986) Exercise haemodynamics and oxygen delivery in human hypertension. Response to verapamil. Hypertension 8: 3–10
16. Aoki K, Sato K, Kondo S, Yamamoto M (1983) Hypotensive effects of diltiazem to normals and essential hypertension. Eur J Clin Pharmacol 25: 475–480
17. Yamakado T, Oonishi N, Nakano T, Takezawa H (1985) Effects of nifedipine and diltiazem on haemodynamic responses at rest and during exercise in hypertensive patients. Jpn Circ J 49: 415–421
18. Klein W, Brandt D, Vrecko K, Härringer M (1983) Role of calcium antagonists in treatment of essential hypertension. Circ Res (Suppl I) 52: I-174–181
19. Safar ME, Simon ACH, Levenson JA, Cazor JL (1983) Haemodynamic effects of diltiazem in hypertension. Circ Res (Suppl I) 52: I-169–173
20. Pool PE, Seagren SC, Salel AF, Skalland ML (1985) Effects of diltiazem on serum lipids, exercise performance and blood pressure: randomized, double-blind, placebo-controlled evaluation for systemic hypertension. Am J Cardiol 56: H86–H91
21. Bühler FR, Bolli P, Erne P, Kiowski W, Müller FB, Hulthén UL, JI BH (1985) Position of calcium antagonists in antihypertensive therapy. J Cardiovasc Pharmacol (Suppl IV) 7: IV-S21–S27
22. Franz I-W, Wiewel D (1984) Antihypertensive effects on blood pressure at rest and during exercise of calcium antagonists, beta-receptor blockers, and their combination in hypertensive patients. J Cardiovasc Pharmacol 6: S1037–S1042
23. Lorimer AR, McAlpine HM, Rae AP, Simpson IA, Barbour MP, Forret EA, Kent Hill TW, Veitch Lawrie TD (1985) Effects of felodipine on rest and exercise heart rate and blood pressure in hypertensive patients. Drugs (Suppl II) 29: II-154–164
24. Herpin D, Amiel A, Boutaud P, Wajman A, Demange J (1985) Effect of a calcium antagonist (1985) verpamil on resting blood pressure and pressor response to dynamic exercise. Acta Cardiol 40: 277–290

25. Lund-Johansen P (1975) Haemodynamic changes at rest and during exercise in long-term prazosin therapy for essential hypertension. Postgrad Med J (Suppl I) 58: I-45–52
26. Lund-Johansen P, Omvik P, Haugland H (1986) Acute and chronic haemodynamic effects of doxazosin in hypertension at rest and during exercise. Br J Clin Pharmacol 21: S45–S54
27. Lund-Johansen P, Omvik P (1984) Long-term haemodynamic effect of enalapril (alone and in combination with hydrochlorothiazide) at rest and during exercise in essential hypertension. J Hypertens (Suppl II) 2: II-49–56
28. Aoki K, Sato K (1989) Hyperdynamic state in normotensives and hypertensives with calcium antagonists at rest. In: Aoki K, Frohlich ED (eds) Calcium in essential hypertension. Academic Press, Tokyo San Diego New York London, pp 537–552
29. Lund-Johansen P (1989) Hemodynamics in essential hypertension: spontaneous changes and effects of calcium antagonists. In: Aoki K, Frohlich ED (eds) Calcium in essential hypertension. Academic Press, Tokyo San Diego New York London, pp 553–574
30. Reid JL, Elliott HL (1989) Calcium antagonists in essential hypertension: clinical pharmacological aspects. In: Aoki K, Frohlich ED (eds) Calcium in essential hypertension. Academic Press, Tokyo San Diego New York London, pp 575–600
31. Doyle AE (1989) An overview of indication and choice for the use of calcium antagonists for the treatment of essential hypertension. In: Aoki K, Frohlich ED (eds) Calcium in essential hypertension. Academic Press, Tokyo San Diego New York London, pp 601–619

V Mechanism of Blood Plessure Elevation in Gene (Essential) Hypertension

Calcium Metabolism in Hypertension: Clinical Evidence and Cellular Hypothesis

LAWRENCE M. RESNICK and JOHN H. LARAGH[1]

Summary. Investigation of calcium metabolism in hypertension has led to a new understanding of the pathophysiology of the hypertensive process. Essential hypertensive subjects exhibit deviations in calcium metabolism, which parallel deviations in the renin-aldosterone system. These linked shifts suggest a molecular-cellular theory of hypertension and a more general scheme by which dietary and drug factors affect blood pressure.

At the cellular level, all hypertension may derive from one of two types of defective cellular calcium handling. Low-renin forms of hypertension are more critically dependent on extracellular calcium, and are thus more responsive to calcium channel blockade. In contrast, in high-renin forms of hypertension, excess angiotensin II mediates release of calcium from intracellular stores. This form of hypertension is much less dependent on calcium influx from the extracellular space, and therefore is less sensitive to the antihypertensive effects of calcium channel blockade.

More generally, renin and calcium regulating hormone systems preside over the uptake and distribution of monovalent and divalent cations between intracellular and extracellular sites and, thus, mediate the blood pressure effects of stimuli normally affecting these hormones, such as dietary salt and calcium intake. Assessment of calcium metabolic indices can be of practical use in the choice of dietary and drug recommendations for subjects with hypertensive disease.

Key words: Calcium—Hypertension—Calciotropic hormones—1,25 Dihydroxy vitamin D—Salt-sensitivity

Introduction

Rapidly accumulating research data attests to an increasing focus on calcium in the pathophysiology of hypertension. This progress has largely proceeded along two parallel lines. At the more basic level, calcium is critical as part of a second messenger system, in transducing cellular responses to a wide variety of stimuli. In at least four organ systems, this appears to be of special relevance to blood pressure homeostasis. First, since the time of Ringer [1], when calcium was shown to be important to cardiac contraction, the generation and regulation of contractile force in all types of muscle is now known to be calcium dependent.

[1]Cardiovascular Center, The New York Hospital-Cornell Medical Center, New York, NY 10021, USA

This is especially true in smooth muscle, where devoid of fast, Na^+-linked channels, the slow, voltage-dependent calcium channels are of predominant importance. Hence, calcium becomes important in considering the normal mechanisms of cardiac output and peripheral vascular resistance and in attempting to define abnormal vascular function in hypertension. Second, calcium also participates in the final common pathway of stimulus-secretion coupling mechanisms and is thus central to endocrine function. Numerous humoral pathways such as the renin-aldosterone system, calcium-regulating hormones, and other adrenal hormones both affect and regulate cardiac function, peripheral vascular tone, and renal excretion. It is understandable, therefore, that much research has also focused on how altered calcium-dependent processes might underlie an endocrine contribution to the pathogenesis of hypertension. These considerations are also true of central and peripheral neural output, representing a special case of calcium-mediated endocrine function in which neurotransmitter release and post-synaptic responses are similarly calcium-dependent processes. Thus, neural factors, together with the peripheral endocrine system, control in an integrated fashion cardiac, peripheral vascular, and renal aspects of pressure and volume homeostasis.

Lastly, abnormalities of renal endocrine and excretory function and, in particular, of sodium handling and renin secretion also seem to characterize the hypertensive process. Not only are tubular sodium reabsorption as well as renin secretion each influenced by altered cellular calcium handling, but tubular-glomerular feedback, linking excretory signals to intrarenal vascular and renin responses, is also calcium dependent. Much attention has thus been appropriately focused on defining key renal abnormalities in hypertension, and a hypothesis to account for linked alterations of renin secretion and tubular sodium handling has recently been suggested [2]. The challenge confronting calcium research at the basic, mechanistic level is to develop a more unified view, to identify common defects in cellular calcium handling which manifest themselves as the wide variety of organ-defined abnormalities rapidly being catalogued.

A second, previously separate line of investigation has proceeded more slowly and empirically, but has increasingly called attention to the involvement of calcium in hypertension at the epidemiological and clinical levels. The data often appear contradictory and support both a positive, pathogenetic role for excess calcium in the hypertensive process as well as alternatively suggesting an ameliorative role for calcium in hypertension, thus implying that hypertension is somehow associated with a calcium deficiency. Specifically, while excess cytosolic free calcium has been demonstrated in a variety of cell types in hypertension and the height of the blood pressure is usually found to be positively correlated with circulating calcium levels, serum ionized calcium levels have been reported as lower and urinary calcium excretion as higher in both clinical and experimental hypertensive states. Additionally, epidemiologic studies, while failing to show within-population correlations between dietary sodium intake and the height of the blood pressure, do show significant, inverse relationships between blood pressure and dietary calcium intake. More provocative still have been studies demonstrating the ability of dietary calcium supplementation to lower blood pressure in many forms of experimental hypertension and to a variable degree in human essential hypertension as well [3]. The challenge confronting these types of calcium-related research is first to resolve the apparent paradox of calcium both contributing to and protecting from hypertensive disease [4]. More importantly, however, must be the challenge of how to integrate the knowledge obtained at these two different levels of investigation, at the clinical-environmental level and at the molecular-cellular level. How do the cells know that environmental changes in dietary mineral intake have occurred? What determines their response to these signals? How do the molecular-cellular alterations characteristic of hypertension either contribute to or result from (or both) clinically relevant alterations of calcium mineral balance?

We have attempted to elucidate these issues, adopting an approach that attempts to bridge this cellular-clinical gap and that emphasizes both the biochemical and clinical heterogeneity of the hypertensive process. Utilizing this approach, we have studied calcium metabolism in both experimental and clinical forms of hypertensive disease and have found broad-based deviations of ionic and hormonal circulating calcium indices. Indeed, divalent cation metabolism appears to be shifted in both directions away from average normotensive values among different types of hypertension, paralleling shifts in the activity of the hormonal system involved in regulating monovalent cation metabolism, the renin-aldosterone system. Furthermore, these renin-linked shifts in calcium ionic and hormonal metabolism are clinically relevant in understanding the mechanism by which dietary intake, e.g., salt or calcium, may affect blood pressure. Indeed, assessment of calcium metabolism and of plasma renin activity may help target specific subgroups of hypertensive individuals for whom calcium supplementation may provide a non-pharmacologic alternative to dietary salt restriction, especially in elderly and black hypertensive populations. These same considerations seem to apply in predicting the utility of a calcium channel blocking drug as the drug of first choice in the pharmacologic therapy of hypertension [5, 6].

This type of analysis, emphasizing the heterogeneous nature of human hypertension, has allowed us to formulate a working cellular hypothesis in which all hypertensive disease consists of two calcium dependent forms, each contributing to a greater or lesser extent. One form is critically dependent on cellular exchange with extracellular calcium, while the other type of mechanism is more critically dependent on agonist-mediated intracellular calcium release and uptake mechanisms.

Furthermore, our results suggest a more general scheme in which calcium regulating hormones, coordinately with the renin-aldosterone system, translate environmental dietary mineral (and even drug) signals at the cellular level and thus determine the effects of these inputs on blood pressure. Hence the different metabolic set points of these systems, themselves both genetically and environmentally defined, determine the divergent and often opposite biological responses observed among different hypertensive populations to the same dietary or drug maneuver. Analysis of these monovalent and divalent hormonal mediators may not only help us to understand the underlying heterogeneity of hypertension, but may also serve to bridge the purely cellular vs clinical-environmental approaches previously utilized and provide a perspective with which to plan future research.

Calcium Metabolism is Altered in Human Hypertension

A wide range of alterations in calcium metabolism have been reported in hypertension. Early epidemiologic evidence suggested that circulating calcium and blood pressure were directly related—the higher the serum total calcium, the higher the blood pressure [7]. More recently, a similar, strikingly positive relationship was demonstrated between intracellular cytosolic free calcium in platelets and blood pressure [8]. Once again, the higher the cytosolic free calcium, the greater the blood pressure. The hypercalcemia of primary hyperparathyroidism is routinely associated with increases in blood pressure [9], as is iatrogenic hypercalcemia, whether as acute intravenous loads or chronically as in vitamin D intoxication [10]. Conversely, hypocalcemia may be associated with hypotension [11]. These data collectively suggest that alterations in calcium metabolism may exist in blood pressure disorders and that a calcium surfeit contributes to its underlying pathophysiology. However, equally suggestive are data indicating exactly the opposite, that a calcium deficit characterizes the hypertensive state. Epidemiologic surveys of diverse ethnically and geographically defined groups indicate a rather consistent pattern—an inverse relationship between assessed dietary calcium intake and concurrent blood pressure. The higher the measured blood pressure, the lower the ingested calcium [12, 13]. At the same time relative to renal sodium excretion, calcium excretion is excessive among hypertensive subjects, a greater loss of calcium resulting at any given level of sodium intake [14]. Lower serum ionized calcium levels have also been measured in some hypertensives [15]. Lastly, dietary calcium supplementation in various rat models of hypertension, and in human hypertensives as well, may in at least some circumstances lower blood pressure [16–18]. Hence, a deficit of calcium has also been claimed as central to the pathogenesis of hypertension. These apparently paradoxical data have made the contribution of calcium metabolism to hypertension still undefined and controversial.

We believe that analyzing hypertension as if it were of uniform underlying pathophysiology, considering all hypertensives as one homogeneous group is mistaken and contributes to ongoing artificial controversies, obscuring the central role calcium plays in the pathophysiology of hypertension. We began investigating calcium metabolism in hypertension by measuring circulating levels of ionized calcium and of the calcium regulating hormones, parathyroid hormone (PTH), calcitonin (CT), and 1,25 dihydroxy vitamin D (1,25 D) in subjects with primary, essential hypertension [19, 20]. Distinct pathophysiologic subgroups were indentified by means of renin-sodium profiling. The renal pressor hormone, renin, released by the juxtaglomerular apparatus in response to a variety of ionic, neural, humoral, and direct pressure-stretch stimuli, both reflects and contributes to pressure and volume homeostasis. Of interest, renin, like parathyroid hormone, is unusual among hormones, increasing intracellular calcium levels associated with suppression of cellular hormone secretion. Approximately one-third of hypertensive individuals have renin levels inappropriately suppressed compared to normotensive subjects at the same average level of urinary sodium excretion. These subjects are more commonly found in black and elderly hypertensive populations in whom sodium-volume factors may pre-

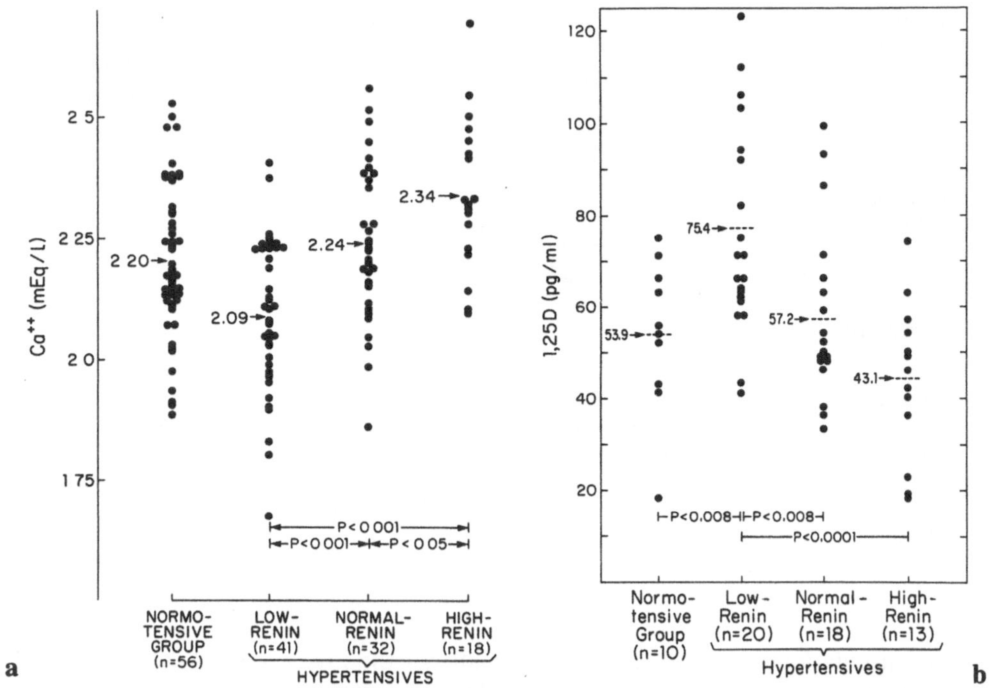

Fig. 1. **a** Serum ionized calcium levels (*Ca^{++}*) and **b** serum 1,25 dihydroxyvitamin D (*1,25 D*) levels in normotensive and renin subgroups of essential hypertensive subjects [19, 20]

dominate. On the other hand, inappropriately high levels of renin usually signify angiotensin II dependent hypertension, in which sodium-volume mechanisms appear to be less clinically relevant. Thus, renin-sodium profiling may allow for a more pathophysiologically oriented way of categorizing hypertensive subjects and has allowed us to more logically fashion appropriate antihypertensive treatment regimens [21].

When patients were categorized in this manner, serum ionized calcium levels in hypertensives were within the normal range but were distinguishably different among different renin subgroups [19] (Fig. la). Low-renin hypertensives had lower average serum ionized calcium values while high-renin subjects had values higher than other hypertensive or normotensive control subjects. With considerable overlap observed between individual subjects within renin subgroups, we sought to establish the pathophysiologic rather than just the statistical significance of the small ionized calcium changes within the normal range by measuring the calcium regulating hormones PTH, CT, and 1,25 D. What we found was that each of the calcium regulating hormonal species measured was likewise shifted [20]. Like serum ionized calcium values themselves, however, each hormone was deviated in opposite directions away from average normotensive values, appropriate for and, thus, presumably secondary to the altered circulating concentrations of calcium (Table 1). Thus, PTH values were consistently

Table 1. Calcium regulating hormones in essential hypertension

Group	n	Parathyroid hormone	Calcitonin	1,25 Dihydroxy vitamin D
Normotensive	10	244 ± 21 (254 ± 42)	68.0 ± 6.5	53.9 ± 5.3
Hypertensive				
Low renin	20	433 ± 38** (398 ± 31)**	56.4 ± 1.8	75.4 ± 5.5***
Normal renin	18	318 ± 22* (307 ± 33)†	66.2 ± 4.0#	57.2 ± 4.3††
High renin	13	292 ± 37 (311 ± 39)	84.3 ± 3.9*** ††††	43.1 ± 4.6†††

* $P < 0.05$; ** $P < 0.01$; *** $P < 0.008$ vs normotensive
† $P < 0.05$; †† $P < 0.008$; ††† $P < 0.0001$; †††† $P < 0.00001$ vs low renin hypertensive
\# $P < 0.005$ vs high-renin hypertensive
* and † for values in parentheses are mean adjusted for covariance of urinary sodium excretion
P-values for calcitonin and 1,25 D are for pooled variance-t statistics (Bonferroni)

elevated in low-renin subjects, appropriate for the lower ionized calcium values found in these patients. Similarly, 1,25 D values were elevated in these subjects while calcitonin levels were lower, again exactly what one would expect if calcium levels were physiologially sensed as being significantly lower in these individuals (Fig. 1b). Conversely, high-renin subjects had elevated CT levels and suppressed values for both PTH and 1,25 D compared to the other hypertensive subjects, again consistent with the higher average circulating ionized calcium levels observed in these subjects. Low-renin essential hypertensive subjects are thus acting as if they have a "calcium deficit" while high-renin hypertensive subjects, as a group, act as if a calcium surfeit was present. It may be that apparently paradoxical calcium-related abnormalities of hypertensive subjects previously reported may at least partly reflect different hypertensive populations, with calcium metabolism shifted in each in different directions away from average normotensive values.

Uniform Versus Heterogeneous Defects in Calcium Metabolism: A Cellular Hypothesis

If, as first demonstrated by Swiss workers, higher blood pressures are uniformly associated with higher levels of free cytosolic calcium in proportion to the elevation of the pressure [8], then how can one explain the opposite extracellular deviations in ionized calcium, some being lower, other higher than average normotensive values? We have developed a hypothesis to reconcile this intracellular uniformity in the face of extracellular diversity (Fig. 2). This hypothesis also predicts clinical consequences that we have tested.

Analyzing hypertension with respect to calcium and disregarding magnesium,

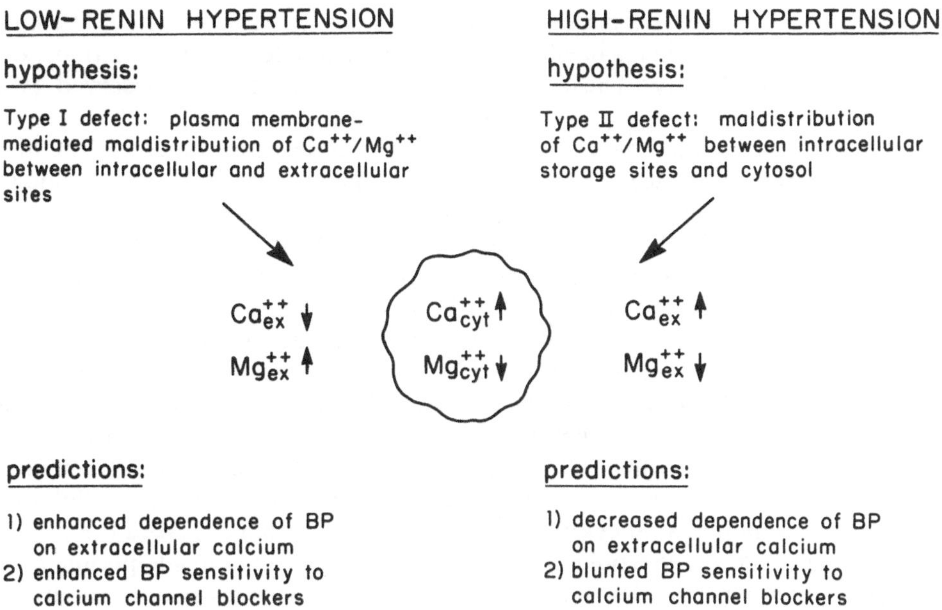

LOW-RENIN HYPERTENSION
hypothesis:
Type I defect: plasma membrane-mediated maldistribution of Ca++/Mg++ between intracellular and extracellular sites
HIGH-RENIN HYPERTENSION
hypothesis:
Type II defect: maldistribution of Ca++/Mg++ between intracellular storage sites and cytosol
Ca++ex ↓
Mg++ex ↑
Ca++cyt ↑
Mg++cyt ↓
Ca++ex ↑
Mg++ex ↓
predictions:
1) enhanced dependence of BP on extracellular calcium
2) enhanced BP sensitivity to calcium channel blockers
predictions:
1) decreased dependence of BP on extracellular calcium
2) blunted BP sensitivity to calcium channel blockers

Fig. 2. Cellular calcium hypothesis of hypertension that reconciles heterogeneous extracellular divalent cation measurements with presumed uniform intracellular concentrations. Ca^{++}_{ex}, extracellular calcium; Ca^{++}_{cyt}, cytosolic free calcium; Mg^{++}_{ex}, extracellular magnesium; Mg^{++}_{cyt}, cytosolic free magnesium; *BP*, blood pressure

also the subject of recent investigation [22], let us assume that there is an elevation of cytosolic free calcium in proportion to the elevation in blood pressure. At the same time, we have also observed that in low-renin hypertensive patients, serum ionized calcium levels are suppressed. Notice that the intracellular abnormality appears as the mirror image of the extracellular abnormality. The implied hypothesis would be a defect (call it type I defect) in the plasma membrane separating intracellular from extracellular calcium pools. In the steady-state, the plasma membrane incorrectly partitions calcium with more calcium accumulating on the "inside," coming from the "outside." The result is the calcium metabolic pattern of the low-renin state: lower serum ionized calcium and high cytosolic free calcium levels. This hypothesis is perfectly general and not dependent on any particular one of a variety of membrane defects already postulated in hypertension including defects of sodium-potassium-adenosine triphosphatase ATPase, sodium-potassium co-transport, sodium-sodium countertransport, sodium-calcium exchange, and calcium-ATPase as well as humorally mediated calcium uptake [23–28]. Furthermore, this type I, plasma membrane-dependent defect implies certain testable predictions. First, this kind of hypertension should be exquisitely sensitive to calcium channel blockade, since the excess accumulated intracellular calcium must come from the extracellular calcium pool. Second, in accord with current molecular hypotheses regarding salt-sensitive hypertension, all of which link the ability of dietary salt to raise pressure with its ability to somehow increase the cellular uptake of calcium, we would predict that hypertensive patients with a low-renin type I defect should be more salt-sensitive than others, having a greater tendency to transport extracellular calcium intracellularly. These predictions have indeed been observed (vide infra).

On the other hand, how can we reconcile the observations made in the high-renin hypertensive state? Once again, intracellular calcium levels are assumed to be increased, as demonstrated in all hypertensive patients [8]. Yet, remember that in patients with high-renin hypertension levels of ionized calcium outside the cell were also higher than in other people. Rather than the mirror image of low-renin hypertension, the same distribution of free calcium seems to be going on inside and outside the cell. Here, we postulate the problem is one of "domestic affairs"—let us call it a type II defect. Most cells have organelles within which calcium is in steady-state equilibrium with cytoplasmic free calcium. The endoplasmic reticulum, the mitochondria, various cytoplasmic proteins, and calmodulin may all be involved. We hypothesize that an abnormal steady-state partitioning between stored and/or bound vs free intracellular calcium results in increased cytoplasmic free calcium. If that is indeed the primary abnormality, then one would expect an otherwise normal outside plasma membrane to pump out the excess accumulated free cytoplasmic calcium, resulting in increased extracellular calcium levels. Indeed, angiotensin II recruits intracellular calcium stores, even to the extent of total cellular calcium depletion [29]. Here, abnormal increases in intracellular calcium do not result from, but rather are the cause of, the increase in extracellular calcium. Once again, this hypothesis is independent of the various molecular mechanisms by which this might occur, such as endoplasmic reticulum or mitochondrial pump defects, abnormal calcium-binding protein interactions, or alterations of calcium-calmodulin binding.

Regardless of how this "domestic affairs" type II defect might arise, certain consequences can be predicted. First, since intracellular calcium is, in this model, less dependent on calcium entering from outside the cell, a patient with type II, high-renin hypertension should be less sensitive to the blood pressure lowering effects of calcium channel blockade. Second, since excess intracellular free cytoplasmic calcium can itself gate its own (calcium) channels, patients with high-renin hypertension would be predicted to be less sensitive to the hypertensive influences of dietary salt. These predictions are also consistent with published observations (vide infra).

The Relevance of Altered Calcium Metabolism in Clinical Hypertension

Dietary Salt Sensitivity in Hypertension

Salt provokes altered calcium metabolism and exacerbates hypertension.

What might be the potential clinical significance, if any, of these subtle shifts in calcium metabolism among different renin-defined subgroups of essential hypertension? We reasoned that if these calcium metabolic shifts contributed to the pathogenesis of the hypertension, then maneuvers which were known to exacerbate hypertension ought to necessarily provoke and/or exacerbate those same alterations in calcium metabolism. We therefore studied the short- and longer-term effects of dietary salt loading on blood pressure and calcium metabolism. Essential hypertensive subjects were randomly allocated to metabolic balance diets containing 10 and 200 mEq of sodium chloride for 5 days each. Longer-term studies among essential hypertensives provided for low-sodium chloride (< 50 mEq/day) and high-sodium chloride (> 200 mEq/day) diets for 1 month each under outpatient supervision [30, 31].

In each group of studies, consistent relationships were observed between the ability of salt to raise blood pressure and its ability to alter calcium metabolism. Salt raised pressure significantly in approximately one-half of the hypertensive subjects in both short- and long-term studies, while pressure actually declined in some patients with salt loading. Regardless of the magnitude of the blood pressure change, continuous relationships were observed between the change in diastolic pressure on high vs low dietary salt and salt-induced changes in both serum ionized calcium (short-term: $r = -0.72$, $P < 0.001$; long-term: $r = -0.78$, $P < 0.001$) and 1,25 D (short-term: $r = 0.82$, $P < 0.001$; long-term: $r = 0.71$, $P < 0.001$) (Fig. 3). Salt raised pressure the most in those individuals in whom it most lowered ionized calcium levels and most elevated levels of 1,25 D—provoking exactly those calcium metabolic changes characteristic of and originally observed in low-renin hypertension. It thus appears that the originally observed "calcium deficient" ionic and hormonal pattern observed in low-renin patients is of pathophysiologic significance for hypertension, contributing to and/or reflective of dietary salt sensitivity. Interestingly, preliminary studies by others also suggested that salt loading in normotensive subjects may elevate PTH and 1,25 D levels, although pressure measurements were not reported in that study [32]. Conversely, when extremes of dietary salt intake did not result in significant changes in blood pressure, significant changes in serum ionized calcium were also not observed [33].

Oral Calcium Supplementation in Hypertension

Calcium reverses altered calcium metabolic indices and reverses hypertension.

Although the above results indicated to us that the ability of salt to raise blood pressure was associated with its ability to alter calcium metabolism, the two phenomena need not have been causally linked; the salt-induced blood pressure rise was perhaps not dependent on the salt-induced calcium metabolic shifts. We thus proceeded to supplement hypertensive patients with 2 calcium carbonate daily in 4 divided doses under a variety of conditions: (a) in short-term studies of hypertensive patients on metabolic balance diets, (b) in longer-term outpatient studies, and (c) on high vs low dietary salt intakes. In each instance, calcium supplementation lowered blood pressure preferentially among those subjects with lower plasma renin activity [34] (Fig. 4), lower initial serum ionized calcium levels [35] (Fig. 5), and among those who were salt-sensitive [36] (Fig. 6). Of note, while dietary calcium supplementation did not elevate circulating ionized

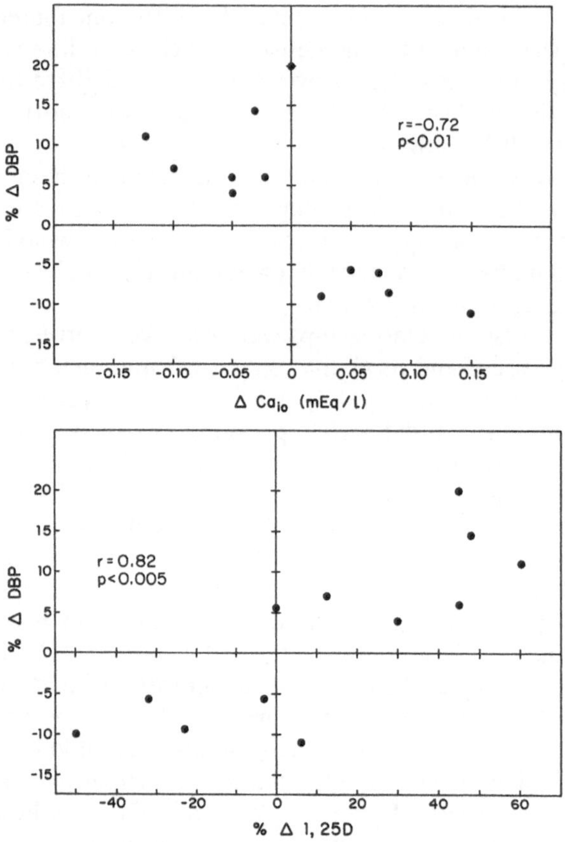

Fig. 3a,b. Dependence of salt-induced blood pressure effects on salt-induced changes in calcium metabolism. **a** *ΔCa_{io}*, salt-induced change in serum ionized calcium; **b** *%Δ1,25 D*, percent change in 1,25 dihydroxyvitamin D [30]

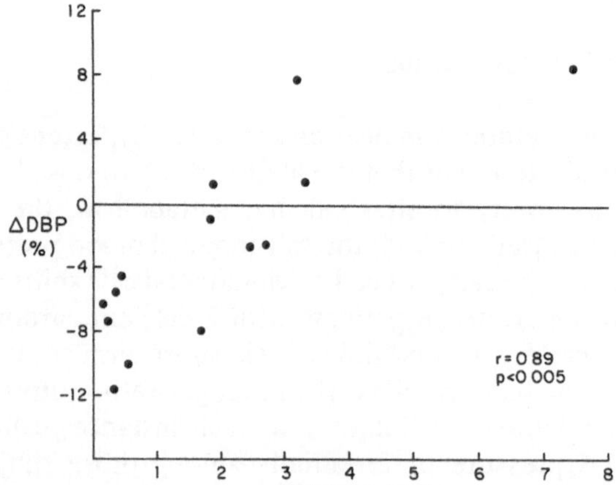

Fig. 4. The effect of short-term calcium supplementation on diastolic blood pressure (*ΔDBP*) in relation to initial plasma renin activity (*PRA*) [35]

calcium levels themselves, it always reversed the elevated PTH and 1,25 D levels characteristic of the low-renin, salt-sensitive patient. Indeed, it was only among salt-sensitive subjects that calcium supplementation significantly lowered 1,25 D levels, and only among salt-sensitive patients did calcium supplementation blunt salt-induced elevations in blood pressure. Interestingly, pressor responses to oral calcium supplementation were also observed in subjects with high-renin and higher initial ionized calcium levels and among those who were salt-insensitive. Once again, these results reflect the different and often opposite clinical responses to the same dietary maneuver. These data thus emphasize the heter-

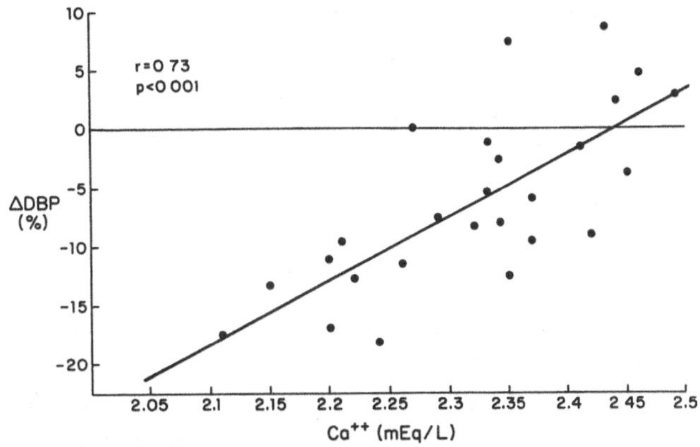

Fig. 5. The effect of long-term (6 month) oral calcium supplementation on diastolic blood pressure (*ΔDBP*) in relation to pretreatment serum ionized calcium (*Ca*++) [35]

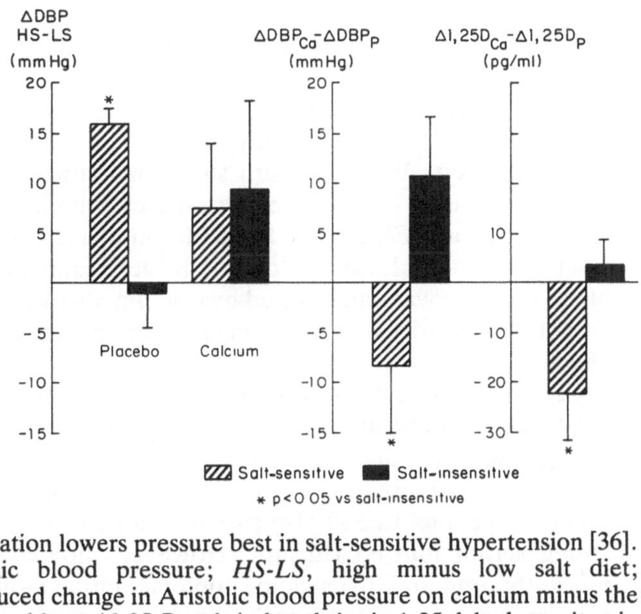

Fig. 6. *Calcium* supplementation lowers pressure best in salt-sensitive hypertension [36]. *ΔDBP*, change in diastolic blood pressure; *HS-LS*, high minus low salt diet; $\Delta DBP_{Ca} - \Delta DBP_{P}$, salt-induced change in Aristolic blood pressure on calcium minus the change observed on placebo tablets; *Δ1,25 D*, salt-induced rise in 1,25 dehydroxyvitamin D on calcium versus on placebo tablets

ogeneity of hypertensive mechanisms among different hypertensive subjects and suggest the relevance of calcium metabolic indices in identifying and perhaps underlying this heterogeneity.

Hence, the subtle shifts in calcium metabolism initially observed among different renin subgroups of essential hypertensives and induced by dietary salt loading in salt-sensitive individuals, are indeed of clinical relevance to hypertension. Specifically, sodium-volume dependent hypertension is a "calcium-dependent" hypertension in which salt induces the "calcium metabolic deficit" characteristic of the low-renin state: lower serum ionized calcium and elevated 1,25 D levels. It is especially in this form of hypertension that calcium supplementation reverses and/or blunts both the blood pressure and these same calcium variables that sodium loading exacerbates.

The Relationship of Calcium Regulating Hormones to the Renin-Aldosterone System in Hypertension

Why does calcium supplementation lower blood pressure? What is the role of 1,25 D vs PTH?

We wondered whether the observed "seesawlike" opposing effects of dietary salt vis-à-vis dietary calcium on blood pressure were due to opposite salt- and calcium-induced effects on circulating calcium ion levels per se, or to induced alterations in calcium regulating hormones, or both. We reasoned that adding 1,25 D therapy to oral calcium supplementation would, by enhancing calcium absorption, provide a more positive calcium balance than calcium supplementation alone. If total body calcium balance and/or circulating calcium levels were themselves critical to the observed dietary mineral-induced blood pressure effects, then the addition of 1,25 D to calcium itself would potentiate calcium effects on blood pressure. Interestingly, exactly the opposite was observed [37, 38]

Essential hypertensive patients on metabolic balance diets, already receiving 2 g supplemental calcium daily, were then given 0.25 μg 1,25 D daily for 3 additional days. While calcium supplementation lowered blood pressure preferentially among low-renin hypertensives, who had lower initial levels of serum ionized calcium (Figs. 4, 5), the addition of 1,25 D to calcium reversed the calcium effects and elevated pressure in these same patients (Fig. 7). Conversely, blood pressure was ameliorated by calcium plus 1,25 D in those high-renin, higher initial ionized calcium patients in whom calcium alone exerted a pressor effect.

Comparing the effects of these two calcium maneuvers on circulating hormone levels was also revealing (Table 2). Calcium supplementation alone suppressed endogenous PTH and 1,25 D levels and elevated plasma renin activity without significantly elevating serum ionized calcium levels. Calcium given with additional exogenous 1,25 D also suppressed PTH levels while not significantly elevating serum ionized calcium. However, circulating 1,25 D levels rose significantly compared to calcium supplementation alone, and plasma renin activity was suppressed. These opposite hormonal effects parallel their opposite blood pressure effects. It thus appears that:

1. PTH is probably of little significance in mediating the blood pressure effects of oral calcium supplementation, since it was suppressed by both calcium-related maneuvers, each having opposite blood pressure consequences.
2. At least part of the mechanism by which increased oral calcium intake influences blood pressure is by virtue of its ability to suppress circulating 1,25 D levels.

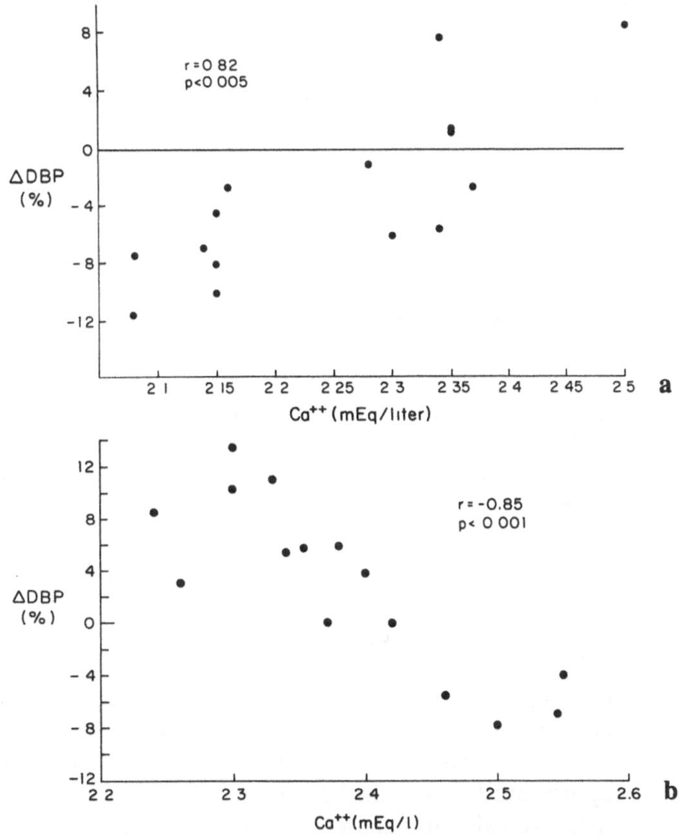

Fig. 7. Blood pressure effects of **a** calcium supplementation alone compared with **b** co-administration of calcium with 1,25 dihydroxyvitamin D. Blood pressure changes (*ΔDBP*) were compared on the basis of initial serum ionized calcium levels ($Ca^{++}{}_{io}$) [38]

Table 2. Metabolic effects of calcium with and without 1,25 dihydroxyvitamin D (1,25 D)

	Δ Ca^{2+} (%)	*Δ* PTH (%)	*Δ* 1,25 D (%)	*Δ* PRA (%)
Ca^{2+}	3.2 ± 3.0	−35.7 ± 11.6	−53 ± 15	38 ± 8.3
Ca^{2+} + 1,25 D	2.3 ± 1.1	− 20 ± 7.7	63 ± 21*	−31 ± 9.5*

PTH, parathyroid hormone; PRA, plasma renin activity
* $P < 0.001$ vs Ca^{2+} alone

3. The inverse relation of plasma renin activity of 1,25 D observed in steady-state screening of hypertensive populations is also consistently observed dynamically. Calcium supplementation, in suppressing endogenous 1,25 D, elevates plasma renin activity. This further strengthens the linkage between the renin-aldosterone system and calcium regulating hormones.
4. The fact that calcium supplementation lowers blood pressure best among those subjects in whom it reverses those very renin and calcium hormonal deviations initially observed in low-renin and salt-sensitive hypertensive subjects emphasizes the clinical relevance of these hormonal deviations in the pathogenesis of the hypertensive process.

Primary Aldosteronism Alters the Relation of PTH to 1,25 D

Is there a role for mineralocorticoids in regulating renal 1,25 D?

In essential hypertension, although alterations in calcium regulating hormones may be relevant to the expression of the elevated blood pressure, they do not seem to be primary defects. Indeed, the hyperparathyroidism and the associated elevations of circulating 1,25 D observed in low-renin and salt-sensitive forms of hypertension were appropriate for and, thus, presumably secondary to the lower serum ionized calcium levels measured in those hypertensive states. Conversely, the calcium hormonal patterns observed in high-renin, salt-insensitive hypertensives, suggestive of a calcium surfeit, are exactly what one would expect for the higher than average normotensive values for circulating ionized calcium measured in these subjects. These altered extracellular distributions of serum ionized calcium are themselves as yet unexplained, and we have hypothesized how they may reflect primary underlying abnormalities of cellular calcium handling (vide supra) [4]. Nevertheless, these calcium hormonal changes appear to be necessary for the expression of the elevated blood pressure and are linked to the activity of the renin-aldosterone system as described.

Unlike essential hypertension, however, in primary aldosteronism, relationships between calcium hormones and circulating calcium levels are abnormal and suggest primary effects of mineralocorticoid excess on the renal production of 1,25 D independently of standard calcium signals. Although it has been appreciated that mineralocorticoid excess causes renal calcium wasting and negative calcium balance [39, 40], only recently has it been clinically recognized that the syndrome of primary aldosteronism is routinely associated with secondary hyperparathyroidism [41]. Serum ionized calcium levels in primary aldosteronism, a syndrome with the greatest degree of renin suppression, are even lower than in low-renin essential hypertensive subjects. Appropriately, PTH levels are even higher than in low-renin essential hypertension. At the same time, however, serum 1,25 D levels are paradoxically suppressed in primary aldosteronism compared with low renin essential hypertension. This has also been found in desoxycorticosterone-saline hypertensive rats compared with sodium chloride-loaded control rats—lower calcium levels with concurrent paradoxically lower 1,25 D levels (Fig. 8). Furthermore, the normal, direct, positive relationship between PTH and 1,25 D is reversed in this syndrome. The higher the 1,25 D, the lower the PTH, implying perhaps an intact feedback relationship between 1,25 D and PTH, but at the same time implying that neither calcium

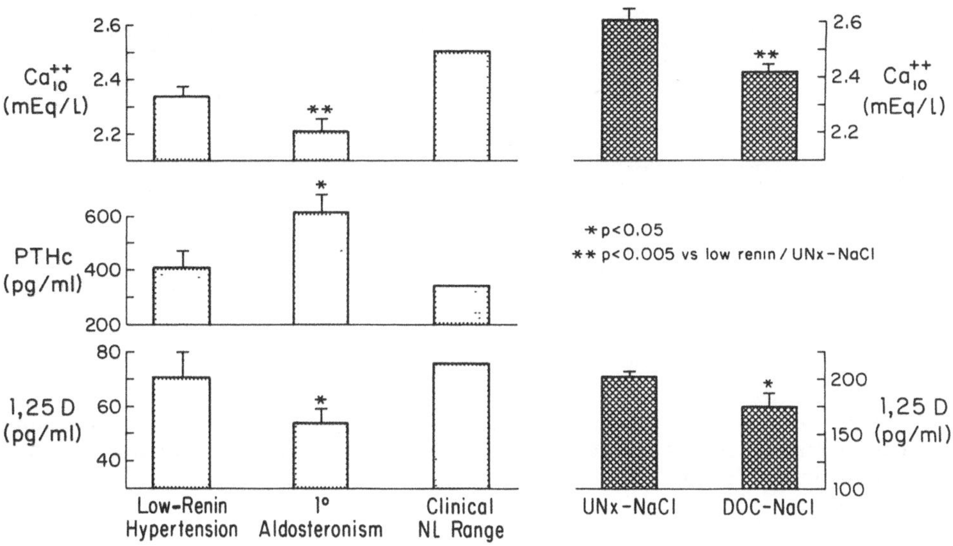

Fig. 8. Calcium metabolic indices in primary aldosteronism and experimental mineralocorticoid excess. Ca^{++}_{io}, serum ionized calcium; PTH_c, C-terminal PTH levels; *1,25 D*, 1,25 dihydroxy vitamin D; *NL*, normal; *UNx*, uninephrectomized; DOC, desoxycorticosterone [42]

itself nor PTH are primary influences on circulating 1,25 D levels in primary aldosteronism. We hypothesize that mineralocorticoid hormones, here in opposition to physiologically "appropriate" calcium-related inputs, participate in the control of 1,25 D synthesis [42].

Overall Hypothesis

What emerges from these data is an appreciation of a critical role for calcium regulating hormones coordinately and reciprocally with renin-aldosterone system linking the humoral control of monovalent and divalent cation metabolism in transducing dietary mineral signals at the cellular level. The resultant steady-state alterations in the distribution of calcium and magnesium intracellularly and between intracellular and extracellular compartments in a variety of different tissues directly influences (1) cardiac hemodynamic function, (2) central nervous and peripheral vasoactive hormone release, and (3) peripheral smooth muscle vasoconstrictor tone, and hence, the resultant blood pressure. This pattern of relationships is illustrated in Fig. 9. This scheme, in which the ability of an altered dietary mineral balance to affect blood pressure is necessarily mediated by and in turn affects mono- and divalent cation regulating hormones, provides a perspective for better understanding both the biochemical and clinical heterogeneity of hypertension. Specifically, this may help to resolve the bothersome observations that (1) a variety of genetically inherited ion transport defects found in hypertension are also found in normotensive family relatives and, hence, cannot by themselves explain elevated blood pressure, and (2) chronic

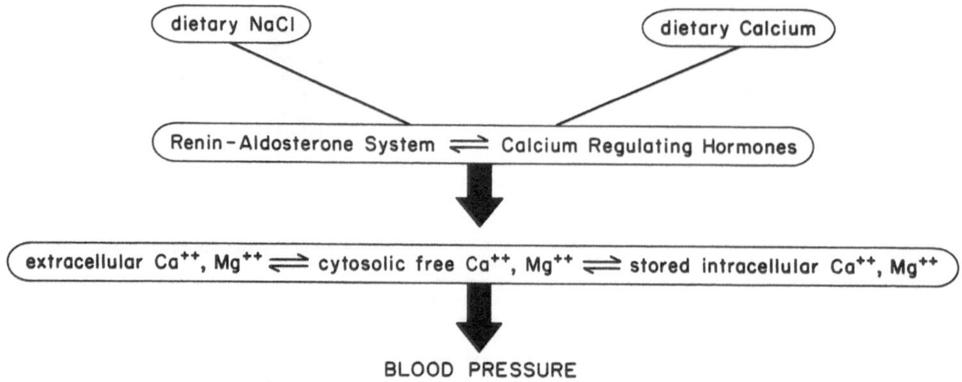

Fig. 9. Hypothetical scheme in which the blood pressure effects of environmental dietary-mineral signals are mediated by the renin-aldosterone system and calcium regulating hormones [20]

dietary sodium excess and/or calcium deficiency also do not usually result in clinical hypertension. Therefore, in neither case can these factors be considered "causes" of hypertension [43, 44], although they may each represent necessary but not sufficient condition.

In essential hypertension for instance, different primary genetically inherited or acquired alterations in cellular ion transport systems may result in the heterogeneous distribution of serum ionized calcium we have observed. These in turn, create the metabolic set points of these two hormone systems, i.e., the renin-aldosterone system and calcium regulating hormones, which would lead to hypertension only when access to dietary minerals such as sodium chloride, potassium, calcium, and magnesium is also shifted significantly with each mineral either enhancing or suppressing these pathophysiologic hormone shifts. A high dietary salt intake in the setting of high endogenous 1,25 D and secondary hyperparathyroidism would result in accelerated calcium transport intracellularly and, thus, the vasoconstriction, enhanced cardiovascular function, and increased central neural hormonal release often observed in salt-sensitive hypertension. In the absence of this skewed metabolic profile, dietary salt loading would result in no elevation in blood pressure. Furthermore, by helping to offset the calcium metabolic profile characterizing the low-renin and salt-sensitive state, dietary calcium supplementation blunts the pressor effects of salt and, in already salt-loaded people, is significantly antihypertensive. These considerations also appear to help explain the heterogeneous blood pressure responses to a variety of different pharmacologic agents, including newer drug classes such as calcium channel antagonists and converting enzyme inhibitors [5, 6, 21].

Knowing how these hormonal systems operate to preside co-ordinately over blood pressure homeostasis will not only afford additional meaningful insights into the pathophysiology of hypertension, but will allow us diagnostically to identify subgroups of hypertensives for whom more physiologic, individualized therapy can be constructed. Ultimately, long-term prophylactic strategies to prevent the onset of hypertensive disease might be developed.

References

1. Ringer S (1883) A third contribution regarding the infusion of the inorganic constituents of the blood on the ventricular contraction. J Physiol (Lond) 4: 222–225
2. Sealey JE, Blumenfeld JD, Bell GM, Pecker MS, Sommers SC, Laragh JH (1988) On the renal basis of essential hypertension: Nephron heterogeneity with discordant renin secretion and sodium excretion causing a hypetensive vaso-constriction-volume relationship. J Hypertens 6: 763–777
3. Addison WLT (1924) The use of calcium chloride in arterial hypertension. Can Med Assoc J 14: 1059–1061
4. Resnick LM (1987) Uniformity and diversity of calcium metabolism in hypertension: A conceptual framework. Am J Med (Suppl I) 82: I-B16–B26
5. Resnick LM, Nicholson JP, Laragh JH (1987) Calcium, the renin-angiotensin system, and the hypertensive response to nifedipine. Hypertension 10: 254–258
6. Resnick LM, Nicholson JP, Laragh JH (1988) The antihypertensive effects of calcium channel blockade: Role of sodium and calcium metabolism. J Cardiovasc Pharmacol (Suppl VI) 12: VI-S114–S116
7. Kesteloot H, Geboers J (1982) Calcium and blood pressure. Lancet 1: 813–315
8. Erne P, Bolli P, Burgissen E, Buhler FR (1984) Correlation of platelet calcium with blood pressure: Effect of antihypertensive therapy. N Engl J Med 319: 1084–1088
9. Brinton GS, Jubiz W, Lagerquist LD (1975) Hypertension in primary hyperparathyroidism: The role of the renin-angiotensin system. J Clin Endocrinol Metab 41: 1025–1029
10. Weidmann P, Massry SG, Coburn JW, Maxwell MH, Atleson J, Kleeman CR (1972) Blood pressure effects of acute hypercalcemia: studies in patients with chronic renal failure. Ann Intern Med 76: 741–745
11. Clowes GHA, Simeone FA (1957) Acute hypocalcemia in surgical patients. Ann Surg 145: 530–540
12. McCarron D, Morris CD, Henry JH, Standon JL (1984) Blood pressure and nutrient intake in the United States. Science 224: 1392–1398
13. Achley S, Connor-Barrett E, Suary L (1983) Dairy products, calcium and blood pressure. Am J Clin Nutr 38: 457–461
14. Strazzullo P, Nunziata V, Cirillo M, Giannattasio R, Ferrara LA, Mattioli PL, Mancini M (1983) Abnormalities of calcium metabolism in essential hypertension. Clin Sci 65: 137–141
15. McCarron DA (1982) Low serum concentrations of ionized calcium in patients with hypertension. N Engl J Med 307: 226–228
16. Ayachi S (1979) Increased dietary calcium lowers blood pressure in the spontaneous hypertensive rat. Metabolism 28: 1234–1238
17. Belizan JM, Villar J, Pineda O, Gonzalez AE, Soing E, Garrera G, Sibrian R (1983) Reduction of blood pressure with calcium supplementation in young adults. JAMA 249: 1161–1165
18. McCarron DA, Morris CD (1985) Blood pressure response to oral calcium in persons with mild to moderate hypertension. Ann Intern Med 103: 825–831
19. Resnick LM, Laragh JH, Sealey JE, Alderman MH (1983) Divalent cations in essential hypertension: Relations between serum ionized calcium, magnesium and plasma renin activity. N Engl J Med 309: 888–891
20. Resnick LM, Müller FB, Laragh JH (1986) Calcium regulating hormones in essential hypertension: Relation to plasma renin activity and sodium metabolism. Ann Intern Med 105: 649–654
21. Resnick LM, Laragh JH (1985) Renin, calcium metabolism and the pathophysiologic basis of antihypertensive therapy. Am J Cardiol 56: H68–H74
22. Resnick LM, Gupta RK, Laragh JH (1984) Intracellular free magnesium in erythrocytes of essential hypertension: Relation to blood pressure and serum divalent cations. Proc Natl Acad Sci USA 81: 6511–6516

23. Kuriyama H, Yushi I, Suzuki H, Kitamura K, Itoh T (1982) Factors modifying contraction-relation cycle in vascular smooth muscles. Am J Physiol 243: H641–H662
24. Robertson BR (1984) Alterated calcium handling as a cause of primary hypertension. Hypertension 2: 453–460
25. Hall CE, Hungerford S (1983) Prevention of DOCA-salt hypertension with the calcium blocker nitrendipine. Clin Exp Hypertens (A) 5: 721–728
26. Resnick LM, Nicholson JP, Laragh JH (1984) Calcium metabolism, blood pressure and salt intake in essential hypertension. Circulation (Suppl II) 70: II–A1
27. Bible DD, Rasmussen H (1978) A biochemical model for the ionic control of 25-hydroxyvitamin D_3 1α-hydroxylase. J Biol Chem 253: 3042–3048
28. McCarron DA (1985) Is calcium more important than sodium, in the pathogenesis of essential hypertension? Hypertension 7: 607–627
29. Resnick LM, Laragh JH (1983) The hypotensive effect of short-term oral calcium loading in essential hypertension. Clin Res 31: A334
30. Resnick LM, Nicholson JP, Laragh JH (1985) Alterations in calcium metabolism mediate dietary salt sensitivity in essential hypertension. Trans Assoc Am Physicians 98: 313–321
31. Resnick LM, Nicholson JP, Laragh JH (1985) Sodium sensitivity and calcium regulating hormones in essential hypertension. J Am Coll Cardiol 5: 436
32. Breslau NA, McCurie JL, Zerwith JE, Pak CYC (1982) The role of dietary sodium on renal excretion and intestinal absorption of calcium and on vitamin D metabolism. J Clin Endocrinol Metab 55: 369–372
33. McCarron DA, Rankin LI, Bennett MLU, Krutzik S, McClung MR, Luft FC (1981) Urinary calcium excretion at extremes of sodium intake in normal man. Am J Nephrol 1: 84
34. Resnick LM, Sealey JE, Laragh JH (1983) Short and long-term oral calcium alters blood pressure (BP) in essential hypertension. Fed Proc (Suppl III) 42: III 300
35. Resnick LM, Nicholson JP, Laragh JH (1986) Calcium metabolism and essential hypertension—relationship to altered renin system activity. Fed Proc 45: 2739–2745
36. Resnick LM, Di Fabio B, Marion RM, James GD, Laragh JH (1986) Dietary calcium modifies the pressor effects of dietary salt intake in essential hypertension. J Hypertens 4 (Suppl VI): VI-S679–S681
37. Resnick LM, Laragh JH (1984) Does dihydroxyvitamin D (1,25 D) cause low renin hypertension? Hypertension 6: 792
38. Resnick LM (1987) Calcium and vitamin D metabolism in the pathophysiology of human hypertension. In: Levander O (ed) American Inst Nutrition Proceedings—Nutrition 1987. AIN, Bethesda, MD, pp 110–115
39. Luft R, Sjogren B (1953) Some aspects of the metabolic effect of deoxycorticosterone acetate. Metabolism 2: 313–321
40. Massry S, Coburn JW, Chapman LW, Kleeman CR (1968) The effect of long-term deoxycorticosterone acetate administration on the renal excretion of calcium and magnesium. J Lab Clin Med 71: 212–219
41. Resnick LM, Laragh JH (1985) Calcium metabolism and parathyroid function in primary aldosteronism. Am J Med 78: 385–390
42. Resnick LM, Gertner JM, Laragh JH (1987) Abnormal vitamin D metabolism in primary aldosteronism and experimental mineralcorticoid excess. J Hypertens 5 (Suppl V): V-S99–S101
43. Aoki K (1989) The three-way classification of hypertension: essential hypertension, environment hypertension, and disease hypertension. In: Aoki K, Frohlich ED (eds) Calcium in essential hypertension. Academic Press, Tokyo San Diego New York London, pp 9–36
44. Aoki K (1989) The calcium membrane theory of essential hypertension. In: Aoki K, Frohlich ED (eds) Calcium in essential hypertension. Academic Press, Tokyo San Diego New York London, pp 623–655

The History of the Calcium Membrane Theory of Gene (Essential) Hypertension

KYUZO AOKI[1]

Summary. High blood pressure (hypertension) is one of symptoms of many different diseases. The hypertension is classified into three types: (1) gene (essential) hypertension, (2) environment hypertension, and (3) disease hypertension. The calcium membrane theory of gene (essential) hypertension is based on the concept that functional defects of arterial smooth muscle membrane, *i.e.*, abnormal transmembrane calcium movement, cause a rise in blood pressure. The process of the rise in blood pressure is as follows: The membrane defects increase the calcium concentration of cytosol in arterial muscle. The raised calcium concentration increases the overlap of actin and myosin filaments, which results in over-contraction of the smooth muscle. The over-contraction brings about narrowing of arterial lumen, leading to elevation of vascular resistance. In conclusion gene (essential) hypertension develops and persists through the membrane defects in animals as well as in humans.

Key words: Calcium membrane theory—Gene hypertension—Essential hypertension—Calcium antagonist—Calcium agonist

Introduction

Development of Concept of Essential Hypertension

High blood pressure (hypertension) is a symptom of disease. Thus, hypertension is observed in different diseases which are developed by different causes as well as different pathophysiologies. These complicated hypertensive diseases will be simply classified by following process: (1) accurate description of the evidence, (2) arrangement and recognition of evidence, (3) invention of a hypothesis to explain the relationship between the cause and evidence, and (4) the hypothesis is proved by experiment, thus it achieves the status of a theory, and conclusively, the dignity of a law [1].

Regarding the history of hypertension, Bright (1827) [2] described a series of cases in which albuminous urine was associated with edema as hypertensive dis-

[1] Second Department of Internal Medicine, Nagoya City University Medical School, Mizuho-ku, Nagoya 467, Japan

ease. Bright (1836) [3] wrote "The obvious structural changes in the heart consisted chiefly of hypertrophy with or without valvular disease". He proposed a hypothesis in which the hypertrophy of the heart led us to look for the unusual and causal efforts to which the heart were impelled. Then the two hypotheses of the mechanisms of hypertension appeared to be, either that the altered quality of the blood affords unusual stimulus to the heart, or that the altered blood affects the small arteries and capillary vessels, so as to render greater action necessary to force the blood through the vascular system.

The idea that the alteration of a pressor substance in the blood afforded an unusual stimulus to the heart and arterial vessels, leading to hypertension, stimulated the investigation of the endocrine function of the kidney [4]. Tigerstedt and Bergman (1898) [5] found that saline extracts of kidney have a pressor response. They named this pressor substance renin. In 1934, Goldblatt et al. [6] found that a persistent rise in blood pressure is produced by constriction of a renal artery in dogs. The following hypothesis was proposed: renin is secreted from the kidney into the blood leading to the rise in blood pressure. Pickering and Prinzmetal (1938) [7] confirmed a rise in blood pressure by intravenous injection of renin. Both renin-induced and arterial-clip hypertension confirmed the hypothesis of the renal origin of essential hypertension.

The next landmark for research on hypertension was the paper of Gull and Sutton (1872) [8] on "arterio-capillary fibrosis". They concluded that chronic Bright's disease is associated with contracted kidney, in which the alteration is characterized by 'hyalin-fibrinoid' formation in the arterioles. Huchard (1889) [9] observed the elevation of arterial pressure in patients in the absence of albuminuria and concluded that hypertension occurs without nephritis.

Traube [see 4] suggested that arteriosclerosis and hypertrophy of the heart are the coordinated effects of the primary rise in arterial blood pressure. These studies promoted the development of the hypothesis of arterial origin of hypertension. Ewald (1877) [10] confirmed that hypertrophy of the media of arteries correlated with hypertrophy of the heart and regarded this to be a consequence of hypertension [11]. von Basch (1893) [12] called, what we now call essential hypertension, "latent arteriosclerosis". In 1915, Allbutt [13, 14] gave the name "hyperpiesis" to the following disease: Firstly, a malady in which blood pressure rises excessively at or towards middle life; secondly, a malady with a distinctive course which deserves of the name of a disease; and thirdly, at one other disease class, the rise in blood pressure is not typically associated with age, the rise is in later life. Allbutt gave this third class the name "decrescent" arteriosclerosis, what we now call "elderly" arteriosclerosis, elderly hypertension, or systolic arteriosclerotic hypertension in the elderly. Allbutt's classification and concept of essential hypertension have been expanded. However, hyperpiesis never became widely used. On the other hand, Frank's term "*essentielle Hypertonie*" (1911) [15] was widely used. The term "*essentielle Hypertonie*", which corresponds to hyperpiesis, has been translated into English as essential hypertension. Frank (1911) [15] described as a hypothesis of the mechanism of hypertension that "*Der erhöhte Tonus der arteriolen Ringmuskular bei der essentiellen Hypertonie bleibt ätiologische Zunächst ganz unklar.*"

A History of Modern Classification of Hypertension

One classification of hypertension dates from Volhard and Fahr's study [16]. It should be recognized that the classification was the subdivision of renal diseases into three groups, but not a classification of hypertension [16]. The classification was in 1914 as follows:

1. Degenerative diseases: Nephrosis
2. Inflammatory disease: Nephritis
3. Arteriosclerotic disease: Renal arteriosclerosis
   A. Pure arteriosclerosis in the kidney = simple benign hypertension
   B. The combination form of arteriosclerosis with nephritis = malignant hypertension

Pickering (1955) [4] suggested that the benign and malignant phase of hypertension depend on the severity of hypertension. Pickering [4] proposed the two-way classification of hypertension in 1955.

1. Systolic hypertension, in which the only systolic pressure is elevated
2. Hypertension, in which both systolic and diastolic pressures are elevated
   A. Classification by type
      a. Secondary hypertension: hypertension occurring as a manifestation of a disease
      b. Essential hypertension: hypertension occurring polygenically as an interaction of a genetic factor, environmental factor, and aging effect
   B. Classification by degree
      a. Benign phase: diastolic blood pressure below 130 mmHg
      b. Malignant phase: diastolic blood pressure over 130 mmHg with papilledema

Three-way Classification of Hypertension

It is clear that hypertension is not a single disease, but a symptom. Furthermore, hypertension is a biophysical sign, like fever, the specific cause of which must be determined in each individual patient. The hypothesis of hypertension as a symptom is supported by the demonstration of a different cause of hypertension in animals and humans which suggests that a variety of hypertension diseases could be classified according to the specific evidence into three types in animals as well as in humans.

Firstly, renovascular hypertension, perinephritis hypertension, mineralocorticoid- and glucocorticoid-induced hypertension, and renal failure hypertension in dogs and rats are considered as models for the study of human secondary hypertension, in which hypertension is a sign of a disease. Secondary, Dahl's salt-sensitive rats [17] develop hypertension by the interaction of two factors, the accessory (minor) hypertension gene and the environmental factor, high-salt intake. Dahl's salt-sensitive rats are considered a tool for studying Kawasaki's "salt-sensitive" human hypertension [18, 19]. Thirdly, the genetic animal model, Aoki spontaneously hypertensive rat (Aoki SHR) [20–24], is a tool for studying gene essential hypertension in humans [9–15, 24–28]. On the basis of the three types of animal hypertension, in 1985 Aoki [24, 26, 27] classified hypertension into the following three types:

1. Gene (essential) hypertension (Aoki)
   = Frank (Allbutt) essential hypertension
   = Aoki spontaneous gene hypertension in rats (SHR)
2. Environment hypertension
   = Salt hypertension
   = Obesity hypertension
   = Alcohol hypertension
   = Dahl salt hypertension in rats
3. Disease hypertension (= Pickering secondary hypertension)
   = Renal arterial hypertension
   = Pheochromocytoma
   = Aldosteronism
   = Goldblatt renovascular hypertension in dogs
   = Skelton adrenal regeneration hypertension in rats

This three-way classification used in this chapter is described in detail on pp 9–33.

Mechanisms of Vasoconstriction and Vasodilation of Arteries

Changes in the Arterial Wall and Arterial Smooth Muscle during Vasoconstriction and Vasodilation

Since 1911, it has been accepted that gene (essential) hypertension is caused by an increased contraction of arterial muscle [8–15]. In addition, it has been demonstrated that the elevation of total peripheral vascular resistance is the primary hemodynamics for the pressure rise in essential hypertension [25]. This evidence has supported the hypothesis that the pressure rise in essential hypertension may be due to abnormalities in the vasoconstriction and vasodilation of arteries through an altered contractile mechanism of arterial muscle [26–46]. Therefore, to clarify the mechanisms of the development of hypertension, the mechanisms of vasoconstriction and vasodilation of arteries, as well as those of contraction and relaxation of arterial muscle should be well understood.

The arterial wall is composed of arterial muscle and connective tissue. The wall is subdivided into three sections, intima, media, and adventitia. The media, the middle layer, has a spindle-shaped arterial muscle arranged in concentric circles. Contraction of the arterial muscle causes arterial vasoconstriction which induces a reduction in the lumen of the arteries. The contracted muscles are thickened and rounded, or as described another way, the contracted muscles have a decreased long-axis and an increased short-axis. Thus, by contraction, the cell form changes from an elongated cylinder to a spheroid configuration. The configuration brings about the widening of wall thickness and the reduction of lumen size of arteries, which is the state of arterial vasoconstriction (Fig. 1b). The elongated configuration of arterial muscles leads to the narrowing of wall thickness and the increase in lumen size of arteries, which is the state of vasodilation (Fig. 1a).

Calcium movement during contraction and relaxation of Arterial Smooth Muscle

Contraction of the muscle is initiated by a rise in the concentration of activator calcium in the cytoplasma. The cytoplasmic-free calcium comes from the extracellular space by calcium influx through the calcium channels and from membrane storage sites by calcium release. In addition, sodium-free solution brings about the contraction of arterial muscle by calcium influx throughout voltage-dependent calcium channels, because sodium efflux induces a depolarization of the cell membrane of arterial muscle in sodium-free medium [47–49]

The calcium concentration of the intracellular space is on the order of 10^{-8} *M* at relaxation, about 10^{-7} *M* at threshold for activation of the contractile proteins, and about 10^{-3} *M* at full activation of the proteins at maximum contraction. Arterial muscle contains the contractile proteins of actin filament and myosin filament, which lie parallel to the long axis. An increase in calcium in the cytoplasm causes the interaction of actin and myosin filaments. An increase in this interaction is the contraction of muscle, and a reduction of this interaction is muscle relaxation (Fig. 1). Thus, the contraction and relexation of muscle are controlled by cytosolic concentration of calcium via membrane regulation of calcium movement [47].

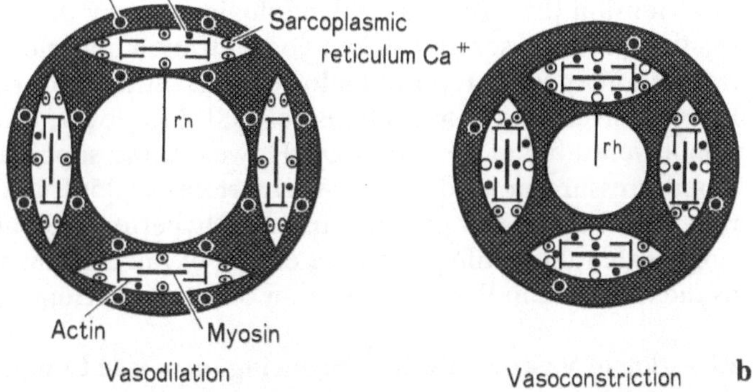

a b

Fig. 1a,b. Cross-sectional schema of arteries during vasodilation (**a**) and vasoconstriction (**b**). Extracellular calcium and intracellular storage calcium become intracellular free calcium through calcium influx and calcium releases, which raise the calcium concentration in the cytosol of arterial muscles. An increase in the cytosolic calcium concentration leads to the overlap of actin and myosin filaments, leading from the elongated spindle muscle cell to a round configuration. The round muscle brings about thickening of the arterial wall, which reduces the lumen of arteries. The reduction of arterial lumen is vasoconstriction (**b**). From Aoki [39, 40]

Calcium in Essential Hypertension

Calcium in Pathogenesis of Essential Hypertension

Although many hypotheses of the mechanisms of blood pressure elevation in essential hypertension have been proposed, the cause of high blood pressure is not clear. One of the reasons for the complexities and controversies is that there is no true consensus regarding the concept and definition of essential hypertension [26, 27]. However, most investigators agree that the rise of total peripheral vascular resistance is responsible for pressure elevation in essential hypertension [25]. One possible cause of the elevation of vascular resistance is that the calcium influx through calcium channels is increased in arterial muscles. The increased calcium influx increases intracellular-free calcium. This would cause arterial muscle contraction, thereby decreasing the diameter of arterial vessels.

The increase in calcium channel activity could be the result of an intrinsic modification of the channel protein or the result of membrane depolarization. This hypothesis is in fact supported by the following evidence. First, arterial muscle cells in spontaneously hypertensive rats are slightly depolarized [48, 49]. Second, voltage-dependent calcium channels are altered [49]. Third, calcium antagonists that inhibit the voltage-sensitive calcium channel activity, cause a reduction in high blood pressure through vasodilation [35–38]. Therefore, understanding the properties of the calcium channels in arterial muscle is a prerequisite to the elucidation of their role in the mechanism of hypertension.

In view of the role of calcium in the regulation of contraction and relaxation of arterial muscle, the possibility that the rise of blood pressure in essential hypertension reflects an abnormally high level of cytoplasmic calcium has been suggested for at least two decades [47], and remains the subject of intensive studies [28–43, 48–54]. Recently, the causal relationship between calcium and hypertension is a topic for the mechanisms of the rise of blood pressure in essential hypertension [54–60]. An epidemiological survey demonstrated a positive correlation between serum calcium concentration and blood pressure [57]. In contrast, another study reported a lower concentration of serum ionized calcium in hypertensives than in normotensives [58]. Moreover, oral calcium supplementation lowered blood pressure [59]. However, the supplementation did not lower blood pressure in mild essential hypertensives [56], and calcium supplementation raised the blood pressure in some hypertensive patients [56]. The reports suggest that the antihypertensive effects of the calcium supplementation as well as the increases in the concentration of serum calcium remain controversial.

Elevation of Vascular Resistance during Increased Concentration of Serum Calcium

It is suggested that an increased concentration of serum calcium through intravenous infusion of calcium may induce entry of calcium into the arterial wall and then into the cytosol of arterial muscle. Therefore, the infused calcium leads to the increased concentration of cytosolic calcium, which brings about arterial vasoconstriction through contraction of arterial muscle. The vasoconstriction

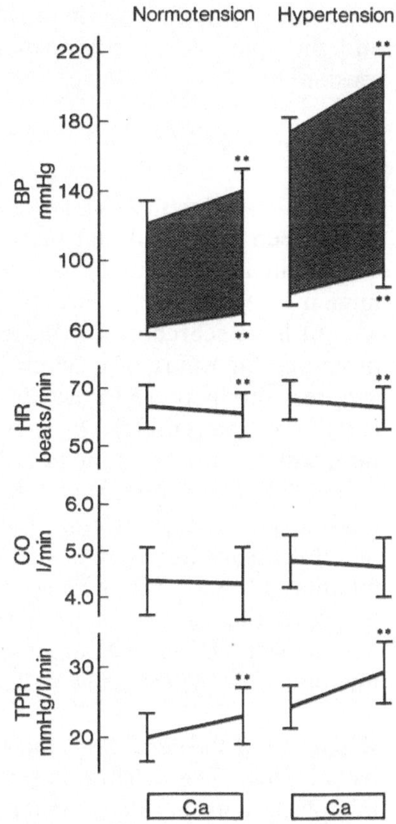

Fig. 2. Intravenous calcium (*Ca*) infusion causes a rise in blood pressure (*BP*) with an elevation of vascular resistance (*TPR*) in normotensives ($n = 6$) and hypertensives ($n = 6$). The rise in blood pressure and vascular resistance are inhibited by the calcium antagonist, verapamil (see p 304). *HR*, heart rate; *CO*, cardiac output. Values are mean ± SD. Compared with values before calcium infusion, $^*P < 0.05$, $^{**}P < 0.01$. From Suzuki and Aoki [57]

causes an elevation of vascular resistance, leading to hypertension. This hypothesis may be proved by the pressor effect of acute hypercalcemia by intravenous infusion of calcium. The finding may confirm the calcium membrane theory of essential hypertension as well as the role of calcium in the rise of blood pressure. Suzuki and Aoki [60] in 1988, have demonstrated that an intravenous infusion of calcium raises blood pressure in normotensives and hypertensives, and that the elevated blood pressure caused by the elevation of vascular resistance, is accompanied by a parallel increase in the serum concentration of calcium (Fig. 2). The elevation of both blood pressure and vascular resistance during calcium infusion are inhibited by calcium antagonists (see p 302) [60].

In view of the hypertensive action of serum calcium, it is suggested that a specific initiating factor to induce hypertension through increasing vascular resistance could be the cause of the increase in cellular calcium through a defect in the calcium binding protein or the calcium channel protein of the cell membrane and sarcoplasmic reticulum membrane of arterial muscle in essential hypertension.

Inhibition of Calcium Influx in Arterial Smooth Muscle

Hypotensive Effects of Calcium Influx Inhibition

It has been suggested that there is an increased influx and a decreased efflux of calcium in arterial muscle of spontaneously hypertensive rats (SHR) [28–30] and human essential hypertension. Therefore, from 1968 Aoki and others [28, 29, 35, 36] have searched for the substance which inhibits the entry of calcium or promotes the efflux of calcium in arterial muscle, as these substances possibly suppress the increase in calcium concentration in the cytosol, which causes relaxation of the muscle. The muscle relaxation brings about an arterial vasodilation, which causes a reduction of blood pressure in hypertension [35, 36].

In 1973, we learned that a newly synthesized antagonist, nifedipine, has an inhibitory action on calcium influx. It was thought that nifedipine may reduce blood pressure in patients with essential hypertension. The calcium antagonist nifedipine (kindly supplied by Yoshitaka Hirakawa) was given to hypertensives to prove the calcium hypothesis in 1973 (Fig. 3). Aoki et al. reported the hypotensive action and increased plasma renin activity by calcium antagonist, nifedipine, in hypertensive patients [35].

Reduction of Vascular Resistance by Calcium Influx Inhibition

Nifediphine. The calcium antagonist nifedipine inhibits calcium influx into the myoplasm and its release from calcium storage sites in the membrane system, which prevents the interaction of actin and myosin filaments [35, 36]. Thereby,

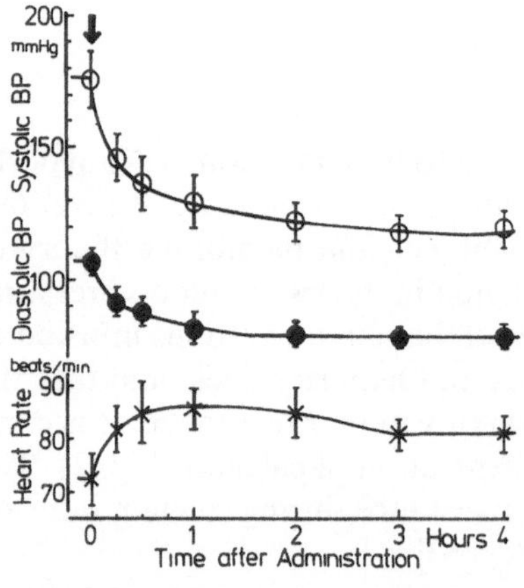

Fig. 3. The calcium antagonist, nifedipine, decreased systolic (○) and diastolic (●) blood pressure, and increased heart rate (×) in essential hypertensives. Values are mean ± SD. From Aoki et al. [35]

the drug reduces vascular resistance through a decrease in the contraction of arterial muscle.

The action of nifedipine was hemodynamically examined to demonstrate a reduction of vascular resistance in normotensives, and in borderline, and established essential hypertensives. Forty-five Japanese men, including twelve healthy normotensive volunteers, eleven patients with borderline stage and twenty-two patients with established stage essential hypertension, were investigated to demonstrate the reduction in both vascular resistance and blood pressure after the administration of nifedipine. All subjects rested in the supine position for a 1-h equilibration period and a 1-h testing period. Nifedipine (30 mg, Adalat) was sublingually administered. Before and 1 h after nifedipine administration, blood pressure, heart rate, and cardiac output were measured. Blood pressure of the upper right arm was measured by auscultation, using a mercury sphygmomanometer (WHO Memorandum, 1983). Heart rate was continuously measured with a heart rate monitor. Cardiac output was determined by the ear-piece dye dilution method, using a bedside cardiac computer (Eram Dyemac EW 90-A) [25]. Total peripheral vascular resistance (TPR) was calculated as follows:

Mean blood pressure (BP) (mmHg)
diastolic BP + 1/3 (systolic BP—diastolic BP)

Total peripheral vascular resistance (TPR) (mmHg/liter per minute)
mean BP (mmHg)/cardiac output (liter/min)

Nifedipine decreased vascular resistance in both normotensives and hypertensives, but it did not reduce the blood pressure in the normotensives. The drug increased the heart rate and cardiac output in both groups (Fig. 4). In spite of the reduction of vascular resistance, nifedipine did not alter blood pressure in normotensives.

The vascular resistance was calculated from the mean blood pressure divided by cardiac output. Therefore, blood pressure was dependent on the balance between cardiac output and vascular resistance. It was observed that nifedipine reduced vascular resistance (35%) and increased cardiac output (54%) in normotensives. The reduction of vascular resistance was compensated by an increase in cardiac output, in which blood pressure did not decrease in the normotensives. On the contrary, a reduction of vascular resistance (44% in borderline and 46% in established) could not compensated for by an increase in cardiac output (56% and 56%, respectively), which induced a fall in blood pressure (14%, and 16%) in hypertensives [37].

A potent vasodilator of nifedipine induced a specific circulatory state of high cardiac output with low vascular resistance, which is a hyperdynamic state in normotensives and hypertensives. The nifedipine-induced hyperdynamic state could be responsible for the induction of angina pectoris and cerebrovascular accidents during administration of excessive doses of nifedipine.

Diltiazem. Diltiazem is a calcium antagonist, which directly acts on the plasma membrane and probably also on the sarcoplasmic reticulum of arterial muscle to inhibit the entry of calcium into the myoplasma through membrane calcium channels and their release from binding sites in the membrane systems [38, 39,

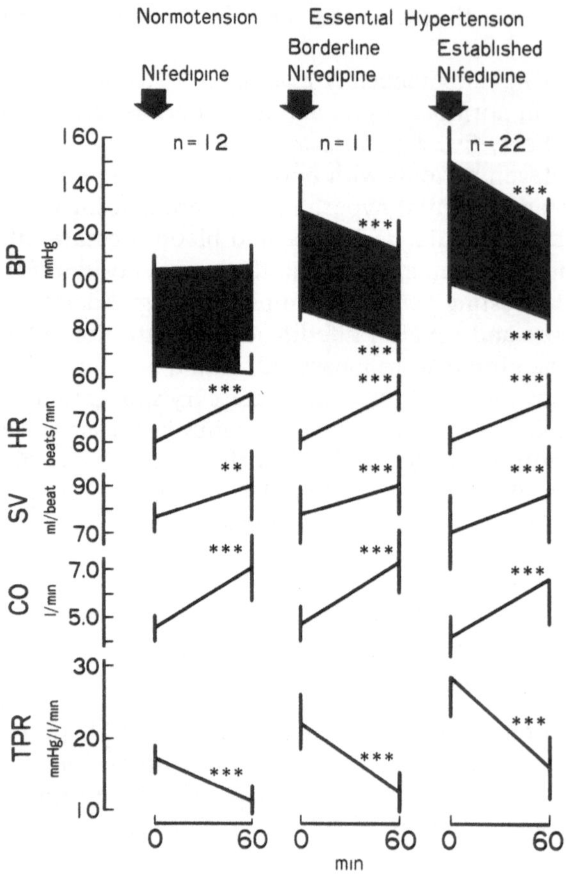

Fig. 4. The calcium antagonist, nifedipine, decreased blood pressure (*BP*) in borderline (*Borderline*) and in established (*Established*) stage of essential hypertension, but not in normotensives (*Normotension*). The drug increased heart rate (*HR*), stroke volume (*SV*), and cardiac output (*CO*). In contrast, nifidipine reduced vascular resistance (*TPR*) in both normotensives and hypertensives. Values are mean ± SD. Compared with values before nifedipine, $^{*}P<0.05$, $^{**}P<0.01$, $^{***}P<0.001$. From Aoki and Sato [37]

61]. An inhibition of the increase in intracellular-free calcium by the calcium antagonist could bring about a reduction of the vascular resistance through arterial vasodilation which decreases the blood pressure in subjects with essential hypertension. To clarify the hypotensive mechanisms, diltiazem (0.5 mg/min, then 1.0 mg/min, each for 30 min) was intravenously infused to mild essential hypertensives ($n = 10$, 35–60 y, average 49 ± 8 y). The infusion rapidly decreased blood pressure from 164/98 to 144/86 mmHg with an increase in heart rate from 66 to 67 beats/min and cardiac output from 37 to 42 liters/min, and it reduced vascular resistance from 32 to 25 mmHg/liters per minute [61] (Fig. 5).

Diltiazem infusion (2 mg/min for 20 min, i.v.) in severe hypertensives decreased blood pressure with a reduction of vascular resistance. The evidence suggests that an intravenous infusion of diltiazem is a choice of treatment for

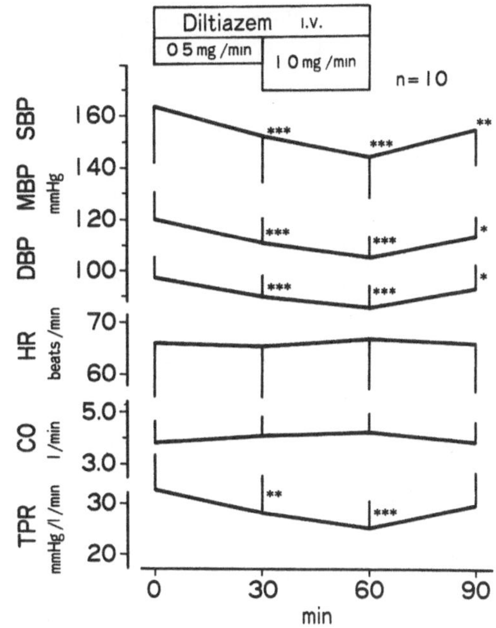

Fig. 5. Intravenous infusion of calcium antagonist, diltiazem, decreased systolic (*SBP*), diastolic (*DBP*) and mean (*MBP*) blood pressure, and vascular resistance (*TPB*), but did not change heart rate (*HR*) or cardiac output (*CO*) in hypertensives. Values are mean ± SD. Compared with values before diltiazem, $*P<0.05$, $**P<0.01$, $***P<0.001$. From Aoki et al. [61]

severe hypertensives [39]. Furthermore, the fact that the infusion decreased blood pressure with a reduction of vascular resistance indicates that the hypotensive action is caused by the reduction of vascular resistance through arterial vasodilation [38, 39, 61].

Verapamil. Verapamil is a calcium antagonist which inhibits the opening of the voltage-dependent calcium channels. Therefore, verapamil may inhibit the calcium influx and release which may cause vasodilation leading to a decrease in blood pressure. Intravenous injection of verapamil (5 mg) decreased blood pressure in hypertensives. The onset of maximal hypotension was observed within a few min, and a lowered pressure persisted throughout the infusion. The drug increased both stroke volume and cardiac output, however, it decreased vascular resistance [38] (Fig. 6).

Hypotensive action of calcium influx inhibition. These studies indicate that the calcium antagonists, nifedipine, diltiazem, and verapamil have a strong hypotensive action associated with reduced vascular resistance in hypertensives. From these results it was concluded that the fall of blood pressure is caused by a reduction of vascular resistance, probably due to vasodilation [39] (Fig. 7).

An elevation of vascular resistance is observed in the majority of essential hypertensives. The findings suggest that the main aim of the treatment of hyper-

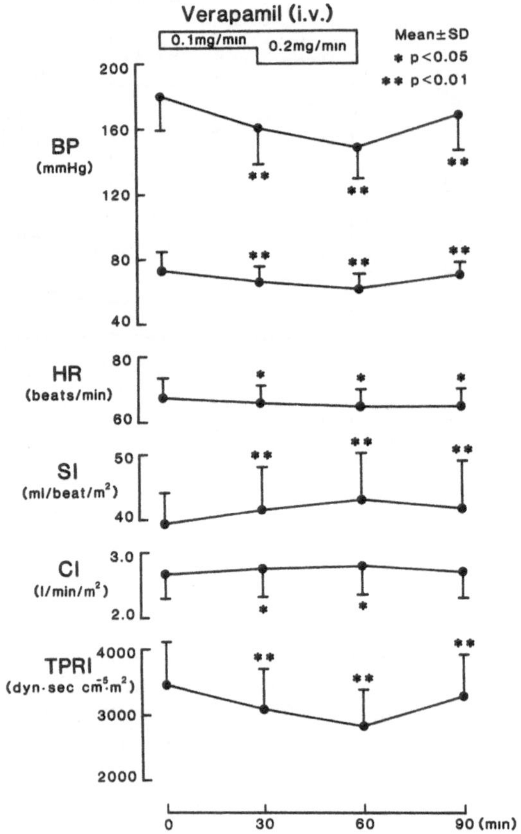

Fig. 6. Intravenous infusion of calcium antagonist, verapamil, decreased blood pressure (*BP*), heart rate (*HR*), and vascular resistance (*TPRI*), and increased stroke volume (*SI*) and cardiac output (*CI*) in hypertensives ($n = 12$). Values are mean ± SD. From Suzuki, Sato, Aoki [92]

tension is to decrease the elevated resistance, because the reduction of elevated resistance leads to lower blood pressure in essential hypertensives. Therefore, a calcium antagonist is the first and most logical choice for the treatment of hypertension [38].

Elevated Vascular Resistance in Essential Hypertension

The product of cardiac output and total vascular resistance determines the level of blood pressure (blood pressure = cardiac output × total vascular resistance). An increase in either cardiac output or vascular resistance results in high blood pressure. Therefore, it can be understood that hypertension is induced by an elevation of vascular resistance through an increased vasoconstriction. This idea brings up the hypothesis that an elevated vascular resistance due to vasoconstric-

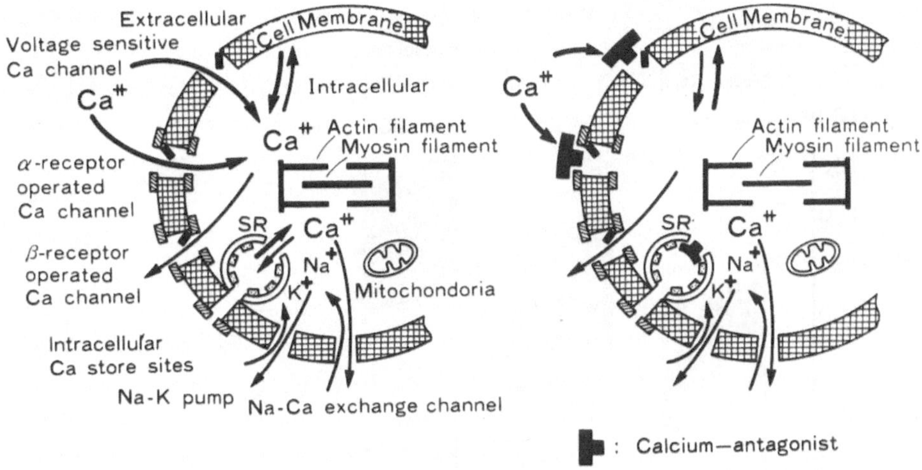

Fig. 7. Schema of cellular and molecular basis of arterial muscle. The cell membrane is characterized as a barrier between extracellular and intracellular spaces. The membrane has voltage-sensitive and receptor-operated calcium channels, Na-K pump and calcium binding sites, and it may also have Na-Ca exchange channels. The sarcoplasmic reticulum (*SR*) has calcium channels and calcium binding sites. Calcium antagonists (▶) inhibit calcium influx through calcium channels and calcium release from calcium binding sites. These effects depress the rise in the intracellular free calcium, inhibiting the overlapping of actin and myosin filaments. Thus, the calcium antagonist inhibits contraction of arterial muscle, which induces a widening of arterial lumen. The widening of the lumen decreases vascular resistance, leading to a fall in blood pressure in hypertensives. From Aoki [42, 43]

tion causes an increase in blood pressure in essential hypertension. To prove this hypothesis, cardiac output and blood pressure were measured in both normotensives and hypertensives. A majority of the studies from 1960 to 1980 demonstrated that cardiac output and heart rate were about 15% higher in young mild hypertensives than in normotensives, and that stroke volume and vascular resistance did not differ in normotensives and hypertensives [62]. However, recently Hofman et al. [63] have reported that cardiac output is lower in patients with early stage hypertension than in normotensives, but the vascular resistance is elevated in the hypertensives. Ohlsson [64] has reported that the heart rate and cardiac output are not greater in the offspring of hypertensive parents than in those of normotensive parents; however, the vascular resistance is slightly elevated in the offspring of hypertensive parents. Lund-Johansen [62] reported that the elevated blood pressure in the resting supine position in both children and adolescents with hypertensive parents is caused by an elevated vascular resistance rather than by an increased cardiac output.

Aoki and Sato [25] demonstrated that a resting supine position decreases blood pressure, heart rate, stroke volume, and cardiac output, whereas it elevates vascular resistance (Fig. 8). The vascular resistance during the supine position was greater in both borderline and established stage essential hypertensives than in normotensives. A positive correlation between blood pressure and vascular resistance was observed in the subjects (Fig. 9). Blood pressure was

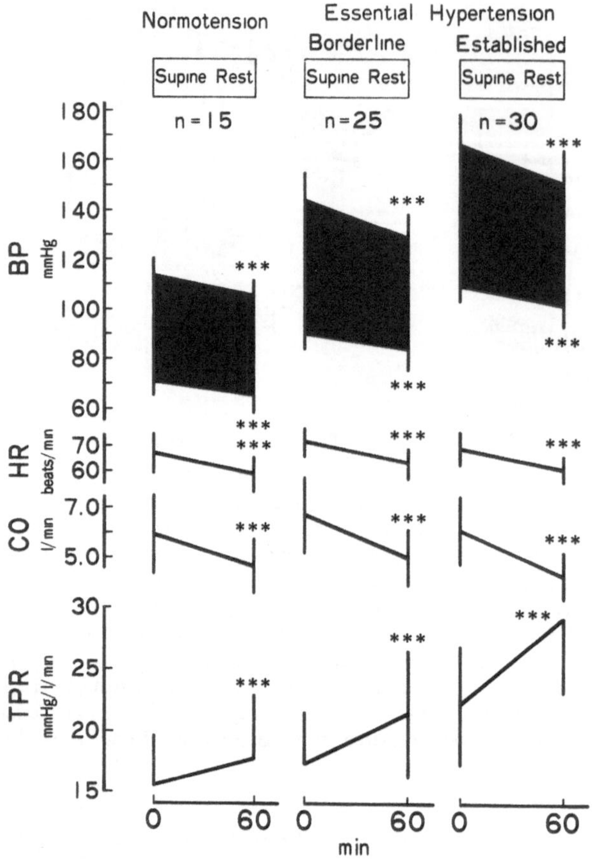

Fig. 8. The supine resting position decreased blood presure (*BP*), heart rate (*HR*), and cardiac output (*CO*), whereas it elevated vascular resistance (*TPR*). The decrease and elevation were greater in hypertensives than in normotensives. From Aoki and Sato [25]

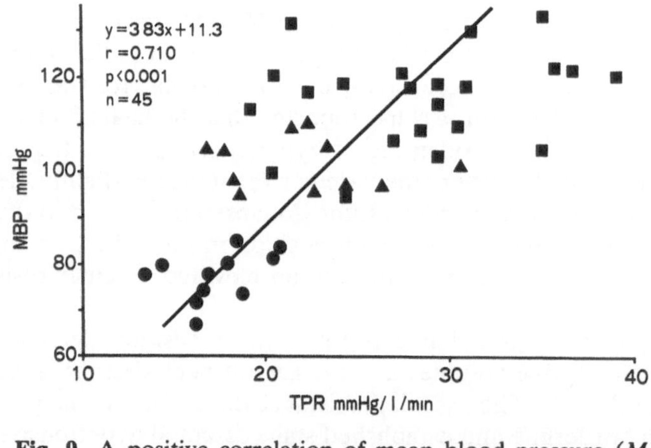

Fig. 9. A positive correlation of mean blood pressure (*MBP*) and vascular resistnace (*TPR*) was demonstrated in normotensives (●), borderline (▲), and in established (■) phase of essential hypertension. From Aoki and Sato [25]

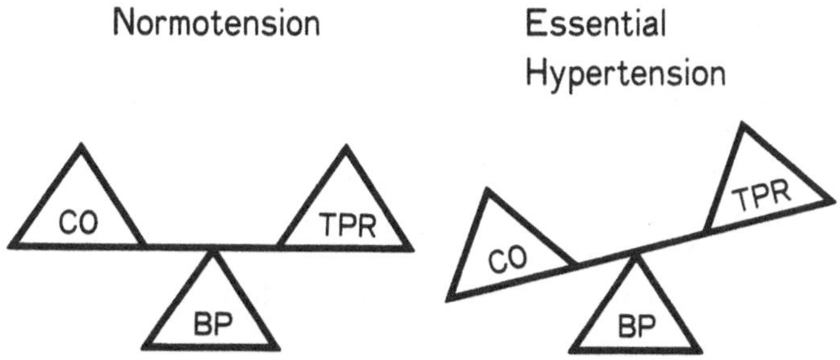

Fig. 10. Schema of hemodynamic balance in rise of blood pressure (*BP*) in established essential hypertensives. The elevated vascular resistance (*TPR*) causes the rise of blood pressure in essential hypertensives, but not cardiac output (*CO*). From Aoki [25, 42, 43]

significantly correlated with vascular resistance, but not with cardiac output. These findings indicate that the rise in blood pressure is caused by the elevation of vascular resistance in essential hypertension [25, 63, 64] (Fig. 10).

Calcium and Contraction in Arterial Smooth Muscle

Excitation-Contraction Coupling

Contraction and relaxation of arterial muscle is regulated by the intracellular calcium concentration, which determines arterial vascular tone. In view of the primary role of calcium in the regulation of muscle contractility, there is a possibility that an elevated vascular resistance in hypertension reflects on abnormally high level of cytoplasmic-free calcium [42, 43]. Changes in the cytoplasmic level of calcium are due to the calcium influx through the pathway which is controlled by neurotransmitters, vasoactive drugs, the membrane potential, and local physio-chemical conditions. Both intrinsic and external stimuli are translated into mechanical activation of arteries. Changes in the membrane potential, such as action potentials or graded depolarizations, are the events which lead to contraction in arterial muscle. Agonists contract the muscle by a mechanism which does not involve a change in the membrane potential. These findings indicate the presence of electro- and pharmaco-mechanical couplings [65].

Two different calcium-extrusion mechanisms have been proposed. The first mechanism drives the calcium out of the cell by the use of the Na-K pump, which interacts with Na-Ca exchange. Na-deficient solution and inhibition of Na-K pump, as cardiac glycosides or exposure to K-free solutions, induce contraction of arterial muscle. The contraction causes arterial vasoconstriction. However, the effects of Na-free solution on ^{45}Ca efflux is similar to the effects of potassium depolarization, which suggest that Na-free solution opens the voltage-sensitive channels [65, 66]. Recently, Sato and Aoki [67] demonstrated that a Na-K pump inhibitor of ouabain induces an arterial contraction, which is inhibited by a voltage-sensitive calcium channel blocker of nifedipine. In addition, contraction induced by Na-free solutions was inhibited by nifedipine [68]. These findings support the hypothesis that the function of Na-Ca exchange does not occur in the membrane of the muscle. The second type of extrusion mechanism is an ATP-dependent Ca pump. Numerous studies have demostrated ATP-dependent calcium accumulation into the microsomal fraction prepared from arterial muscle. The ATP-dependent calcium accumulation is more important for the decrease in cytosolic calcium than the Na-Ca exchange in the muscle [65, 66]

Altered Calcium Concentration of Arterial Smooth Muscle

Either the increased number of calcium channels or the decreased number of calcium binding sites may raise the calcium concentration in the intracellular space of arterial muscle. The increase in calcium concentration may be responsible for the rise of blood pressure through over-contraction of the muscle [28–32].

The alteration of membrane calcium handling has been demonstrated as follows [28–30]. The calcium binding constant and binding sites of microsomal fractions from the membrane of arterial muscle were measured in SHR and normotensive rats. The maximum uptake (number of binding sites) and apparent binding constant (binding affinity) were estimated by double reciprocal plots of free and bound calcium to the microsomal fraction. The ATPase activity in the fraction was increased compared to that of normotensives [28]. The calcium uptake of the sarcoplasmic reticulum was lower in SHR than in normotensive rats (Fig. 11a). The calcium binding constant did not differ in the two groups (Fig. 11b). This impaired calcium sequestration in the membranes, may possibly induce the binding of a large amount of available calcium to the regulatory protein, calmodulin and then to contractile proteins. The increased available calcium may result in the over-contraction of arterial muscle, which leads to high blood pressure.

The calcium content of the isolated membrane fraction was measured in a solution containing ATP and Mg (A), ATP, Mg, and EGTA (B), and EGTA, ATP- and Mg-free (C). The calcium content of A-solution was increased in SHR, which may consist of intracellular free calcium, ATP-dependent uptake calcium, and tightly bound calcium. After washing out calcium from the fraction with the B-solution, the calcium content did not differ between the two groups. The calcium may consist of ATP-dependent uptake calcium and tightly bound calcium. After washing with the C-solution, the calcium content was greater in SHR than in normotensive rats, thus the calcium may be tightly bound (unmobilizable) calcium. Intracellular-free calcium, calculated as the difference between the calcium content in A and B, was greater in SHR than in normotensive rats. The total mobilizable calcium, calculated as the difference between the calcium content in A and C, did not differ in the two groups. In addition, the amount of ATP-dependent uptake binding (calcium binding sites), the difference between the calcium content in B and C, was smaller in SHR, suggesting a decreased number of ATP-dependent calcium binding sites (Table 1). These findings indicate that the available calcium for contractile proteins may be greater in the arterial muscle of SHR. The calcium may cause an increase in arterial tone.

Fig. 11a,b. Decrease in calcium uptake (**a**) and in maximum calcium binding (**b**) by arterial smooth muscle membrane (*sarcoplasmic reticulum*) fraction isolated from SHR. Values are means ± SD. SHR (●) and normotensive rat (○). From Aoki et al. [28–30]

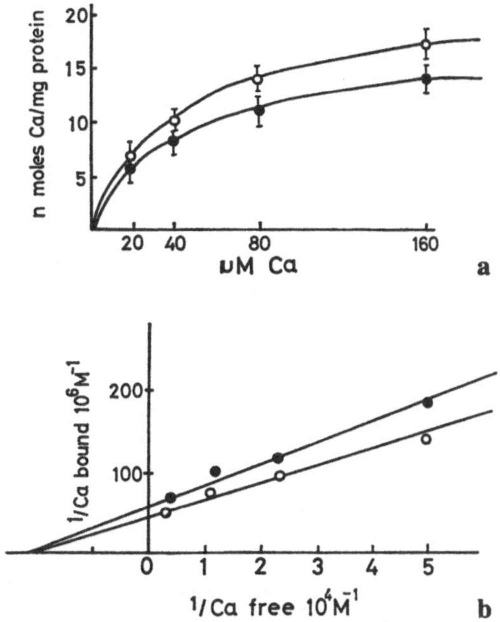

Table 1. Intracellular free calcium, mobilizable calcium, and ATP-dependent binding calcium in arterial muscle spontaneously hypertensive rats (SHR) and normotensive rats

Study	Normotensive rats	SHR
	(Calcium content 10^{-8} mol/mg)	
Sample in standard buffer containing ATP and Mg (a)	2.43 ± 0.22	2.72 ± 0.28
Sample washed with buffer containing ATP, Mg, and EGTA (b)	1.87 ± 0.23	1.83 ± 0.30
Sample washed with buffer containing EGTA (unmobilizable calcium) (c)	1.20 ± 0.22	1.47 ± 0.30
Intracellular free calcium (a–b)	0.56	0.89
Mobilizable calcium (a–c)	1.23	1.25
ATP-dependent binding calcium (b–c)	0.70	0.40

ATP 3m*M*, Mg 3m*M*, EGTA 5m*M*

Increase in Contraction of Arterial Smooth Muscle

An altered contractile response of arterial muscle may be a possible mechanism for the rise of blood pressure in essential hypertensives as well as in SHR. However, there are conflicting results regarding the muscle response of SHR compared with that of normotensive rats. It has been reported that the greater sensitivity and smaller contractile response of the muscle from SHR [50–52] and

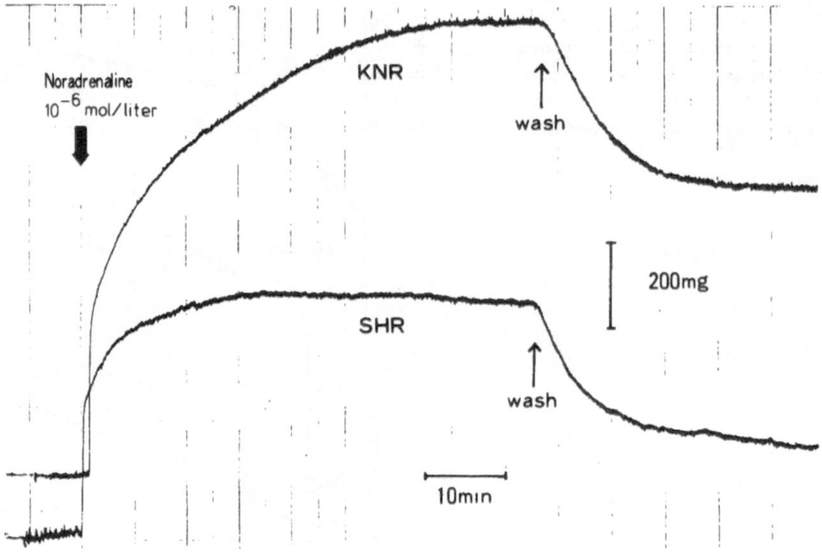

Fig. 12. A recording of noradrenaline (10^{-6} *M*)-induced tension in aortic strips in solution containing calcium (1.6×10^{-3} *M*). The norepinephrine-induced maximum tension was smaller in the strips from *SHR* than in the strips from normotensive rats (*KNR*). From Aoki et al. [71]

the hyperreactivity of SHR arteries are due to the increased wall to lumen ratio [69, 70]. Regarding the aspect of a calcium handling abnormality of the muscle, we have measured the tension development of arterial strips by norepinephrine and calcium in the absence and presence of calcium in the medium. The maximum tension induced by norepinephrine in solution containing calcium was significantly smaller in the strips of SHR than in those of normotensive rats [71] (Fig. 12).

After removal of both intra- and extracellular mobilizable calcium from the tissue by treatment of the strips with calcium-free EGTA solution, the strips developed tension with the addition of 1.5 m*M* calcium. The calcium-induced tension of the SHR strips was greater than that of normotensive strips, suggesting that the arterial muscle of SHR has strong spontaneous tone from an increase in calcium permeability of the cell membrane and an increase in the calcium release from sarcoplasmic reticulum in the resting condition [71] (Fig. 13).

Femoral arterial strips from SHR were more sensitive to exogenously applied KCl than those from normotensive rats [50]. The contractile response of the muscle to potassium was dependent on the calcium influx through voltage-dependent calcium channels. Transmembrane influx of calcium via the channel appears to induce a spontaneous contraction of the muscle. The calcium agonist, Bay K 8644 causes calcium influx through the voltage-dependent calcium channels. It was reported in 1986 that (1) this calcium agonist induces a greater contractile response in SHR arteries, (2) SHR arteries are more sensitive to K-depolarization than normotensive arteries, and (3) the calcium channel antagonist, nifedipine, produces a relaxation response from the resting contraction in

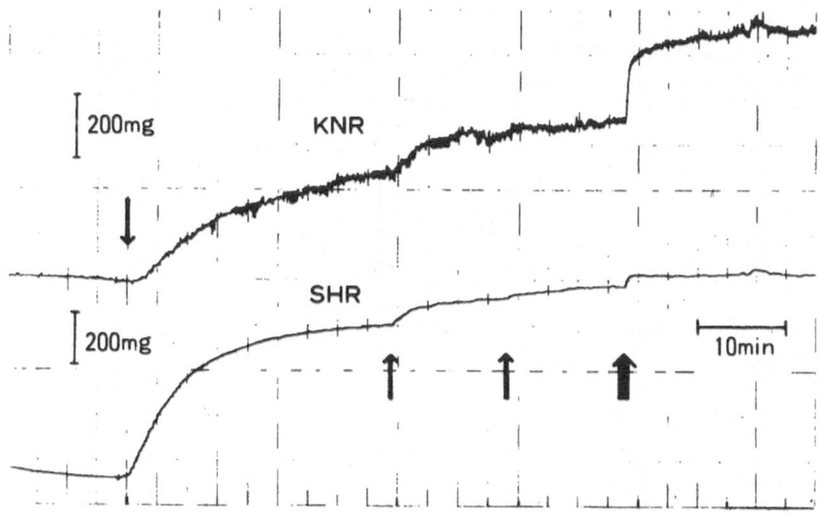

Fig. 13. A recording of calcium induced tension in aortic strips in the zero-calcium EGTA solution after calcium being washed out from the strips by the EGTA solution. The calcium induced tension was greater in SHR than in normotensive rat. Addition of calcium (*thin arrows*) and norepinephrine (10^{-6} *M*) (*thick arrow*) to the medium. From Aoki et al. [71]

SHR arteries, but not in normotensive arteries (Fig. 14) [34]. These findings suggest that an abnormality of the voltage-dependent calcium channel may be one of the membrane abnormalities of arterial muscle in SHR.

Membrane Potential, Electrogenic Sodium Pump, and Na-Ca Exchange

Early investigations indicated that arterial muscle from SHR is more permeable to ions of sodium, potassium, and chloride than that from normotensive rats [66]. Rencently, it was observed that the sodium flux in arterial muscle of SHR and WKY do not differ, and that intracellular sodium is similar in SHR and WKY [72]. In addition, the intracellular potassium concentration has been estimated to be unchanged in SHR [73].

The permeability to monovalent ions in arterial muscle may be partly offset by an electrogenic transport of sodium and potassium by sodium-potassium ATPase. When sodium pump activity is inhibited by removing the potassium from the extracellular environment, depolarization is greater in the muscle from SHR [73, 74]. Helical strips from SHR showed greater relaxation responses to potassium than did those from WKY [72]. A functional index of sodium-pump activity could be obtained as the magnitude of contractile responses to ouabain- or potassium-free solution. The rate of contraction of the muscle from SHR to ouabain- or potassium-free solution is faster than in that from WKY. This enhanced contractile response of SHR arterial muscle to sodium-pump inhibition may have relevance to the vascular reactivity in the whole animal [75–77].

There are two mechanisms from vascular contraction by sodium-pump inhibi-

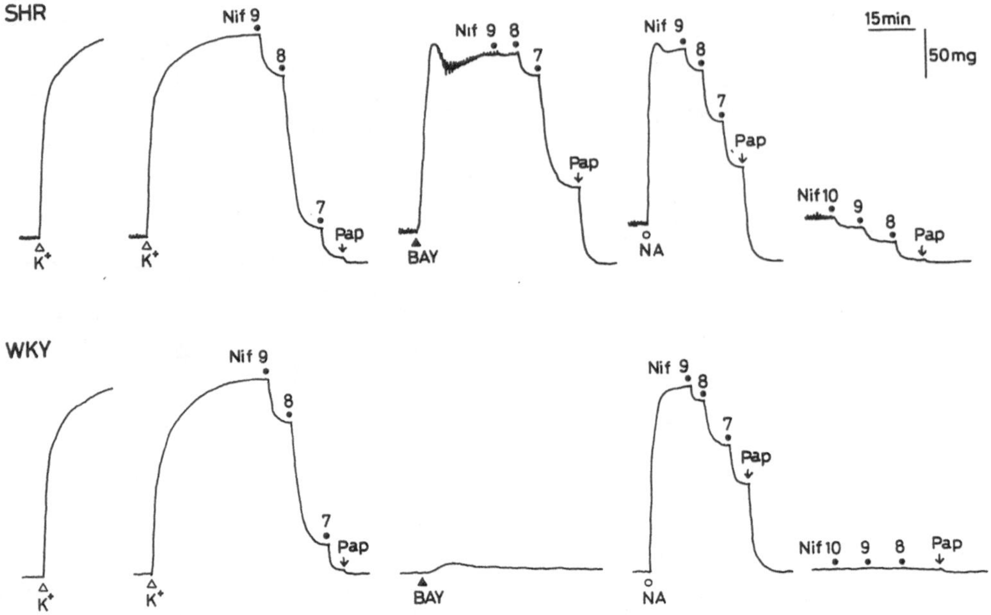

Fig. 14. Responses to calcium antagonist (*nifedipine*, *Nif*) and calcium agonist (Bay K 8644, *BAY*) in femoral arterial strips from 6-week-old *SHR* and age matched *WKY*. The strips were contracted with K (60 m*M*, K^+), Bay K (10^{-7} *M*, BAY), and noradrenaline (10^{-5} *M*, *NA*) plus timolol (3×10^{-7} *M*). The calcium agonist of Bay K induced contraction of the SHR strips, but not that of WKY. The agonist induced contraction was cancelled by the antagonist of nifedipine (*upper-middle-column*). The basal tone of SHR strips was reduced by calcium antagonist (*right-upper column*). From Aoki et al. [34]

tion. First, intracellular sodium concentration may increase after sodium-pump inhibition and cause a decrease in the calcium extrusion via sodium-calcium exchange. Second, since the sodium-pump is electrogenic, pump inhibition may cause depolarization and calcium influx through potential-operated channels [66]. It is not clear how intracellular sodium concentration leads to contraction. One possible explanation is the decrease in the sodium-calcium exchange. However, evidence for a sodium-calcium exchange has not been found in the muscle. Since the sodium-calcium exchange is a passive but electrogenic process (3 or 4 sodium ions move in for every 1 calcium ion that moves out), one would predict that inhibition of this exchange would cause contraction associated with membrane hyperpolarization [66]. This hyperpolarization-contraction does not occur in arterial muscle, suggesting that this is not the mechanism by which sodium-pump inhibition causes contraction. Another possible mechanism for ouabain-induced contraction is calcium influx through potential-operated channels secondary to depolarization. Lamb et al [77] have shown that exposure of aortic strips to verapamil produce a 75% reduction in the contractile response to ouabain and potassium-free solution. Sato and Aoki [67] have observed in 1988 that early contraction of ouabain-induced biphasic contraction of human umbilical arteries is inhibited by the calcium antagonist, nifedipine, and disappears in the calcium-free solution. These observations suggest that calcium entry through potential-operated channels is altered in SHR. This could be a mechanism contributing to the difference in the response to ouabain between SHR and WKY rats [76, 77].

The alteration in membrane potential electrogenesis in SHR appears to consist not only of an increased dependence of membrane potential on the activity of the sodium-pump, but also consists of a smaller electrochemical gradient for potassium. Hermsmeyer [73, 74] and Harder [74] have provided evidence that ionized intracellular potassium concentration is approximatly 20 m*M* lower in SHR than WKY tail arteries. The difference in intracellular potassium concentration causes membrane potential to be approximately 3 mV less at each value of extracellular potassium concentration. Hermsmeyer proposed that this difference in the potassium equilibrium potential between tail arteries of SHR and WKY rats contributes to the greater depolarization of SHR cells to norepinephrine. The greater sensitivity of arterial muscle from SHR to norepinephrine-induced depolarization has also been reported [73].

Receptors of Arterial Smooth Muscle in Hypertension

Few studies have attempted to determine whether or not the altered vascular reactivity in SHR lies at the adrenoceptor level. Knorr et al. [78] concluded that there is an increase in vascular α_2-receptor sensitivity in SHR. Hicks et al. [79] found that pressor responses to norepinephrine, phenylephrine, and angiotensin II were enhanced in SHR, in contrast the pressor responses to the α_2-agonists were not enhanced. They concluded that the difference in the potency of α_1- and α_2-antagonists in SHR and WKY most likely reflects a difference at the post-receptor level, involving excitation-contraction coupling mechanisms. Weiss et al. [80] suggested that the alterations in post-receptor activation mechanisms also may play a role in the altered vascular response to α_2-agonists in SHR.

It is known that β-adrenoceptor-mediated relaxation of arterial muscle is diminished in SHR arteries [81–83]. Fujimoto et al. [83] demonstrated that β_1-adrenoceptor-mediated relaxation of the muscles is diminished in SHR. The diminished relaxation is related to the distal and proximal activation of adenylate cyclase or cAMP formation. It was reported that cAMP-dependent protein kinase activity is decreased in SHR aorta [84]. There was no difference in the cAMP-dependent protein kinase response to forskolin between SHR and WKY aortae [85]. These results suggested that the diminished relaxation response of SHR arterial muscle to forskolin is due to an abnormality in the distal activation of cAMP-dependent protein kinase [86]. In addition, the phosphorylation of membraneous protein by protein kinase is followed by transmembrane efflux of calcium. The efflux and calcium uptake into the sarcoplasmic reticulum lead to decreased amounts of calcium available for maintenance of muscle tonicity. The cAMP-stimulated phosphorylation and calcium uptake by sarcoplasmic reticulum from the muscle were decreased in SHR [84–86]. These results suggest that increased amounts of calcium available for the contraction of the muscle may contribute to the initiation of the rise of blood pressure in gene (essential) hypertension.

The Calcium Membrane Theory of Gene (Essential) Hypertension

Evidence for the Calcium Membrane Theory

Hinke [87] demonstrated that norepinephrine and potassium induces contraction of arterial segments. The following findings were observed: (1) Hypertensive arteries are contracted further than normotensive arteries. (2) The arterial contraction in hypertensives is prolonged in zero-calcium perfusion. (3) Reestablishment of the contraction after abolished contraction in zero-calcium perfusion requires less calcium in the hypertensive arteries. These findings suggest that the hypertensive arteries are hyperresponsive for contraction to norepinephrine and potassium, and are hyperresponse for contraction to calcium. The following mechanisms may be responsible for the hyperresponsiveness of arteries of hypertensives: (1) there is a greater calcium influx through the cell membrane and a greater calcium release from the calcium storage sites of arterial muscle, and (2) The calcium utilization is more effective on the contractile mechanism of arterial muscle.

Bruner et al. [66] proposed two theories to explain the observations of the reduced membrane stabilization effects of calcium on the contractility in arterial muscle of SHR, which may cause an elevation of vascular resistance in hypertension. Both of these theories are based on the early work of Frankenhaeuser and Hodgkin [88] in 1957. The first is called "the divalent gating-particle theory" and the second, "the surface-potential theory." In the divalent gating-particle theory, calcium ions bind to a specific component of the voltage-dependent gating channel causing the pore to close in the cell membrane. When the membrane is depolarized, the calcium ions are stripped off, causing the ion port to open. This plugging action of calcium ions may be reduced in cell membranes of arterial muscle from SHR as indicated by the reduced calcium-binding capacity in the membrane fractions [66].

The second theory is based on the idea that negatively charged ions in the membrane can create local potentials. In the absence of calcium, the outer surface of the membrane bears a net negative charge due to the ionized groups of molecular components. This situation sets up a local negative surface potential that alters the electric field within the membrane. Presumably, this electrical field alteration is "sensed" locally as membrane depolarization, causing to open the calcium channels [66]. In hypertension, the number of calcium binding sites appears to be reduced, providing an explanation for the relative inability of elevated calcium to decrease monovalent ion turnover. The opening of the calcium channels can induce the influx of calcium, which is responsible for the elevation of arterial tone in hypertension.

Aoki et al. [27–30] have observed that the ATPase activity of the membrane fraction from SHR is elevated compared with that of normotensive rats,and the calcium uptake of the fraction is lower in SHR than in normotensive rats; also the apparent binding constant did not differ between SHR and normotensive rats. The findings suggest that there is a decreased number of binding sites in the membrane fraction. These results have been confirmed by several investigators [31, 32, 89, 90]. From the measurements of the calcium content in the isolated membrane fraction, mobilizable (intracellular free) calcium was greater in SHR than in normotensives, and ATP-dependent calcium binding was decreased in SHR [28–32]. These results support the calcium theory that a primary defect in the membrane causes the elevation of vascular resistance through over-contraction of arterial muscle.

It is suggested from these findings that the membrane calcium handling abnormality can induce an increase in contractile response of arterial muscle in SHR as well as in human essential hypertension. The hypothesis is supported by the demonstration that the maximum tension induced by norepinephrine in the solution containing calcium is less in the arterial strips of SHR, and in contrast, the tension induced by calcium in the calcium-free solution is greater in the strips of SHR [33], and calcium agonist-induced maximum tension is greater in SHR arteries [34]. In addition, Suzuki and Aoki [60] have demonstrated that hypercalcemia induced by intravenous infusion of calcium can cause an increase in vascular resistance leading to hypertension, and the increase in vascular resistance and blood pressure is greater in hypertensives than in normotensives. A significant positive relationship between the increase in serum concentration of calcium and the elevation of blood pressure has also been demonstrated in a recent study of Aoki and Miyagawa [91]. The calcium-induced hypertension is lowered by calcium antagonists [60].

Bohr et al. [66, 75–76], Hermsmeyer et al. [73, 74], Aoki et al. [40–43], and many other investigators [44–54] believe that the defect in both the cell membrane and the sarcoplasmic reticulum is the primary cause of the over-contraction of arterial muscle in gene hypertension of SHR and human essential hypertension.

The Calcium Membrane Theory of Essential Hypertension

Many findings concerning the abnormalities of membrane calcium handling have been reported in arterial muscle from Aoki spontaneously hypertensive rat (SHR) as well as in human essential hypertension. Thus, the calcium membrane theory of gene (essential) hypertension has been proposed as a mechanism of the development and persistance of high blood pressure in essential hypertension.

The findings of the membrane defect are expressed as an increase in passive permeability of calcium ions, a decrease in calcium-binding ability, and a depolarization of the cell membrane, probably in the sarcoplasmic reticulum and calcium channels in essential hypertension. These membrane defects cause a rise in the calcium concentration of the cytosol in arterial muscle. The rise in calcium concentration induces an increase in the overlap of actin and myosin filaments, which results in over-contraction and incomplete relaxation. The over-contraction of the muscle brings about shortening of the long length and widening of the short length of the muscle, leading to a reduction in the lumen of arteries, which causes an elevation of vascular resistance. The elevation raises blood pressure, causing the development and persistance of hypertension in humans and animals with essential hypertension (Fig. 15). The abnormality in the handling of membrane calcium is the primary factor, which leads to high blood pressure in gene (essential) hypertension (Fig. 16). This hypothesis of the mechanism of the development of gene hypertension is called the calcium membrane theory of essential hypertension [42, 43, 92].

In Inheritance:

Major hypertension gene

↓

In Arterial Muscle:

Genetic abnormalities of cell membrane and sarcoplasmic reticulum

↓

Increase in calcium influx and release

↓

Increase in calcium concentration in cytosol

↓

Increase in overlap of actin and myosin filaments
(overcontraction and incomplete relaxation)

↓

Shortening of long diameter and widening of short diameter of arterial muscle

↓

In Arteries:

Thickening of arterial wall

↓

Narrowing of arterial lumen (vasoconstriction)

↓

In Systemic Circulation:

Elevation of total peripheral vascular resistance

↓

Rise in blood pressure

↓

Hypertension

Fig. 15. Mechanism of the rise of blood pressure in gene (essential) hypertension. Major hypertension genes induce abnormalities in the cell membrane and sarcoplasmic reticulum in arterial muscle, which increase the calcium influx and release. The increased influx and release of calcium causes over-contraction and incomplete relaxation of the muscle, which leads to reduction of arterial lumen. The lumen reduction of arteries brings about an elevation of vascular resistance, causing the development and persistance of hypertension. From Aoki [42, 43]

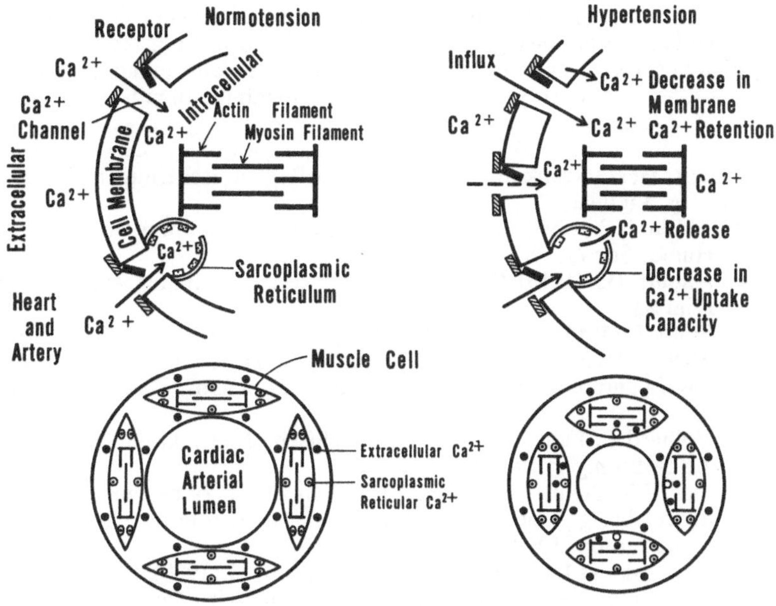

Fig. 16. A schema of the calcium membrane theory of gene (essential) hypertension. Membrane abnormalities, consisting of a decreased calcium binding and uptake, and an increased calcium influx through calcium channels, induces an increase in intracellular calcium. The increased calcium causes contraction of the arterial muscle, which leads to an elevation of vascular resistance. The elevation raises blood pressure in gene (essential) hypertension. From Aoki [42, 43]

References

1. Pickering GW (1955) Introduction. In: High blood pressure. Grune & Stratton, New York, J & A Churchill, London, pp 1–7
2. Bright R (1827) Reports of medical cases, selected with a view of illustrating the symptoms and cure of diseases by a reference to morbid anatomy. Longman, London
3. Bright R (1836) Cases illustrative of the effects produced when the arteries and brain are diseased: selected chiefly with a view to the diagnosis in such affections. Guy's Hosp Rep 1
4. Pickering GW (1955) The classification of hypertension. In: High blood pressure. Grune and Stratton, New York, J and A Churchill, London, pp 122–130
5. Tigerstedt R, Bergman PG (1898) Niere und Kreislauf, Scand Arch Physiol 8: 223–240
6. Goldblatt H, Lynch J, Hanzal RF, Summerville WW (1934) Studies on experimental hypertension. I. The production of persistent elevation of systolic blood pressure by means of renal ischemia. J Exp Med 59: 347–379
7. Pickering GW, Prinzmetal M (1938) Some observations on renin, a pressor substance contained in normal kidney, together with a method for its biological asaay. Clin Sci 3: 211–227
8. Gull WW, Sutton HG (1872) On the pathology of the morbid state commonly called chronic Bright's disease with contracted kidney: Arterio-capillary fibrosis. Med Chir Trans 55: 273–326
9. Huchard H (1889) Maladies du coeur et des vaisseaux. Doin, Paris

10. Ewald CA (1877) Ueber die Veränderungen Kleiner Gefässe bei Morbus Brightii und die darauf bezüglichen Theorien. Virchows Arch 71: 453
11. Johnson G (1868) On certain points in the anatomy and pathology of Bright's disease of the kidney. II. On the influence of the minute blood-vessels upon the circulation. Med Chir Trans 51: 57
12. von Basch S (1893) Ueber latent Arteriosclerose und deren Beziehung zu Fettleibigkeit, Herzerkrankungen und anderen Begleiterscheinugen. Urban and Schwartzenberg, Vienna.
13. Allbutt TC (1895) Senile plethora or high arterial pressure in elderly persons. Trans Hunter Soc 77: 38
14. Allbutt TC (1915) Diseases of the arteries, including angina pectoris, Macmillan, London.
15. Frank E (1911) Bestehen Beziehungen zwischen chromaffinen System und der chronischen Hypertonie des Menschen? Ein kritischer Beitrag zu der Lehre von der physio-pathologischen Bedeutung des Adrenalins. Dtsch Arch Klin Med 103: 397–412
16. Volhard F, Fahr T (1914) Die Brightsche Nierenkrankheit: Klinik Pathologie und Atlas. Springer, Berlin.
17. Dahl LK, Heine M, Tassinari L (1962) Effects of chronic excess salt ingestion. Evidence that genetic factors play an important role in susceptibility to experimental hypertension. J Exp Med 115: 1173–1190
18. Kawasaki T, Delea CS, Bartter FC, Smith H (1978) The effect of high sodium and low sodium intake of arterial pressure and other related variables in humans subjects with ideopathic hypertension. Am J Med 64: 193–198
19. Weinberger MH, Miller JZ, Luft FC, Grim CE, Fineberg NS (1986) Definitions and characteristics of sodium sensitivity and blood pressure resistance. Hypertension 8 (Suppl II): II-127–134
20. Aoki K (1989) The spontaneously hypertensive rat: Evidence of the genetic hypothesis in essential hypertension. In: Aoki K, Frohlich ED (eds), Calcium in essential hypertension. Academic Press, Tokyo San Diego New York London, pp 3–8
21. Aoki K (1986) Discovery of the spontaneously hypertensive rat. In: Aoki K (ed) Essential hypertension: calcium mechanisms and treatment. Springer, Tokyo Berlin Heidelberg New York London Paris, pp 3–7
22. Okamoto K, Aoki K (1963) Development of a strain of spontaneously hypertensive rats. Jpn Circ J 27: 282–293
23. Aoki K, Yamori Y, Ooshima A, Okamoto K (1972) Effects of high or low sodium intake in spontaneously hypertensive rats. Jpn Circ J 36: 539–545
24. Aoki K (1985) Essential hypertension and secondary hypertension in humans and rats. Asian Med J 28: 529–548
25. Aoki K, Sato K (1986) Decrease in blood pressure and increase in total peripheral vascular resistance in supine resting subjects with normotension or essential hypertension. Jpn Heart J 27: 467–474
26. Aoki K (1986) Etiological classification of hypertension: Essential hypertension, environment hypertension, and disease hypertension. In: Aoki K (ed) Essential hypertension: calcium mechanisms and treatment. Springer, Tokyo Berlin Heidelberg New York London Paris, pp 11–24
27. Aoki K (1989) The three-way classification of hypertension: Essential hypertension, environment hypertension, and disease hypertension. In: Aoki K, Frohlich ED (eds) Calcium in essential hypertension. Academic Press, Tokyo San Diego New York London, pp 9–36
28. Aoki K, Ikeda N, Yamashita K, Tazumi K, Sato I, Hotta K (1974) Cardiovascular contraction in spontaneously hypertensive rat: Ca^{2+} interaction of myofibrils and subcellular membrane of heart and arterial smooth muscle. Jpn Cir J 38: 1115–1121
29. Aoki K, Ikeda N, Yamashita K, Hotta K (1974) ATPase activity and Ca^{2+} interaction of myofibrils and sarcoplasmic reticulum isolated from the heart of spontaneously hypertensive rats. Jpn Heart J 15: 475–484

30. Aoki K, Yamashita K, Hotta K (1976) Calcium uptake by subcellular membranes from vascular smooth muscle of spontaneously hypertensive rats. Jpn J Pharmacol 26: 624–627
31. Moore L, Hurwitz L, Davenport GR, Landon EJ (1975) Energy-dependent calcium uptake activity of microsomes from the aorta of normal and hypertensive rats. Biochim Biophys Acta 413: 432–443
32. Wei JW, Janis RA, Daniel EE (1976) Calcium accumulation and enzymatic activities of subcellular fractions from aortae and ventricles of genetically hypertensive rats. Circ Res 39: 133–140
33. Aoki K, Mochizuki A, Hotta K (1981) Noradrenaline and calcium-induced tension in aortic strips of normotensive and spontaneously hypertensive rats. Jpn Circ J 45: 547–551
34. Aoki K, Asano M (1986) Effects of Bay K 8644 and nifedipine on femoral arteries of spontaneously hypertensive rats. Br J Pharmacol 88: 221–230
35. Aoki K, Yoshida T, Kato S, Tazumi K, Sato I, Takikawa K, Hotta K (1976) Hypotensive action and increased plasma renin activity by Ca^{2+} antagonist (nifedipine) in hypertensive patients. Jpn Heart J 17: 479–484
36. Aoki K, Kondo S, Mochizuki A, Yoshida T, Kato S, Kato K, Takikawa K (1978) Antihypertensive effect of cardiovascular Ca^{2+}-antagonist in hypertensive patients in the absence and presence of beta-adrenergic blockade. Am Heart J 96: 218–226
37. Aoki K, Sato K (1989) Hyperdynamic state in normotensives and hypertensives with calcium antagonists at rest. In: Aoki K, Frohlich ED (eds) Calcium in essential hypertension. Academic Press, Tokyo San Diego New York London, pp 537–552
38. Aoki K (1989) Hypotensive mechanism of calcium antagonists and treatment of essential hypertension. In: Aoki K, Frohlich ED (eds) Calcium in essential hypertension. Academic Press, Tokyo San Diego New York London, pp 485–509
39. Aoki K, Sato K, Kawaguchi Y, Yamamoto M (1982) Acute and long-term hypotensive effects and plasma concentrations of nifedipine in patients with essential hypertension. Eur J Clin Pharmacol 23: 197–201
40. Aoki K, Kondo S, Mochizuki A, Sato K, Yoshida T, Kato S, Kato K (1979) Ca^{2+}-antagonist therapy for hypertension in combination with beta-blockade: A new concept of essential hypertension. In: Yamori Y, Lovenberg W, Freis ED (eds) Prophylactic approach to hypertensive disease. Raven, New York, pp 377–386
41. Aoki K, Asano M, Sato K, Kondo S, Mochizuki A, Kawaguchi Y, Yamamoto M (1983) Calcium antagonists on the vascular smooth muscle of spontaneously hypertensive rat and human essential hypertension: A calcium membrane theory of essential hypertension. In: Beven JA, et al. (eds) Vascular neuroeffector mechanisms: 4th International Symposium. Raven, New York, pp 295–299
42. Aoki K (1986) Calcium membrane theory of essential hypertension. In: Aoki K (ed), Essential hypertension: calcium mechanisms and treatment, Springer, Tokyo Berlin Heidelbery New York London Paris, pp 223–242
43. Aoki K (1989) The calcium membrane theory of essential hypertension. In: Aoki K, Frohlich ED (eds), Calcium in essential hypertension. Academic Press, Tokyo San Diego New York London, pp 623–655
44. Robinson BF (1984) Altered calcium handling as a cause of primary hypertension. J Hypertens 2: 453–460
45. Postnov YV, Orlov SN (1984) Cell membrane alteration as a source of primary hypertension. J Hypertens 2: 1–6
46. Beilin LJ (1988) Epitaph to essential hypertension—a preventable disorder of known aetiology? J Hypertens 6: 85–94
47. Somlyo AP, Somlyo AV (1968) Vascular smooth muscle—normal structure, pathology, biochemistry and biophysics. Pharmacol Rev 20: 197–272
48. Hermsmeyer K, Harder D (1976) Electrogenesis of increased norepinephrine sensitivity of arterial vascular muscle in hypertension. Circ Res 38: 362–367
49. Rusch NJ, Hermsmeyer K (1988) Calcium currents are altered in the vascular smooth muscle cell membrane of spontaneously hypertensive rats. Circ Res 63: 997–1002

50. Holloway ET, Bohr DF (1973) Reactivity of vascular smooth muscle in hypertensive rats. Circ Res 33: 678–685
51. Bohr DF (1974) Reactivity of vascular smooth muscle from normal and hypertensive rats: effects of several cations. Fed Proc 33: 127–132
52. Hansen TR, Bohr DF (1975) Hypertension, transmural pressure, and vascular smooth muscle response in rats. Circ Res 36: 590–598
53. Bohr DF, Webb RC (1984) Vascular smooth muscle function and its changes in hypertension. Am J Med (Suppl IV) 77: IV-3–16
54. Postnov YV, Orlov SN (1985) Ion transport across plasma membrane in primary hypertension. Physiol Rev 65: 904–945
55. Singh RB, Singh NK, Mehta PJ, Pastogi SS (1987) Does calcium aggravate and cause hypertension? Acta Cardiol 42: 445–467
56. Xoccali C, Mallamaci F, Delfino D, Ciccarelli M, Parlong S, Lellamo D, Moscato D, Maggiore Q (1988) Double-blind randomized, cross-over trial of calcium supplementation in essential hypertension. J Hypertens 6:451–455
57. Kesteloot H, Geboers J, Math L, van Hoof R (1983) Epidemiological study of the relationship between calcium and blood pressure. Hypertension (Suppl II) 5: II-52–56
58. McCarron DA (1982) Low serum concentration of ionized calcium in patients with hypertension. New Engl J Med 307: 226–228
59. McCarron DA, Morris CD (1985) Blood pressure response to oral calcium in persons with mild to moderate hypertension. Ann Intern Med 103: 825–831
60. Suzuki T, Aoki K (1988) Hypertensive effects of calcium infusion in subjects with normotension and hypertension. J Hypertension 6: 1003–1008
61. Aoki K, Sato K, Yamamoto M (1983) Hypotensive effects of diltiazem to normals and essential hypertension. Eur J Clin Pharmacol 25: 475–480
62. Lund-Johansen P (1983). Haemodynamics in early essential hypertension—still an area of controversy. J Hypertens 1: 209–213
63. Hofman A, Ellison RC, Newbuger J, Miettinen OS (1982) Blood pressure and haemodynamics in teenagers. Br Heart J 48: 377–380
64. Ohlsson O (1987) Haemodynamic studies of relatives to hypertensive patients. Univ Malmo
65. Droogmans G (1989) Excitation-contraction coupling in arterial smooth muscle: Electro- and pharmacomechanical coupling. In: Aoki K, Frohlich ED (eds) Calcium in essential hypertension. Academic Press, Tokyo San Diego New York London, pp 65–83
66. Bruner CA, Webb RC, Bohr DF (1989) Vascular reactivity and membrane stabilizing effect of calcium in spontaneously hypertensive rats. In: Aoki K, Frohlich ED (eds) Calcium in essential hypertension, Academic Press, Tokyo San Diego New York London, pp 275–306
67. Sato K, Aoki K (1988) Biphasic contraction induced by ouabain in human umbilical arteries. Eur J Pharmacol 158: 299–302
68. Makita Y, Kanmura Y, Itoh T, Suzuki H, Kuriyama H (1983) Effects of nifedipine derivatives on smooth muscle cells and neuromuscular transmission in the rabbit mesenteric artery. Naunyn-Schniedebergs Arch Pharmacol 324: 302–312
69. Folkow B, Hallbäck M, Lundgren Y, Weiss L (1970) Structurally based increase of flow resistance in spontaneously hypertensive rats. Acta Physiol Scand 79: 373–378
70. Folkow B (1982) Physiological aspects of primary hypertension. Physiol Rev 62: 347–504
71. Aoki K, Kawaguchi Y, Sato K, Kondo S, Yamamoto M (1982) Clinical and pharmacological properties of calcium antagonists in essential hypertension in humans and spontaneously hypertensive rats. J Cardiovasc Pharmacol 4: S298–S302
72. Aalkjaer C, Kjeldsen K, Norgaared A, Clausen T, Mulvany MJ (1985) Ouabain binding and Na content in resistance vessels and skeletal muscle of spontaneously hypertensive and K^+-depleted rats. Hypertension 7: 277–286
73. Hermsmeyer K (1984) Altered arterical muscle ion transport mechanism in the spontaneously hypertensive rat. J Cardiovasc Pharm 6: S10–S15

74. Hermsmeyer K, Harder D (1986) Membrane ATPase mechanism of K^+ relaxation in arterial muscle of stroke-prone SHR and WKY. Am J Physiol 250: C557–C562
75. Webb RC, Bohr DF (1978) Membrane stabilization by calcium in vascular smooth muscle. Am J Physiol 235: C227–C232
76. Webb RC, Bohr DF (1979) Potassium relaxation of vascular smooth muscle from spontaneously hypertensive rats. Blood Vessels 16: 71–79
77. Lamb FS, Moreland RS, Webb RC (1983) Inhibition of sodium pump activity and vascular responsiveness in SHR and WKY. Clin Res 31: A814
78. Knorr A, Muller B, Kazda (1983) Increased sensitivity of alpha-2 adrenoceptors mediating pressor response in spontaneously hypertensive rats. Biochem Pharmacol 32: 2639–2642
79. Hicks PE, Nahorski SR, Cook N (1983) Postsynaptic alpha-adrenoceptors in the hypertensive rat: Studies on vascular reactivity in vivo and receptor binding in vitro. Clin Exp Hypertens 3: 401–427
80. Weiss RJ, Webb RC, Smith CB (1984) Comparison of alpha-2 adrenoceptors on arterial smooth muscle and brain homogenates from spontaneously hypertensive and Wistar-Kyoto normotensive rats. Hypertension 2: 249–255
81. Cohen ML, Berkowitz BA (1976) Decreased vascular relaxation in hypertension. J Pharmacol Exp Ther 196: 396–406
82. Cheng JB, Shibata S (1981) Vascular relaxation in the spontaneously hypertensive rat. J Cardiovasc Pharmacol 3: 1126–1140
83. Fujimoto S, Dohi Y, Aoki K, Matsuda T (1988) Beta-1 and beta-2 adrenoceptor-mediated relaxation responses in peripheral arteries from spontaneously hypertensive rats at prehypertensive and early hypertensive stages. J Hypertens 6: 543–550
84. Bhalla RC, Webb RC, Singh D, Ashley T, Brock T (1978) Calcium fluxes, calcium binding, and adenosine cyclic 3′,5′-mono-phosphate-dependent protein kinase activity in the aorta of spontaneously hypertensive and Kyoto Wistar normotensive rats. Mol Pharmacol 14: 468–477
85. Coquil JF, Hamet P (1980) Activity of cyclic AMP-dependent protein kinase in heart and aorta of spontaneously hypertensive rat. Proc Soc Exp Biol Med 164: 569–575
86. Silver PJ, Michalak RJ, Kocmund SM (1985) Role of cyclic AMP protein kinase in decreased arterial cyclic AMP responsiveness in hypertension. J Pharmacol Exp Ther 232: 595–601
87. Hinke JAM (1966) Effect of Ca^{++} upon contractility of small arteries from DOCA-hypertensive rats.Circ Res 18, (Suppl I) 19: I-23–24
88. Frankenhaeuser B, Hodgkin AL (1957) The action of calcium on the electrical properties of squid axon. J Physiol (London) 137: 218–244
89. Kwan C-Y, Daniel ED (1989) Calcium handling by membranes isolated from vascular smooth muscle in hypertension. In: Aoki K, Frohlich ED (eds), Calcium in essential hypertension. Academic Press, Tokyo San Diego New York London, pp 201–230
90. Kwan C-Y (1985) Dysfunction of calcium handling by smooth muscle in hypertension. Can J Physiol Pharmacol 63: 366–374
91. Aoki K, Miyagawa K (to be published) Correlation of increase in serum calcium with elevation of blood pressure and vascular resistance during calcium infusion in normotensive man
92. Aoki K, Frohlich ED (eds) (1989) Calcium in essential hypertension. Academic Press, Tokyo San Diego New York London

Index of Key Words